MW00845868

AIDED NAVIGATION

GPS with High Rate Sensors

About the Author

Jay A. Farrell received B.S. degrees (1986) in physics and electrical engineering from Iowa State University, and M.S. (1988) and Ph.D. (1989) degrees in electrical engineering from the University of Notre Dame. At Charles Stark Draper Lab (1989–1994), he was principal investigator on projects involving intelligent and learning control systems for autonomous vehicles. Dr. Farrell received the Engineering Vice President's Best Technical Publication Award in 1990, and Recognition Awards for Outstanding Performance and Achievement in 1991 and 1993. He is a Professor and former Chair of the Department of Electrical Engineering at the University of California, Riverside. He has served as Vice President of Finance and Vice President of Technical Activities for the IEEE Control Systems Society. He is a Fellow of the IEEE (2008) and author of over 150 technical publications. He is author of the book *Aided Navigation: GPS with High Rate Sensors* (McGraw-Hill, 2008). He is also co-author of the books *The Global Positioning System and Inertial Navigation* (McGraw-Hill, 1998) and *Adaptive Approximation Based Control: Unifying Neural, Fuzzy and Traditional Adaptive Approximation Approaches* (John Wiley, 2006).

AIDED NAVIGATION

GPS with High Rate Sensors

Jay A. Farrell

New York Chicago San Francisco
Lisbon London Madrid Mexico City
Milan New Delhi San Juan
Seoul Singapore Sydney Toronto

The McGraw-Hill Companies

Library of Congress Cataloging-in-Publication Data

Aided Navigation: GPS with High Rate Sensors

1234567890 DOC DOC 0198

ISBN 978-0-07-149329-1
MHID 0-07-149329-8

Sponsoring Editor
Wendy Rinaldi

Editorial Supervisor
Patty Mon

Acquisitions Coordinator
Mandy Canales

Production Supervisor
Jean Bodeaux

Composition
Jay A. Farrell

Art Director, Cover
Jeff Weeks

Cover Designer
Pattie Lee

Cover Illustration
12E Design

To my students and colleagues, without whom this book would not have been possible.

Contents

Preface

Technological innovations over recent decades have enabled an ever increasing array of inexpensive, high accuracy, aided navigation systems suitable for student projects, research test-bed usage, and commercial and military applications. These innovations include: small, low cost, low power sensors; GPS technology; and, high performance, low cost, computational equipment for embedded processing.

Aided navigation involves two categories of sensors. The output signals from sensors in the first category are integrated using a kinematic model of the system. The result of this integration provides a reference trajectory. Example kinematic input sensors include inertial measurement units, Doppler radar or sonar, and wheel encoders. Elements of the second category of sensors are used to estimate the error between this reference trajectory and trajectory of the vehicle. In the estimation process, these aiding sensors may also determine various calibration parameters to improve the future performance of the system. The advent of Global Navigation Satellite Systems (GNSS) has provided an accurate and reliable aiding measurement source for navigation systems in certain outdoor applications. A prototypical example of a GNSS sensor suitable for aided operations is the Global Positioning System (GPS).

Aided navigation is motivated when an application, such as an autonomous vehicle, requires accurate high bandwidth information about the vehicle state reliably at a high sample rate. The aiding sensors may satisfy the accuracy requirement, but not the reliability, bandwidth, and high sample rate specifications. The reference trajectory computed from the kinematic input sensors may satisfy the reliability, bandwidth, and high sample rate specifications; however, due to its integrative nature, errors that accumulate over time can cause the unaided system to eventually fail to meet the accuracy specification.

The aiding sensors and the reference trajectory computed from the kinematic input sensors have complementary characteristics; therefore, it is natural to consider their implementation within an integrated approach.

Objective

My main objective in writing this book is to provide a self-contained reference on aided navigation system design that is appropriate for use in a classroom setting. The methodology presented herein is an industry standard approach. It has been used by many persons in various applications with a high-level of success.

The text is written with the expectation that the reader has the standard background of a senior in a bachelor-of-science (BS) engineering program.

I have taken special care in presenting the material at the BS level. For example, the topic of optimal filtering is derived both in the least squares context and in the standard stochastic processes context. Examples are included throughout the text to relate theoretical concepts back to applications, to motivate the importance of specific concepts, and to illustrate the tradeoffs between alternative techniques. In addition, detailed derivations are included. The intended audience for the book includes engineers, students, researchers, scientists, and project managers who may be interested in either designing or using an aided navigation system in a given application.

Outline

The text presents a systematic method for designing and analyzing aided navigation systems along with the essential theory to support that methodology. The book is divided into two parts.

Part I contains seven chapters. Chapter 1 motivates the design and analysis methodology. This chapter introduces simple examples and comparisons to illustrate the method and to stimulate interest in the theoretical information presented in Chapters 2–6. Chapter 2 defines frames-of-reference, transformations between frames-of-reference, and methods for maintaining the transformations between rotating frames-of-reference. Chapters 3 and 4 focus on model development. An accurate mathematical model is a necessary precondition to being able to make accurate quantitative statements about navigation system performance during the design process. Chapter 3 presents various concepts from (deterministic) systems theory. Chapter 4 introduces certain necessary concepts from the theory of stochastic processes. Chapter 5 introduces and derives optimal state estimation methods (i.e., Kalman filtering). Knowledge of optimal and sub-optimal state estimation are essential to the implementation and performance analysis of navigation systems. Chapter 6 discusses methods for performance analysis. Chapter 7 returns to the methodology suggested in Chapter 1; however, now with the tools provided in Chapters 2-6 the same examples can be rigorously and quantitatively analyzed. Examples are used extensively throughout Chapters 1-7.

Part II of the text provides in-depth discussion of several specific navigation applications. Chapter 8 provides a self-contained description of the basic GPS solution, differential GPS, Doppler processing, and various carrier phase processing techniques. This presentation of the various GPS techniques using unified notation greatly facilitates the understanding of the techniques and their relative tradeoffs. For clarity of the presentation and comparison of methods within Chapter 8, all techniques presented in that chapter consider only point-wise data processing (i.e., no Kalman filtering). GPS is discussed herein as a prototype of the various GNSS which are now or soon will be available. Chapters 9–12 each focus on a specific aided navigation application. Each chapter derives the kinematic model and presents the navigation mechanization equations based on the kinematic model; derives and analyzes the dynamic model for the navigation error state; and, presents equations for predicting the aiding measurements and for modeling the residual aiding measurements. Different forms of performance analysis are included in each of the chapters including covariance analysis, observability analysis, and data analysis from application or simulation. Chapter 9 discusses aided encoder-based dead-reckoning. Chapter 10 discusses an attitude and heading reference system that uses gyros as inputs to the kinematic model and accelerometers as gravity sensors for error correction. Chapter 11 discusses aided inertial navigation. Chapter 12 discusses a specific application of an aided inertial system. Together these chapters provide examples of methods that the reader can modify to fit their particular application needs.

Four appendices are included. Appendix A defines the notation and constants used in the text. Appendix B reviews various linear algebra concepts that are used in the main body of the text. Appendix C presents material from the GPS interface specification that is necessary to process GPS pseudorange, phase, and Doppler measurements. Example calculations are also included. Appendix D presents a short tutorial on quaternions. This appendix discusses their definition, operations, and kinematic model. Example calculations are included in each appendix.

Motivation

This book is motivated in part by the reader response to my prior book [48]. That response requested two major changes. First, that the material be restructured to support the book's use in a classroom setting. Second, that the subject matter be widened to the broader topic of aided navigation. Therefore, while the present book draws a significant amount of its source material from [48], the objectives of this book are distinct enough and the amount of new material is significant enough to necessitate the new title.

The present book discusses the general topic of aided navigation system design. Chapters 1 and 7 present the aided navigation system design methodology. Chapters 2–6 present the theoretical material required to understand and implement that method. Chapter 8 provides a detailed discussion of GPS. Chapters 9–12 provide detailed discussion of the aided navigation system design and analysis methodology.

This book is appropriate for use as a textbook in a senior or first year graduate level course. As such, the book includes extensive use of examples and end-of-chapter problems throughout Part I. As a textbook, the lectures could cover Part I, with a student end-of-course project to design and implement an aided navigation system. The systems in Chapters 9, 10, and 11 are particularly appropriate for senior design projects.

Many engineering programs require senior-level students to complete a substantial design project. A challenge for faculty involved in directing such projects is leading the student team through the quantitative definition of the specification and analysis of system performance relative to the specifications. A main reason that the aided navigation system design and analysis methodology presented herein is widely used in commercial and military applications is that it is easily amenable to such quantitative analysis of performance relative to specifications. Performance analysis is discussed in Chapter 6.

Book Website

Associated with this book is the publisher hosted website. The website contains various resources related to this book:

- The source code used to create examples using MATLAB;
- Data sets to support examples;
- Data sets that are requested by readers and that are reasonable for the author to provide;
- An errata list;
- Clarifications as requested by readers.

The official web site for this text can be found at

www.mhprofessional.com

then searching by isbn, title, or author. The book web site that results will contain a download section containing the above material.

Errata

While I have worked to ensure that this book is free from errors, previous experience has taught me that this is unlikely to occur. An errata list will be made available through the publisher website and through my university homepage, currently

www.ee.ucr.edu/~farrell

Readers who detect errors not already on the errata list are encouraged to report them electronically to the author. My e-mail address is currently

farrell@ee.ucr.edu

Identification of errors, or suggestions of either additional material (e.g., examples, exercises, or topics), is greatly appreciated.

Acknowledgements

First, I appreciate the many readers of [48] who communicated to me interesting ideas, corrections, and useful suggestions. Those communications helped to motivate and define this book. Second, I am grateful to acknowledge the help and collaboration of my students and colleagues in the development and proofreading of this book. The first set of reviewers included Anning Chen, Licheng Luo, Angello Pozo, and Arvind Ramanandan. The second set of reviewers included Jinrong Cheng, Wenjie Dong, Yu Lu, Paul Miller, Rolf Rysdyk, and Peng Xie. In addition, Wenjie Dong collaborated in the writing of Section 4.9.1; Arvind Ramanandan collaborated in the writing of Section 4.9.2; Yunchun Yang provided information essential to the writing of Chapter 10; Yu Lu collaborated in the writing of Section C.4 and provided source material helpful in the writing of Chapter 9; and, collaboration with Paul Miller was essential to the writing of Chapter 12. Of course, I take responsibility for any errors or omissions in the final presentation of all the material contained in the book. Third, I appreciate the help and support of the production team, mainly Wendy Rinaldi and Jean Bodeaux. Finally, I appreciate the patience of my family throughout this project.

This book was typeset using LaTeX. The examples and graphs for figures were implemented in MATLAB and SIMULINK. Drawings for figures were created using Macromedia FreeHand MX.

Jay A. Farrell
Department of Electrical Engineering
College of Engineering
University of California, Riverside
February 23, 2008

Part I

Theory

Part I Overview

This portion of the book presents a methodology for the design of aided navigation systems along with the theory necessary for implementation of the method. The methodology is briefly presented in Chapter 1 to motivate the study of the material in Chapters 2–6. Chapter 7 uses the material from Chapters 2–6 to reconsider the design and analysis approach in significantly more depth.

Part II of the book will use the skills developed in Part I to work through the design and analysis of several navigation systems.

Chapter 1

Introduction

Throughout history, the science of navigation has played an important role for mankind. Individuals who could reliably travel to and return from distant locations were successful both militarily and commercially [140].

Increasing levels of vehicle autonomy have been demonstrated in recent decades. A fundamental capability for an autonomous vehicle is navigation. In the autonomous vehicle literature, navigation may have one of two meanings:

1. accurate determination of the *vehicle state* (e.g., position, velocity, and attitude); or

2. planning and execution of the maneuvers necessary to move between desired locations.

The first capability is necessary to accurately achieve the second. This book focuses on the methods to implement the first capability; therefore, herein, the term *navigation* is used to refer to the process of estimating the vehicle state in realtime as the vehicle maneuvers along a trajectory. Both manned and unmanned vehicles require navigation capabilities. The vehicle state may be necessary for automatic control, real-time planning, data logging, simultaneous location and mapping, or operator communications. is referred to as *navigation.*

A typical systems approach to the larger guidance, navigation, and control (GNC) problem is illustrated in Figure 1.1. The objective of the GNC problem is to cause the vehicle, weapon, or robot (referred to below as the *plant*) to achieve a useful set of tasks requiring translational motion. The plant alone is an actuated connection of mechanical components. To move purposefully, the hardware must be augmented with sensors and a control system. The inputs to the control system are the estimated and commanded values of key system variables. When properly designed, the

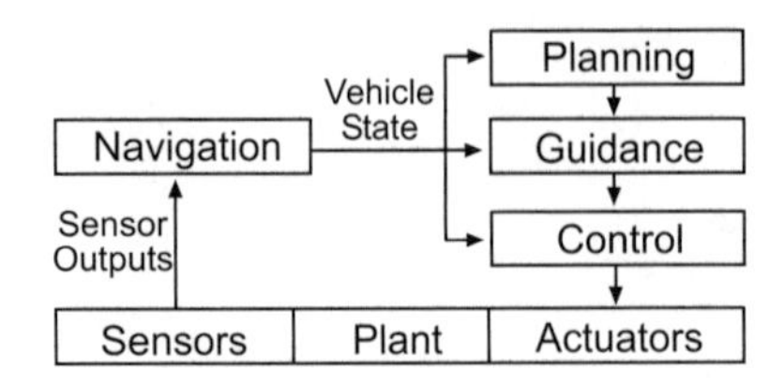

Figure 1.1: Block Diagram of a Typical GNC System.

output actuation signals from the control system force the system variables to their desired values [43, 53, 114]. If the system is to be capable of high performance translational motion, then it must be capable of accurately determining either its local (i.e., landmark relative) or world coordinates. This is the purpose of a navigation system. Based on knowledge of the desired objectives and the estimate of the vehicle state from the navigation system, the planning and guidance systems determine and output a trajectory in the appropriate coordinate system for input to the control system.

A classical approach to vehicle state estimation is to equip the vehicle with inertial sensors capable of measuring the acceleration and angular rate of the vehicle. With proper calibration and initialization, integration of the angular rates provides an estimate of the attitude, while integration of acceleration provides estimates of velocity and position. The integrative nature of this approach has both positive and negative aspects. On the positive side, integration will smoothed out the high-frequency errors (e.g., sensor noise). On the negative side, integration of low frequency errors due to biases, scale factor error, or misalignment will cause increasing error between the true and estimated vehicle state. The vehicle state estimate computed by integration of the data from the high rate sensors can be corrected using measurements from suitably selected low rate sensors. Examples of current and potential future aiding sensors include camera imagery, BeiDou (or Compass), Galileo, GLONASS, laser range finders, LORAN, Navstar global positioning system (GPS), pseudolites, radar, and star sensors. Navigation systems wherein low rate sensors are used to correct the state estimate produced by integration of the outputs from high rate sensors are referred to as *aided* or *integrated navigation systems*.

The main objective of this book is to systematically present the standard tools and methods used in aided navigation system design and analysis. The book is divided into two parts. Part I presents the theoretical issues that an engineer must understand to design and analyze aided navigation systems. The material is presented in a practical fashion with numerous illustrative examples. Derivations are presented with sufficient detail to allow the development of the theory in a step by step fashion. Part II presents various applications of the theory from Part I to navigation systems. The applications have been selected based on commercial, military, student project, and research interests. The GPS system is discussed in a chapter of its own. GPS is currently the dominant aiding system for outdoor applications. GPS will be used as the aiding system in two of the subsequent chapters of Part II. Each aided navigation system in Part II will be presented and analyzed using a consistent methodology and notation. In each case, the chapter will present a complete kinematic model with sufficiently detailed sensor error models to allow the reader to extend that portion of the model to fit the circumstances of their own application.

With the advent of multiple Global Navigation Satellite Systems (GNSS), it is likely that future commercial and student projects will use information from other GNSS in addition to GPS. In the author's experience, if a student learns the theory well and is able to develop and analyze a single aided navigation system, then the student is able to extend that approach to use alternative aiding sensors. Therefore, this book focuses on GPS aided systems to allow a thorough presentation and exemplification of system design and analysis without getting bogged down in the specific details of the many different possible aiding sensors.

The next section presents a very simple example of the navigation system design and analysis methodology that is used throughout this text. This section is intended to serve as a motivation for the various theoretical issues that will be considered in Part I of this text. Section 1.2 provides a summary of the material in Part I with discussion of the relation of that material to aided navigation system design and analysis. Section 1.3 provides a summary of the applications considered in Part II.

1.1 Method Overview

The method presented in this text is a classical approach that has been applied successfully in many applications for several decades. The method can best be understood in relation to a specific example system. The system is intentionally simple and low dimensional to focus on the main theoretical ideas without large amounts of algebra. The intent of the discussion is to motivate the theoretical coverage in Part I and the general approach used in Part II. The realistic systems of Part II are considerably more complex.

Most readers will not be familiar with all of the ideas presented in the following discussion, for example, differential equations with random inputs. Such readers should not be intimidated, the purpose of the example is to illustrate the necessity for the study of such topics in relation to aided navigation system design and analysis. The reader should use the example to identify and distinguish between the topics in which they are already knowledgable and the topics in which they are not. Then, the reader can use the discussion of Section 1.2 to identify a reading plan for Part I of the book. This example is revisited in greater depth in Chapter 7.

1.1.1 Methodology Example

Consider a point that is moving on a line. The position p, velocity v, and acceleration a are related by the kinematic model

$$\begin{aligned} \dot{p}(t) &= v(t) \\ \dot{v}(t) &= a(t). \end{aligned}$$

Kinematics is the study of motion without discussion of the masses or forces involved in creating the motion. Note that there is no uncertainty in the kinematic model of the system: no unknown parameters and no random signals.

If a sensor is available that provides a continuous-time measurement $\tilde{a}(t)$ of the acceleration $a(t)$, then a navigation system could integrate this measurement through the kinematic model to provide an estimate of the system state vector $\mathbf{x} = [p, v]^\top$. Throughout this book, the equations for computing the state estimate based on the integration of sensor signals through the vehicle kinematics will be referred to as the *mechanization equations*. Denoting the estimate of the state as $\hat{\mathbf{x}} = [\hat{p}, \hat{v}]^\top$, the mechanization equations for the navigation system would be

$$\left.\begin{aligned} \dot{\hat{p}}(t) &= \hat{v}(t) \\ \dot{\hat{v}}(t) &= \tilde{a}(t). \end{aligned}\right\} \tag{1.1}$$

Assuming that initial conditions $[\hat{p}(0), \hat{v}(0)]$ are available, numerical integration of the mechanization equations from time zero forward in time would provide an estimate of the system state for any positive value of time. Throughout the text, when the meaning is clear, the time argument of signals will be eliminated to simplify the notation.

To study the accuracy of the system we define the error state vector $\delta\mathbf{x} = \mathbf{x} - \hat{\mathbf{x}}$. The components of this vector in this simplified example would be $\delta p = p - \hat{p}$ and $\delta v = v - \hat{v}$. We can find the model for the time variation of the error state by subtracting the above equations:

$$\begin{aligned} \delta\dot{p} &= \dot{p} - \dot{\hat{p}} \\ &= v - \hat{v} \\ &= \delta v \end{aligned}$$

and

$$\begin{aligned} \delta\dot{v} &= \dot{v} - \dot{\hat{v}} \\ &= a - \hat{a} \end{aligned}$$

where

$$\hat{a} = \tilde{a} + \hat{b} \tag{1.2}$$

when a bias estimate $\hat{b}$ is available. For this simple example, $\hat{b} = 0$. Summarizing the results of the above analysis, the model of the time variation of the navigation error is

$$\left.\begin{array}{rcl} \delta\dot{p} & = & \delta v \\ \delta\dot{v} & = & a - \hat{a}. \end{array}\right\} \tag{1.3}$$

The navigation error state changes with time as dictated by eqn. (1.3). This model shows that the time variation of the error state is driven by the error in the measurement of a. Note that while the navigation system actually integrates eqns. (1.1) during the operation of the system, eqns. (1.3) cannot be integrated to determine the error state because the input $(a - \hat{a})$ is not known. Instead, eqns. (1.3) are analyzed theoretically to quantify and to identify the means to improve the performance of the system.

To characterize the performance of the navigation system, we must start with a characterization of the accuracy of the instruments that are used as inputs to the kinematic model of eqn. (1.1). For the purpose of this discussion, assume that the acceleration sensor measurement is accurately modeled as

$$\tilde{a}(t) = a(t) - b(t) - \omega_1(t). \tag{1.4}$$

The signals ω_1 and b represent imperfections in the sensor that can be modeled as stochastic processes. The signal ω_1 represents random measurement noise. The quantity b represents a sensor bias. This bias is unknown and varies slowly with time. The sensor manufacturer provides the following model of the bias time variation

$$\dot{b}(t) = -\lambda_b b(t) + \omega_b(t)$$

where λ_b is a positive constant and ω_b is another a random process. The sensor manufacturer also provides time-invariant statistical descriptions of the random signals $\omega_1(t)$ and $\omega_b(t)$.

Based on eqn. (1.4),

$$a(t) - \tilde{a}(t) = b(t) + \omega_1(t).$$

With this equation, the model for the system error becomes

$$\left.\begin{array}{rcl} \delta\dot{p} & = & \delta v \\ \delta\dot{v} & = & b + \omega_1 \\ \dot{b} & = & -\lambda_b b + \omega_b \end{array}\right\} \tag{1.5}$$

where the accelerometer error state dynamics have been included. This is a stochastic process model for the system error state. It is a linear system with two random inputs that, in this particular example, results in a time invariant model. It should be clear that the study of linear systems with random inputs will be critical to the characterization of the performance of navigation systems. Variations of this example are considered throughout Part I of the text. Readers curious about the nature of the error accumulation and the benefits of aiding could glance at Sections 4.9.3–4.9.4.

If a second sensor, referred to as an aiding sensor, is available that provides a measurement of some portion of the state at discrete instants of time. For example, a sensor might provide measurements of the position p at a 1 Hz rate. We might be interested in using the measured value of p to correct the estimated state of the navigation system. This can be formulated as a state estimation process. Then, important practical questions concern the design of, and performance available from, an optimal estimator.

1.1.2 Methodology Summary

In general, the procedure can be described as follows.

1. A kinematic model is derived for the system of interest in the form

$$\dot{\mathbf{x}} = \mathbf{f}(\mathbf{x}, \mathbf{u}) \tag{1.6}$$

where a sensor is available for the kinematic input variable $\mathbf{u}$.

2. The navigation mechanization equations are defined by

$$\dot{\hat{\mathbf{x}}} = \mathbf{f}(\hat{\mathbf{x}}, \tilde{\mathbf{u}}) \tag{1.7}$$

where $\tilde{\mathbf{u}}$ represents the measured value of the signal $\mathbf{u}$. Samples of $\tilde{\mathbf{u}}$ are available at the sample rate f_1, where f_1 is large relative to the bandwidth of the system of interest.

3. Models are developed for the $\tilde{\mathbf{u}}$ sensor and for any aiding sensors. In the following, the aiding sensors will be denoted by the variable $\mathbf{y}$. The aiding signals may be asynchronous or be sampled at different rates, but for simplicity of discussion, assume that they are all sampled at a rate f_2. It will typically be the case that $f_1 >> f_2$. It may be the case that f_2 is smaller than the bandwidth of the system of interest.

4. A linearized model is developed for the navigation error state vector. The goal for this step is to obtain a set of linear differential equations with inputs that are defined by stochastic processes with known statistics. This model is the basis for the subsequent two steps.

5. A state estimator is designed to stabilize the error state system.

6. The system performance is analyzed.

The design objective is to provide an optimal estimate of $\mathbf{x}$ for the given set of sensors. Additionally, the designer is often interested in performing quantitative analysis to answer questions such as:

- What values of f_1 and f_2 are necessary to achieve a specified performance?
- What quality of kinematic input sensor $\mathbf{u}$ is sufficient to achieve a specified performance?
- What quality of aiding sensor $\mathbf{y}$ is sufficient to achieve a specified performance?
- How do these design specifications interact? For example, if the quality of $\mathbf{y}$ is increased, how much can the quality of $\mathbf{u}$ be decreased?
- What algorithm should be selected to compute the error state estimate?
- Are there implementation simplifications that can be made without significantly affecting the performance relative to the specifications?

The example of Section 1.1.1 exhibits several standard characteristics of many navigation problems.

- The kinematic model of the actual system is deterministic and dynamic; however, the sensors used in the navigation system cause the model of the overall system to contain stochastic variables. Therefore, we will be interested in the behavior of deterministic and stochastic dynamic systems.
- The system dynamics naturally evolve in continuous-time; however, the sensed variables are only available at discrete-time instants. Therefore, we will be interested in equivalent discrete-time models for continuous-time systems.
- The sensors have different sampling rates with the quantities to be integrated (i.e., $\mathbf{u}$) sampled at a much higher rate than the quantities that result from the integration (i.e., $\mathbf{y}$).

With these design questions and model characteristics in mind, additional more theoretical questions are of interest.

- What is meant by optimal estimation?
- What is the optimal estimation algorithm?

- What are properties of the optimal estimator?
- To what form of model does the optimal estimator apply?
- What model structural properties are required for an optimal estimator to exist?
- How can the navigation problem be manipulated into a suitable form?
- How is the performance of the estimator analyzed?
- How can the performance analysis methods be used to analyze system design tradeoffs?

The example of Section 1.1.1 and previous discussion of this section motivates discussion of several topics critical to the design and analysis of aided navigation systems. Part I will present various theoretical topics that should be well understood by the navigation system designer and analyst. At the conclusion of Part I, the navigation system design and analysis methodology is reviewed with greater detail in Section 7.1. Part II will consider the use of these theoretical topics in the design and analysis of navigation systems. The following sections overview the structure of this text.

1.2 Overview of Part I: Theory

This section provides an overview of the material in Chapters 2-7.

1.2.1 Reference Frames

A clear understanding of reference frames is necessary in navigation system design. For example, Newton's equations apply in inertial reference frames, while the GPS ephemeris equations are typically solved in a specific Earth-centered Earth-fixed frame-of-reference. Since the Earth rotates about its own axis and orbits the sun, the Earth is not an inertial reference frame.

As sensors are added to a navigation system and equations are derived, the navigation system designer must make careful note and definition of any new reference frames that are introduced. To make comparisons between quantities that are naturally defined in different frames-of-reference, transformations between these frames-of-reference will be required. Such transformations may be fixed or time-varying. In either case, the accuracy of the transformation is critical to the accuracy of the transformed quantities.

As an example, consider the vehicle and Earth reference frames. Sensors onboard the vehicle may resolve their data in the vehicle reference frame;

however, the user of the vehicle is interested in knowing the vehicle state (i.e., position and velocity) in the Earth reference frame. The vehicle frame is attached to and rotates with the vehicle. The Earth frame is attached to and rotates with the Earth. Because the vehicle is free to translate and rotate with respect to the Earth, the vehicle and Earth frames can arbitrarily translate and rotate with respect to each other. Therefore, the navigation system must maintain accurate estimates of the quantities necessary to allow transformation of variables between these reference frames.

Definition of various important reference frames and transformations between rotating reference frames is addressed in Chapter 2.

1.2.2 Deterministic Systems

The kinematics of the vehicle naturally evolve in continuous-time. Therefore, the analysis and design process typically begins with the development of a *continuous-time* kinematic model for the vehicle. Various continuous-time model formats exist. The majority of this text will focus on the state space model format. Since navigation systems and optimal estimators are conveniently implemented in discrete-time on a computer, we will require discrete-time equivalent state space models of continuous-time systems.

To accommodate modeling of sensor alignment and calibration parameters, it may be desirable to *augment* the vehicle state variables with additional alignment and calibration variables. In the simple example in Section 1.1, p and v are the vehicle model states and b is a sensor calibration parameter. If we are able to estimate this parameter accurately, then the estimated value can be removed from the sensed quantity to produce a more accurate estimate of the acceleration a: $\hat{a}(t) = \tilde{a}(t) + \hat{b}(t)$, which could be used in the mechanization equations instead of the raw sensor output $\tilde{a}$.

Chapter 3 will review various important concepts related to continuous and discrete-time dynamic systems. Since all engineering students take ordinary differential equations, physics, and some dynamics courses, the underlying concepts should be familiar; however, not all such courses are formulated within the state variable framework. Therefore, the two main goals of that chapter are to review state space analysis and state estimation.

1.2.3 Stochastic Processes

By their senior year at college, most students are fairly comfortable with modeling continuous-time deterministic systems. The idea that we integrate acceleration to find velocity and integrate velocity to find position is now rather innate. However, for many readers derivation and analysis of stochastic models remains somewhat esoteric.

Even in the simple example in Section 1.1, the model has (at least) three stochastic variables: ω_1, b and ω_b. In fact, we will see that any quantities computed from these variables will also be stochastic variables. Therefore, accurate stochastic system modeling and analysis is critical to navigation system design. Accurate characterization of model and variable properties and analysis of their affect on the system defines the analysis of the overall system performance.

When purchasing or designing sensors, the designer naturally searches for the *best* sensors that fit within the project budget. However, any use of an adjective such as "best" requires further consideration of its definition in this context. For example, for the acceleration sensor, the best sensor would have no noise (i.e., $\omega_1(t) = 0$) and no bias (i.e., $b(t) = 0$); however, this is typically not possible under application conditions. Instead, thermal fluctuations and component variation will cause both quantities to be nonzero. To surpass the competition, each sensor manufacturer has the incentive to remove, to the extent possible, all the predictable components of sensor error (e.g., nonlinearities, temperature dependent characteristics). Therefore, the remaining sensor error is non-deterministic and will be accommodated within the navigation system design by stochastic process modeling and state estimation. Knowing this, the sensor manufacturer designs the sensor so that the characteristics that define the random nature of the sensor are time-invariant. Also, the sensor manufacturer will typically supply the parameters that quantify the stochastic nature of the sensor. A brief discussion of sensor specifications is presented in Section 4.9.2.

Chapter 4 discusses the various topics from the theory of random variables and stochastic processes that must be understood for the design and analysis of navigation systems. This discussion includes definitions of basic stochastic process concepts, discussion of dynamic systems with stochastic inputs, definition of various stochastic process models that regularly appear in navigation systems, and discussion of the time variation of the mean and variance of the state of a linear stochastic state space system. Chapter 4 concludes with a discussion of state estimation as it relates to linear stochastic state space systems and the characterization of their state estimation accuracy through covariance analysis. The example of Section 1.1.1 is considered again in Section 4.9.3-4.9.4 using the quantitative methods of Chapter 4.

1.2.4 Optimal State Estimation

Based on the considerations of the previous subsections, aided navigation system design will be organized as a state estimation problem for a stochastic dynamic system. Due to the stochastic nature of the measured variables, the designer cannot expect to achieve a state estimation error that converges exactly to zero; however, the designer has the means to quantify the performance of the state estimation process. Therefore, a natural question to consider is what is the optimal level of performance that can be achieved and what is the estimator that will provide that level of performance.

Chapter 5 derives the Kalman filter and discusses its properties along with key implementation considerations. Under certain assumptions, the Kalman filter is the optimal state estimator for a linear system with stochastic inputs. These assumptions and their relation to navigation applications will be considered.

1.2.5 Performance Analysis

The analysis in Chapter 5 assumes that the estimator design model is an exact model of the navigation system error dynamics. That assumption is optimistic. In the design process, a series of modeling tradeoffs may be considered for various reasons including computational loading. Due to such tradeoffs and the fact that any mathematical model is only an approximation to the actual physical system, the estimator design model will not exactly match the real system. Therefore, the analyst should be interested in analyzing the robustness of the system performance to discrepancies between the actual and model systems. In addition, the designer may be interested in considering how the system performance would change under different sets of design choices, for example, changing the quality of a sensor.

Chapter 6 presents both a method to study the robustness of the system performance to model errors and a method that allows the designer to consider each source of system error separately to determine its contribution to the overall error budget.

1.2.6 Aided Navigation System Design and Analysis

Chapters 2–6 will have presented various theoretical topics which can be combined to provide a classical theory for the design and analysis of aided navigation systems. Chapter 7 combines together the various theoretical concepts within simple navigation examples to illustrate the key concepts of the approach. This understanding is the stepping off point for Part II of the text. It is critical to understand the basic theory and overall approach prior to attempting more complex and higher dimensional real world navigation systems.

Chapter 7 will discuss the complementary filter with applications and exercises. The complementary filter framework relates well to aided navigation system design. Within this framework it is natural to focus on key ideas such as the complementary nature of sensors, observability, and linearized models for stochastic error variables.

Researchers might be tempted to investigate more advanced concepts such as particle filters, neural networks, or fuzzy estimators. Such advanced topics are sometimes proposed and may, in some applications, be appropriate. However, the classical approach built on the theory in Part I of this text, outlined in Chapter 7, and discussed for several specifics applications in Part II is a complete and rigorous approach to the design and analysis of navigation systems. It is highly recommended that the researcher understand this classical approach, its strengths, and its weaknesses before moving on to more advanced concepts.

1.3 Overview of Part II: Applications

Part I of this text has presented the essential components of the theory necessary for the design and analysis of navigation systems. Chapter 7 has brought those concepts together to present the classical approach to aided navigation system design. Chapter 7 focused on a rather simple example using the idea of a complementary filter. The purpose of Part II is to extend that methodology to the design of various navigation systems suitable for real world application. Prior to considering aided navigation systems, Part II first introduces the GPS system as a typical aiding sensor. The coverage of each chapter is discussed in the following subsections.

1.3.1 GPS

Chapter 8 discusses the Navstar Global Positioning System. This chapter includes discussion of the GPS observables, measurement models, error models, position solution methods, and differential GPS.

In this text GPS is used as the prototypical aiding sensor. Aiding with Compass, Galileo, GLONASS, combined GNSS signals, or pseudolites would be similar in nature, but distinct in specific integration details.

Similarly, underwater applications are denied access to satellite signals; however, a variety of acoustic ranging sensor systems exist. Understanding of the use of GPS as an aiding signal enables extension to underwater acoustic ranging systems. However, the speed of sound in water is significantly slower and more variable than the speed of light in Earth's atmosphere; therefore, depending on the desired level of performance characterization of the speed of sound in water and accommodation of the time of travel of the ranging signal can be critical.

1.3.2 Aided Navigation Systems

Chapters 9 and 11 will discuss the design and analysis of GPS aided Encoder and GPS aided INS, respectively. Chapter 10 discusses the design of an attitude and heading reference system. Chapter 12 discusses the design of a

Doppler and acoustic ranging aided inertial navigation system for an underwater vehicle. Each chapter will present the system kinematic model, basic sensor models, measurement prediction equations, measurement residual equations, an error state dynamic model, observability analysis, and example covariance analysis. Each chapter includes enough discussion to serve as a detailed starting point for applications or student projects.

1.4 Overview of Appendices

Four appendices are included. Appendix A is provided as a summary of notation used in the book. Appendix B reviews various linear algebra concepts and includes a few more advanced topics. It is intended to serve as a convenient point of reference to make the text more self-contained. Appendix C presents the GPS ephemeris and ionospheric models. The appendix includes an example set of calculations and equations for computing the velocity of the GPS satellites. Appendix D discusses the important topic of quaternions. The main body of the text discusses mainly the Euler angle and direction cosine attitude representations. Even though quaternions are often the preferred attitude representation, since students often find quaternions difficult and they are not essential for implementation, quaternions have been included as an appendix. The intent is to allow the main body of the text to flow through the theory presented in Part I and the applications of Part II. Once these concepts are familiar and at least one application is successfully functioning, the topic of numeric efficiency and singularity free representations is an extension that often will naturally occur.

Chapter 2

Reference Frames

Navigation systems require the transformation of measured and computed quantities between various frames-of-reference. The purpose of this chapter is to define the various frames and the methods for transforming the coordinates of points and representation of vectors between frames. Before proceeding to the main body of the chapter, the next paragraph steps through the measurements and computations of a strapdown inertial navigation system (INS). The goals of this brief introduction are to define notation and to illustrate how the different frames-of-reference come into play in navigation applications.

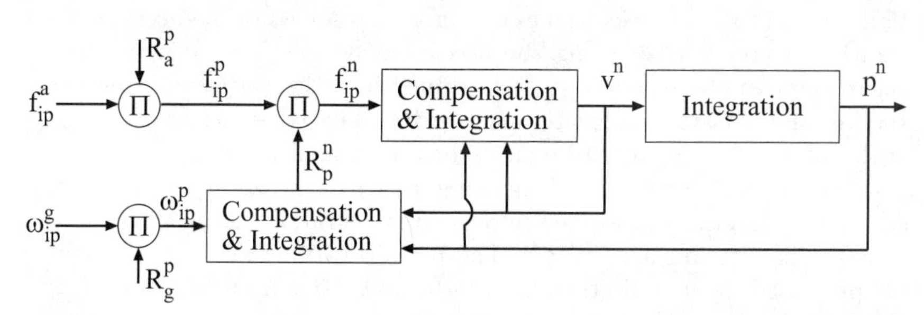

Figure 2.1: High level block diagram of an inertial navigation system. For a vector, the superscript defines the frame-of-reference in which the vector is represented. For a matrix transformations between two frames-of-reference, the subscript defines the origination frame and the superscript defines the destination frame. The encircled symbol Π represents a product.

A high level block diagram for a typical strapdown INS is shown in Figure 2.1. Before discussing its operation, it is necessary to briefly discuss notational conventions. As will be discussed in this chapter, vectors have distinct coordinate representations in distinct reference frames. Therefore, a notation is required to record the frame of the representation. The reference frame in which a vector is represented is indicated by a superscript, for example $\mathbf{v}^n$ is the velocity vector represented in the n or navigation frame. In the text of the book, vectors and matrices will be represented in boldface. Rotational transformations will be represented by an $\mathbf{R}$ with a subscript and superscript indicating the origin and destination frames-of-reference. For example, $\mathbf{R}_p^n$ represents the rotational transformation from the p or platform frame to the n or navigation frame. The various reference frames are defined in Section 2.2. Finally, certain quantities such as angular rates measure the rotational rate of one frame relative to another in addition to being represented in a specific reference frame. For example, $\boldsymbol{\omega}_{ip}^p$ should be read as the angular rate of frame p with respect to frame i represented in frame p. With this notation, it is true that

$$\boldsymbol{\omega}_{ip}^p = -\boldsymbol{\omega}_{pi}^p.$$

Throughout the book, all angular rates are defined to be positive in the right-handed sense. This means that if the thumb of the right hand points along the direction of the angular rate vector, then the fingers of the right hand will indicate the physical sense of the rotation. Alternatively, this can be stated that looking up along the angular rate vector, from the tail toward the head, the sense of the rotation is clockwise.

The INS in Figure 2.1 uses variables related to five different coordinate systems. The accelerometers measure $\mathbf{f}_{ip}^a$ which is the platform acceleration relative to an *inertial* frame-of-reference, resolved in the *accelerometer* frame-of-reference (i.e., along the accelerometer sensitive axis). The accelerometer measurements are transformed into the *platform* frame using the (usually) constant calibration matrix $\mathbf{R}_a^p$. The gyros measure $\boldsymbol{\omega}_{ip}^g$ which is the platform angular rate relative to the inertial frame-of-reference, as resolved in *gyro frame* (i.e., along the gyro sensitive axis). The gyro measurements are also transformed into the platform frame using the (usually) constant calibration matrix $\mathbf{R}_g^p$. The platform frame gyro measurements are processed to maintain the platform-to-*navigation* frame rotation matrix $\mathbf{R}_p^n$. Finally, the platform-to-navigation frame rotation matrix $\mathbf{R}_p^n$ is used to transform the accelerometer measurements into the navigation frame where they are processed to determine navigation frame velocity and position.

This chapter discusses the definition of, properties of, and transformation between frames-of-reference. Section 2.1 provides a short summary of some properties of reference frames and coordinate systems to simplify the discussion of the subsequent sections. Section 2.2 defines each frame-of-reference along with at least one definition of a suitable coordinate frame. Section 2.3 focuses on definition of and transformations between common coordinates systems for the important Earth Centered Earth Fixed frame-of-reference. Section 2.4 discusses the general approach to transforming points and vectors between rectangular coordinate systems. Section 2.5 discusses plane rotations with derivations of specific rotation matrices that are useful in navigation applications. Since navigation systems are freely moving, one navigation system may rotate relative to another. Therefore, Section 2.6 discusses the properties of time derivatives of vectors represented in and rotation matrices representing transformations between rotating reference frames. Section 2.7 discusses methods to compute the direction cosine matrix based on measurements of the relative angular rate between two reference frames.

2.1 Reference Frame Properties

As previously motivated, navigation systems involve variables measured in various frames-of-reference. To insure system interoperability and clear communication between engineers working with the system each frame and its properties should be clearly defined.

As exemplified in the introduction to this chapter, throughout this book, vectors and matrices will be written in boldface. When a vector is represented relative to a specific reference frame, the frame-of-reference will be indicated by a superscript. For example, $\boldsymbol{\omega}^a$ is the vector $\boldsymbol{\omega}$ represented with respect to the a frame-of-reference. The rotation matrix for transforming a vector representation from frame a to frame b is written as $\mathbf{R}_a^b$ where

$$\boldsymbol{\omega}^b = \mathbf{R}_a^b \boldsymbol{\omega}^a. \tag{2.1}$$

This equation is derived in Section 2.4. Specific rotation matrices $\mathbf{R}_a^b$ will be defined in Section 2.5. Appendix A contains summary tables of the various coordinate frame, notation, variable, and constant definitions used throughout this chapter and the remainder of the book.

Unless otherwise stated, all rectangular frames-of-reference are assumed to have three axes defined to be orthonormal and right-handed. Let the unit vectors along the coordinate axes of a reference frame be $\mathbf{I}$, $\mathbf{J}$, and $\mathbf{K}$. Orthonormality requires that

$$\begin{array}{ccc} \mathbf{I}\cdot\mathbf{I}=1 & \mathbf{I}\cdot\mathbf{J}=0 & \mathbf{I}\cdot\mathbf{K}=0 \\ \mathbf{J}\cdot\mathbf{I}=0 & \mathbf{J}\cdot\mathbf{J}=1 & \mathbf{J}\cdot\mathbf{K}=0 \\ \mathbf{K}\cdot\mathbf{I}=0 & \mathbf{K}\cdot\mathbf{J}=0 & \mathbf{K}\cdot\mathbf{K}=1 \end{array}$$

where $\mathbf{K}\cdot\mathbf{I}$ denotes the inner product between $\mathbf{K}$ and $\mathbf{I}$. Right-handed implies that $\mathbf{I}\times\mathbf{J}=\mathbf{K}$.

Due to the gravitational effects near the surface of the Earth, it is often convenient to consider ellipsoidal coordinate systems in addition to rectangular coordinate systems. The Earth's geoid is a hypothetical equipotential surface of the Earth's gravitational field that coincides with the Earth's mean sea level. By definition the gravity vector is everywhere perpendicular to the geoid. For analytic tractability the Earth's geoid is usually approximated by a reference ellipsoid that is produced by rotating an ellipse around its semi-minor axis. Figure 2.2 depicts an ellipse that has much greater flatness than the that of the Earth, but the ellipse is useful for making a few important points. Most importantly, the normal to the ellipse, when extended towards the interior of the ellipse does not intersect the center of the Earth. See exercise 2.1. Since the gravity vector is (nominally) perpendicular to the ellipsoid, the gravity vector does not point to the center of the Earth. Also, as shown in the left portion of the figure, the latitude of a point P can be defined in two different ways. The geodetic latitude ϕ is the angle from the equator to the outward pointing normal vector with positive angles in the northern hemisphere. The geocentric latitude ϕ_c is the angle from the equator to the vector from the center of the Earth to P. Therefore, we can have geodetic or geocentric ellipsoidal coordinates. Also, we must be careful to clearly distinguish whether a local coordinate systems aligns with the geodetic or the geocentric normal.

There are also two common classes of methods for defining the origin O of a reference system on the surface of the Earth. As shown in the left portion of Figure 2.2, the origin of a reference frame can be defined as the projection of the point P onto the reference ellipsoid. In this case, the reference frame origin O moves as dictated by the horizontal portion of the motion of P. Alternatively, as shown in the right portion of the figure, the origin O may be fixed to the surface of the Earth at a point convenient for local reference, e.g., the end of a runway. In this case, the origin of the local reference frame does not move with P.

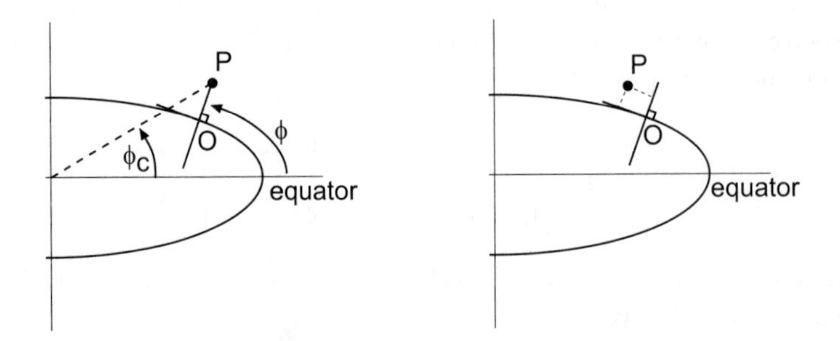

Figure 2.2: Left — Exaggerated ellipse depicting the difference between geodetic and geocentric latitude. Right — Depiction of a local frame-of-reference.

2.2 Reference Frame Definitions

This section defines various frames-of-reference that are commonly used in navigation system applications.

2.2.1 Inertial Frame

An inertial frame is a reference frame in which Newton's laws of motion apply. An inertial frame is therefore not accelerating, but may be in uniform linear motion. The origin of the inertial coordinate system is arbitrary, and the coordinate axis may point in any three mutually perpendicular directions. All inertial sensors produce measurements relative to an inertial frame, resolved along the instrument sensitive axis.

For discussion purposes it is sometimes convenient to define an Earth centered inertial (ECI) frame which at a specified initial time has its origin coincident with the center of mass of the Earth, see Figure 2.3. At the same initial time, the inertial x and z axes point toward the vernal equinox and along the Earth spin axis, respectively. The y-axis is defined to complete the right-handed coordinate system. The axes define an orthogonal coordinate system. Note that the ECEF frame, defined in Section 2.2.2, rotates with respect to this ECI frame with angular rate ω_{ie}; therefore, in the ECI frame the angular rate vector is $\boldsymbol{\omega}_{ie}^i = [0, 0, \omega_{ie}]^\top$.

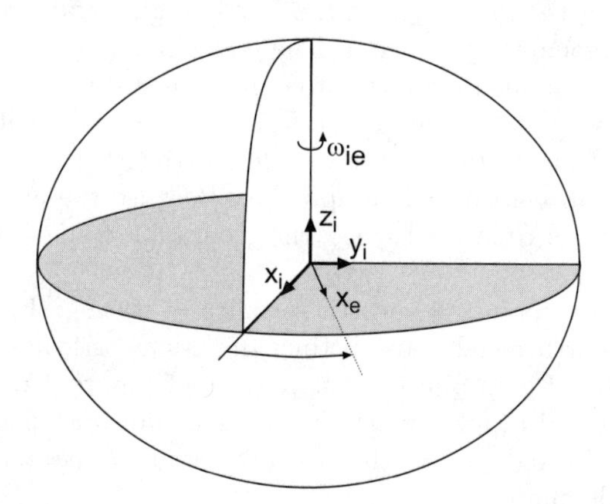

Figure 2.3: Rotation of the ECEF frame with respect to an Earth-centered inertial frame. The vectors $\mathbf{x}_i$, $\mathbf{y}_i$, and $\mathbf{z}_i$, define the axes of the ECI frame. The vector $\mathbf{x}_e$ defines the x-axis of the ECEF frame.

2.2.2 Earth Centered Earth Fixed (ECEF) Frames

This frame has its origin fixed to the center of the Earth. Therefore, the axes rotate relative to the inertial frame with frequency

$$\omega_{ie} \approx \left(\frac{1 + 365.25 \text{ cycle}}{(365.25)(24) \text{ hr}}\right)\left(\frac{2\pi \text{ rad/cycle}}{3600 \text{ sec/hr}}\right) = 7.292115 \times 10^{-5} \frac{rad}{sec} \quad (2.2)$$

due to the 365.25 daily Earth rotations per year plus the one annual revolution about the sun. Relative to inertial frame, the Earth rotational rate vector expressed relative to the ECEF axes is $\boldsymbol{\omega}_{ie}^{e} = [0, 0, 1]^{\top} \omega_{ie}$.

The Earth's geoid is usually approximated as an ellipsoid of revolution about its minor axis. A consistent set of Earth shape (i.e., ellipsoid) and gravitation model parameters must be used in any given application. Therefore, the value for ω_{ie} in eqn. (2.2) should only be considered as an approximated value. Earth shape and gravity models are discussed in Section 2.3. Due to the Earth rotation, the ECEF frame-of-reference is *not* an inertial reference frame. Two common coordinate systems for the ECEF frame-of-reference are discussed in Section 2.3.

2.2.3 Geographic Frame

The geographic frame is defined locally, relative to the Earth's geoid. The origin of the geographic frame moves with the system and is defined as the projection of the platform origin P onto the reference ellipsoid, see the left portion of Figure 2.2. The geographic z-axis points toward the interior of the ellipsoid along the ellipsoid normal. The x-axis points toward true north (i.e., along the projection of the Earth angular rate vector $\boldsymbol{\omega}_{ie}$ onto the plane orthogonal to the z-axis). The y-axis points east to complete the orthogonal, right-handed rectangular coordinate system.

Since the origin of the geographic frame travels along with the vehicle, the axes of the frame rotate as the vehicle moves either north or east. The rotation rate is discussed in Example 2.6. Because the geographic frame rotates with respect to inertial space, the geographic frame is not an inertial frame.

Two additional points are worth specifically stating. First, true north and magnetic north usually are distinct directions. Second, as illustrated by the exaggerated ellipse in the left portion of Figure 2.2, the normal to a reference ellipsoid (approximate Earth geoid) does not pass through the center of the ellipsoid, unless the platform origin P is at the equator or along the Earth spin axis.

2.2.4 Geocentric Frame

Closely related to the geographic frame-of-reference is the geocentric frame. The main distinction is that the geocentric z-axis points from the system location towards the Earth's center. The x-axis points toward true north in the plane orthogonal to the z-axis. The y-axis points east to complete the orthogonal, right-handed rectangular coordinate system. Like the geographic frame, the axes of the geocentric frame rotate as the vehicle moves north or east; therefore, the geocentric frame is not an inertial frame.

2.2.5 Local Geodetic or Tangent Plane

The local geodetic frame is the north, east, down rectangular coordinate system we often refer to in our everyday life (see Figure 2.4). It is determined by fitting a tangent plane to the geodetic reference ellipse at a point of interest. The tangent plane is attached to a fixed point on the surface of the Earth at some convenient point for local measurements. This point is the origin of the local frame. The x-axis points to true north. The z-axis points toward the interior of the Earth, perpendicular to the reference ellipsoid. The y-axis completes the right-handed coordinate system, pointing east.

For a stationary system, located at the origin of the tangent frame, the geographic and tangent plane frames coincide. When a system is in motion, the tangent plane origin is fixed, while the geographic frame origin is the projection of the platform origin onto the reference ellipsoid of the Earth. The tangent frame system is often used for local navigation (e.g., navigation relative to a runway).

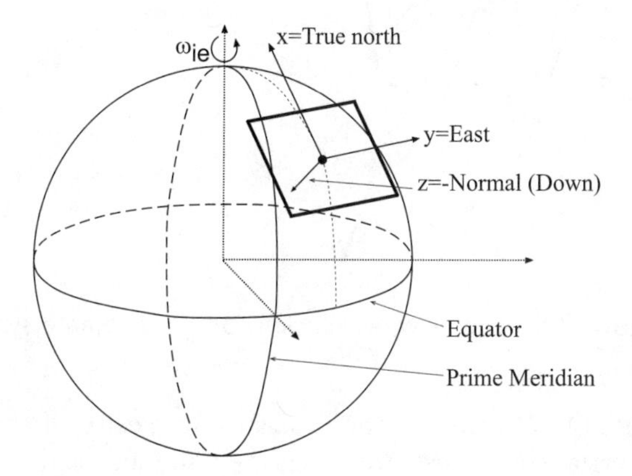

Figure 2.4: Local geodetic or tangent plane reference coordinate system in relation to the ECEF frame.

2.2.6 Body Frame

In navigation applications, the objective is to determine the position and velocity of a vehicle based on measurements from various sensors attached to a platform on the vehicle. This motivates the definition of vehicle and instrument frames-of-reference and their associated coordinate systems.

The body frame is rigidly attached to the vehicle of interest, usually at a fixed point such as the center of gravity. Picking the center of gravity as the location of the body frame origin simplifies the derivation of the kinematic equations [90] and is usually convenient for control system design. The u-axis is defined in the forward direction of the vehicle. The w-axis is defined pointing to the bottom of the vehicle and the v-axis completes the right-handed orthogonal coordinate system. The axes directions so defined (see Figure 2.5) are not unique, but are typical in aircraft and underwater vehicle applications. In this text, the above definitions will be used. In addition, the notation $[u, v, w]$ for the vehicle axes unit vectors has been used instead of $[x, y, z]$, as the former is more standard.

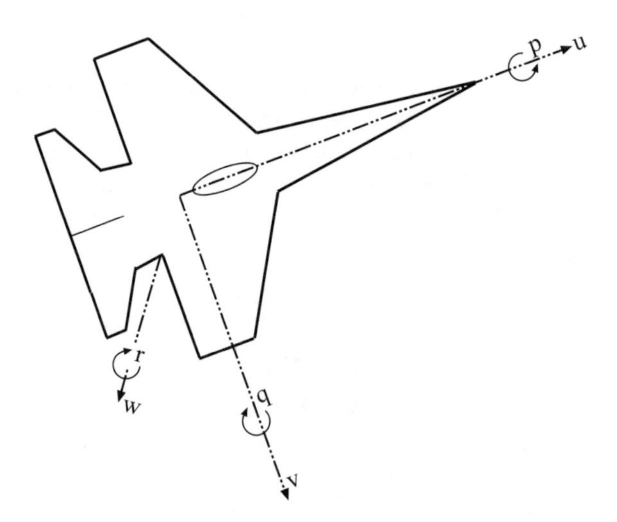

Figure 2.5: Top view of vehicle (body) coordinate system.

As indicated in Figure 2.5, the rotation rate vector of the body frame relative to inertial space, resolved along the body axis is denoted by $\boldsymbol{\omega}_{ib}^{b} = [p, q, r]^{\top}$ where p is the angular rate about the u-axis (i.e., roll rate), q is the angular rate about the v-axis (i.e., pitch rate), and r is the angular rate about the w-axis (i.e., yaw rate). Each angular rate is positive in the right-hand sense. The body frame is not an inertial frame-of-reference.

2.2.7 Platform Frame

This book will only discuss applications where the sensors are rigidly attached to the vehicle. For inertial navigation, such systems are referred to as *strap-down* systems. Although the sensor platform is rigidly attached to the vehicle, for various reasons the origin of the platform frame may be offset or rotated with respect to the origin of the body frame. The origin of the platform coordinate frame is at an arbitrary point on the platform. The platform frame axes are defined to be mutually orthogonal and right-handed, but their specific directions are application dependent. Often the rotation matrix $\mathbf{R}_p^b$ is constant and determined at the design stage.

2.2.8 Instrument Frames

Typically an instrument frame-of-reference is defined by markings on the case of the instrument. Sensors within the instrument resolve their measurements along the sensitive axes of the instruments. Ideally, the instrument sensitive axes align with the instrument frame-of-reference; however, perfect alignment will not be possible under realistic conditions. In addition, the sensitive axes may not be orthogonal. Instrument manufacturers may put considerable effort into in-factory calibration and orthogonalization routines in addition to temperature, linearity, and scale factor compensation. Such manufacturer defined compensation algorithms will be factory preset and programmed into the instrument. Depending on the desired level of performance, the navigation system designed may have to consider additional instrument calibration either at system initialization or during operation. Methods to accomplish this are a major item of discussion throughout the subsequent chapters of this book.

Even though the transformation of sensor data from the instrument frame to the platform frame is not necessarily a rotation, for consistency, we will use a similar notation. For example, the transformation of accelerometer data to platform frame will be denoted by $\mathbf{R}_a^p$.

2.2.9 Summary

A natural question at the beginning of this section is why are the various coordinate systems required? The answer is that each sensor provides measurements with respect to a given reference frame. The fact that the sensor-relevant coordinate frames are typically distinct from the navigation frame can now be made more concrete.

The GPS system (described in Chapter 8) provides estimates of an antenna position in the ECEF coordinate system, vision and radar provide distance measures in a local instrument-relative coordinate system, accelerometers and gyros provide inertial measurements expressed relative to their instrument axes. Given that different sensors provide measurements relative to different frames, the measurements in different frames are only comparable if there are convenient means to transfer the measurements between the coordinate systems. For example, a strap-down GPS aided INS system performing navigation relative to a fixed tangent plane frame-of-reference will typically:

1. transform acceleration and angular rate measurements to platform coordinates;

2. compensate the platform angular rate measurement for navigation frame rotation;

3. integrate the compensated platform frame angular rates to maintain an accurate vector transformation from platform to navigation coordinates;

4. transform platform frame accelerations to tangent plane using the transformation from step 3;

5. integrate the (compensated) tangent plane accelerations to calculate tangent plane velocity and position;

6. use the position estimate to predict the GPS observables;

7. make GPS measurements, compute the residual error between the predicted and measured GPS observables, and use these measurement residuals to estimate and correct errors in the sensed and calculated INS quantities;

8. transform the vehicle inertial measurements and state variables that are estimated above to frames-of-reference (e.g., body) that might be desired by other vehicle systems (e.g., control or mission planning).

A similar procedure is followed for other aided navigation systems. The necessary coordinate system transformations and the algorithms for their on-line calculation are derived in the following sections.

2.3 ECEF Coordinate Systems

Two different coordinate systems are common for describing the location of a point in the ECEF frame. These rectangular and geodetic coordinate systems are defined in the following two subsections. Subsection 2.3.3 discusses the transformation between the two types of coordinates.

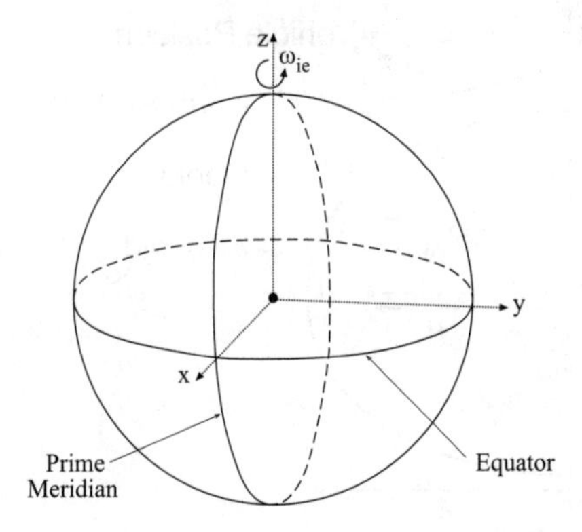

Figure 2.6: ECEF rectangular coordinate system.

2.3.1 ECEF Rectangular Coordinates

The usual rectangular coordinate system $[x, y, z]^e$, herein referred to as the ECEF coordinate system, has its origin at the Earth's center of mass with its x-axis extend through the intersection of the prime meridian (0 deg longitude) and the equator (0 deg latitude). The z-axis extends through the true north pole (i.e., coincident with the Earth spin axis). The y-axis completes the right-handed coordinate system, passing through the equator and 90 deg longitude, as shown in Figure 2.6.

2.3.2 The Earth Geoid and Gravity Model

The following sections briefly discuss the need for and definition of the Earth geoid and the gravity model relative to this geoid.

The gravity vector is the vector sum of the gravitational force of the Earth mass and the centrifugal force due to Earth rotation. The relative size and orientation of these two forces is dependent on the point of evaluation. For example, the centrifugal force is maximum at the equator and zero along the Earth axis of rotation. The geodetic surface of the Earth is defined to be everywhere normal to the gravity vector (i.e., an equipotential surface relative to the force of gravity). The geodetic surface is distinct from the actual irregular shape of the surface, but may be imagined as the mean shape that the Earth would take if the solid surface of the Earth were completely covered with sea water.

2.3.2.1 Earth Geoid

Figure 2.7 illustrates various surfaces useful for understanding the actual shape of the Earth and analytic representations of that shape. The geoid is

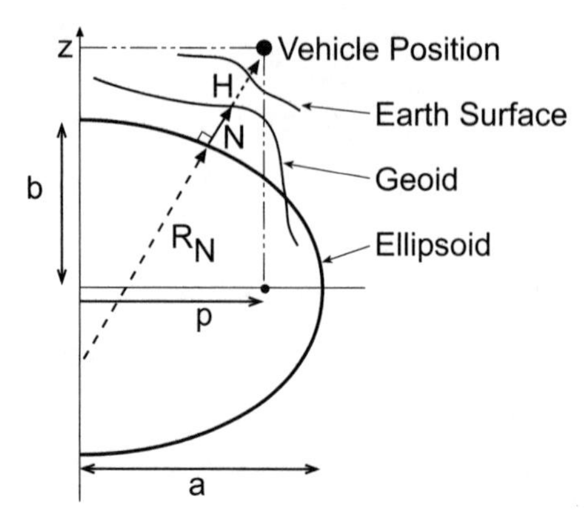

Figure 2.7: Definitions of Earth surfaces related to geodetic frame.

the mean sea level of the Earth. The actual Earth surface fluctuates relative to the geoid. For analytical convenience the geoid is approximated by an ellipsoid created by rotation of an ellipse around its minor axis as shown in Figure 2.8. The center of the ellipsoid is coincident with the center of mass of the Earth. The minor axis of the ellipse is coincident with the mean rotational axis of the Earth.

Most charts and navigation tools are expressed in the (ϕ, λ, h) geodetic coordinates, as depicted in Figure 2.8, where ϕ denotes latitude, λ denotes longitude, and h denotes altitude above the reference ellipsoid. *Latitude* is the angle in the meridian plane from the equatorial plane to the ellipsoidal normal N. Note that the extension of the normal towards the interior of the ellipsoid will not intersect the center of the Earth except in special cases such as $\phi = 0$ or $\pm 90°$. See Exercise 2.1. *Longitude* is the angle in the equatorial plane from the prime meridian to the projection of the point of interest onto the equatorial plane. *Altitude* is the distance along the ellipsoidal normal, away from the interior of the ellipsoid, between the surface of the ellipsoid and the point of interest.

A geodetic model is defined by specification of four constants. The eccentricity of the ellipsoid that approximates the geoid is determined by the gravitational attraction and angular rotation rate of the Earth. Therefore, the defining parameters for a geodetic system must be defined consistently to determine the ellipsoid and the gravity model. The parameters of a geodetic model are determined by least squares fitting to experimental data. Various geodetic parameter sets have been defined by different users at different times. The model accuracy has increased along with the associated instrumentation accuracy. In particular,improved accuracy was made possible by the analysis of the trajectories of artificial satellites. This book will use the WGS84 geodetic system. The WGS84 defining constants are

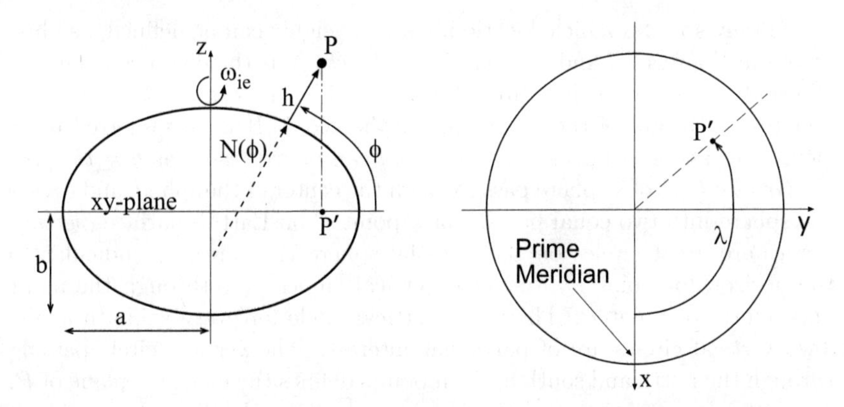

Figure 2.8: Geodetic reference coordinate system. Left - Cross section of the ellipsoid containing the rotational axis and the projection of the point P onto the equatorial plane. Right - Cross section of the ellipsoid containing the equatorial plane.

Name	Sysmbol	Value	Units
Equatorial radius	a	6378137	m
Reciprocal flattening	$\frac{1}{f}$	298.257223563	
Angular rate	ω_{ie}	7.292115×10^{-5}	$\frac{rad}{s}$
Gravitational constant	GM	$3.986004418 \times 10^{14}$	$\frac{m^3}{s^2}$

Table 2.1: WGS84 defining parameters.

contained in Table 2.1 [12].

Two constants define the reference ellipse. The ellipse defining parameters are typically the length of the semimajor axis a and one of the following:

$$\begin{aligned} \text{semiminor axis:} \quad & b \\ \text{eccentricity:} \quad & e = \sqrt{\tfrac{a^2-b^2}{a^2}} = \sqrt{f(2-f)} \\ \text{flatness:} \quad & f = \tfrac{a-b}{a}. \end{aligned}$$

Based on the defining constants in Table 2.1, the following ellipse parameter values can be derived:

$$f = 0.00335281, \tag{2.3}$$

$$b = a(1-f) = 6356752.314m, \tag{2.4}$$

$$e = \sqrt{f(2-f)} = 0.08181919. \tag{2.5}$$

Replication of the above computations should be completed using machine precision, not the precision displayed in the above results.

For any specific vehicle location, various heights can be defined, as illustrated in Figures 2.7 and 2.8. The *geoid height* N is the distance along the ellipsoid normal from the ellipsoid to the geoid. The *orthometric height* H denotes the height of the vehicle above the geoid. It is also referred to as elevation. The altitude or *geodetic height* can be expressed as $h = H + N$.

Great circles of a sphere pass through the center of the sphere and divide the sphere into two equal parts. For a point P on Earth's surface, *vertical circles* are great circles on the celestial sphere that are perpendicular to the horizon for point P. As such, vertical circles pass through the nadir and zenith for point P. There is a vertical circle for every azimuth angle. Two vertical circles are of particular interest. The vertical circle passing through the north and south horizon points defines the meridian plane of P. The intersection of that plane with the Earth ellipsoid defines the *meridian*, which is an ellipse of constant longitude (modulo 180 degrees). Let ϕ denote the latitude of P. At P the meridian ellipse has radius of curvature $R_M(\phi)$ referred to as the *meridian radius* or the *radius of the ellipse*

$$R_M(\phi) = \frac{a(1-e^2)}{\left(1 - e^2\sin^2(\phi)\right)^{\frac{3}{2}}}. \tag{2.6}$$

The vertical circle through the east and west horizon points is the prime vertical. The prime vertical also defines a plane. The radius of curvature at P of the ellipse defined by the intersection of this prime-vertical plane with the Earth ellipsoid is the *normal radius*

$$R_N(\phi) = \frac{a}{\left(1 - e^2\sin^2(\phi)\right)^{\frac{1}{2}}}. \tag{2.7}$$

It is also referred to as the *prime vertical radius of curvature* or the *great normal.* For $\phi \neq \pm\frac{\pi}{2}$, the normal radius is the length of the normal to the ellipsoid, from its intersection with the ECEF z-axis to the surface of the ellipsoid. On the reference ellipsoid, *parallels* are closed curves with constant latitude. Such curves are circles with radius $R_N\cos(\phi)$.

Exercise 2.2 steps through the derivation of eqns. (2.6) and (2.7). Exercise 2.6 asks the user to derive formulas related to eqns. (2.6) and (2.7) that are useful in subsequent derivations.

2.3.2.2 Earth Gravity Model

The gravity vector varies as a function of position due to the gravitational attraction being a function of geocentric radius, the centripetal acceleration being a function of latitude and radius, and the non-uniform Earth mass distribution.

The WGS84 geodetic and gravity model parameters are defined in Table 2.1. In the geographic reference frame, a local gravity vector model developed by the Defense Mapping Agency that is accurate on the surface of the WGS84 ellipsoid is

$$\mathbf{g}^g = \begin{bmatrix} 0 \\ 0 \\ \gamma(\phi) \end{bmatrix} + \begin{bmatrix} \zeta_g \\ -\eta_g \\ \delta_g \end{bmatrix} \tag{2.8}$$

where ϕ is the latitude, $\gamma_e = 9.7803267715 \frac{m}{s^2}$ is the equatorial effective gravity and

$$\gamma(\phi) = \gamma_e \frac{1 + 0.001931851353 \sin^2(\phi)}{\sqrt{1 - 0.0066943800229 \sin^2(\phi)}}.$$

The vector $[\zeta_g, -\eta_g, \delta_g]^\top$ represents local perturbations in the gravity vector relative to the ellipsoidal normal vector. Both $\mathbf{g}^g$ and $[\zeta_g, -\eta_g, \delta_g]^\top$ are functions of position, but the argument is not shown to simplify the notation.

2.3.3 ECEF Transformations

This section presents algorithms to transform between the geodetic (i.e., $(\phi, \lambda, h)^e$) and rectangular (i.e., $(x, y, z)^e$) ECEF coordinates. The WGS-84 ellipsoid parameters are used throughout the discussion.

When the geodetic coordinates are known, the coordinates in the ECEF frame is calculated as

$$x = (R_N + h)\cos(\phi)\cos(\lambda) \tag{2.9}$$
$$y = (R_N + h)\cos(\phi)\sin(\lambda) \tag{2.10}$$
$$z = [R_N(1 - e^2) + h]\sin(\phi). \tag{2.11}$$

Example 2.1 *The geodetic position of a point in the vicinity of Los Angeles, CA is* $\phi = 34$ *degrees,* 0 *minutes,* 0.00174 *seconds North,* $\lambda = 117$ *degrees,* 20 *minutes,* 0.84965 *seconds West, and* $h = 251.702$ *m. These coordinates convert to* $\phi = 0.5934119541$ *rad and* $\lambda = -2.0478571082$ *rad. The signs and precision are critical, see Exercise 2.3. Then* $R_N = 6.384823214 \times 10^6$ *m and*

$$[x, y, z]^e = [-2430601.828, -4702442.703, 3546587.358] \ m. \tag{2.12}$$

Readers unfamiliar with the degrees, minutes, seconds notation should see Exercises 2.3 and 2.4. △

The transformation from $[x, y, z]^e$ to geodetic coordinates is more involved, but is important in GPS applications where the ECEF rectangular coordinates are directly determined from range measurements and the geodetic coordinates are often desired. Longitude can be found explicitly from eqns. (2.9-2.10) as $\lambda = arctan2(y, x)$. Solution for h and ϕ can be computed by iteration as follows [92]:

1. Initialization: Let

$$\begin{aligned} h &= 0 \\ R_N &= a \\ p &= \sqrt{x^2 + y^2}. \end{aligned}$$

2. Perform the following iteration until convergence:

$$\begin{aligned} \sin(\phi) &= \frac{z}{(1-e^2)R_N + h} \\ \phi &= \text{atan}((z + e^2 R_N \sin(\phi))/p) \\ R_N(\phi) &= \frac{a}{\sqrt{1 - e^2 \sin^2(\phi)}} \\ h &= \frac{p}{\cos(\phi)} - R_N \end{aligned}$$

where the eccentricity e is a defining constant of the geodetic system as discussed in Section 2.3.2.1. Although it is conceptually simpler to take the inverse sine in the second step, the presented algorithm has much faster convergence, as shown in the following example.

Example 2.2 *Consider the inversion of the previous example in which*

$$[x, y, z]^e = [-2430601.828, -4702442.703, 3546587.358]\ m.$$

Starting with an initial estimate of $h = 0.0$ m, the algorithm as written converges to centimeter accuracy (1×10^{-9} rads.) in four iterations:

iteration	*1*	*2*	*3*	*4*	*5*
lat: deg	*34*	*34*	*34*	*34*	*34*
lat: min	*0*	*0*	*0*	*0*	*0*
lat: sec	*0.027145*	*0.003159*	*0.001738*	*0.001740*	*0.001740*
h	*237.6519*	*251.7325*	*251.7020*	*251.7020*	*251.7020*

The same algorithm with the same initial estimate for h, but using an inverse sine instead of an arctan function in the second calculation of the iteration has still not converged to meter level accuracy after eight iterations:

iteration	*1*	*2*	*3*	*4*	*5*	*6*	*7*	*8*
lat:deg	*34*	*33*	*34*	*33*	*34*	*33*	*34*	*33*
lat: min	*2*	*58*	*0*	*59*	*0*	*59*	*0*	*59*
lat: sec	*31.41*	*50.62*	*31.77*	*45.45*	*6.67*	*56.95*	*1.40*	*59.36*
h	*3401*	*-1189*	*912*	*-50*	*390*	*188*	*280*	*238*

Convergence to centimeter accuracy requires 25 iterations. Generally, the arctan-based iterative technique converges quickly, especially when a good initial guess at the solution (e.g., from a previous measurement) is available. △

Various alternative approximate and closed form solutions exist in the literature [24, 66, 68, 76, 117].

2.4 Reference Frame Transformations

This section presents methods for transforming points and vectors between rectangular coordinate systems. The axes of each coordinate system are assumed to be right-handed and orthogonal. Three dimensions are used throughout the discussion; however, the discussion is equally valid for $\mathbb{R}^n$.

2.4.1 The Direction Cosine Matrix

Let ϕ_1 represent a right-handed orthogonal coordinate system. Let $\mathbf{v}_1$ be a vector from the origin O_1 of the ϕ_1 frame to the point P. The representation of the vector $\mathbf{v}_1$ with respect to frame ϕ_1 is

$$\mathbf{v}_1 = x_1\mathbf{I}_1 + y_1\mathbf{J}_1 + z_1\mathbf{K}_1 \tag{2.13}$$

where $\mathbf{I}_1, \mathbf{J}_1, \mathbf{K}_1$ are unit vectors along the ϕ_1 axes and

$$\begin{aligned} x_1 &= (P - O_1) \cdot \mathbf{I}_1 \\ y_1 &= (P - O_1) \cdot \mathbf{J}_1 \\ z_1 &= (P - O_1) \cdot \mathbf{K}_1. \end{aligned}$$

The vector $[\mathbf{v}_1]^1 = [x_1, y_1, z_1]^\top$ contains the coordinates of the point P with respect to the axes of ϕ_1 and is the representation of the vector $\mathbf{v}_1$ with respect to ϕ_1. The physical interpretations of the coordinates is that they are the projections of the vector $\mathbf{v}_1$ onto the ϕ_1 axes. For the two-dimensional $x - y$ plane, the discussion of this paragraph is depicted in Figure 2.9.

A vector $\mathbf{v}$ can be defined without reference to a specific reference frame. When convenient, as discussed above, the representation of $\mathbf{v} \in \mathbb{R}^3$ with respect to the axes of frame ϕ_1 is

$$[\mathbf{v}]^1 = \begin{bmatrix} \mathbf{I}_1 \cdot \mathbf{v} \\ \mathbf{J}_1 \cdot \mathbf{v} \\ \mathbf{K}_1 \cdot \mathbf{v} \end{bmatrix} = \begin{bmatrix} \mathbf{I}_1^\top \\ \mathbf{J}_1^\top \\ \mathbf{K}_1^\top \end{bmatrix} \mathbf{v}. \tag{2.14}$$

Eqn. (2.14) is used in derivations later in this subsection.

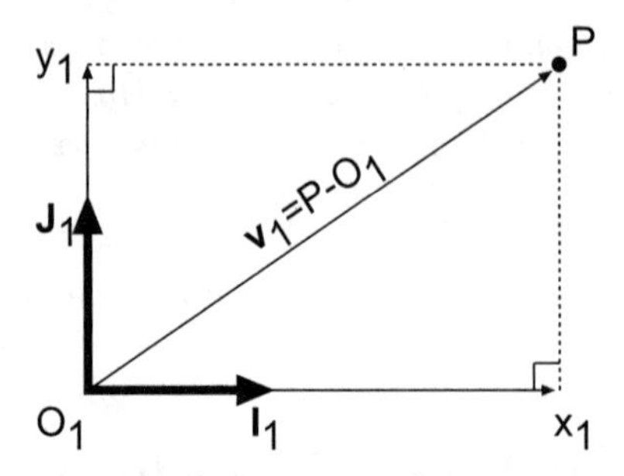

Figure 2.9: Two dimensional representation of the determination of the coordinates of a point P relative to the origin O_1 of reference frame ϕ_1.

With two distinct reference frames ϕ_1 and ϕ_2, the same point can be represented by a different sets of coordinates in each reference frame. For two dimensions with both reference frames defined within the plane of the page, the situation is depicted in Figure 2.10 where the projections and coordinate definitions of the previous paragraph are performed with respect to each frame-of-reference. The remainder of this section discusses the important question of how to use the coordinates of a point in one frame-of-reference to compute the coordinates of the same point with respect to a different frame-of-reference. The transformation of point coordinates from one frame-of-reference to another will require two operations: *translation* and *rotation*.

From the above discussion, $\mathbf{v}_1 = P - O_1$ is the vector from O_1 to P and $\mathbf{v}_2 = P - O_2$ is the vector from O_2 to P. Define $\mathbf{O}_{12} = O_2 - O_1$ as the vector from O_1 to O_2. Therefore, we have that

$$\mathbf{v}_1 = \mathbf{O}_{12} + \mathbf{v}_2.$$

This equation must hold whether the vectors are represented in the coordinates of the ϕ_1 frame or the ϕ_2 frame.

Denote the components of vector $\mathbf{v}_1$ relative to the ϕ_1 frame as $[\mathbf{v}_1]^1 = [x_1, y_1, z_1]^1$, the components of $\mathbf{v}_2$ relative to the ϕ_2 frame as $[\mathbf{v}_2]^2 = [x_2, y_2, z_2]^2$ and the components of $\mathbf{O}_{12}$ relative to the ϕ_1 frame as $[\mathbf{O}_{12}]^1 = [x_O, y_O, z_O]^1$. Assume that $[\mathbf{v}_2]^2$, $[\mathbf{O}_{12}]^1$ and the relative orientation of the two reference frames are known. Then, the position of P with respect to the ϕ_1 frame can be computed as

$$\begin{bmatrix} x_1 \\ y_1 \\ z_1 \end{bmatrix}^1 = \begin{bmatrix} x_O \\ y_O \\ z_O \end{bmatrix}^1 + [\mathbf{v}_2]^1. \tag{2.15}$$

Because it is the only unknown term in the right hand side, the present

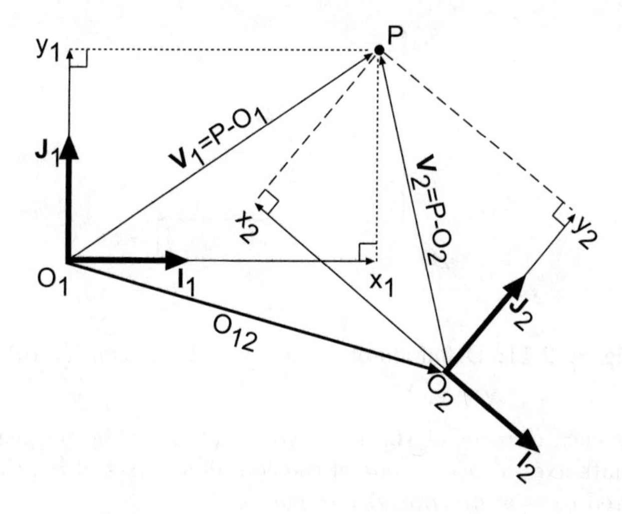

Figure 2.10: Definition of the coordinates of a point P with respect to two frames-of-reference ϕ_1 and ϕ_2.

question of interest is how to calculate $[\mathbf{v}_2]^1$ based on the available information.

Let $\mathbf{I}_2, \mathbf{J}_2, \mathbf{K}_2$ represent the unit vectors along the ϕ_2 axes. As discussed relative to eqn. (2.14), vectors $\mathbf{v}_I$, $\mathbf{v}_J$, and $\mathbf{v}_K$ defined as

$$\mathbf{v}_I = \begin{bmatrix} \mathbf{I}_1 \cdot \mathbf{I}_2 \\ \mathbf{J}_1 \cdot \mathbf{I}_2 \\ \mathbf{K}_1 \cdot \mathbf{I}_2 \end{bmatrix}, \mathbf{v}_J = \begin{bmatrix} \mathbf{I}_1 \cdot \mathbf{J}_2 \\ \mathbf{J}_1 \cdot \mathbf{J}_2 \\ \mathbf{K}_1 \cdot \mathbf{J}_2 \end{bmatrix}, \text{ and } \mathbf{v}_K = \begin{bmatrix} \mathbf{I}_1 \cdot \mathbf{K}_2 \\ \mathbf{J}_1 \cdot \mathbf{K}_2 \\ \mathbf{K}_1 \cdot \mathbf{K}_2 \end{bmatrix}$$

represent the unit vectors in the direction of the ϕ_2 coordinate axes that are resolved in the ϕ_1 reference frame. Since $\mathbf{I}_2, \mathbf{J}_2, \mathbf{K}_2$ are orthonormal, so are $\mathbf{v}_I, \mathbf{v}_J, \mathbf{v}_K$. Therefore, the matrix

$$\mathbf{R}_2^1 = [\mathbf{v}_I, \mathbf{v}_J, \mathbf{v}_K]$$

is an orthonormal matrix (i.e., $\left(\mathbf{R}_2^1\right)^\top \mathbf{R}_2^1 = \mathbf{R}_2^1 \left(\mathbf{R}_2^1\right)^\top = \mathbf{I}$).

Each element of $\mathbf{R}_2^1$ is the cosine of the angle between one of $\mathbf{I}_1, \mathbf{J}_1, \mathbf{K}_1$ and one of $\mathbf{I}_2, \mathbf{J}_2, \mathbf{K}_2$. To see this, consider the element in the third row second column:

$$\begin{aligned} \left[\mathbf{R}_2^1\right]_{3,2} &= \mathbf{K}_1 \cdot \mathbf{J}_2 \\ &= \|\mathbf{K}_1\|_2 \|\mathbf{J}_2\|_2 \cos(\beta_3) \\ &= \cos(\beta_3) \end{aligned}$$

where β_3 is the angle between $\mathbf{K}_1$ and $\mathbf{J}_2$ and we have used the fact that $\|\mathbf{K}_1\|_2 = \|\mathbf{J}_2\|_2 = 1$.

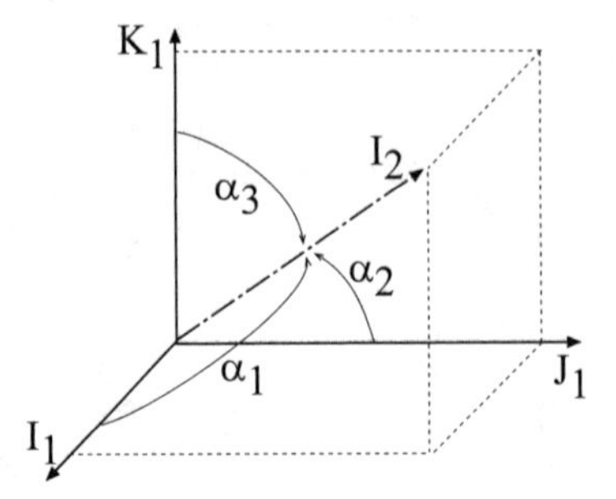

Figure 2.11: Definition of α_i for $i = 1, 2, 3$ in eqn. (2.16).

Because each element of $\mathbf{R}_2^1$ is the cosine of the angle between one of the coordinate axes of ϕ_1 and one of the coordinate axes of ϕ_2, the matrix $\mathbf{R}_2^1$ is referred to as a *direction cosine matrix*:

$$\mathbf{R}_2^1 = \begin{bmatrix} \cos(\alpha_1) & \cos(\beta_1) & \cos(\gamma_1) \\ \cos(\alpha_2) & \cos(\beta_2) & \cos(\gamma_2) \\ \cos(\alpha_3) & \cos(\beta_3) & \cos(\gamma_3) \end{bmatrix}. \tag{2.16}$$

Figure 2.11 depicts the angles α_i for $i = 1, 2, 3$ that define the first column of $\mathbf{R}_2^1$. The β_i and γ_i angles are defined similarly. When the relative orientation of two reference frames is known, the direction cosine matrix $\mathbf{R}_2^1$ is unique and known.

Although the direction cosine matrix has nine elements, due to the three orthogonality constraints and the three normality constraints, there are only three degrees of freedom. Alternative parameterizations are discussed in Section 2.5 and Appendix D.

Analogous to eqn. (2.13)

$$\mathbf{v}_2 = \mathbf{I}_2 x_2 + \mathbf{J}_2 y_2 + \mathbf{K}_2 z_2$$

where $[\mathbf{v}_2]^2 = [x_2, y_2, z_2]^\top$ and by eqn. (2.14)

$$[\mathbf{v}_2]^1 = \begin{bmatrix} \mathbf{I}_1^\top \\ \mathbf{J}_1^\top \\ \mathbf{K}_1^\top \end{bmatrix} \mathbf{v}_2.$$

Therefore,

$$\begin{aligned} [\mathbf{v}_2]^1 &= \begin{bmatrix} \mathbf{I}_1^\top \\ \mathbf{J}_1^\top \\ \mathbf{K}_1^\top \end{bmatrix} (\mathbf{I}_2 x_2 + \mathbf{J}_2 y_2 + \mathbf{K}_2 z_2) \\ &= \begin{bmatrix} \mathbf{I}_1 \cdot \mathbf{I}_2 \\ \mathbf{J}_1 \cdot \mathbf{I}_2 \\ \mathbf{K}_1 \cdot \mathbf{I}_2 \end{bmatrix} x_2 + \begin{bmatrix} \mathbf{I}_1 \cdot \mathbf{J}_2 \\ \mathbf{J}_1 \cdot \mathbf{J}_2 \\ \mathbf{K}_1 \cdot \mathbf{J}_2 \end{bmatrix} y_2 + \begin{bmatrix} \mathbf{I}_1 \cdot \mathbf{K}_2 \\ \mathbf{J}_1 \cdot \mathbf{K}_2 \\ \mathbf{K}_1 \cdot \mathbf{K}_2 \end{bmatrix} z_2 \end{aligned}$$

$$= \mathbf{R}_2^1 [\mathbf{v}_2]^2 . \tag{2.17}$$

2.4.2 Point Transformation

When eqn. (2.17) is substituted into eqn. (2.15) it yields the desired equation for the transformation of the coordinates of P with respect to frame 2, as represented by $[\mathbf{v}_2]^2$, to the coordinates of P with respect to frame 1, as represented by $[\mathbf{v}_1]^1$:

$$\begin{aligned} \begin{bmatrix} x_1 \\ y_1 \\ z_1 \end{bmatrix}^1 &= \begin{bmatrix} x_O \\ y_O \\ z_O \end{bmatrix}^1 + \mathbf{R}_2^1 [\mathbf{v}_2]^2 \\ [\mathbf{v}_1]^1 &= [\mathbf{O}_{12}]^1 + \mathbf{R}_2^1 [\mathbf{v}_2]^2 . \end{aligned} \tag{2.18}$$

The reverse transformation is easily shown from eqn. (2.18) to be

$$\begin{aligned} \begin{bmatrix} x_2 \\ y_2 \\ z_2 \end{bmatrix}^2 &= \mathbf{R}_1^2 \left(\begin{bmatrix} x_1 \\ y_1 \\ z_1 \end{bmatrix}^1 - \begin{bmatrix} x_O \\ y_O \\ z_O \end{bmatrix}^1 \right) \\ [\mathbf{v}_2]^2 &= \mathbf{R}_1^2 \left([\mathbf{v}_1]^1 - [\mathbf{O}_{12}]^1 \right) \end{aligned} \tag{2.19}$$

where we have used the fact that $\mathbf{R}_1^2 = \left(\mathbf{R}_2^1\right)^{-1} = \left(\mathbf{R}_2^1\right)^\top$, where the last equality is true due to the orthonormality of $\mathbf{R}_2^1$. Note that the point transformation between reference systems involves two operations: translation to account for separation of the origins, and rotation to account for non-alignment of the axis.

2.4.3 Vector Transformation

Consider two points P_1 and P_2. Let the vector $\mathbf{v}$ denote the directed line segment from P_1 to P_2. Relative to ϕ_1, $\mathbf{v}$ can be described as

$$\begin{aligned} \mathbf{v} = \begin{bmatrix} x_2 - x_1 \\ y_2 - y_1 \\ z_2 - z_1 \end{bmatrix}^1 &= \begin{bmatrix} x_2 \\ y_2 \\ z_2 \end{bmatrix}^1 - \begin{bmatrix} x_1 \\ y_1 \\ z_1 \end{bmatrix}^1 \\ &= \left(\begin{bmatrix} x_O \\ y_O \\ z_O \end{bmatrix}^1 + \mathbf{R}_2^1 \begin{bmatrix} x_2 \\ y_2 \\ z_2 \end{bmatrix}^2 \right) \\ &- \left(\begin{bmatrix} x_O \\ y_O \\ z_O \end{bmatrix}^1 + \mathbf{R}_2^1 \begin{bmatrix} x_1 \\ y_1 \\ z_1 \end{bmatrix}^2 \right) \end{aligned}$$

$$\begin{aligned}
&= \mathbf{R}_2^1 \left(\begin{bmatrix} x_2 \\ y_2 \\ z_2 \end{bmatrix}^2 - \begin{bmatrix} x_1 \\ y_1 \\ z_1 \end{bmatrix}^2 \right) \\
\mathbf{v}^1 &= \mathbf{R}_2^1 \mathbf{v}^2. \qquad (2.20)
\end{aligned}$$

Eqn. (2.20) is the vector transformation between coordinate systems. This relation is valid for any vector quantity. As discussed in detail in [95], it is important to realized that vectors, vector operations, and relations between vectors are invariant relative to any two particular coordinate representations as long as the coordinate systems are related through eqn. (2.20). This is important, as it corresponds to the intuitive notion that the physical properties of a system are invariant no matter how we orient the coordinate system in which our analysis is performed.

In the discussion of this section, the two frames have been considered to have no relative motion. Issues related to relative motion will be critically important in navigation systems and are discussed in subsequent sections.

Throughout the text, the notation $\mathbf{R}_a^b$ will denote the rotation matrix transforming vectors from frame a to frame b. Therefore,

$$\mathbf{v}^b = \mathbf{R}_a^b \mathbf{v}^a \qquad (2.21)$$

$$\mathbf{v}^a = \mathbf{R}_b^a \mathbf{v}^b \qquad (2.22)$$

where $\mathbf{R}_b^a = \left(\mathbf{R}_a^b\right)^\top$.

2.4.4 Matrix Transformation

In some instances, a matrix will be defined with respect to a specific frame of reference. Eqns. (2.21–2.22) can be used to derive the transformation of such matrices between frames-of-reference.

Let $\mathbf{\Omega}^a \in \mathbb{R}^{3\times3}$ be a matrix defined with respect to frame a and $\mathbf{v}_1^a$, $\mathbf{v}_2^a \in \mathbb{R}^3$ be two vectors defined in frame a. Let b be a second frame of reference. If $\mathbf{v}_1^a$ and $\mathbf{v}_2^a$ are related by

$$\mathbf{v}_1^a = \mathbf{\Omega}^a \mathbf{v}_2^a,$$

then eqns. (2.21–2.22) show that

$$\mathbf{R}_b^a \mathbf{v}_1^b = \mathbf{\Omega}^a \mathbf{R}_b^a \mathbf{v}_2^b$$

or

$$\mathbf{v}_1^b = \mathbf{\Omega}^b \mathbf{v}_2^b$$

where

$$\mathbf{\Omega}^b = \mathbf{R}_a^b \mathbf{\Omega}^a \mathbf{R}_b^a \qquad (2.23)$$

is the representation of the matrix $\mathbf{\Omega}$ with respect to frame b.

2.5 Specific Vector Transformations

In Section 2.4, the transformation of vectors from frame a to frame b is shown to involve an orthonormal matrix denoted by $\mathbf{R}_a^b$. The elements of this matrix, called the direction cosine matrix, are the cosines of the angles between the coordinate axes of the two frames-of-reference. Although this appears to allow nine independent variables to define $\mathbf{R}_a^b$, orthonormality restrictions result in only three independent quantities. Section 2.5.1 introduces the concept of a plane rotation. Sections 2.5.2–2.5.5 will use plane rotations to define the transformations between specific pairs of reference systems.

In addition to the direction cosine and Euler angle representations of the relative orientations of two reference frames, various other representations exist [120]. Advantages of alternative representations may include efficient computation, lack of singularities, or compact representation. One popular representation of relative attitude is the quaternion. Quaternions offer accurate and efficient computation methods without singularities. Often quaternions are preferred over both direction cosine and Euler angle methods. Nonetheless, their discussion is a topic in and of itself. To maintain the flow of the book, the discussion of quaternions has been placed in Appendix D. It is recommended that designers read the main body of the text first, to understand the role and issues related to attitude representation; however, they should understand and consider quaternions prior to implementation of their first system.

2.5.1 Plane Rotations

A plane rotation is a convenient means for mathematically expressing the rotational transformation of vectors between two coordinate systems where the second coordinate system is related to the first by a rotation of the first coordinate system by an angle x around a vector $\mathbf{v}$. In the special case where the vector $\mathbf{v}$ is one of the original coordinate axes, the plane rotation matrix takes on an especially simple form. In the following, a rotation of the first coordinate system by x radians[1] around the i-th axis will be expressed as $[x]_i$. Using this notation,

$$[x]_1 = \begin{bmatrix} 1 & 0 & 0 \\ 0 & \cos(x) & \sin(x) \\ 0 & -\sin(x) & \cos(x) \end{bmatrix} \tag{2.24}$$

$$[x]_2 = \begin{bmatrix} \cos(x) & 0 & -\sin(x) \\ 0 & 1 & 0 \\ \sin(x) & 0 & \cos(x) \end{bmatrix} \tag{2.25}$$

[1]In all of the above equations, as will be true through out this book, the radian rotation x represents a positive rotation as defined by the right-hand rule.

$$[x]_3 = \begin{bmatrix} \cos(x) & \sin(x) & 0 \\ -\sin(x) & \cos(x) & 0 \\ 0 & 0 & 1 \end{bmatrix}. \tag{2.26}$$

Each of these plane rotation matrices is an orthonormal matrix. For a rotation of x radians about the i-th axis of the first coordinate system, the components of vector $\mathbf{z}$ in each coordinate system are related by

$$\mathbf{z}^2 = [x]_i \mathbf{z}^1. \tag{2.27}$$

When two coordinate systems are related by a sequence of rotations, then the corresponding rotation matrices are multiplied in the corresponding order. For example, continuing from the last equation, if a third frame is defined by a rotation of y radians about the j-th axis of the second frame, then the representation of the vector $\mathbf{z}$ in this frame is

$$\begin{aligned} \mathbf{z}^3 &= [y]_j \mathbf{z}^2 \\ &= [y]_j [x]_i \mathbf{z}^1. \end{aligned}$$

The order of the matrix multiplication is critical. Since matrix multiplication is not commutative, neither is the order of rotation. For example, a 90 degree rotation about the first axis followed by a 90 degree rotation about the resultant second axis results in a distinct orientation from a 90 degree rotation about the second axis followed by a 90 degree rotation about the resultant first axis. The following two sections use plane rotations to determine the Euler angle representations of a few useful vector transformations.

2.5.2 Transformation: ECEF to Tangent Plane

Let

$$\Delta \hat{\mathbf{x}}^e = [x, y, z]^e - [x_o, y_o, z_o]^e \tag{2.28}$$

where $[x_o, y_o, z_o]^e$ are the ECEF coordinates of the origin of the local tangent plane. Then $\Delta \hat{\mathbf{x}}^e$ is a vector from the local tangent plane origin to an arbitrary location $P^e = [x, y, z]^e$, with the vector and point coordinates each expressed relative to the ECEF axis.

The transformation of vectors from ECEF to tangent plane (TP) can be constructed by two plane rotations, as depicted in Figure 2.12. First, a plane rotation about the ECEF z-axis to align the rotated y-axis (denoted y') with the tangent plane east axis; second, a plane rotation about the new y'-axis to align the new z-axis (denoted z'') with tangent plane inward pointing normal vector. The first plane rotation is defined by

$$\mathbf{R}_e^1 = [\lambda]_3 = \begin{bmatrix} \cos(\lambda) & \sin(\lambda) & 0 \\ -\sin(\lambda) & \cos(\lambda) & 0 \\ 0 & 0 & 1 \end{bmatrix} \tag{2.29}$$

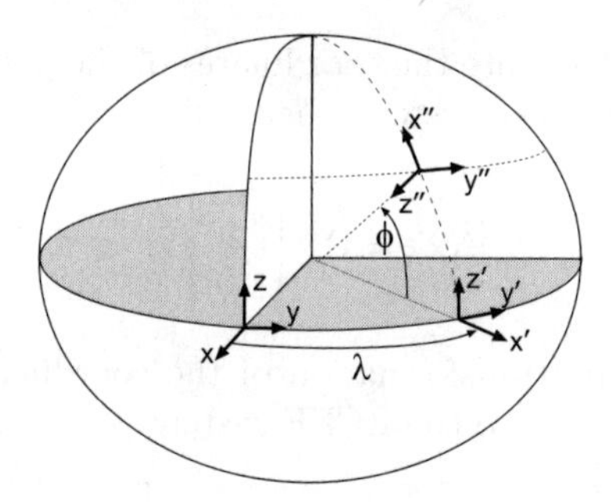

Figure 2.12: Variables for derivation of $\mathbf{R}_e^t$.

where λ is the longitude of the point $[x_o, y_o, z_o]^e$. The second plane rotation is defined by

$$\mathbf{R}_1^t = \left[-\left(\phi + \frac{\pi}{2}\right)\right]_2 = \begin{bmatrix} \cos\left(\phi + \frac{\pi}{2}\right) & 0 & \sin\left(\phi + \frac{\pi}{2}\right) \\ 0 & 1 & 0 \\ -\sin\left(\phi + \frac{\pi}{2}\right) & 0 & \cos\left(\phi + \frac{\pi}{2}\right) \end{bmatrix} \quad (2.30)$$

$$= \begin{bmatrix} -\sin(\phi) & 0 & \cos(\phi) \\ 0 & 1 & 0 \\ -\cos(\phi) & 0 & -\sin(\phi) \end{bmatrix}, \quad (2.31)$$

where ϕ is the latitude of the point $[x_o, y_o, z_o]^e$.

The overall transformation for vectors from ECEF to tangent plane representation is then $\mathbf{v}^t = \mathbf{R}_e^t \mathbf{v}^e$ where

$$\mathbf{R}_e^t = \mathbf{R}_1^t \mathbf{R}_e^1 = \begin{bmatrix} -\sin(\phi)\cos(\lambda) & -\sin(\phi)sin(\lambda) & \cos(\phi) \\ -\sin(\lambda) & \cos(\lambda) & 0 \\ -\cos(\phi)\cos(\lambda) & -\cos(\phi)sin(\lambda) & -\sin(\phi) \end{bmatrix}. \quad (2.32)$$

The inverse transformation for vectors from tangent plane to ECEF is $\mathbf{v}^e = \mathbf{R}_t^e \mathbf{v}^t$ where $\mathbf{R}_t^e = (\mathbf{R}_e^t)^\top$.

Example 2.3 *The angular rate of the ECEF frame with respect to the inertial frame represented in the ECEF frame is* $\boldsymbol{\omega}_{ie}^e = [0, 0, 1]^\top \omega_{ie}$. *Therefore, the angular rate of the ECEF frame with respect to the inertial frame represented in the tangent frame is*

$$\boldsymbol{\omega}_{ie}^t = \mathbf{R}_e^t \boldsymbol{\omega}_{ie}^e = \begin{bmatrix} \omega_{ie}\cos(\phi) \\ 0 \\ -\omega_{ie}\sin(\phi) \end{bmatrix}.$$

△

Let $P^t = [n, e, d]^\top$ denote the coordinates of the point P represented in the tangent plane reference system, then

$$\Delta\hat{\mathbf{x}}^e = \mathbf{R}_t^e \begin{bmatrix} n \\ e \\ d \end{bmatrix}^t .$$

Using eqn. (2.28), the transformation of the coordinates of a point from the tangent plane system to the ECEF system is

$$\begin{bmatrix} x \\ y \\ z \end{bmatrix}^e = \begin{bmatrix} x_o \\ y_o \\ z_o \end{bmatrix}^e + \mathbf{R}_t^e \begin{bmatrix} n \\ e \\ d \end{bmatrix}^t . \tag{2.33}$$

Example 2.4 *For the ECEF position given in the Example 2.1 on page 33, the matrix*

$$\mathbf{R}_e^t = \begin{bmatrix} 0.2569 & 0.4967 & 0.8290 \\ 0.8882 & -0.4594 & 0.0000 \\ 0.3808 & 0.7364 & -0.5592 \end{bmatrix}$$

transforms vectors from the ECEF coordinate system to tangent plane coordinates. The point transform is defined as

$$\begin{bmatrix} x \\ y \\ z \end{bmatrix}^t = \mathbf{R}_e^t \left(\begin{bmatrix} x \\ y \\ z \end{bmatrix}^e - \begin{bmatrix} -2.430601 \\ -4.702442 \\ 3.546587 \end{bmatrix} \times 10^6 \right)$$

where the origin location $[x_o, y_o, z_o]^e$ *is defined in eqn. (2.12).*

The inverse transformations are easily derived from the preceding text. For example, the local unit gravity vector which is assumed to be $(0, 0, 1)^\top$ *in tangent plane coordinates, transforms to* $(0.3808, 0.7364, -0.5592)$ *in ECEF coordinates.* △

2.5.3 Transformation: ECEF to Geographic

The geographic frame has a few points that distinguish it from the other frames. First, because the origin of the geographic frame moves with the vehicle and is the projection of vehicle frame origin onto the reference ellipsoid, the position of the vehicle in the geographic frame is $\mathbf{x}^g = [0, 0, -h]^\top$. The latitude ϕ and longitude λ define the position of the geographic frame origin (vehicle frame projection) on the reference ellipsoid. Second, $\frac{d}{dt}\mathbf{x}^g = [0, 0, -\dot{h}]^\top$; which is not the velocity vector for the vehicle. The Earth relative velocity vector represented in the ECEF frame is $\mathbf{v}_e^e = \frac{d}{dt}\mathbf{x}^e$. This vector can be represented in the geographic frame as

$$\mathbf{v}_e^g = \mathbf{R}_e^g \mathbf{v}_e^e . \tag{2.34}$$

The vector $\mathbf{v}_e^g$ is not the derivative of the geographic frame position vector $\mathbf{x}^g$. The components of the Earth relative velocity vector represented in the geographic frame are named as $\mathbf{v}_e^g = [v_n, v_e, v_d]^\top$ which are the north, east, and down components of the velocity vector along the instantaneous geographic frame axes.

The rotation matrix $\mathbf{R}_e^g$ has the exact same form as $\mathbf{R}_e^t$. The distinction is that $\mathbf{R}_e^g$ is computed using the latitude ϕ and longitude λ defined by the position of the vehicle at the time of interest whereas $\mathbf{R}_e^t$ is a constant matrix defined by the fixed latitude and longitude of the tangent plane origin. It should be clear that $\boldsymbol{\omega}_{et} = \mathbf{0}$, while $\boldsymbol{\omega}_{eg} = \boldsymbol{\omega}_{ig} - \boldsymbol{\omega}_{ie}$, where $\boldsymbol{\omega}_{eg}$ and $\boldsymbol{\omega}_{ig}$ are discussed in eqns. (2.56) and (2.57), respectively.

Eqns. (2.9–2.11) provide the relationship between $\mathbf{x}^e = [x, y, z]^\top$ and (ϕ, λ, h) which is repeated below

$$x = (R_N + h)\cos(\phi)\cos(\lambda) \tag{2.35}$$

$$y = (R_N + h)\cos(\phi)\sin(\lambda) \tag{2.36}$$

$$z = [R_N(1 - e^2) + h]\sin(\phi) \tag{2.37}$$

and will be used to relate $\mathbf{v}_e^g$ to $(\dot{\phi}, \dot{\lambda}, \dot{h})$. First, we note that

$$\mathbf{v}_e^e = \frac{\partial \mathbf{x}^e}{\partial \phi}\dot{\phi} + \frac{\partial \mathbf{x}^e}{\partial \lambda}\dot{\lambda} + \frac{\partial \mathbf{x}^e}{\partial h}\dot{h} = \left[\frac{\partial \mathbf{x}^e}{\partial \phi}, \frac{\partial \mathbf{x}^e}{\partial \lambda}, \frac{\partial \mathbf{x}^e}{\partial h}\right]\begin{bmatrix}\dot{\phi}\\ \dot{\lambda}\\ \dot{h}\end{bmatrix}.$$

Next, using eqns. (2.78–2.78) it is straightforward to show that

$$\frac{\partial \mathbf{x}^e}{\partial \phi} = (R_M + h)\begin{bmatrix}-\sin(\phi)\cos(\lambda)\\ -\sin(\phi)\sin(\lambda)\\ \cos(\phi)\end{bmatrix},$$

$$\frac{\partial \mathbf{x}^e}{\partial \lambda} = (R_N + h)\begin{bmatrix}-\cos(\phi)\sin(\lambda)\\ \cos(\phi)\cos(\lambda)\\ 0\end{bmatrix}, \text{ and } \frac{\partial \mathbf{x}^e}{\partial h} = \begin{bmatrix}\cos(\phi)\cos(\lambda)\\ \cos(\phi)\sin(\lambda)\\ \sin(\phi)\end{bmatrix}.$$

Therefore,

$$\left[\frac{\partial \mathbf{x}^e}{\partial \phi}, \frac{\partial \mathbf{x}^e}{\partial \lambda}, \frac{\partial \mathbf{x}^e}{\partial h}\right] = \mathbf{R}_g^e\begin{bmatrix}(R_M + h) & 0 & 0\\ 0 & (R_N + h)\cos(\phi) & 0\\ 0 & 0 & -1\end{bmatrix}.$$

With this expression and eqn. (2.34) it is straightforward to show that

$$\mathbf{v}_e^g = \begin{bmatrix}v_n\\ v_e\\ v_d\end{bmatrix} = \begin{bmatrix}(R_M + h)\dot{\phi}\\ \cos(\phi)(R_N + h)\dot{\lambda}\\ -\dot{h}\end{bmatrix} \tag{2.38}$$

and

$$\begin{bmatrix} \dot{\phi} \\ \dot{\lambda} \\ \dot{h} \end{bmatrix} = \begin{bmatrix} \frac{v_n}{R_M+h} \\ \frac{v_e}{\cos(\phi)(R_N+h)} \\ -v_d \end{bmatrix}. \tag{2.39}$$

2.5.4 Transformation: Vehicle to Navigation Frame

Consider the situation shown in Figure 2.13, which depicts two coordinate systems. The first coordinate system, denoted by (n, e, d), is the geographic frame. The second coordinate system, denoted by (u, v, w) is the vehicle body frame which is at an arbitrary orientation relative to the geographic frame[2].

The relationship between vectors in the body and geographic reference frames can be completely described by the rotation matrix $\mathbf{R}_b^g$. This rotation matrix can be defined by a series of three plane rotations involving the *Euler angles*[3] (ϕ, θ, ψ) where ϕ represents roll, θ represents pitch, and ψ represents yaw angle[4].

In Figures 2.14–2.16, the axes of the geographic frame are indicated by the **I**, **J**, and **K** unit vectors.

The first rotation, as shown in Figure 2.14, rotates the geographic coordinate system by ψ radians about the geographic frame d-axis (i.e., **K** unit vector). This rotation aligns the new $\mathbf{I}'$-axis with the projection of the vehicle u-axis onto the tangent plane to the ellipsoid. The plane rotation for this operation is described as

$$\begin{bmatrix} x' \\ y' \\ z' \end{bmatrix} = \mathbf{R}_g^1 \begin{bmatrix} x \\ y \\ z \end{bmatrix}^t \tag{2.40}$$

where $\mathbf{R}_g^1 = \begin{bmatrix} \cos(\psi) & \sin(\psi) & 0 \\ -\sin(\psi) & \cos(\psi) & 0 \\ 0 & 0 & 1 \end{bmatrix} = [\psi]_3$. The resultant $\mathbf{I}'$ and $\mathbf{J}'$-axes still lie in the north-east tangent plane.

The second rotation, as shown in Figure 2.15, rotates the coordinate system that resulted from the previous yaw rotation by θ radians about the

[2]In applications where navigation is performed in the local tangent frame, the method of this section could be used to define the rotation matrix between the body and local tangent plane frames-of-reference.

[3]Although represented as a three-tuple, (ϕ, θ, ψ) is not a true vector. For example, addition of Euler angles is not commutative.

[4]While the angles ϕ and θ are referred to almost universally as the roll and pitch angles, various authors use different terms for the angle ψ (e.g., heading, yaw, azimuth). In addition, different terms are used for the angle of the velocity vector with respect to north in the local tangent plane (e.g., heading angle or course angle). This majority of this text will follow the practice recommended for aircraft from Section 1.3.3.1 in [5] where ψ is referred to as the yaw or azimuth angle.

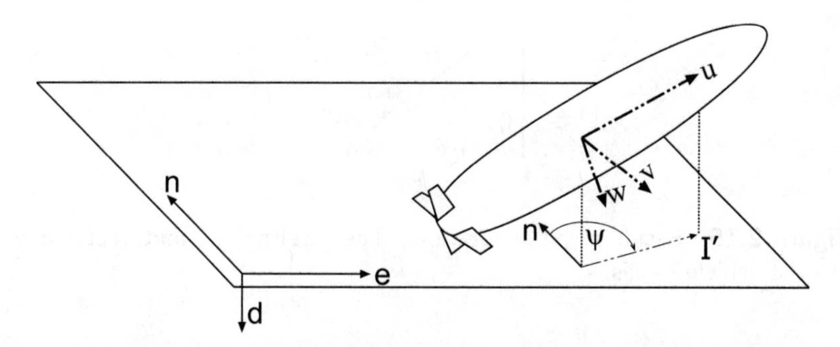

Figure 2.13: Relation between vehicle and navigation frame coordinate systems. For navigation in the geographic frame, the origin of the ned coordinate axes would be at the projection of the vehicle frame origin onto the ellipsoid. For navigation in a local tangent frame, the origin of the ned coordinate axes would be at some convenient point near the area of operation.

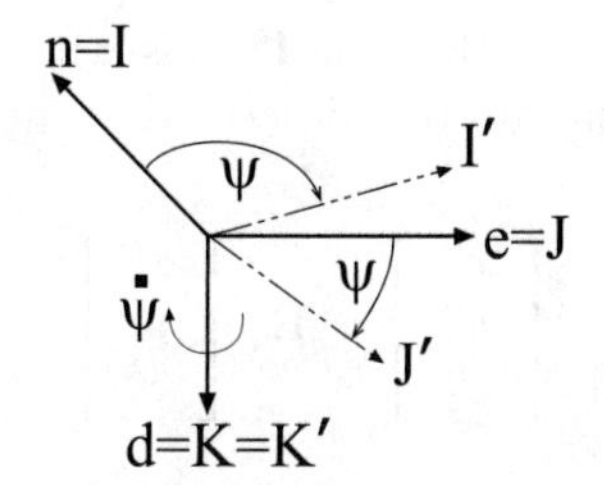

Figure 2.14: Result of yaw rotation. The initial $\mathbf{I}$, $\mathbf{J}$, $\mathbf{K}$ unit vectors align with the tangent frame (n, e, d) directions.

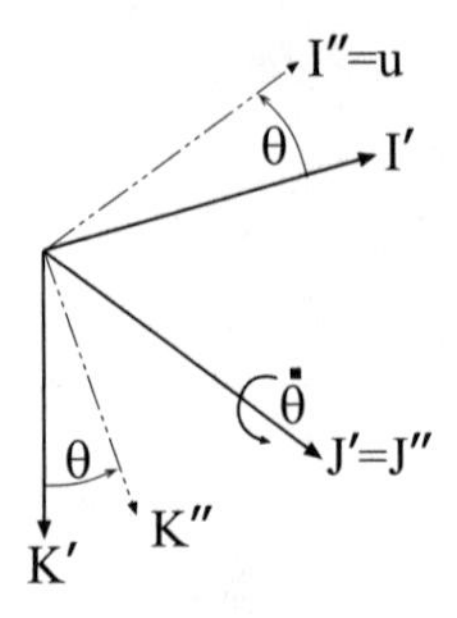

Figure 2.15: Result of pitch rotation. The resultant $\mathbf{I}''$ unit vector aligns with the vehicle u-axis.

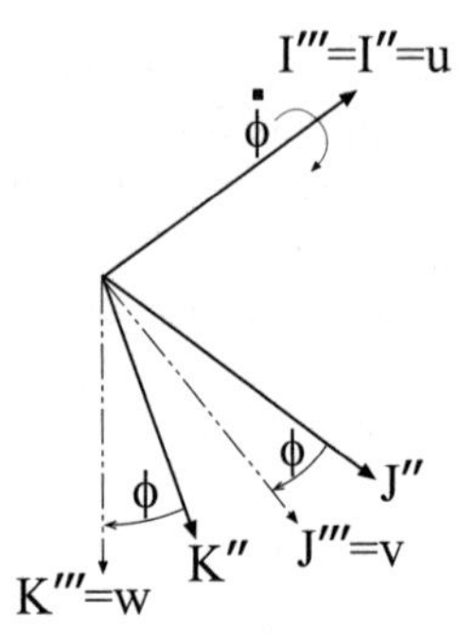

Figure 2.16: Result of roll rotation. The final $\mathbf{I}'''$, $\mathbf{J}'''$, $\mathbf{K}'''$ unit vectors align with the vehicle u, v, w directions.

$\mathbf{J}'$-axis. This rotation aligns the new $\mathbf{I}''$-axis with the vehicle u-axis. The plane rotation for this operation is described as

$$\begin{bmatrix} x'' \\ y'' \\ z'' \end{bmatrix} = \mathbf{R}_1^2 \begin{bmatrix} x' \\ y' \\ z' \end{bmatrix} \tag{2.41}$$

where $\mathbf{R}_1^2 = \begin{bmatrix} \cos(\theta) & 0 & -\sin(\theta) \\ 0 & 1 & 0 \\ \sin(\theta) & 0 & \cos(\theta) \end{bmatrix} = [\theta]_2.$

The third rotation, as shown in Figure 2.16, rotates the coordinate system that resulted from the previous pitch rotation by ϕ radians about the $\mathbf{I}''$-axis. This rotation aligns the new $\mathbf{J}'''$ and $\mathbf{K}'''$ axes with the vehicle v and w axes, respectively. The plane rotation for this operation is described as

$$\begin{bmatrix} u \\ v \\ w \end{bmatrix}_b = \mathbf{R}_2^b \begin{bmatrix} x'' \\ y'' \\ z'' \end{bmatrix} \tag{2.42}$$

where $\mathbf{R}_2^b = \begin{bmatrix} 1 & 0 & 0 \\ 0 & \cos(\phi) & \sin(\phi) \\ 0 & -\sin(\phi) & \cos(\phi) \end{bmatrix} = [\phi]_1$.

Therefore, vectors represented in geographic frame (or in the local tangent plane) can be transformed into a vehicle frame representation by the series of three rotations: $[\phi]_1[\theta]_2[\psi]_3$.

$$\begin{aligned}
\mathbf{v}^b &= \begin{bmatrix} 1 & 0 & 0 \\ 0 & c\phi & s\phi \\ 0 & -s\phi & c\phi \end{bmatrix} \begin{bmatrix} c\theta & 0 & -s\theta \\ 0 & 1 & 0 \\ s\theta & 0 & c\theta \end{bmatrix} \begin{bmatrix} c\psi & s\psi & 0 \\ -s\psi & c\psi & 0 \\ 0 & 0 & 1 \end{bmatrix} \mathbf{v}^g \\
\mathbf{v}^b &= \begin{bmatrix} c\psi c\theta & s\psi c\theta & -s\theta \\ -s\psi c\phi + c\psi s\theta s\phi & c\psi c\phi + s\psi s\theta s\phi & c\theta s\phi \\ s\psi s\phi + c\psi s\theta c\phi & -c\psi s\phi + s\psi s\theta c\phi & c\theta c\phi \end{bmatrix} \mathbf{v}^g \\
\mathbf{v}^b &= \mathbf{R}_g^b \mathbf{v}^g
\end{aligned} \tag{2.43}$$

where the notation $cx = \cos(x)$ and $sx = \sin(x)$. The inverse vector transformation is

$$\begin{aligned}
\mathbf{v}^g &= \mathbf{R}_b^g \mathbf{v}^b \\
&= \left(\mathbf{R}_g^b\right)^\top \mathbf{v}^b.
\end{aligned} \tag{2.44}$$

Example 2.5 *The velocity of a vehicle in the body frame is measured to be* $\mathbf{v}^b = [50, 0, 0]^\top \frac{m}{s}$. *The attitude of the vehicle is* $(\phi, \theta, \psi) = (0, 45, 90)$ deg. *What is the instantaneous rate of change of the vehicle position in the local (geographic) tangent plane reference frame?*

In this case,

$$\mathbf{R}_b^g = \begin{bmatrix} 0 & -1 & 0 \\ .707 & 0 & .707 \\ -.707 & 0 & .707 \end{bmatrix}.$$

Therefore, the vehicle velocity relative to the tangent plane reference system is $\mathbf{v}^g = [v_n, v_e, v_d]^\top = [0, 35.35, -35.35]^\top \frac{m}{s}$. △

Once a sequence of rotations (in this case zyx) is specified, the rotation angle sequence to represent a given relative rotational orientation is unique except at points of singularity. Note for example that for the zyx sequence of rotations, the rotational sequence $[x]_1 \left[\frac{\pi}{2}\right]_2 [x]_3$ yields the same orientation for any $|x| \leq \pi$. This demonstrates that the zyx sequence of rotations is singular points at $\theta = \pm\frac{\pi}{2}$. These are the only points of singularity of the zyx sequence of rotations.

The above zyx rotation sequence is not the only possibility. Other rotation sequences are in use due to the fact that the singularities will occur at different locations. The zyx sequence is used predominantly in this book. In land or sea surface-vehicle applications the singular point (hopefully) does not occur. In other applications, alternative Euler angle sequences may be used. Also, singularity free parameterizations, such as the quaternion, offer attractive alternatives.

When the matrix R_b^g is known, the Euler angles can be determined, for control or planning purposes, by the following equations

$$\theta = -\text{atan}\left(\frac{\mathbf{R}_b^g[3,1]}{\sqrt{1-(\mathbf{R}_b^t[3,1])^2}}\right) \tag{2.45}$$

$$\phi = \text{atan2}\left(\mathbf{R}_b^g[3,2], \mathbf{R}_b^g[3,3]\right) \tag{2.46}$$

$$\psi = \text{atan2}\left(\mathbf{R}_b^g[2,1], \mathbf{R}_b^g[1,1]\right)) \tag{2.47}$$

where atan2(y, x) is a four quadrant inverse tangent function and the numbers in square brackets refer to a specific element of the matrix. For example, $\mathbf{A}[i,j] = a_{ij}$ is the element in the i-th row and j-th column of matrix $\mathbf{A}$.

This section has alluded to the fact that the rotation matrices $\mathbf{R}_b^g$ and $\mathbf{R}_b^t$ would have the same form. This should not be interpreted as meaning that the matrices or the Euler angles are the same. The Euler angles relative to a fixed tangent plane will be distinct from the Euler angles defined relative to the geographic frame. Also, the angular rates $\boldsymbol{\omega}_{bt}$ and $\boldsymbol{\omega}_{bg}$ are distinct. We have that

$$\begin{aligned} \boldsymbol{\omega}_{bt} &= \boldsymbol{\omega}_{it} - \boldsymbol{\omega}_{ib} \\ \boldsymbol{\omega}_{bg} &= \boldsymbol{\omega}_{ig} - \boldsymbol{\omega}_{ib}. \end{aligned}$$

The vector $\boldsymbol{\omega}_{it} = \mathbf{0}$, while $\boldsymbol{\omega}_{ig}$ can be non-zero as discussed in Example 2.6.

2.5.5 Transformation: Orthogonal Small Angle

Subsequent discussions will frequently consider small angle transformations. A small angle transformations, is the transformation between two coordinate systems differing infinitesimally in relative orientation. For example, in discussing the time derivative of a direction cosine matrix, it will be convenient to consider the small angle transformation between the direction cosine matrices valid at two infinitesimally different instants of time. Also, in analyzing INS error dynamics it will be necessary to consider transformations between physical and computed frames-of-reference, where the error (at least initially) is small. In contrast, Euler angles define finite angle rotational transformations.

Consider coordinate systems a and b where frame b is obtained from frame a by the infinitesimal rotations $\delta\theta_3$ about the third axis of the a frame, $\delta\theta_2$ about the second axis of the resultant frame of the first rotation, and $\delta\theta_1$ about the first axis of the resultant frame of the second rotation. Denote this infinitesimal rotation by $\delta\boldsymbol{\theta} = [\delta\theta_1, \delta\theta_2, \delta\theta_3]^\top$. The vector transformation from frame a to frame b is defined by the series of three rotations: $[\delta\theta_1]_1[\delta\theta_2]_2[\delta\theta_3]_3$. Due to the fact that each angle is infinitesimal (which implies that $\cos(\delta\theta_i) \approx 1$, $\sin(\delta\theta_i) \approx \delta\theta_i$, and $\delta\theta_i\delta\theta_j \approx 0$ for $i, j = 1, 2, 3$) the order in which the rotations occur will not be important. The matrix representation of the vector transformation is

$$\begin{aligned}
\mathbf{v}^b &= \begin{bmatrix} 1 & 0 & 0 \\ 0 & 1 & \delta\theta_1 \\ 0 & -\delta\theta_1 & 1 \end{bmatrix} \begin{bmatrix} 1 & 0 & -\delta\theta_2 \\ 0 & 1 & 0 \\ \delta\theta_2 & 0 & 1 \end{bmatrix} \begin{bmatrix} 1 & \delta\theta_3 & 0 \\ -\delta\theta_3 & 1 & 0 \\ 0 & 0 & 1 \end{bmatrix} \mathbf{v}^a \\
\mathbf{v}^b &= \begin{bmatrix} 1 & \delta\theta_3 & -\delta\theta_2 \\ -\delta\theta_3 & 1 & \delta\theta_1 \\ \delta\theta_2 & -\delta\theta_1 & 1 \end{bmatrix} \mathbf{v}^a \\
\mathbf{v}^b &= (\mathbf{I} - \delta\boldsymbol{\Theta})\,\mathbf{v}^a
\end{aligned} \tag{2.48}$$

where $\delta\boldsymbol{\Theta} = [\delta\boldsymbol{\theta}\times]$ is the skew symmetric representation of $\delta\boldsymbol{\theta}$ as defined in eqn. (B.15) of Appendix B. To first order, the inverse rotation is

$$\mathbf{v}^a = (\mathbf{I} + \delta\boldsymbol{\Theta})\,\mathbf{v}^b. \tag{2.49}$$

2.6 Rotating Reference Frames

As discussed relative to Figure 2.1, reference frames may be free to rotate arbitrarily with respect to one another. Consider for example the body frame moving with respect to the ECEF frame. The following subsections are concerned with frames-of-reference rotating with respect to one another.

2.6.1 Direction Cosine Kinematics

Section 2.5 showed that the transformation of vectors between two coordinate systems could be represented by an appropriately defined direction cosine matrix. In subsequent sections, it is necessary to calculate derivatives of direction cosine matrices for coordinate systems experiencing relative rotation. Such derivatives are the subject of this present section.

The definition of the derivative of the rotation matrix from frame a to frame b is

$$\dot{\mathbf{R}}_a^b(t) = \lim_{\delta t \to 0} \frac{\mathbf{R}_a^b(t + \delta t) - \mathbf{R}_a^b(t)}{\delta t}. \tag{2.50}$$

For small δt the rotation $\mathbf{R}_a^b(t+\delta t)$ can be considered as the rotation from frame a to frame b at time t followed by the rotation from frame b at time t to frame b at time $t+\delta t$:

$$\mathbf{R}_a^b(t+\delta t) = \mathbf{R}_{b(t)}^{b(t+\delta t)} \mathbf{R}_a^b(t). \tag{2.51}$$

Because the instantaneous angular velocity $\boldsymbol{\omega}_{ab}^b$ of frame b with respect frame a represented in frame b is finite and δt will be approaching zero, the rotation matrix $\mathbf{R}_{b(t)}^{b(t+\delta t)}$ represents the small angle rotation $\delta\boldsymbol{\theta} = \boldsymbol{\omega}_{ab}^b \delta t$ which by eqn. (2.48) is

$$\mathbf{R}_{b(t)}^{b(t+\delta t)} = \mathbf{I} - \boldsymbol{\Omega}_{ab}^b \delta t \tag{2.52}$$

where $\boldsymbol{\Omega}_{ab}^b = \left[\boldsymbol{\omega}_{ab}^b \times\right]$. Substituting eqns. (2.51) and (2.52) into eqn. (2.50) yields

$$\begin{aligned} \dot{\mathbf{R}}_a^b(t) &= \lim_{\delta t \to 0} \frac{\left(\mathbf{I} - \boldsymbol{\Omega}_{ab}^b \delta t\right) \mathbf{R}_a^b(t) - \mathbf{R}_a^b(t)}{\delta t} \\ &= -\boldsymbol{\Omega}_{ab}^b \mathbf{R}_a^b(t). \end{aligned} \tag{2.53}$$

Using the facts that $\boldsymbol{\Omega}_{ab}^b = \mathbf{R}_a^b \boldsymbol{\Omega}_{ab}^a \mathbf{R}_b^a$ and that $\boldsymbol{\Omega}_{ab}^a = -\boldsymbol{\Omega}_{ba}^a$, eqn. (2.53) can also be expressed as

$$\dot{\mathbf{R}}_a^b = \mathbf{R}_a^b \, \boldsymbol{\Omega}_{ba}^a. \tag{2.54}$$

Example 2.6 *Let a represent the geographic frame and b represent the ECEF frame. Using the fact the $\dot{\mathbf{R}}_g^e = \mathbf{R}_g^e \boldsymbol{\Omega}_{eg}^g$, we can compute $\boldsymbol{\omega}_{eg}^g$ as the vector form of $\boldsymbol{\Omega}_{eg}^g = \mathbf{R}_e^g \dot{\mathbf{R}}_g^e$.*

For $\mathbf{R}_g^e$ as defined in Section 2.5.3, we can compute $\dot{\mathbf{R}}_g^e$ as

$$\dot{\mathbf{R}}_g^e = \frac{\partial}{\partial \phi} \mathbf{R}_g^e \dot{\phi} + \frac{\partial}{\partial \lambda} \mathbf{R}_g^e \dot{\lambda}.$$

After algebra to simplify the result, this product yields

$$\boldsymbol{\Omega}_{eg}^g = \begin{bmatrix} 0 & \dot{\lambda}\sin(\phi) & -\dot{\phi} \\ -\dot{\lambda}\sin(\phi) & 0 & -\dot{\lambda}\cos(\phi) \\ \dot{\phi} & \dot{\lambda}\cos(\phi) & 0 \end{bmatrix} \tag{2.55}$$

which gives

$$\boldsymbol{\omega}_{eg}^g = [\dot{\lambda}\cos(\phi), -\dot{\phi}, -\dot{\lambda}\sin(\phi)]^\top. \tag{2.56}$$

This, combined with $\boldsymbol{\omega}_{ie}^g = [\omega_{ie}\cos(\phi), 0, -\omega_{ie}\sin(\phi)]^\top$ and $\boldsymbol{\omega}_{ig}^g = \boldsymbol{\omega}_{ie}^g + \boldsymbol{\omega}_{eg}^g$, gives

$$\boldsymbol{\omega}_{ig}^g = \begin{bmatrix} (\dot{\lambda} + \omega_{ie})\cos(\phi) \\ -\dot{\phi} \\ -(\dot{\lambda} + \omega_{ie})\sin(\phi) \end{bmatrix}. \tag{2.57}$$

△

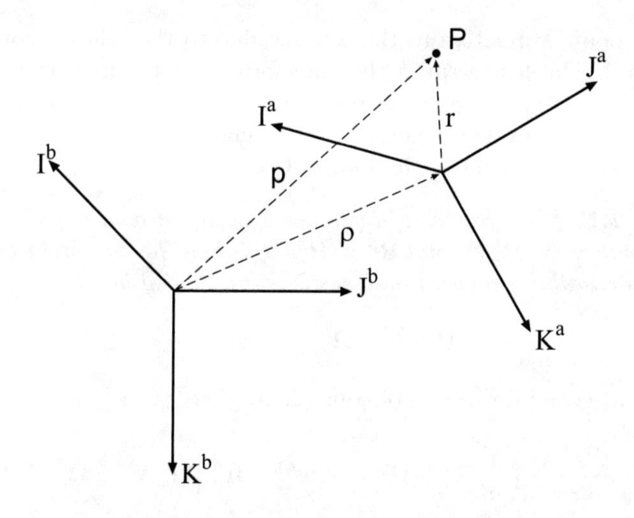

Figure 2.17: Rotating coordinate frames.

2.6.2 Derivative Calculations in Rotation Frames

As shown in Figure 2.17, let $\mathbf{p}$ be the vector from the b frame origin to point P, $\boldsymbol{\rho}$ be the vector from the b frame origin to the a frame origin, and $\mathbf{r}$ be the vector from the a frame origin to point P. These vectors are related by

$$\mathbf{p} = \boldsymbol{\rho} + \mathbf{r}.$$

If $\boldsymbol{\rho}^b$, $\mathbf{R}_a^b$, and $\mathbf{r}^a$ are known, then from Section 2.4, the representation of $\mathbf{p}$ in frame b can be computed:

$$\mathbf{p}^b = \boldsymbol{\rho}^b + \mathbf{R}_a^b \mathbf{r}^a. \tag{2.58}$$

If the a frame is rotating with respect to the b frame, the rate of change of $\mathbf{p}^b$ can be expressed as in eqn. (2.59):

$$\begin{aligned}
\frac{d\mathbf{p}^b}{dt} &= \frac{d\boldsymbol{\rho}^b}{dt} + \left(\frac{d}{dt}\mathbf{R}_a^b\right)\mathbf{r}^a + \mathbf{R}_a^b \frac{d\mathbf{r}^a}{dt} \\
\frac{d\mathbf{p}^b}{dt} &= \frac{d\boldsymbol{\rho}^b}{dt} + \mathbf{R}_a^b \boldsymbol{\Omega}_{ba}^a \mathbf{r}^a + \mathbf{R}_a^b \mathbf{v}^a \\
\frac{d\mathbf{p}^b}{dt} &= \frac{d\boldsymbol{\rho}^b}{dt} + \mathbf{R}_a^b \left(\boldsymbol{\Omega}_{ba}^a \mathbf{r}^a + \mathbf{v}^a\right).
\end{aligned} \tag{2.59}$$

The first term on the right accounts for the relative instantaneous linear velocity of the two reference frames. The second term is the instantaneous velocity of point P relative to the b frame due to the relative rotation of the a frame. The last term is the transformation to the b frame of the instantaneous velocity of point P relative to the origin of the a frame.

This equation can be considered as a special case of the following theorem, which is a statement of the Law of Coriolis.

Theorem 2.1 *If two frames-of-reference experience relative angular rotation* $\mathbf{\Omega}^a_{ba}$ *with* $\mathbf{v}^b = \mathbf{R}^b_a\mathbf{v}^a$ *and* $\dot{\mathbf{R}}^b_a = \mathbf{R}^b_a\mathbf{\Omega}^a_{ba}$, *then the time rate of change of the vector in the two coordinate systems are related by*

$$\mathbf{R}^a_b\dot{\mathbf{v}}^b = [\mathbf{\Omega}^a_{ba}\mathbf{v}^a + \dot{\mathbf{v}}^a]. \tag{2.60}$$

Taking a second derivative of eqn. (2.59) gives

$$\begin{aligned}\frac{d^2\mathbf{p}^b}{dt^2} &= \frac{d^2\boldsymbol{\rho}^b}{dt^2} + \left(\frac{d}{dt}\mathbf{R}^b_a\right)[\mathbf{\Omega}^a_{ba}\mathbf{r}^a + \mathbf{v}^a] + \mathbf{R}^b_a\left[\mathbf{\Omega}^a_{ba}\mathbf{v}^a + \dot{\mathbf{\Omega}}^a_{ba}\mathbf{r}^a + \frac{d^2\mathbf{r}^a}{dt^2}\right]\\ \frac{d^2\mathbf{p}^b}{dt^2} &= \frac{d^2\boldsymbol{\rho}^b}{dt^2} + \mathbf{R}^b_a\left[2\mathbf{\Omega}^a_{ba}\mathbf{v}^a + \mathbf{\Omega}^a_{ba}\mathbf{\Omega}^a_{ba}\mathbf{r}^a + \dot{\mathbf{\Omega}}^a_{ba}\mathbf{r}^a + \frac{d^2\mathbf{r}^a}{dt^2}\right]. \qquad (2.61)\end{aligned}$$

Note the following points regarding the derivation of eqn. (2.61): the equation is exact; the equation is applicable between any two coordinate systems; and, the equation is linear in the position and velocity vectors. This equation is the foundation on which a variety of navigation systems are built.

2.7 Calculation of the Direction Cosine

The previous sections have motivated the necessity of maintaining accurate direction cosine matrices. The following two subsections will consider two methods for maintaining the direction cosine matrix as the two reference frames experience arbitrary relative angular motion. Each technique relies on measuring the relative angular rate and integrating it (via different methods). Initial conditions for the resulting differential equations are discussed in Sections 10.3 and 11.7. Prior to integration, the angular rates should be properly compensated for biases and navigation frame rotation as discussed in Section 11.3.

The measurement of angular rates followed by integration to determine angle has the disadvantage that measurement errors will accumulate during the integration process. However, the approach has several distinct advantages. The angular rates are measurable via inertial measurements. Inertial measurements do not rely on the reception of any signal exterior to the sensor itself. Therefore, the accuracy of the measurement and integration processes will only be limited by the accuracy inherent to the instrument and of the integration process. Both of these quantities can be accurately calibrated so that the system accuracy can be reliably predicted. Also, the error accumulation through the integration of inertial quantities is a slow process that can be corrected via aiding sensors. Alternatively, level sensors and compass type instruments could be used to directly measure the Euler angles. However, level (gravity vector) sensors are sensitive to acceleration as well as attitude changes; and, yaw/heading sensors are sensitive to local magnetic fields. The resulting measurement errors are difficult to quantify accurately and reliably at the design stage.

2.7.1 Direction Cosine Derivatives

Eqn. (2.53) provides a differential equation which can be integrated to maintain the direction cosine matrix. Such an approach could numerically integrate each of the direction cosine elements separately; however, since the equation is linear, a closed form solution should be obtainable. Direct numeric integration of eqn. (2.53) does not enforce the orthogonality nor normality constraints on the direction cosine matrix; hence, additional normalization calculations would be required (see Chapter 2 in [27] or Section B.9 of Appendix B).

Let $t_k = kT$ and select T sufficiently small that $\boldsymbol{\omega}^a_{ba}$ can be considered constant $\forall t \in [t_{k-1}, t_k]$. To derive a closed form solution to eqn. (2.53) with initial condition $\mathbf{R}^a_b(t_{k-1})$ for $t \in [t_{k-1}, t_k)$, define the integrating factor

$$\Xi(t) = e^{\int_{t_{k-1}}^{t} \boldsymbol{\Omega} d\tau} \tag{2.62}$$

where $\boldsymbol{\Omega} = [\boldsymbol{\omega}\times]$ has been used as a shorthand notation for $\boldsymbol{\Omega}^a_{ba}$ to decrease the complexity of the notation. The $[\boldsymbol{\omega}\times]$ notation is defined in eqn. (B.15).

Multiplying the integrating factor into eqn. (2.53) and simplifying, yields

$$\begin{aligned}
\dot{\mathbf{R}}^a_b(t) &= -\boldsymbol{\Omega}\mathbf{R}^a_b(t) \\
e^{\int_{t_{k-1}}^{t} \boldsymbol{\Omega} d\tau}\dot{\mathbf{R}}^a_b(t) + e^{\int_{t_{k-1}}^{t} \boldsymbol{\Omega} d\tau}\boldsymbol{\Omega}\mathbf{R}^a_b(t) &= \mathbf{0} \\
\frac{d}{dt}\left(e^{\int_{t_{k-1}}^{t} \boldsymbol{\Omega} dt}\mathbf{R}^a_b(t)\right) &= \mathbf{0}
\end{aligned}$$

where the last step is valid given the assumption that T is sufficiently small so that $\boldsymbol{\omega}$ is constant over the period of integration. Integrating both sides, over the period of integration yields

$$\begin{aligned}
e^{\int_{t_{k-1}}^{t} \boldsymbol{\Omega} d\tau}\mathbf{R}^a_b(t) - \mathbf{R}^a_b(t_{k-1}) &= \mathbf{0} \\
\mathbf{R}^a_b(t) &= e^{-\int_{t_{k-1}}^{t} \boldsymbol{\Omega} d\tau}\mathbf{R}^a_b(t_{k-1})
\end{aligned} \tag{2.63}$$

for $t \in [t_{k-1}, t_k)$.

To simplify this expression, let

$$\boldsymbol{\Upsilon}(t) = \int_{t_{k-1}}^{t} \boldsymbol{\Omega} d\tau \tag{2.64}$$

where $\upsilon_i(t) = \int_{t_{k-1}}^{t} \omega_i(\tau)d\tau$ and $\boldsymbol{\Upsilon} = [\boldsymbol{\upsilon}\times]$. Powers of the matrix $\boldsymbol{\Upsilon}$ reduce as follows:

$$\boldsymbol{\Upsilon}^2 = \begin{bmatrix} -(\upsilon_2^2 + \upsilon_3^2) & \upsilon_1\upsilon_2 & \upsilon_1\upsilon_3 \\ \upsilon_1\upsilon_2 & -(\upsilon_1^2 + \upsilon_3^2) & \upsilon_2\upsilon_3 \\ \upsilon_1\upsilon_3 & \upsilon_2\upsilon_3 & -(\upsilon_1^2 + \upsilon_2^2) \end{bmatrix} \tag{2.65}$$

$$\boldsymbol{\Upsilon}^3 = -\|\boldsymbol{\upsilon}\|^2\boldsymbol{\Upsilon} \tag{2.66}$$

$$\boldsymbol{\Upsilon}^4 = -\|\boldsymbol{\upsilon}\|^2\boldsymbol{\Upsilon}^2 \tag{2.67}$$

$$\boldsymbol{\Upsilon}^5 = +\|\boldsymbol{\upsilon}\|^4\boldsymbol{\Upsilon} \tag{2.68}$$

which are verified in Exercise 2.10. Therefore, using eqn. (B.42),

$$\begin{aligned} e^{-\int_{t_{k-1}}^{t}\boldsymbol{\Omega}d\tau} &= \mathbf{I} - \boldsymbol{\Upsilon} + \frac{1}{2}\boldsymbol{\Upsilon}^2 - \frac{1}{3!}\boldsymbol{\Upsilon}^3 + \dots \\ &= \mathbf{I} - \boldsymbol{\Upsilon} + \frac{1}{2}\boldsymbol{\Upsilon}^2 + \frac{1}{3!}\|\upsilon\|^2\boldsymbol{\Upsilon} - \frac{1}{4!}\|\upsilon\|^2\boldsymbol{\Upsilon}^2 - \frac{1}{5!}\|\upsilon\|^4\boldsymbol{\Upsilon} + \dots \\ &= \mathbf{I} - \left(1 - \frac{1}{3!}\|\boldsymbol{\upsilon}\|^2 + \frac{1}{5!}\|\boldsymbol{\upsilon}\|^4 + \dots\right)\boldsymbol{\Upsilon} \\ &\quad + \left(\frac{1}{2} - \frac{1}{4!}\|\boldsymbol{\upsilon}\|^2 + \dots\right)\boldsymbol{\Upsilon}^2 \\ &= \mathbf{I} - \frac{\sin(\|\boldsymbol{\upsilon}\|)}{\|\boldsymbol{\upsilon}\|}\boldsymbol{\Upsilon} + \frac{1-\cos(\|\boldsymbol{\upsilon}\|)}{\|\boldsymbol{\upsilon}\|^2}\boldsymbol{\Upsilon}^2. \end{aligned} \tag{2.69}$$

Substituting eqn. (2.69) into eqn. (2.63) yields

$$\mathbf{R}_b^a(t_k) = \left(\mathbf{I} - \frac{\sin(\|\boldsymbol{\upsilon}\|)}{\|\boldsymbol{\upsilon}\|}\boldsymbol{\Upsilon} + \frac{1-\cos(\|\boldsymbol{\upsilon}\|)}{\|\boldsymbol{\upsilon}\|^2}\boldsymbol{\Upsilon}^2\right)\mathbf{R}_b^a(t_{k-1}). \tag{2.70}$$

where $\boldsymbol{\Upsilon} = \boldsymbol{\Upsilon}(t_k) = [\boldsymbol{\upsilon}(t_k)\times]$ and $\upsilon_i(t_k) = \int_{t_{k-1}}^{t_k} \omega_i(\tau)d\tau$. Eqn. (2.70) is properly defined theoretically, even as $\|\boldsymbol{\upsilon}\| \to 0$, but must be implemented numerically with care.

The designer must ensure that the interval $[t_k, t_{k-1}]$ is sufficiently small (i.e., the sample frequency is sufficiently fast) to satisfy the assumption above eqn. (2.62) that each ω_i can be considered constant over each period of integration.

2.7.2 Euler Angle Derivatives

Section 2.5 defined the Euler angles (ϕ, θ, ψ) and described their use to determine the direction cosine matrix $\mathbf{R}_b^g$. This method of determining the direction cosine matrix requires the navigation system to measure or compute the Euler angles. This section presents the method for computing the Euler angle derivatives $(\dot{\phi}, \dot{\theta}, \dot{\psi})$ from the angular rate vector $\boldsymbol{\omega}_{gb}^b$. The navigation system could then integrate the Euler angle derivatives, to determine the Euler angles from which the direction cosine matrix can be computed.

Note that the three-tuple $(\dot{\phi}, \dot{\theta}, \dot{\psi})$ is *not* related to $\boldsymbol{\omega}_{gb}^b$ by a rotational transform because each of the components of $(\dot{\phi}, \dot{\theta}, \dot{\psi})$ is defined in a different reference frame. Instead, we have that

$$\boldsymbol{\omega}_{gb}^b = \boldsymbol{\omega}_{g1}^b + \boldsymbol{\omega}_{12}^b + \boldsymbol{\omega}_{2b}^b. \tag{2.71}$$

Figures 2.14–2.16 define the yaw, pitch, and roll rotations. In addition, the figures show that $\dot{\psi}$ is the rate of rotation about the $\mathbf{K}$ or $\mathbf{K}'$ axes, which are coincident; that $\dot{\theta}$ is the rate of rotation about the $\mathbf{J}'$ or $\mathbf{J}''$ axes, which are coincident; and, that $\dot{\phi}$ is the rate of rotation about the $\mathbf{I}''$ or $\mathbf{I}'''$ axes, which are coincident. From the facts in the previous sentence we have that

$$\boldsymbol{\omega}_{g1}^1 = \begin{bmatrix} 0 \\ 0 \\ 1 \end{bmatrix} \dot{\psi}, \quad \boldsymbol{\omega}_{12}^2 = \begin{bmatrix} 0 \\ 1 \\ 0 \end{bmatrix} \dot{\theta}, \quad \boldsymbol{\omega}_{2b}^b = \begin{bmatrix} 1 \\ 0 \\ 0 \end{bmatrix} \dot{\phi}.$$

From eqn. (2.71) we have

$$\boldsymbol{\omega}_{gb}^b = \mathbf{R}_1^b \boldsymbol{\omega}_{g1}^1 + \mathbf{R}_2^b \boldsymbol{\omega}_{12}^2 + \boldsymbol{\omega}_{2b}^b. \tag{2.72}$$

From eqn. (2.42),

$$\mathbf{R}_2^b \boldsymbol{\omega}_{12}^2 = \begin{bmatrix} 0 \\ \cos(\phi) \\ -\sin(\phi) \end{bmatrix} \dot{\theta};$$

and from eqns. (2.41–2.42),

$$\mathbf{R}_1^b \boldsymbol{\omega}_{g1}^1 = \mathbf{R}_2^b \mathbf{R}_1^2 \boldsymbol{\omega}_{g1}^1 = \begin{bmatrix} -\sin(\theta) \\ \sin(\phi)\cos(\theta) \\ \cos(\phi)\cos(\theta) \end{bmatrix} \dot{\psi}.$$

Therefore, eqn. (2.72) yields

$$\boldsymbol{\omega}_{gb}^b = \begin{bmatrix} 1 & 0 & -\sin(\theta) \\ 0 & \cos(\phi) & \sin(\phi)\cos(\theta) \\ 0 & -\sin(\phi) & \cos(\phi)\cos(\theta) \end{bmatrix} \begin{bmatrix} \dot{\phi} \\ \dot{\theta} \\ \dot{\psi} \end{bmatrix}. \tag{2.73}$$

The inverse transformation is

$$\begin{aligned} \begin{bmatrix} \dot{\phi} \\ \dot{\theta} \\ \dot{\psi} \end{bmatrix} &= \begin{bmatrix} 1 & \sin(\phi)\tan(\theta) & \cos(\phi)\tan(\theta) \\ 0 & \cos(\phi) & -\sin(\phi) \\ 0 & \frac{\sin(\phi)}{\cos(\theta)} & \frac{\cos(\phi)}{\cos(\theta)} \end{bmatrix} \boldsymbol{\omega}_{gb}^b \\ &= \boldsymbol{\Omega}_E^{-1} \boldsymbol{\omega}_{gb}^b \end{aligned} \tag{2.74}$$

where $\mathbf{\Omega}_E = \begin{bmatrix} 1 & 0 & -\sin(\theta) \\ 0 & \cos(\phi) & \sin(\phi)\cos(\theta) \\ 0 & -\sin(\phi) & \cos(\phi)\cos(\theta) \end{bmatrix}$

and $\mathbf{\Omega}_E^{-1} = \begin{bmatrix} 1 & \sin(\phi)\tan(\theta) & \cos(\phi)\tan(\theta) \\ 0 & \cos(\phi) & -\sin(\phi) \\ 0 & \frac{\sin(\phi)}{\cos(\theta)} & \frac{\cos(\phi)}{\cos(\theta)} \end{bmatrix}$.

Neither $\mathbf{\Omega}_E$ nor its inverse is a rotation matrix. Eqns. (2.73) and (2.74) do not represent vector transformations between frames-of-reference. The matrix $\mathbf{\Omega}_E$ is singular when $\theta = \frac{\pi}{2}$.

Eqn. (2.74) is a nonlinear ordinary differential equation for the Euler angles. Numeric integration of eqn. (2.74) provides estimates of the Euler angles from which the direction cosine matrix can by computed using eqn. (2.43). This approach requires numerous trigonometric operations at the high rate of the attitude portion of the INS system. The vector $\boldsymbol{\omega}_{gb}^b$ is not directly measured; instead it is computed as $\boldsymbol{\omega}_{gb}^b = \boldsymbol{\omega}_{ib}^b - \boldsymbol{\omega}_{ig}^b$ where $\boldsymbol{\omega}_{ib}^b$ is measured by the body mounted gyros and $\boldsymbol{\omega}_{ig}^b = \mathbf{R}_g^b \boldsymbol{\omega}_{ig}^g$ with $\boldsymbol{\omega}_{ig}^g$ expressed in eqn. (2.57).

2.8 References and Further Reading

The main references for this chapter are [27, 37, 50, 73]. The reference frames are presented very precisely in [99]. The main sources for the material related to gravity models and definition of geodetic reference systems are [27, 37, 67, 73, 102, 127].

Alternative attitude representations are reviewed in for example [120]. An introduction and review of the quaternion representation is presented in Appendix D.

2.9 Exercises

Exercise 2.1 The purpose of this exercise is for the reader to demonstrate that the extension of the normal to an ellipse does not in general intersect the center of the ellipse.

Consider the ellipse $\left(\frac{x}{a}\right)^2 + \left(\frac{y}{b}\right)^2 = 1$ which is centered at the origin.

1. Show that the outward pointing normal to the ellipse is $N = \left[\frac{2x}{a^2}, \frac{2y}{b^2}\right]$.

2. Select a point on the ellipse for which neither coordinate is zero, for example $\left[\frac{\sqrt{3}}{2}a, \frac{1}{2}b\right]$. Determine the equation for the line through the selected point in the direction of the normal vector.

3. Find the intersection of this line with both the x axis. Then find the intersection of this line with the y axis.

4. Show that these intersections coincide with the origin only if $a = b$.

This problem can be easily implemented in a MATLAB program to illustrate the solution.

Exercise 2.2 This exercise guides the reader through the derivation of eqns. (2.7) and (2.6) using the notation defined in Figure 2.18 and the method from Chapter 4 in [113].

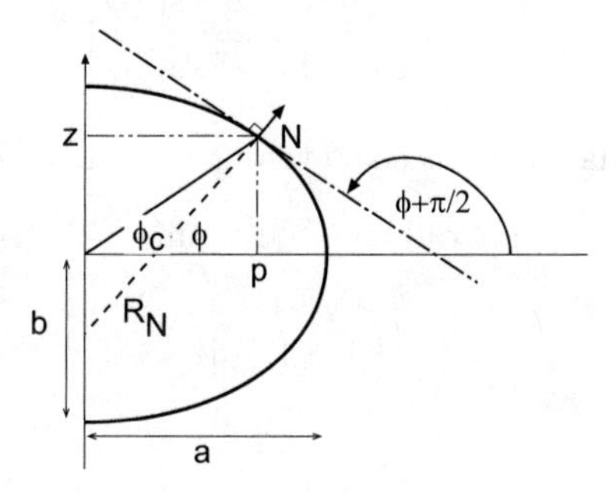

Figure 2.18: Ellipse with notation for Exercise 2.2.

The equation for the ellipse in Figure 2.18 is

$$\frac{p^2}{a^2} + \frac{z^2}{b^2} = 1$$

where (p, z) is a point on the ellipse.

1. Show that for a point on the ellipse,

$$\begin{aligned} \frac{dz}{dp} &= -\frac{p}{z}\frac{b^2}{a^2} \\ \frac{dz^2}{dp^2} &= -\frac{b^4}{a^2}\frac{1}{z^3}. \end{aligned}$$

2. Show that the line tangent to the ellipse at (p, z) has slope

$$\frac{dz}{dp} = \tan\left(\phi + \frac{\pi}{2}\right) = -\frac{1}{\tan\phi}.$$

3. Using the results from Steps 1 and 2 show that

$$\frac{z}{p} = (1 - e^2)\tan\phi \tag{2.75}$$

where e is the eccentricity of the ellipse. Combine this result with the fact that $p^2 = a^2\left(1 - \frac{z^2}{b^2}\right)$ to derive

$$p = \frac{a\cos\phi}{\sqrt{1 - e^2\sin^2\phi}}. \tag{2.76}$$

Substitute eqn. (2.76) into eqn. (2.75) to obtain

$$z = \frac{a\left(1 - e^2\right)\sin\phi}{\sqrt{1 - e^2\sin^2\phi}}. \tag{2.77}$$

Use the fact that $p = R_N\cos\phi$ to derive eqn. (2.7).

4. Use the expression for radius of curvature

$$\frac{\left(1 + \left(\frac{dz}{dp}\right)^2\right)^{\frac{3}{2}}}{\pm\frac{d^2z}{dp^2}}$$

and the fact that $b^2 = -(e^2 - 1)a^2$ to derive

$$R_M = \frac{\left(z^2 + p^2(1 - e^2)\right)^{\frac{3}{2}}}{b^2(1 - e^2)}.$$

Finally, use eqns. (2.76–2.77) to derive eqn. (2.6).

Exercise 2.3 This exercise discusses alternative representations of latitude and longitude angles. For simplicity, consider a spherical Earth model with radius $R = 6.4 \times 10^6$ m.

1. Let ϕ represent the radian latitude of a point on a sphere of radius R. The relation between angular and rectangular error is approximately $dx = Rd\phi$. What precision of representation should be used for ϕ if the loss of precision in computing x is desired to be less than 1.0×10^{-3} m?

2. If instead ϕ was represented in degrees, what precision of representation should be used for ϕ?

3. If the representation of ϕ is deg:min where deg represents the integer number of degrees and min is the real number of minutes, what precision of representation should be used for *min*.

4. If the representation of ϕ is deg:min:sec where deg represents the integer number of degrees, min is the integer number of minutes and sec is the real number of seconds, what precision of representation should be used for *sec*.

Note that when using the deg:min:sec notation, the user must be careful with respect to which hemisphere contains the point of interest. For latitude, S and N represent the northern and southern hemispheres, respectively. For longitude, E and W represent the eastern and western hemispheres, respectively.

Exercise 2.4 Write a program to compute the ECEF rectangular coordinates for a given set of ECEF geodetic coordinates. Test the program with the following input data.

City	Country	Latitude, d:m:s	Longitude, d:m:s	height, m
Pretoria	S. Africa	26:08:42.20 S	28:03:00.92 E	1660.86
Sydney	Australia	33:53:28.15 S	151:14:57.07 E	86.26
Minsk	Belarus	53:53:58.61 N	27:34:21.10 E	197.21
Apia	Samoa	13:49:53.05 S	171:45:06.71 W	8.53

The d:m:s notation in this table is defined in Exercise 2.3.

Exercise 2.5 Write a program to compute the ECEF geodetic coordinates for a given set of ECEF rectangular coordinates. Test the program with the following input data.

City	x $\times 10^6$ m	y $\times 10^6$ m	z $\times 10^6$ m
Pretoria	5.057 590 377	2.694 861 463	-2.794 229 000
Sydney	-4.646 678 571	2.549 341 033	-3.536 478 881
Minsk	3.338 602 399	1.743 341 362	5.130 327 709
Apia	-6.130 311 688	-0.888 649 276	-1.514 877 991

Present your final results in the degree:minute:second notation and correctly indicate the hemisphere of the location.

Exercise 2.6 For $R_M(\phi)$ and $R_N(\phi)$ as defined in eqns. (2.6) and (2.7) show the validity of the following:

$$\frac{d}{d\phi}\left[(R_N + h)\cos(\phi)\right] = -(R_M + h)\sin(\phi) \tag{2.78}$$

$$\frac{d}{d\phi}\left[\left(R_N\left(1 - e^2\right) + h\right)\sin(\phi)\right] = (R_M + h)\cos(\phi). \tag{2.79}$$

Exercise 2.7 As was done in Example 2.4, find the vector and point transformation from ECEF to tangent plane for each of the locations listed in Problem 2.4.

Exercise 2.8 In body frame, the four corners of a vehicle are given in the following table.

Name	x m	y m	z m
FL	0.50	-0.25	0.00
FR	0.50	0.25	0.00
RR	-0.75	0.25	0.00
RL	-0.75	-0.25	0.00

Assuming that the offset from the tangent frame origin to the vehicle frame origin is $[100, 50, -7]^\top m$, for each of the following attitudes, compute the tangent plane location for each corner of the vehicle.

Attitude, deg		
ϕ	θ	ψ
45.00	0.00	0.00
0.00	45.00	-30.00
45.00	45.00	0.00
45.00	-45.00	0.00

Exercise 2.9 Use direct multiplication to show that to first order, the product of $(\mathbf{I} - \delta\boldsymbol{\Theta}^a_{ba})$ and $(\mathbf{I} + \delta\boldsymbol{\Theta}^a_{ba})$ is the identity matrix.

Exercise 2.10 Use eqn. (B.15) to form the skew symmetric matrix $\boldsymbol{\Upsilon}$. By direct matrix multiplication, confirm eqns. (2.65–2.68).

Exercise 2.11 Section 2.7.2 derived the relationship between the Euler attitude rates $\dot{\boldsymbol{\theta}} = (\dot{\phi}, \dot{\theta}, \dot{\psi})^\top$ and the body-frame inertial angular rate vector $\boldsymbol{\omega}^b_{ib}$.

1. Use a similar approach to show that the relationship between the Euler attitude rates and the angular rate of the body-frame relative to the inertial-frame represented in geodetic frame is $\boldsymbol{\omega}^g_{ib} = \boldsymbol{\Omega}_T \dot{\boldsymbol{\theta}}$ where

$$\boldsymbol{\Omega}_T = \begin{bmatrix} \cos(\psi)\cos(\theta) & -\sin(\psi) & 0 \\ \sin(\psi)\cos(\theta) & \cos(\psi) & 0 \\ -\sin(\theta) & 0 & 1 \end{bmatrix}. \tag{2.80}$$

2. Check the above result by confirming that $\mathbf{R}^g_b = \boldsymbol{\Omega}_T \boldsymbol{\Omega}_E^{-1}$.

This matrix $\boldsymbol{\Omega}_T$ is useful for relating the attitude error $\boldsymbol{\rho}$ represented in geodetic frame (defined in Section 10.5) to the Euler angle errors $\delta\boldsymbol{\theta} = (\delta\phi, \delta\theta, \delta\psi)^\top$. The relation $\delta\boldsymbol{\rho} = \boldsymbol{\Omega}_T \delta\boldsymbol{\theta}$ is used in Sections 10.5.5 and 12.5.1.

Exercise 2.12 Figure 9.1 shows the axle of a vehicle. Denote the length of the axle by L. The body frame velocity and angular rate are $\mathbf{v} = [u, v, 0]^\top$ and $\omega = [0, 0, 1]^\top$. Use the law of Coriolis to find the tangent plane velocity of each wheel.

Chapter 3

Deterministic Systems

The quantitative analysis of navigation systems will require *analytic system models.* Models can take a variety of forms. For finite dimensional linear systems with zero initial conditions that evolve in continuous-time, for example, the ordinary differential equation, transfer function, and state space models are equivalent. The dynamics of the physical systems of interest in navigation applications typically evolve in continuous-time, while the equations of the navigation system itself are often most efficiently implemented in discrete-time. Therefore, both difference and differential equations are of interest. This chapter present several essential concepts from linear and nonlinear systems theory.

3.1 Continuous-Time Systems Models

Three equivalent model structures for continuous-time systems are discussed in this section. The models are equivalent in the sense that each contains the same basic information about the system. However, some forms of analysis are more convenient in one model format than in another. Models for physical systems derived from basic principles often result in ordinary differential equation or state space models. Frequency response analysis and frequency domain system identification techniques utilize the transfer function representation. Optimal state estimation techniques are most conveniently presented and implemented using the state space approach. A major objective of this section is to provide the means to translate efficiently and accurately between these three model representations.

3.1.1 Ordinary Differential Equations

Many systems that evolve dynamically as a function of a continuous-time variable can be modeled effectively by a set of n-th order ordinary differential equations (ODE). When the applicable equations are linear, each n-th order differential equation is represented as

$$\begin{aligned} y^{(n)}(t) + d_1(t)y^{(n-1)}(t) + \ldots + d_n(t)y(t) \\ = n_1(t)u^{(n-1)}(t) + \ldots + n_n(t)u(t) \end{aligned} \tag{3.1}$$

where the notation $()^{(j)}$ denotes the j-th time derivative of the term in parenthesis. In this general form, the coefficients of the differential equation are time-varying. In applications involving time-invariant systems, the coefficients are constants.

Eqn. (3.1) represents a single-input single-output dynamical system. The input signal is represented by $u(t)$. The output signal is represented by $y(t)$. Eqn. (3.1) is referred to as the input-output ordinary differential equation.

When the applicable equations are nonlinear, the n-th order differential equation is represented in general as

$$y^{(n)}(t) = f(y^{(n-1)}(t), \ldots, y(t), u^{(n-1)}(t), \ldots, u(t)). \tag{3.2}$$

Taylor series analysis of eqn. (3.2) about a nominal trajectory can be used to provide a linear model described as in eqn. (3.1) for local analysis. See Section 3.3.

To solve an n-th order differential equation for $t \geq t_0$ requires n pieces of information (e.g., initial conditions) which describe the state of the system at time t_0. This concept of *system state* will be made concrete in Section 3.5.

Example 3.1 *Consider a one dimensional frictionless system corresponding to an object with a known external force applied at the center of gravity. The corresponding differential equation is*

$$m\ddot{p}(t) = f(t) \tag{3.3}$$

where m is the mass, $\ddot{p}(t)$ is the acceleration, and $f(t)$ is the external applied force. If the object is also subject to linear friction and restoring forces, then this is the classic forced mass-spring-damper example, The resulting differential equation is

$$m\ddot{p}(t) + b\dot{p}(t) + kp(t) = f(t) \tag{3.4}$$

where b represents the linear coefficient of friction and k is the coefficient of the linear restoring force. △

Example 3.2 *The differential equation for a single channel of the INS horizontal position error due to gyro measurement error is*

$$p^{(3)}(t) + \frac{g}{R_e} p^{(1)}(t) = g\, \epsilon_g(t) \tag{3.5}$$

where p is position in meters, $p^{(i)}$ is the i-th derivative of position, g is gravitational acceleration, R_e is the Earth radius, and ϵ_g represents gyro measurement error. △

3.1.2 Transfer Functions

If a system described by eqn. (3.1) has constant coefficients, then frequency domain or Laplace analysis can aid the analyst's understanding of the system performance. If the initial conditions in eqn. (3.1) are assumed to be zero, then the transfer function representing the linear system is found as follows:

$$\begin{aligned} \mathcal{L}\{y^{(n)}(t) + d_1 y^{(n-1)}(t) + \dots d_n y(t)\} &= \mathcal{L}\{n_1 u^{(n-1)}(t) + \dots + n_n u(t)\} \\ Y(s)s^n + d_1 Y(s)s^{n-1} + \dots + d_n Y(s) &= n_1 U(s)s^{n-1} + \dots + n_n U(s) \\ G(s) = \frac{Y(s)}{U(s)} &= \frac{n_1 s^{n-1} + \dots + n_n}{s^n + d_1 s^{n-1} + \dots + d_n} \end{aligned} \tag{3.6}$$

where $X(s) = \mathcal{L}\{x(t)\}$ denotes the Laplace transform of $x(t)$ and s denotes the Laplace variable. The transfer function for a linear time-invariant (LTI) system is the Laplace transform of the output divided by the Laplace transform of the input. When there is no pure delay in the LTI system, the transfer function can be represented as the ratio of two polynomials in s with constant coefficients. Note that the coefficients of the transfer function polynomials are identical to the input-output differential equation coefficients. The order of the transfer function is n. The system has n poles which are the roots of the denominator polynomial

$$s^n + d_1 s^{n-1} + \dots + d_n = 0.$$

The transfer function (finite) zeros are the roots of the numerator polynomial

$$n_1 s^{n-1} + \dots + n_n = 0.$$

Example 3.3 *For the system described by eqn. (3.3), the transfer function from the forcing function to the position is*

$$G(s) = \frac{P(s)}{F(s)} = \frac{1/m}{s^2}. \tag{3.7}$$

For the system described by eqn. (3.4), the transfer function from the forcing function to the position is

$$G(s) = \frac{P(s)}{F(s)} = \frac{1/m}{s^2 + \frac{b}{m}s + \frac{k}{m}}. \tag{3.8}$$

A transfer function is always defined in reference to a stated input and output. The transfer functions between different inputs and outputs are distinct, but (before pole zero cancellations) the denominators are always the same for all transfer functions related to a given system. For the system described by eqn. (3.4), the transfer function from the forcing function to the velocity is

$$G(s) = \frac{V(s)}{F(s)} = \frac{s/m}{s^2 + \frac{b}{m}s + \frac{k}{m}}. \tag{3.9}$$

△

Example 3.4 *The transfer function corresponding to eqn. (3.5) is*

$$G(s) = \frac{P(s)}{E_g(s)} = \frac{g}{s(s^2 + \frac{g}{R_e})}. \tag{3.10}$$

The response of the systems depends on the form of the gyro error ϵ_g. The system includes a pure integrator, represented by the $\frac{1}{s}$ factor in the transfer function, and an undamped oscillator with frequency $\omega_s = \sqrt{\frac{g}{R_e}}$. This particular value of the natural frequency is referred to as the Schuler frequency. The transfer function from gyro error to velocity error is

$$G(s) = \frac{V(s)}{E_g(s)} = \frac{g}{s^2 + \frac{g}{R_e}}. \tag{3.11}$$

Note that a constant gyro error would result in an oscillatory velocity error at the Schuler frequency. △

3.1.3 State Space

The state space representation converts each n-th order differential equation into n coupled first order differential equations. Such a representation is often desirable for ease of implementation and analysis, as the state space representation allows the use of vector and matrix techniques. State space techniques relevant to both continuous and discrete-time systems are discussed in Section 3.2–3.5.

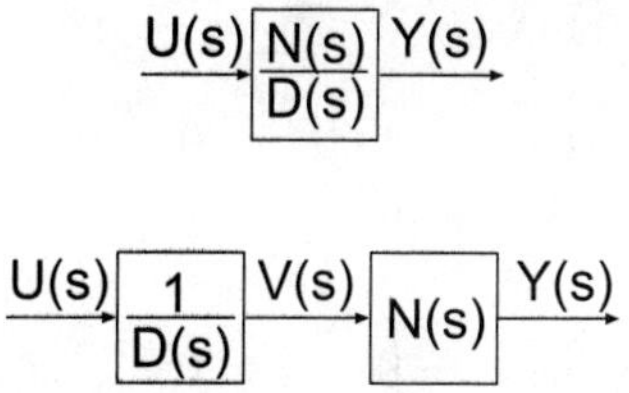

Figure 3.1: Decomposition of a transfer function in eqns. (3.12–3.13).

For analysis only, let the transfer function of eqn. (3.6) be decomposed into two separate filtering operations:

$$
\begin{aligned}
V(s) &= \frac{1}{s^n + d_1 s^{n-1} + \ldots + d_n} U(s) && (3.12)\\
Y(s) &= (n_1 s^{n-1} + \ldots + n_n) V(s) && (3.13)
\end{aligned}
$$

as depicted in Figure 3.1 where $N(s) = n_1 s^{n-1} + \ldots + n_n$ and $D(s) = s^n + d_1 s^{n-1} + \ldots + d_n$. For the n-th order differential equation corresponding to eqn. (3.12), define a state vector $\mathbf{x}$ such that $\mathbf{x} = [v, v^{(1)}, \ldots, v^{(n-1)}]^\top$. Then,

$$
\dot{\mathbf{x}} = \begin{bmatrix} \dot{x}_1 \\ \vdots \\ \dot{x}_{n-1} \\ \dot{x}_n \end{bmatrix} = \begin{bmatrix} v^{(1)} \\ \vdots \\ v^{(n-1)} \\ v^{(n)} \end{bmatrix} = \begin{bmatrix} x_2(t) \\ \vdots \\ x_n(t) \\ u(t) - \sum_{i=1}^n d_i x_{n-i+1} \end{bmatrix}, \quad (3.14)
$$

where the details are provided in Exercise 3.1. Taking the inverse Laplace transform of eqn. (3.13) and using the state vector definition, the system output is represented as

$$
\begin{aligned}
y(t) &= \sum_{i=1}^{n} n_i v^{(n-i)}(t) \\
&= \sum_{i=1}^{n} n_i x_{n-i+1}(t).
\end{aligned}
$$

In matrix notation, the system can be described as

$$
\left.
\begin{aligned}
\dot{\mathbf{x}}(t) &= \mathbf{F}\mathbf{x}(t) + \mathbf{G}\mathbf{u}(t) \\
y(t) &= \mathbf{H}\mathbf{x}(t),
\end{aligned}
\right\} \quad (3.15)
$$

where

$$
\mathbf{F} = \begin{bmatrix} 0 & 1 & 0 & \dots & 0 \\ 0 & 0 & 1 & \dots & 0 \\ & \vdots & & & \vdots \\ 0 & 0 & 0 & \dots & 1 \\ -d_n & -d_{n-1} & -d_{n-2} & \dots & -d_1 \end{bmatrix},
$$

$$
\mathbf{G} = \begin{bmatrix} 0 \\ 0 \\ \vdots \\ 0 \\ 1 \end{bmatrix}, \text{ and } \mathbf{H} = [n_n, n_{n-1}, \dots, n_1].
$$

In such state space representations, $\mathbf{F}$ is referred to as the *system matrix*, $\mathbf{G}$ is referred to as the *input matrix*, and $\mathbf{H}$ is referred to as the *output matrix*. This particular state space implementation is referred to as the *controllable canonical form*. For any system, there exist an infinite number of equivalent state space representations. The transformation between equivalent state space representations is discussed in Section 3.5.1.

Example 3.5 *A state space representation corresponding to the system describe by eqn. (3.3) is*

$$
\dot{\mathbf{v}}(t) = \begin{bmatrix} 0 & 1 \\ 0 & 0 \end{bmatrix} \mathbf{v}(t) + \begin{bmatrix} 0 \\ 1 \end{bmatrix} f(t) \tag{3.16}
$$

$$
p(t) = \begin{bmatrix} 1/m & 0 \end{bmatrix} \mathbf{v}(t) \tag{3.17}
$$

where in this and the subsequent equations of this example

$$
\mathbf{v} = m[p(t), \dot{p}(t)]^\top.
$$

A state space representation corresponding to the system describe by eqn. (3.4) is

$$
\dot{\mathbf{x}}(t) = \begin{bmatrix} 0 & 1 \\ -k/m & -b/m \end{bmatrix} \mathbf{x}(t) + \begin{bmatrix} 0 \\ 1 \end{bmatrix} f(t) \tag{3.18}
$$

$$
p(t) = \begin{bmatrix} 1/m & 0 \end{bmatrix} \mathbf{x}(t). \tag{3.19}
$$

If in addition to position being an output, velocity was also an output, only the output matrix would change to

$$
\mathbf{y}(t) = \begin{bmatrix} p(t) \\ \dot{p}(t) \end{bmatrix} = \begin{bmatrix} 1/m & 0 \\ 0 & 1/m \end{bmatrix} \mathbf{x}(t). \tag{3.20}
$$

△

Example 3.6 *Corresponding to the transfer function in eqn. (3.10) of Example 3.4, the controllable canonical form state space representation is*

$$\begin{bmatrix} \dot{x}_1 \\ \dot{x}_2 \\ \dot{x}_3 \end{bmatrix} = \begin{bmatrix} 0 & 1 & 0 \\ 0 & 0 & 1 \\ 0 & -\frac{g}{R} & 0 \end{bmatrix} \begin{bmatrix} x_1 \\ x_2 \\ x_3 \end{bmatrix} + \begin{bmatrix} 0 \\ 0 \\ g \end{bmatrix} \epsilon_g \tag{3.21}$$

with state vector $\mathbf{x} = [p, \dot{p}, \ddot{p}]^\top$ *and* $y(t) = [1, 0, 0]\mathbf{x}$. *Note that this state space representation is slightly different from the control canonical form. It has been modified, by shifting the gain* g *from the output to the input matrix, to cause each element of the state vector to have a clear physical meaning. Exercise 3.2 confirms that the input-output relationship is correct. Alternative state vector definitions are considered further in Section 3.5.1.* $\triangle$

3.2 State Augmentation

In navigation applications, it will often be the case that we have multiple interconnected systems, each of which is modeled by a set of state space equations. For design and analysis, we want to develop a single state space model for the overall interconnected system. This is achieved by the process of state augmentation. This section first describes the state augmentation process for two systems interconnected in the series configuration of Figure 3.2. From that discussion, the process should become clear and the reader should be able to extend the method to a larger number of systems interconnected in more general ways (e.g., parallel, feedback).

In Figure 3.2, the first system is represented as M_1 and the second system as M_2. Let the first system have the state space model

$$\left.\begin{aligned} \dot{\mathbf{x}} &= \mathbf{F}_1\mathbf{x} + \mathbf{G}_1 u_1 \\ y_1 &= \mathbf{H}_1\mathbf{x} \end{aligned}\right\} \tag{3.22}$$

and the second system have the state space model

$$\left.\begin{aligned} \dot{\mathbf{v}} &= \mathbf{F}_2\mathbf{v} + \mathbf{G}_2 u_2 \\ y_2 &= \mathbf{H}_2\mathbf{v}. \end{aligned}\right\} \tag{3.23}$$

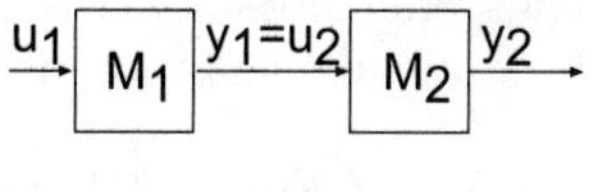

u1 → M1 → y1=u2 → M2 → y2

Figure 3.2: Series interconnection of two state space systems.

By their interconnection, $u_2 = y_1$. Therefore, if we define the augmented state vector $\mathbf{z} = [\mathbf{x}^\top, \mathbf{v}^\top]^\top$, then the state space model for $\mathbf{z}$ is

$$\dot{\mathbf{z}} = \begin{bmatrix} \mathbf{F}_1 & \mathbf{0} \\ \mathbf{G}_2\mathbf{H}_1 & \mathbf{F}_2 \end{bmatrix} \mathbf{z} + \begin{bmatrix} \mathbf{G}_1 \\ \mathbf{0} \end{bmatrix} u_1 \tag{3.24}$$

$$y_2 = \begin{bmatrix} \mathbf{0} & \mathbf{H}_2 \end{bmatrix} \mathbf{z}. \tag{3.25}$$

Other interconnections, such as parallel or feedback, yield different resulting state space formulae.

Example 3.7 *Let the dynamics for* $\mathbf{v}$ *be defined by eqns. (3.16-3.17) in Example 3.5 on page 68. Let the applied force be defined by*

$$\begin{aligned} \dot{x} &= p \\ f &= K_1 x + K_2 v_1 + K_3 v_2. \end{aligned}$$

Note that this is a feedback interconnection. Define the augmented state as $\mathbf{z} = [x, mp, m\dot{p}]^\top$ *and the output as* p. *Then the augmented state model is*

$$\dot{\mathbf{z}}(t) = \left[\begin{array}{c|cc} 0 & 1/m & 0 \\ \hline 0 & 0 & 1 \\ K_1 & K_2 & K_3 \end{array}\right] \mathbf{z}(t) \tag{3.26}$$

$$p(t) = \begin{bmatrix} 0 & 1/m & 0 \end{bmatrix} \mathbf{z}(t). \tag{3.27}$$

△

The state space augmentation technique will be frequently used in navigation system error analysis. The linearized navigation system error dynamics will first be derived. Then error models for the navigation instruments will be obtained either from the manufacturer or through experimentation. Finally, these state space equations will be combined through state space augmentation.

When using the state space augmentation procedure, it is critical that the analyst ensure that neither of the systems being combined *loads* the other system. One system loading another means that the interconnected system behaves significantly different after the interconnection than before. Typically, in navigation systems the augmented state corresponds to instrumentation error variables such as biases and scale factors. By the design of the navigation system and the instruments, the interconnection of the instruments to the navigation system does not change the performance of the instruments; therefore, the state augmentation process works successfully.

Example 3.8 *The left portion of Figure 3.3 shows an example physical system where state augmentation does not lead to a valid system model.*

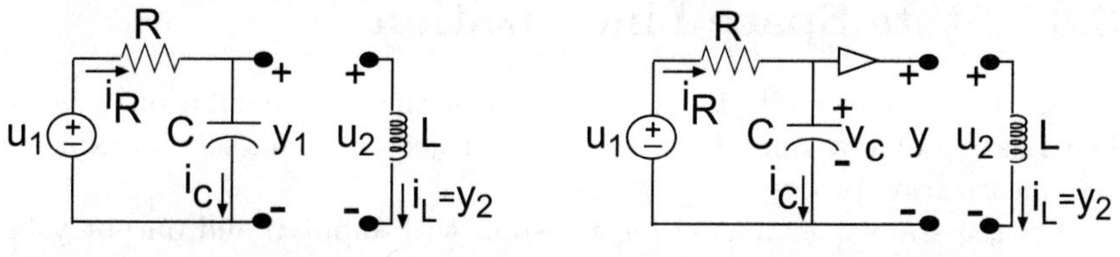

Figure 3.3: Figure for Example 3.8. The sub-circuits are being connected at the nodes as indicated by the adjacent the black dots. Left - Example physical system where the state augmentation process would result in an incorrect model due to loading effects. Right - Example where state augmentation would yield a correct physical model due to the isolation of the loading inductor from the voltage supply by the isolation amplifier depicted by the triangle at the output y_1.

The state space models for the two disconnected systems are

$$\begin{aligned} y_1 &= x_1, & \dot{x}_1 &= -\frac{1}{RC}x_1 + \frac{1}{RC}u_1 \\ y_2 &= x_2, & \dot{x}_2 &= \frac{1}{L}u_2 \end{aligned}$$

where $x_1 = v_C = y_1$ *and* $x_2 = i_L = y_2$. *Direct application of eqns. (3.24–3.25) results in the system model*

$$\begin{bmatrix} \dot{x}_1 \\ \dot{x}_2 \end{bmatrix} = \begin{bmatrix} -\frac{1}{RC} & 0 \\ \frac{1}{L} & 0 \end{bmatrix} \begin{bmatrix} x_1 \\ x_2 \end{bmatrix} + \begin{bmatrix} \frac{1}{RC} \\ 0 \end{bmatrix} u \tag{3.28}$$

which does not apply to the circuit on the left, but does apply to the circuit on the right. In the circuit on the right, an isolation amplifier is inserted at the output of the first circuit. The isolation amplifier has unity gain ($v_c = y_1$) *with high input resistance* ($i_R = i_C$) *and low output impedance which prevents the inductive load from affecting the performance of the capacitive supply.*

The correct model for the circuit on the left is

$$\begin{bmatrix} \dot{x}_1 \\ \dot{x}_2 \end{bmatrix} = \begin{bmatrix} -\frac{1}{RC} & -\frac{1}{C} \\ \frac{1}{L} & 0 \end{bmatrix} \begin{bmatrix} x_1 \\ x_2 \end{bmatrix} + \begin{bmatrix} \frac{1}{RC} \\ 0 \end{bmatrix} u \tag{3.29}$$

due to the constraint that $i_R = i_C + i_L$. *The system with the isolation amplifier described in eqn. (3.28) has two real poles. The system without the isolation amplifier described in eqn. (3.29) has two complex poles.* △

3.3 State Space Linearization

Navigation system analysis, and portions of the implementation, involve linearization of a system about a nominal trajectory. This section explains the linearization process.

Let the state space model for a system with input $\mathbf{u}$ and output $\mathbf{y}$ be described by

$$\dot{\mathbf{x}} = \mathbf{f}(\mathbf{x}, \mathbf{u}) \tag{3.30}$$

$$\mathbf{y} = \mathbf{h}(\mathbf{x}). \tag{3.31}$$

Assume that for a nominal input $\mathbf{u}_o(t)$ a nominal state trajectory $\mathbf{x}_o(t)$ is known which satisfies

$$\begin{aligned} \dot{\mathbf{x}}_o &= \mathbf{f}(\mathbf{x}_o, \mathbf{u}_o) \\ \mathbf{y}_o &= \mathbf{h}(\mathbf{x}_o). \end{aligned}$$

Define the error state vector as $\delta\mathbf{x}(t) = \mathbf{x}(t) - \mathbf{x}_o(t)$. Then,

$$\delta\dot{\mathbf{x}} = \dot{\mathbf{x}} - \dot{\mathbf{x}}_o = \mathbf{f}(\mathbf{x}, \mathbf{u}) - \mathbf{f}(\mathbf{x}_o, \mathbf{u}_o).$$

We can approximate $\mathbf{f}(\mathbf{x}, \mathbf{u})$ using a Taylor series expansion to yield

$$\begin{aligned} \delta\dot{\mathbf{x}} &= \mathbf{f}(\mathbf{x}_o, \mathbf{u}_o) + \left.\frac{\partial \mathbf{f}(\mathbf{x}, \mathbf{u})}{\partial \mathbf{x}}\right|_{\mathbf{x}_o, \mathbf{u}_o} \delta\mathbf{x} \\ &+ \left.\frac{\partial \mathbf{f}(\mathbf{x}, \mathbf{u})}{\partial \mathbf{u}}\right|_{\mathbf{x}_o, \mathbf{u}_o} \delta\mathbf{u} + h.o.t.'s - \mathbf{f}(\mathbf{x}_o, \mathbf{u}_o) \\ \delta\dot{\mathbf{x}} &= \mathbf{F}(t)\delta\mathbf{x} + \mathbf{G}(t)\delta\mathbf{u} + h.o.t.'s \end{aligned} \tag{3.32}$$

where $\delta\mathbf{u} = \mathbf{u} - \mathbf{u}_o$, $\mathbf{F}(t) = \left.\frac{\partial \mathbf{f}}{\partial \mathbf{x}}\right|_{\mathbf{x}_o(t), \mathbf{u}_o(t)}$, and $\mathbf{G}(t) = \left.\frac{\partial \mathbf{f}}{\partial \mathbf{u}}\right|_{\mathbf{x}_o(t), \mathbf{u}_o(t)}$. The resultant perturbation to the system output $\delta\mathbf{y} = \mathbf{y} - \mathbf{y}_o$ is

$$\begin{aligned} \delta\mathbf{y}(t) &= \mathbf{h}(\mathbf{x}(t)) - \mathbf{h}(\mathbf{x}_o(t)) \\ &= \left.\frac{\partial h(\mathbf{x})}{\partial \mathbf{x}}\right|_{\mathbf{x}_o(t), \mathbf{u}_o(t)} \delta\mathbf{x}(t) + h.o.t.'s \\ \delta\mathbf{y}(t) &= \mathbf{H}(t)\delta\mathbf{x}(t) + h.o.t.'s \end{aligned} \tag{3.33}$$

where $\mathbf{H}(t) = \left.\frac{\partial \mathbf{h}}{\partial \mathbf{x}}\right|_{\mathbf{x}_o(t), \mathbf{u}_o(t)}$. By dropping the higher-order terms ($h.o.t$'s), eqns. (3.32) and (3.33) provide the time-varying linearization of the non-linear system:

$$\delta\dot{\mathbf{x}}(t) = \mathbf{F}(t)\delta\mathbf{x}(t) + \mathbf{G}(t)\delta\mathbf{u}(t) \tag{3.34}$$

$$\delta\mathbf{y}(t) = \mathbf{H}(t)\delta\mathbf{x}(t) \tag{3.35}$$

which is accurate near the nominal trajectory (i.e., for small $\|\delta \mathbf{x}\|$ and $\|\delta \mathbf{u}\|$).

Applications of eqns. (3.34) and (3.35) are common in navigation applications. For example, the GPS range equations include the distance from the satellite broadcast antenna effective position to the receiver antenna effective position. When the satellite position is known and the objective is to estimate the receiver location, the GPS measurement is nonlinear with the generic form of eqn. (3.31), but frequently solved via linearization. See Section 8.2.2. Also, the navigation system kinematic equations, discussed in Chapters 9–12, are nonlinear and have the generic form of eqn. (3.30). The following example illustrates the basic state space linearization process.

Example 3.9 *Assume that there is a true system that follows the kinematic equation* $\dot{\mathbf{x}} = f(\mathbf{x}, \mathbf{u})$ *where* $\mathbf{x} = [n, e, \psi]^\top \in \mathbb{R}^3$, $\mathbf{u} = [u, \omega]^\top \in \mathbb{R}^2$, *and*

$$\mathbf{f}(\mathbf{x}, \mathbf{u}) = \begin{bmatrix} u\cos(\psi) \\ u\sin(\psi) \\ \omega \end{bmatrix}. \tag{3.36}$$

For this system, the variables $[n, e]$ *defined the position vector,* ψ *is the yaw angle of the vehicle relative to north,* u *is the body frame forward velocity, and* ω *is the yaw rate. Also, assume that two sensors are available that provide measurements modeled as*

$$\mathbf{y} = \begin{bmatrix} u + e_u \\ \omega + e_\omega \end{bmatrix}.$$

A very simple dead-reckoning navigation system can be designed by integration of $\dot{\mathbf{x}}_o = \mathbf{f}(\mathbf{x}_o, \mathbf{y})$ *for* $t \geq t_o$ *from some initial* $\mathbf{x}_o(t_o)$ *where* $\mathbf{x}_o = [n_o, e_o, \psi_o]^\top$. *The resulting dead-reckoning system state,* $\mathbf{x}_o(t)$ *for* $t \geq 0$, *will be used as the reference trajectory for the linearization process.*

The navigation system equation can be simplified as follows:

$$\begin{aligned} \dot{\mathbf{x}}_o &= \mathbf{f}(\mathbf{x}_o, \mathbf{y}) \\ &= \begin{bmatrix} (u+e_u)\cos(\psi_o) \\ (u+e_u)\sin(\psi_o) \\ (\omega + e_\omega) \end{bmatrix} = \begin{bmatrix} u\cos(\psi_o) \\ u\sin(\psi_o) \\ \omega \end{bmatrix} + \begin{bmatrix} e_u\cos(\psi_o) \\ e_u\sin(\psi_o) \\ e_\omega \end{bmatrix} \\ &= \mathbf{f}(\mathbf{x}_o, \mathbf{u}) + \begin{bmatrix} e_u\cos(\psi_o) \\ e_u\sin(\psi_o) \\ e_\omega \end{bmatrix}. \end{aligned} \tag{3.37}$$

The first order (i.e., linearized) dynamics of the error between the actual and navigation states are

$$\dot{\mathbf{v}} = \begin{bmatrix} 0 & 0 & -\hat{u}\sin(\psi_o) \\ 0 & 0 & \hat{u}\cos(\psi_o) \\ 0 & 0 & 0 \end{bmatrix} \mathbf{v} - \begin{bmatrix} \cos(\psi_o) & 0 \\ \sin(\psi_o) & 0 \\ 0 & 1 \end{bmatrix} \begin{bmatrix} e_u \\ e_\omega \end{bmatrix} \tag{3.38}$$

where $\hat{u} = y_1$ *and* $\mathbf{v} = [\delta n, \delta e, \delta\psi]^\top = \mathbf{x} - \mathbf{x}_o$. *To derive this linearized model, the function* $\mathbf{f}$ *in eqn. (3.36) is expanded using Taylor series. Then eqn. (3.37) is subtracted from the Taylor series expansion of eqn. (3.36) to yield eqn. (3.38). For additional discussion related to this example, see Chapter 9.* △

Error state dynamic equations enable quantitative analysis of the error state itself. Later, the linearized error state equations will be used in the design of Kalman filters to estimate the error state. Prior to that, we must discuss such issues as stochastic modeling of sensor errors which is done in Chapter 4.

3.4 Discrete-Time State Space Notation

It is often advantageous to work with discrete-time equivalent models of continuous-time systems, particularly with the advent of the digital computer. This section discusses only notation definition for state space models of discrete-time systems. The calculation of discrete-time models equivalent to continuous-time systems is discussed in Section 3.5.5.

The standard form for a time-invariant, discrete-time, state space model is

$$\mathbf{x}_{k+1} = \mathbf{\Phi}\mathbf{x}_k + \mathbf{\Gamma}\mathbf{u}_k \tag{3.39}$$

$$\mathbf{y}_k = \mathbf{H}\mathbf{x}_k. \tag{3.40}$$

The above notation is equivalent to

$$\mathbf{x}(k+1) = \mathbf{\Phi}\mathbf{x}(k) + \mathbf{\Gamma}\mathbf{u}(k) \tag{3.41}$$

$$\mathbf{y}(k) = \mathbf{H}\mathbf{x}(k). \tag{3.42}$$

Whenever possible, this book will use the notation of eqns. (3.39-3.40) because it is more compact. If, however, the state space variable names have subscripts, it may be necessary to use the notation of eqns. (3.42-3.42). In either model format, it is understood that $x_k = x(k) = x(kT)$ where T is the time increment of the discrete-time system.

3.5 State Space Analysis

The previous sections have casually referred to the *state* of a system. The following definition formalizes this concept.

Definition 3.1 *The state of a dynamic system is a set of real variables such that knowledge of these variables at time* t_0 *together with knowledge of the system input for* $t \geq t_0$ *is sufficient to determine the system response for all* $t > t_0$.

The definition of state for a discrete-time system is exactly the same with t replaced by k.

If the state variables are independent and organized as a vector, i.e., $\mathbf{x} = [x_1, \ldots, x_n]^\top \in \mathbb{R}^n$, then $\mathbf{x}$ is the state vector and n is the order of the state space model.

The concept of state can be thought of as an extension of the idea of initial conditions. It is well understood that the complete solution of an n-th order differential equation requires specification on n initial conditions. Also, as demonstrated in Section 3.1.3, the state space representation for an n-th order ordinary differential equation entails n state variables with one first-order differential equation for each state variable.

Sections 3.1.3 and 3.4 present the continuous and discrete-time state space models for time-invariant linear systems. In general, the coefficient matrices in these models can be time-varying. The general time-varying linear models are

$$\dot{\mathbf{x}}(t) = \mathbf{F}(t)\mathbf{x}(t) + \mathbf{G}(t)\mathbf{u}(t) \qquad \mathbf{y}(t) = \mathbf{H}(t)\mathbf{x}(t) \tag{3.43}$$

and

$$\mathbf{x}_{k+1} = \mathbf{\Phi}_k\mathbf{x}_k + \mathbf{\Gamma}_k\mathbf{u}_k \qquad \mathbf{y}_k = \mathbf{H}_k\mathbf{x}_k. \tag{3.44}$$

The solution to eqn. (3.43) is involved and important in its own right, so it is discussed in Section 3.5.3. By direct iteration, it is straightforward to show that the solution to eqn. (3.44) is

$$\mathbf{x}_{k_0+n} = \prod_{i=0}^{n-1}\mathbf{\Phi}_{k_0+i}\mathbf{x}_{k_0} + \sum_{j=0}^{n-1}\prod_{i=j+1}^{n-1}\mathbf{\Phi}_{k+i}\mathbf{\Gamma}_{k+j}\mathbf{u}_{k+j} \tag{3.45}$$

$$\mathbf{y}_{k_0+n} = \mathbf{H}_{k+n}\mathbf{x}_{k_0+n}, \tag{3.46}$$

which shows that knowledge of $\mathbf{x}_{k_0}$ and $\mathbf{u}_k$ for $k \geq k_0$ completely specifies the system state and output for all $k \geq k_0$. In eqn. (3.45), the order of the matrix product is important and is interpreted as $\prod_{i=0}^{n-1}\mathbf{\Phi}_{k_0+i} = \mathbf{\Phi}_{k_0+n-1}\cdots\mathbf{\Phi}_{k_0+1}\mathbf{\Phi}_{k_0}$.

3.5.1 Similarity Transformation

Note that Definition 3.1 defines an equivalent class of state vectors of the same dimension. If the set of variables in $\mathbf{x} \in \mathbb{R}^n$ satisfies the definition of state, then so does $\mathbf{v} = \mathbf{P}\mathbf{x}$ for any non-singular matrix $\mathbf{P}$. This is true, since knowledge of $\mathbf{v}$ allows determination of $\mathbf{x}$ according to $\mathbf{x} = \mathbf{P}^{-1}\mathbf{v}$; therefore, $\mathbf{v}$ also satisfies the definition of the system state. The different representations of the state correspond to different scalings of the state variables or different selection of the (not necessarily orthogonal) basis vectors for the state space.

If the state space model corresponding to $\mathbf{x}$ is described by eqn. (3.43) and $\mathbf{P}$ is a constant matrix, then the state space representation for the state vector $\mathbf{v}$ is given in eqns. (3.47–3.48):

$$\begin{aligned}
\dot{\mathbf{x}}(t) &= \mathbf{F}(t)\mathbf{x}(t) + \mathbf{G}(t)\mathbf{u}(t) \\
\mathbf{P}^{-1}\dot{\mathbf{v}}(t) &= \mathbf{F}(t)\mathbf{P}^{-1}\mathbf{v}(t) + \mathbf{G}(t)\mathbf{u}(t) \\
\dot{\mathbf{v}}(t) &= \mathbf{P}\mathbf{F}(t)\mathbf{P}^{-1}\mathbf{v}(t) + \mathbf{P}\mathbf{G}(t)\mathbf{u}(t) \qquad (3.47) \\
\mathbf{y}(t) &= \mathbf{H}(t)\mathbf{P}^{-1}\mathbf{v}(t). \qquad (3.48)
\end{aligned}$$

For specific types of analysis or for unit conversion, it may be convenient to find a nonsingular linear transformation of the original system state which simplifies the subsequent analysis. The use of similarity transformations for discrete-time systems is identical.

Example 3.10 *Example 3.6 on page 69 presented a state space model for a single channel of the error dynamics of the INS with state vector* $\mathbf{x} = [p, \dot{p}, \ddot{p}]^\top$. *An alternative state vector definition is* $\mathbf{v} = [p, \dot{p}, \theta]^\top$ *where* θ *represents error in the computed orientation of the platform frame with respect to the navigation frame. Using the similarity transform* $\mathbf{v} = \mathbf{P}\mathbf{x}$ *with*

$$\mathbf{P} = \begin{bmatrix} 1 & 0 & 0 \\ 0 & 1 & 0 \\ 0 & 0 & -\frac{1}{g} \end{bmatrix},$$

Exercise 3.4 asks the reader to confirm that an equivalent state space model is

$$\begin{bmatrix} \dot{p} \\ \dot{v} \\ \dot{\phi} \end{bmatrix} = \begin{bmatrix} 0 & 1 & 0 \\ 0 & 0 & -g \\ 0 & \frac{1}{R} & 0 \end{bmatrix} \begin{bmatrix} p \\ v \\ \phi \end{bmatrix} + \begin{bmatrix} 0 \\ 0 \\ -1 \end{bmatrix} \epsilon_g \qquad (3.49)$$

with $\mathbf{y} = [1, 0, 0]\mathbf{v}$.

The state space model of eqn. (3.49) is derived from first principles in Exercise 3.12. △

For important applications of the similarity transform, e.g., see Sections 3.6.2–3.6.3.

3.5.2 State Space to Transfer Function

Sections 3.1.3 and 3.4 presented a specific state space representation referred to as the controllable canonical form. This state space format has the advantage that it is easy to compute the associated transfer function and that it is straightforward to determine a state feedback control law, but may not be the model format that results from derivation of a system model based on physical principles. The previous section showed that there are an infinite number of equivalent state space model representations. This section describes how to find the associated transfer function when the available state space model is not in controllable canonical form. Only continuous-time systems are discussed. The analysis is similar for discrete-time systems.

If a time-invariant system is described as

$$\dot{\mathbf{x}}(t) = \mathbf{F}\mathbf{x}(t) + \mathbf{G}\mathbf{u}(t) \tag{3.50}$$

then, assuming that all initial conditions are zero, Laplace transforming both sides results in

$$\begin{aligned} s\mathbf{X}(s) &= \mathbf{F}\mathbf{X}(s) + \mathbf{G}\mathbf{U}(s) \\ (\mathbf{I}s - \mathbf{F})\,\mathbf{X}(s) &= \mathbf{G}\mathbf{U}(s) \\ \mathbf{X}(s) &= (\mathbf{I}s - \mathbf{F})^{-1}\,\mathbf{G}\mathbf{U}(s). \end{aligned}$$

When $\mathbf{y}(t) = \mathbf{H}\mathbf{x}(t)$, then by Laplace transform $\mathbf{Y}(s) = \mathbf{H}\mathbf{X}(s)$ and

$$\mathbf{Y}(s) = \mathbf{H}\,(\mathbf{I}s - \mathbf{F})^{-1}\,\mathbf{G}\mathbf{U}(s). \tag{3.51}$$

To show that all equivalent state space representations have the same input output response, consider the transfer function corresponding to eqns. (3.47) and (3.48):

$$\begin{aligned} \mathbf{Y}(s) &= \mathbf{H}\mathbf{P}^{-1}\mathbf{V}(s) \\ \mathbf{V}(s) &= \left(\mathbf{I}s - \mathbf{P}\mathbf{F}\mathbf{P}^{-1}\right)^{-1}\mathbf{P}\mathbf{G}\mathbf{U}(s) \\ \mathbf{Y}(s) &= \mathbf{H}\mathbf{P}^{-1}\left(\mathbf{I}s - \mathbf{P}\mathbf{F}\mathbf{P}^{-1}\right)^{-1}\mathbf{P}\mathbf{G}\mathbf{U}(s) \\ \mathbf{Y}(s) &= \mathbf{H}\mathbf{P}^{-1}\mathbf{P}\,(\mathbf{I}s - \mathbf{F})^{-1}\,\mathbf{P}^{-1}\mathbf{P}\mathbf{G}\mathbf{U}(s) \\ \mathbf{Y}(s) &= \mathbf{H}\,(\mathbf{I}s - \mathbf{F})^{-1}\,\mathbf{G}\mathbf{U}(s) \end{aligned} \tag{3.52}$$

where we have used the fact that $\left(\mathbf{I}s - \mathbf{P}\mathbf{F}\mathbf{P}^{-1}\right)^{-1} = \mathbf{P}\,(\mathbf{I}s - \mathbf{F})^{-1}\,\mathbf{P}^{-1}$, which can be demonstrated by direct multiplication.

In relation to the previous analysis of this section it is useful to note the following facts. The quantity $(s\mathbf{I} - \mathbf{F})^{-1}$ represents a matrix of transfer functions. Before any pole-zero cancelations, the denominator of each transfer function is the polynomial $d(s) = |s\mathbf{I} - \mathbf{F}|$, which is the determinant of $(s\mathbf{I} - \mathbf{F})$. The system *poles* are the solutions to the equation $d(s) = |s\mathbf{I} - \mathbf{F}| = 0$. The eigenvalues of $\mathbf{F}$ are the solutions of $|s\mathbf{I} - \mathbf{F}| = 0$. Therefore, before pole-zero cancelations, the eigenvalues of $\mathbf{F}$ and the poles of the transfer function from $\mathbf{U}(s)$ to $\mathbf{Y}(s)$ are identical.

Example 3.11 *Consider the state space system*

$$\begin{aligned} \dot{\mathbf{x}} &= \begin{bmatrix} 0 & 1 \\ -6 & -5 \end{bmatrix}\mathbf{x} + \begin{bmatrix} 0 \\ 1 \end{bmatrix}u \\ y &= \begin{bmatrix} 1 & 1 \end{bmatrix}\mathbf{x}. \end{aligned}$$

The eigenvalues of $\mathbf{F}$ *are the roots of* $|s\mathbf{I}-\mathbf{F}|=0$ *where*

$$(s\mathbf{I}-\mathbf{F})=\begin{bmatrix} s & -1 \\ 6 & s+5 \end{bmatrix}.$$

Therefore,

$$|s\mathbf{I}-\mathbf{F}|=s^2+5s+6=(s+3)(s+2)$$

which shows that the eigenvalues are -2 and -3.

The transfer function from u to y is computed as

$$\begin{aligned} \frac{Y(s)}{U(s)}=\mathbf{H}(s\mathbf{I}-\mathbf{F})^{-1}\mathbf{G} &= \begin{bmatrix} 1 & 1 \end{bmatrix}\begin{bmatrix} s & -1 \\ 6 & s+5 \end{bmatrix}^{-1}\begin{bmatrix} 0 \\ 1 \end{bmatrix} \\ &= \begin{bmatrix} 1 & 1 \end{bmatrix}\frac{\begin{bmatrix} s+5 & 1 \\ -6 & s \end{bmatrix}}{(s+3)(s+2)}\begin{bmatrix} 0 \\ 1 \end{bmatrix} \\ &= \frac{(s+1)}{(s+3)(s+2)} \end{aligned}$$

which has a zero at $s=-1$ *and poles at* $s=-2,\ -3$. △

Example 3.12 *Consider the state space system*

$$\begin{aligned} \dot{\mathbf{v}} &= \begin{bmatrix} 0 & 1 & 0 \\ 0 & 0 & 1 \\ 0 & -2 & -2 \end{bmatrix}\mathbf{v}+\begin{bmatrix} 0 \\ 0 \\ 1 \end{bmatrix}u \\ y &= \begin{bmatrix} 0 & 1 & 0 \end{bmatrix}\mathbf{v}. \end{aligned}$$

The eigenvalues are $0,\ 1\pm j$*, because*

$$(s\mathbf{I}-\mathbf{F})=\begin{bmatrix} s & -1 & 0 \\ 0 & s & -1 \\ 0 & 2 & (s+2) \end{bmatrix}$$

which yields the following equation for the eigenvalues

$$|s\mathbf{I}-\mathbf{F}|=s^3+2s^2+2s=s(s+1+1j)(s+1-1j).$$

The transfer function from u to y is computed as follows:

$$\begin{aligned} \frac{Y(s)}{U(s)} &= \mathbf{H}(s\mathbf{I}-\mathbf{F})^{-1}\mathbf{G} \\ &= \begin{bmatrix} 0 & 1 & 0 \end{bmatrix}\begin{bmatrix} s & -1 & 0 \\ 0 & s & -1 \\ 0 & 2 & (s+2) \end{bmatrix}^{-1}\begin{bmatrix} 0 \\ 0 \\ 1 \end{bmatrix} \end{aligned}$$

$$= \begin{bmatrix} 0 & 1 & 0 \end{bmatrix} \frac{\begin{bmatrix} (s^2+2s+2) & (s+2) & 1 \\ 0 & s(s+2) & s \\ 0 & -2s & s^2 \end{bmatrix}}{s(s+1+1j)(s+1-1j)} \begin{bmatrix} 0 \\ 0 \\ 1 \end{bmatrix}$$

$$= \frac{s}{s(s+1+1j)(s+1-1j)}$$

which has a finite zero at $s = 0$ *and* $n = 3$ *poles at* $s = 0,\ 1+j,\ 1-j.$ △

Engineers and analysts should be extremely careful with pole-zero cancelations. In particular, a pole should never be canceled if it is in the right-half complex-plane and only with extreme caution if it is on the $j\omega$-axis of the complex plane. When a pole-zero cancelation occurs it only cancels for a specific transfer function. The order of that transfer function decreases, but the order of the state space representation remains n. The affect of the 'canceled' pole will still appear in other transfer functions and in the transient response.

3.5.3 State Transition Matrix Properties

The homogeneous part of eqn. (3.43) is

$$\dot{\mathbf{x}}(t) = \mathbf{F}(t)\mathbf{x}(t). \tag{3.53}$$

Definition 3.2 *A continuous and differentiable matrix function* $\mathbf{\Phi}(t) : \mathbb{R}^1 \rightarrow \mathbb{R}^{n \times n}$ *is the* fundamental solution *of eqn. (3.53) on* $t \in [0, T]$ *if and only if* $\mathbf{\Phi}(0) = \mathbf{I}$ *and* $\dot{\mathbf{\Phi}}(t) = \mathbf{F}(t)\mathbf{\Phi}(t)$ *for all* $t \in [0, T]$.

The fundamental solution $\mathbf{\Phi}$ is important since it will serve as the basis for finding the solution to both eqns. (3.43) and (3.53).

If $\mathbf{x}(\mathbf{t}) = \mathbf{\Phi}(t)\mathbf{x}(0)$ then $\mathbf{x}(\mathbf{t})$ satisfies the initial value problem corresponding to eqn. (3.53) with initial condition $\mathbf{x}(0)$:

$$\begin{aligned} \dot{\mathbf{x}}(t) &= \frac{d}{dt}\Big(\mathbf{\Phi}(t)\mathbf{x}(0)\Big) \\ &= \dot{\mathbf{\Phi}}(t)\mathbf{x}(0) \\ &= \mathbf{F}(t)\mathbf{\Phi}(t)\mathbf{x}(0) \\ &= \mathbf{F}(t)\mathbf{x}(t). \end{aligned}$$

If $\mathbf{\Phi}(t)$ is nonsingular, then $\mathbf{\Phi}^{-1}(t)\mathbf{x}(t) = \mathbf{x}(0)$ and

$$\begin{aligned} \mathbf{x}(\tau) &= \mathbf{\Phi}(\tau)\mathbf{x}(0) \\ &= \mathbf{\Phi}(\tau)\mathbf{\Phi}^{-1}(t)\mathbf{x}(t) \\ &= \mathbf{\Phi}(\tau, t)\mathbf{x}(t) \end{aligned} \tag{3.54}$$

where $\mathbf{\Phi}(\tau,t)=\mathbf{\Phi}(\tau)\mathbf{\Phi}^{-1}(t)$ is called the *state transition matrix* from time t to time τ. The state transition matrix transforms the solution of the initial value problem corresponding to eqn. (3.53) at time t to the solution at time τ. The state transition matrix has the following properties:

$$\begin{aligned}
\mathbf{\Phi}(t,t) &= \mathbf{I} && (3.55)\\
\mathbf{\Phi}(\tau,t) &= \mathbf{\Phi}^{-1}(t,\tau) && (3.56)\\
\mathbf{\Phi}(\tau,\lambda)\mathbf{\Phi}(\lambda,t) &= \mathbf{\Phi}(\tau,t) && (3.57)\\
\frac{d}{d\tau}\mathbf{\Phi}(\tau,t) &= \mathbf{F}(\tau)\mathbf{\Phi}(\tau,t) && (3.58)\\
\frac{d}{dt}\mathbf{\Phi}(\tau,t) &= -\mathbf{\Phi}(\tau,t)\mathbf{F}(t). && (3.59)
\end{aligned}$$

The general solution to eqn. (3.43) is

$$\mathbf{x}(t) = \mathbf{\Phi}(t,t_0)\mathbf{x}(t_0)+\int_{t_0}^{t}\mathbf{\Phi}(t,\lambda)\mathbf{G}(\lambda)\mathbf{u}(\lambda)d\lambda \qquad (3.60)$$

which can be verified as follows:

$$\begin{aligned}
\frac{d}{dt}\mathbf{x}(t) &= \frac{d}{dt}\Big(\mathbf{\Phi}(t,t_0)\mathbf{x}(t_0)\Big)+\frac{d}{dt}\left(\int_{t_0}^{t}\mathbf{\Phi}(t,\lambda)\mathbf{G}(\lambda)\mathbf{u}(\lambda)d\lambda\right)\\
\dot{\mathbf{x}}(t) &= \mathbf{F}(t)\mathbf{\Phi}(t,t_0)\mathbf{x}(t_0)\\
&\quad+\int_{t_0}^{t}\mathbf{F}(t)\mathbf{\Phi}(t,\lambda)\mathbf{G}(\lambda)\mathbf{u}(\lambda)d\lambda+\mathbf{\Phi}(t,t)\mathbf{G}(t)\mathbf{u}(t)\\
\dot{\mathbf{x}}(t) &= \mathbf{F}(t)\left(\mathbf{\Phi}(t,t_0)\mathbf{x}(t_0)+\int_{t_0}^{t}\mathbf{\Phi}(t,\lambda)\mathbf{G}(\lambda)\mathbf{u}(\lambda)d\lambda\right)+\mathbf{G}(t)\mathbf{u}(t)\\
\dot{\mathbf{x}}(t) &= \mathbf{F}(t)\mathbf{x}(t)+\mathbf{G}(t)\mathbf{u}(t),
\end{aligned}$$

where Leibnitz rule

$$\frac{d}{dt}\int_{a(t)}^{b(t)}f(t,\tau)d\tau=\int_{a(t)}^{b(t)}\frac{\partial}{\partial t}f(t,\tau)d\tau+f(b(t),\tau)\frac{db}{dt}-f(a(t),\tau)\frac{da}{dt}$$

has been used to move derive the second equation from the first.

3.5.4 Linear Time-Invariant Systems

In the case that $\mathbf{F}$ is a constant matrix, it can be shown by direct differentiation that

$$\begin{aligned}
\mathbf{\Phi}(t) &= e^{\mathbf{F}t}, \text{ and} && (3.61)\\
\mathbf{\Phi}(\tau,t) &= e^{\mathbf{F}(\tau-t)}.
\end{aligned}$$

The matrix exponential, its properties, and its computation are discussed in Section B.12. Given this special case of $\mathbf{\Phi}(\tau, t)$ we have that eqn. (3.60) becomes

$$\mathbf{x}(t) = e^{\mathbf{F}(t-t_0)}\mathbf{x}(t_0) + \int_{t_0}^{t} e^{\mathbf{F}(t-\tau)}\mathbf{G}(\tau)\mathbf{u}(\tau)d\tau. \tag{3.62}$$

Assuming that $\mathbf{G}$ is time-invariant, the output is determined by multiplying eqn. (3.62) on the left by the measurement matrix $\mathbf{H}$

$$\mathbf{y}(t) = \mathbf{H}e^{\mathbf{F}(t-t_0)}\mathbf{x}(t_0) + \int_{t_0}^{t} \mathbf{H}e^{\mathbf{F}(t-\tau)}\mathbf{G}\mathbf{u}(\tau)d\tau. \tag{3.63}$$

Defining $\mathbf{m}(t) = \mathbf{H}e^{\mathbf{F}t}\mathbf{G}$, which is the impulse response of the linear system, we see that eqn. (3.63) is the linear combination of the response due to initial conditions and the response due to the input. The response to the input $\mathbf{u}(t)$ is determined as the convolution of the impulse response $\mathbf{m}(t)$ with $\mathbf{u}(t)$. The Laplace transform of the impulse response is the transfer function $\mathbf{M}(s) = \mathbf{H}\left(s\mathbf{I} - \mathbf{F}\right)^{-1}\mathbf{G}$ derived in eqn. (3.51).

An approximate method to compute $\mathbf{\Phi}(t+\tau, t)$ for small τ when $\mathbf{F}(t)$ is slowly time-varying is discussed in Section 7.2.5.2.

3.5.5 Discrete-Time Equivalent Models

It is often the case that a system of interest is naturally described by continuous-time differential equations, but that the system implementation is more convenient in discrete-time. In these circumstances, it is of interest to determine a discrete-time model that is equivalent to the continuous-time model at the discrete-time instants $t_k = kT$ for some fixed value $T > 0$ and $k = 0, 1, 2, \ldots$. Equivalence meaning that the discrete and continuous-time models predict the same system state at the specified discrete-time instants.

If $\mathbf{F}$ is a constant matrix, then from eqn. (3.62)

$$\begin{aligned}
\mathbf{x}_{k+1} &= e^{\mathbf{F}\left((k+1)T-kT\right)}\mathbf{x}_k + \int_{kT}^{(k+1)T} e^{\mathbf{F}\left((k+1)T-\tau\right)}\mathbf{G}(\tau)\mathbf{u}(\tau)d\tau \\
\mathbf{x}_{k+1} &= \mathbf{\Phi}\mathbf{x}_k + \int_{kT}^{(k+1)T} e^{\mathbf{F}\left((k+1)T-\tau\right)}\mathbf{G}(\tau)\mathbf{u}(\tau)d\tau
\end{aligned} \tag{3.64}$$

where $\mathbf{x}_k = \mathbf{x}(kT)$ and $\mathbf{\Phi} = e^{\mathbf{F}T}$. Simplification of the second term on the right hand side is possible under various assumptions. The most common assumption is that $\mathbf{G}(t)$ is a constant vector and that $\mathbf{u}(t)$ has the constant value $\mathbf{u}_k$ for $t \in (kT, (k+1)T]$. With this assumption, eqn. (3.64) reduces to

$$\mathbf{x}_{k+1} = \mathbf{\Phi}\mathbf{x}_k + \mathbf{\Gamma}\mathbf{u}_k \tag{3.65}$$

where $\mathbf{\Gamma} = \int_0^T e^{\mathbf{F}(T-\tau)}\mathbf{G}d\tau$.

Example 3.13 *For the continuous-time system described in eqn. (3.16) on page 68,* $\mathbf{F} = \begin{bmatrix} 0 & 1 \\ 0 & 0 \end{bmatrix}$. *Denoting the sampling time by* T, $\mathbf{\Phi}$ *can be found in closed form via the Taylor series approach described in Section B.12.1. Since* $\mathbf{F}^n = \begin{bmatrix} 0 & 0 \\ 0 & 0 \end{bmatrix}$ *for* $n \geq 2$, *the Taylor series terminates after the second term*

$$\begin{aligned} \mathbf{\Phi} = e^{\mathbf{F}T} &= \mathbf{I} + \mathbf{F}T + 0.5\mathbf{F}^2T^2 + \ldots \\ &= \begin{bmatrix} 1 & T \\ 0 & 1 \end{bmatrix}. \end{aligned}$$

If, in addition, the applied force is held constant at $f(t) = f_k$ *for* $t \in [kT, (k+1)T)$ *for all* $k = 0, 1, 2, \ldots$, *then using the change of variables* $\lambda = \tau - kT$ *we obtain*

$$\begin{aligned} \mathbf{G}_d = \int_{kT}^{(k+1)T} e^{\mathbf{F}((k+1)T-\tau)}\mathbf{G}\mathbf{u}(\tau)d\tau &= \int_0^T \begin{bmatrix} 1 & (T-\lambda) \\ 0 & 1 \end{bmatrix} \begin{bmatrix} 0 \\ 1 \end{bmatrix} f_k d\lambda \\ &= \begin{bmatrix} T & T^2/2 \\ 0 & T \end{bmatrix} \begin{bmatrix} 0 \\ 1 \end{bmatrix} f_k \\ &= \begin{bmatrix} T^2/2 \\ T \end{bmatrix} f_k. \end{aligned}$$

Thus the discrete-time equivalent model to eqn. (3.16) is

$$\begin{aligned} \mathbf{x}_{k+1} &= \begin{bmatrix} 1 & T \\ 0 & 1 \end{bmatrix} \mathbf{x}_k + \begin{bmatrix} T^2/2 \\ T \end{bmatrix} f_k \\ p_k &= \begin{bmatrix} \frac{1}{m} & 0 \end{bmatrix} \mathbf{x}_k. \end{aligned}$$

△

3.6 State Estimation

The state space model format clearly shows that a system may have many internal variables (e.g., states) and fewer outputs (e.g., $\mathbf{y}$). This is true for a variety of reasons including cost, power, or lack of appropriate sensors. If knowledge of the state vector is desired, but is not directly measured, then we have the problem referred to as state estimation. State estimation is useful in control applications, when a few outputs are available, but knowledge of the system state would allow higher performance control. It is also true in navigation systems, where several error variables are of interest, but few external measurements (e.g., ranges or velocities) are available. When

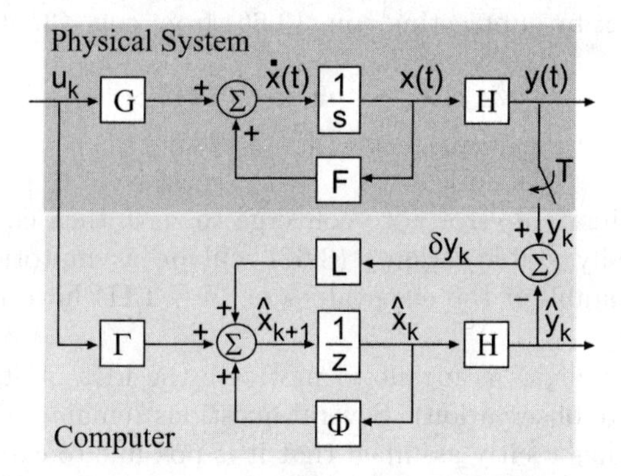

Figure 3.4: State estimator implementation for the continuous-time physical system of eqn. (3.15) with the output sampled every T seconds and the input $u(t)$ constant over each sampling interval. The state estimate is described by eqns. (3.66).

it is advantageous to know the internal state, the question arises of how to estimate the state from the available outputs. This section presents the problem of state estimation for deterministic systems from a stability point of view. Chapter 4 will consider the issue of state estimation performance in applications with stochastic inputs. Chapter 5 will consider the problem of state estimation for stochastic systems from an optimization point of view.

For a discrete-time system described via eqn. (3.39-3.40), consider the following approach:

$$\left.\begin{array}{rcl}\hat{\mathbf{x}}_{k+1} & = & \mathbf{\Phi}\hat{\mathbf{x}}_k + \mathbf{\Gamma}u_k + \mathbf{L}(\mathbf{y}_k - \hat{\mathbf{y}}_k) \\ \hat{\mathbf{y}}_k & = & \mathbf{H}\hat{\mathbf{x}}_k\end{array}\right\} \tag{3.66}$$

where $\hat{\mathbf{x}}_k$ is an estimate of the state $\mathbf{x}_k$ at the k-th instant of time and $\mathbf{L}$ is a design parameter referred to as the *estimator gain vector.* Figure 3.4 depicts the actual system and the state estimator where the symbol z represents a time-advance and $\frac{1}{z}$ represents a time-delay. The figure shows that the state of the physical system evolves in continuous time and is not directly measurable. Instead, output $\mathbf{y}$ is measured at discrete-time instants $t_k = kT$. From the discrete-time measurements of the output, eqn. (3.66) is implemented on the application computer to estimate the state vector. Because the state estimator is implemented computationally, the state estimates are conveniently available on the computer for use by control, planning, or other tasks as deemed appropriate by the designer.

For analysis, we form the difference equation for the state estimation error $\delta \mathbf{x} = \mathbf{x} - \hat{\mathbf{x}}$ by subtracting eqn. (3.66) from eqn. (3.39) to obtain

$$\begin{aligned} \delta \mathbf{x}_{k+1} &= \mathbf{\Phi} \delta \mathbf{x}_k - \mathbf{L}\mathbf{H}\delta \mathbf{x}_k \\ \delta \mathbf{x}_{k+1} &= (\mathbf{\Phi} - \mathbf{L}\mathbf{H})\, \delta \mathbf{x}_k. \end{aligned} \tag{3.67}$$

If the state estimation error is to converge to zero, then eqn. (3.67) must be asymptotically stable. Eqn. (3.67) will be asymptotically (actually exponentially) stable if the eigenvalues of $(\mathbf{\Phi} - \mathbf{L}\mathbf{H})$ have magnitude less than one.

The above analysis is meant to motivate the idea of state estimation (also called state observation). Several questions remain unanswered. The above analysis has tacitly assumed that it is possible to choose the matrix $\mathbf{L}$ so that the eigenvalues of $(\mathbf{\Phi} - \mathbf{L}\mathbf{H})$ have magnitude less than one. This may not always be possible. This issue is related to the question of when the system state can be estimated from a given set of measurements. This issue is discussed in Section 3.6.1. Once it is determined that state estimation is possible for a given system, it is natural to consider whether it is possible to derive an optimal state estimator relative to a given optimality criteria. The above analysis also only discusses time-invariant systems. The time-varying case is more complex and will only be discussed in the context of optimal estimation.

Note that the resulting state estimation algorithm is recursive in nature and can be split into two parts: a measurement update and a time update. At k-th time step prior to using $\mathbf{y}_k$, we have $\hat{\mathbf{x}}_k^-$, $\mathbf{y}_k$, and $\hat{\mathbf{y}}_k^-$. The computation for the k-th time step proceeds as follows:

$$\text{Residual:} \quad \delta \mathbf{y}_k = \mathbf{y}_k - \hat{\mathbf{y}}_k^- \tag{3.68}$$

$$\text{Posterior state estimate:} \quad \hat{\mathbf{x}}_k^+ = \hat{\mathbf{x}}_k^- + \bar{\mathbf{L}}\delta \mathbf{y}_k \tag{3.69}$$

$$\text{A priori state estimate:} \quad \hat{\mathbf{x}}_{k+1}^- = \mathbf{\Phi}\hat{\mathbf{x}}_k^+ + \mathbf{\Gamma} u_k \tag{3.70}$$

$$\text{A priori output estimate:} \quad \hat{\mathbf{y}}_{k+1}^- = \mathbf{H}\hat{\mathbf{x}}_{k+1}^- \tag{3.71}$$

where $\bar{\mathbf{L}} = \mathbf{\Phi}^{-1}\mathbf{L}$. This ordering of the operations is depicted in Figure 3.5. Eqn. (3.68) computes the measurement residual. Eqn. (3.69) can be thought of as a blending of all past measurement information propagated to the present time (as represented by $\hat{\mathbf{x}}_k^-$) with the new information available from the current measurement (as represented by the residual $\delta \mathbf{y}_k$). While $\hat{\mathbf{x}}_k^-$ represents the estimate of $\mathbf{x}_k$ using all measurements up to but not including $\mathbf{y}_k$, the vector $\hat{\mathbf{x}}_k^+$ represents the estimate of $\mathbf{x}_k$ after incorporating all measurements including $\mathbf{y}_k$. Performing the operations in this order minimizes the computation (i.e., delay) between the arrival of $\mathbf{y}_k$ and the availability of $\hat{\mathbf{x}}_k^+$ for use in control (e.g., calculation of u_k), navigation, and planning computations. After those computations are complete, but prior to the arrival of the next measurement, eqns. (3.70-3.71) perform the time

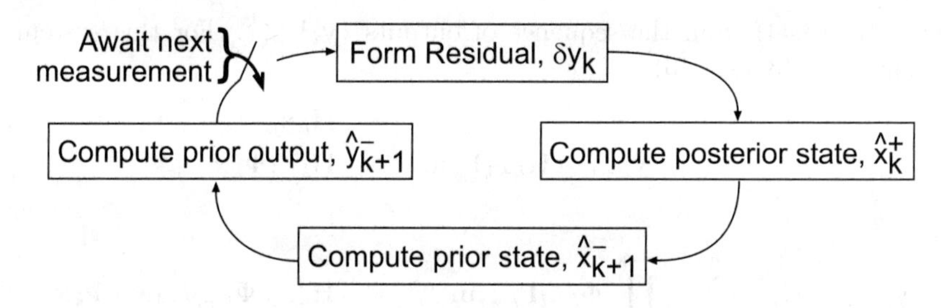

Figure 3.5: State estimator implementation organized to minimize delay between arrival of $\mathbf{y}_k$ and computation of $\mathbf{x}_k$.

propagation necessary to prepare the data for the $(k+1)$-th measurement.

A critical question is proper definition of the state of the system containing the plant and the state estimator. The following variables are open for consideration: $\mathbf{x}$, $\hat{\mathbf{x}}$, and $\delta\mathbf{x}$. A first thought might be to combine all three into a vector of length $3n$; however, this is not a minimal state definition because the three vectors are linearly dependent according to the equation $\delta\mathbf{x} = \mathbf{x} - \hat{\mathbf{x}}$. Knowledge of any two of the three vectors allows calculation of the third. The dimension of system containing both the plant and the state estimator is $2n$. Three valid state vectors are: $[\mathbf{x}, \hat{\mathbf{x}}]$, $[\mathbf{x}, \delta\mathbf{x}]$, and $[\hat{\mathbf{x}}, \delta\mathbf{x}]$.

The question of whether the plant state can be estimated is addressed in Subsection 3.6.1. The question of how to select the gain $\mathbf{L}$ to achieve a stable estimator design is addressed in Subsection 3.6.2. For a linear deterministic system, the state estimator design and analysis work with eqn. (3.67) and performance is independent of the actual state and control signal. Stochastic dynamic systems are introduced in Chapter 4. State estimation for stochastic dynamic systems is discussed in Chapter 5.

3.6.1 Observability

The previous section introduced the idea of estimation of the system state from the available outputs. Prior to trying to design a state estimator, it is important to consider the question of whether the state can be estimated from the available outputs.

This section analyzes the state estimation problem using the inversion-based solution method discussed in Section B.4. It is important for the reader to note that this approach to examining when the state estimation problem is solvable is distinct from the actual method of designing the state estimator. In particular, the vector $\mathbf{Z}$ defined below will never actually be computed directly.

Consider the problem of estimating the state $\mathbf{x}_k$ of the system described by eqn. (3.44) from the sequence of outputs $\{\mathbf{y}_i\}_{i=k}^{N+k}$. For this system, using eqns. (3.45–3.46),

$$\begin{aligned}
\mathbf{y}_k &= \mathbf{H}_k \mathbf{x}_k \\
\mathbf{y}_{k+1} - \mathbf{H}_{k+1}\mathbf{\Gamma}_k \mathbf{u}_k &= \mathbf{H}_{k+1}\mathbf{\Phi}_k \mathbf{x}_k \\
&\vdots \\
\mathbf{y}_{k+N} - \mathbf{H}_{k+N}\sum_{j=0}^{n-1}\prod_{i=j+1}^{n-1} \mathbf{\Phi}_{k+i}\mathbf{\Gamma}_{k+j}\mathbf{u}_{k+j} &= \mathbf{H}_{k+N}\mathbf{\Phi}_{k+N-1}\cdots\mathbf{\Phi}_k \mathbf{x}_k
\end{aligned}$$

where all quantities in the left-hand side of the equation are known. The vector in the left-hand side will be denoted as $\mathbf{Z}$. This set of equations can be written conveniently in matrix notation as

$$\mathbf{Z} = \begin{bmatrix} \mathbf{H}_k \\ \mathbf{H}_{k+1}\mathbf{\Phi}_k \\ \vdots \\ \mathbf{H}_{k+N}\mathbf{\Phi}_{k+N-1}\cdots\mathbf{\Phi}_k \end{bmatrix} \mathbf{x}_k. \tag{3.72}$$

Therefore, $\mathbf{x}_k$ can be estimated from $\{\mathbf{y}_i\}_{i=k}^{N+k}$ if the matrix on the right-hand side of eqn. (3.72) has rank equal to n (i.e., full column rank) for some N. For time-invariant systems this condition reduces to checking the rank of the matrix

$$\mathcal{O}^\top = \left[\mathbf{H}^\top, \mathbf{\Phi}^\top\mathbf{H}^\top, \ldots, \left(\mathbf{\Phi}^\top\right)^{n-1}\mathbf{H}^\top\right]. \tag{3.73}$$

Example 3.14 *Given the state space system*

$$\dot{\mathbf{x}} = \begin{bmatrix} 0 & 1 & 0 \\ 0 & 0 & 1 \\ 0 & 0 & 0 \end{bmatrix} \mathbf{x}(t) \tag{3.74}$$

$$y(t) = \begin{bmatrix} 0 & 0 & 1 \end{bmatrix} \mathbf{x}(t) \tag{3.75}$$

where $\mathbf{x} = [p, \dot{p}, \ddot{p}]^\top$ and only acceleration is being sensed. By the methods of Section 3.5.5, the equivalent discrete-time plant model is

$$\mathbf{x}_{k+1} = \begin{bmatrix} 1 & T & T^2/2 \\ 0 & 1 & T \\ 0 & 0 & 1 \end{bmatrix} \mathbf{x}_k \tag{3.76}$$

$$y_k = \begin{bmatrix} 0 & 0 & 1 \end{bmatrix} \mathbf{x}_k. \tag{3.77}$$

The observability matrix for this system is

$$\mathcal{O} = \begin{bmatrix} \mathbf{H} \\ \mathbf{H}\mathbf{\Phi} \\ \vdots \\ \mathbf{H}\mathbf{\Phi}\cdots\mathbf{\Phi} \end{bmatrix} = \begin{bmatrix} 0 & 0 & 1 \\ 0 & 0 & 1 \\ & \vdots & \\ 0 & 0 & 1 \end{bmatrix}.$$

which has rank one. This should match the reader's intuition regarding this problem, since two constants of integration would be involved in determining position and velocity from acceleration. An accelerometer alone cannot determine these unknown constants. Even if the initial conditions were exactly known, errors in the acceleration measurement could not be detected or removed. △

Example 3.15 *Consider the same dynamic system as described by eqn. (3.74), but with position being measured instead of acceleration. In the system model, eqns. (3.74) and (3.76) would not change, but the output matrix would become* $\mathbf{H} = [1, 0, 0]$. *In this case, the observability matrix is*

$$\mathcal{O} = \begin{bmatrix} \mathbf{H} \\ \mathbf{H\Phi} \\ \mathbf{H\Phi}^2 \end{bmatrix} = \begin{bmatrix} 1 & 0 & 0 \\ 1 & T & T^2/2 \\ 1 & 2T & 2T^2 \end{bmatrix}$$

which has full column rank (for $T \neq 0$*). Again, the mathematics verifies the intuition behind the problem, since at least three samples of position are required to estimate acceleration.* △

The above analysis shows that when the system state is observable it is possible to estimate the state in a minimum of N discrete-time steps; however, this is not usually done. Instead, computational issues, stability, and noise considerations usually motivate estimators which have guaranteed *asymptotic* convergence, as was discussed earlier in this section.

3.6.2 Estimator Design by Pole Placement

Chapter 5 will discuss optimal state estimation. This section discusses estimator design by pole placement (i.e., Ackerman's method) for two reasons. First, it gives the reader a method for selecting the state estimation gain vector for exercises prior to reading Chapter 5. Second, understanding of the concepts of this section will aid the understand of the results derived in Section 3.6.3.

Section 3.1.3 derived the controllable canonical form state space representation for the strictly proper transfer function in eqn. (3.6). For observer design by pole placement, it is also useful to define the *observable canonical form.* In discrete-time, the observable canonical form, state space representation is

$$\hat{\mathbf{v}}_{k+1} = \mathbf{\Phi}_o \hat{\mathbf{v}}_k + \mathbf{\Gamma}_o \mathbf{u}_k + \mathbf{L}_o (\mathbf{y}_k - \hat{\mathbf{y}}_k) \qquad \hat{\mathbf{y}}_k = \mathbf{H}_o \hat{\mathbf{v}}_k. \tag{3.78}$$

where

$$\Phi_o = \begin{bmatrix} -a_1 & 1 & 0 & \dots & 0 \\ -a_2 & 0 & 1 & \dots & 0 \\ \vdots & \vdots & \vdots & \ddots & \vdots \\ -a_{n-1} & 0 & 0 & \dots & 1 \\ -a_n & 0 & 0 & \dots & 0 \end{bmatrix}, \quad \Gamma_o = \begin{bmatrix} b_1 \\ b_2 \\ \vdots \\ b_{n-1} \\ b_n \end{bmatrix},$$

and $\mathbf{H}_o = [1, 0, \dots, 0]$. The observable canonical form will be convenient for selection of the observer gain matrix, assuming that we have method for transforming an arbitrary state space representation to observable canonical form.

The observer gain matrix design process has three steps. First, we find a similarity transform $\mathbf{v} = \mathbf{Ux}$ from the original state space representation

$$\hat{\mathbf{x}}_{k+1} = \Phi\hat{\mathbf{x}}_k + \Gamma\mathbf{u}_k + \mathbf{L}\left(\mathbf{y}_k - \hat{\mathbf{y}}_k\right) \qquad \hat{\mathbf{y}}_k = \mathbf{H}\hat{\mathbf{x}}_k \tag{3.79}$$

to observable canonical form where $\Phi_o = \mathbf{U}\Phi\mathbf{U}^{-1}$, $\Gamma_o = \mathbf{U}\Gamma$, $\mathbf{H}_o = \mathbf{H}\mathbf{U}^{-1}$, and $\mathbf{L}_o = \mathbf{UL}$. Second, we select an observer gain vector $\mathbf{L}_o$ using pole placement for the system in observable canonical form. Third, we use the equation $\mathbf{L} = \mathbf{U}^{-1}\mathbf{L}_o$ to transform the observer gain $\mathbf{L}_o$ back to the original state space representation. Throughout this section, we assume that the system is observable. If the system was not observable, then there would in general be no solution to the full state estimation problem.

The first step is to derive the similarity transform $\mathbf{v} = \mathbf{Ux}$ to transform the original state space representation of eqn. (3.79) to observable canonical form of eqn. (3.78). To achieve this, we begin by computing the observability matrix $\mathcal{O}$ as defined in eqn. (3.73). Because the system is observable, $\mathcal{O}$ is nonsingular; therefore, $\mathcal{O}^{-1}$ exists. Let the last column of $\mathcal{O}^{-1}$ be denoted as $\mathbf{p}$ so that

$$\mathcal{O}^{-1} = [\dots, \mathbf{p}].$$

Using the fact that $\mathcal{O}\mathcal{O}^{-1} = \mathbf{I}$ and focusing only on the last column yields

$$\begin{aligned} \mathbf{Hp} &= 0 \\ \mathbf{H}\Phi\mathbf{p} &= 0 \\ &\vdots \\ \mathbf{H}\Phi^{n-2}\mathbf{p} &= 0 \\ \mathbf{H}\Phi^{n-1}\mathbf{p} &= 1. \end{aligned}$$

Defining $\mathbf{U}^{-1} = [\Phi^{n-1}\mathbf{p}, \Phi^{n-2}\mathbf{p}, \dots, \Phi\mathbf{p}, \mathbf{p}]$, we have $\mathbf{H}\mathbf{U}^{-1} = [1, 0, \dots, 0]$ which shows that the output matrix transforms correctly. Next, consider the rows of

$$\mathbf{U} = \begin{bmatrix} \mathbf{u}_1 \\ \vdots \\ \mathbf{u}_n \end{bmatrix}.$$

Using the fact that

$$\mathbf{I} = \mathbf{U}\mathbf{U}^{-1} = \begin{bmatrix} \mathbf{u}_1\mathbf{\Phi}^{n-1}\mathbf{p} & \mathbf{u}_1\mathbf{\Phi}^{n-2}\mathbf{p} & \dots & \mathbf{u}_1\mathbf{p} \\ \mathbf{u}_2\mathbf{\Phi}^{n-1}\mathbf{p} & \mathbf{u}_2\mathbf{\Phi}^{n-2}\mathbf{p} & \dots & \mathbf{u}_2\mathbf{p} \\ \vdots & \vdots & \ddots & \vdots \\ \mathbf{u}_n\mathbf{\Phi}^{n-1}\mathbf{p} & \mathbf{u}_n\mathbf{\Phi}^{n-2}\mathbf{p} & \dots & \mathbf{u}_n\mathbf{p} \end{bmatrix}$$

we have that for $j \in [1, n]$

$$\mathbf{u}_i\mathbf{\Phi}^{n-j}\mathbf{p} = \begin{cases} 1 & \text{if } j = i \\ 0 & \text{if } j \neq i. \end{cases}$$

Defining $a_i = \mathbf{u}_i\mathbf{\Phi}^n\mathbf{p}$, we are ready to consider the product

$$\begin{aligned} \mathbf{U\Phi U}^{-1} &= \begin{bmatrix} \mathbf{u}_1 \\ \vdots \\ \mathbf{u}_n \end{bmatrix} \mathbf{\Phi}[\mathbf{\Phi}^{n-1}\mathbf{p}, \mathbf{\Phi}^{n-2}\mathbf{p}, \dots, \mathbf{\Phi}\mathbf{p}, \mathbf{p}] \\ &= \begin{bmatrix} \mathbf{u}_1 \\ \vdots \\ \mathbf{u}_n \end{bmatrix} [\mathbf{\Phi}^{n}\mathbf{p}, \mathbf{\Phi}^{n-1}\mathbf{p}, \dots, \mathbf{\Phi}^2\mathbf{p}, \mathbf{\Phi}\mathbf{p}] \\ &= \begin{bmatrix} \mathbf{u}_1\mathbf{\Phi}^{n}\mathbf{p} & \mathbf{u}_1\mathbf{\Phi}^{n-1}\mathbf{p} & \dots & \mathbf{u}_1\mathbf{\Phi}\mathbf{p} \\ \mathbf{u}_2\mathbf{\Phi}^{n}\mathbf{p} & \mathbf{u}_2\mathbf{\Phi}^{n-1}\mathbf{p} & \dots & \mathbf{u}_2\mathbf{\Phi}\mathbf{p} \\ \vdots & \vdots & \ddots & \vdots \\ \mathbf{u}_n\mathbf{\Phi}^{n}\mathbf{p} & \mathbf{u}_n\mathbf{\Phi}^{n-1}\mathbf{p} & \dots & \mathbf{u}_n\mathbf{\Phi}\mathbf{p} \end{bmatrix} \\ &= \begin{bmatrix} -a_1 & 1 & 0 & \dots & 0 \\ -a_2 & 0 & 1 & \dots & 0 \\ \vdots & \vdots & \vdots & \ddots & \vdots \\ -a_n & 0 & 0 & \dots & 1 \end{bmatrix} \end{aligned}$$

which has the desired form for the observable canonical form. Therefore, the similarity transform defined by $\mathbf{U}$ transforms the original observable system to observable canonical form.

In the observable canonical form, it is straightforward to design the observer gain $\mathbf{L}_o$, because

$$(\mathbf{\Phi}_o - \mathbf{L}_o\mathbf{H}_o) = \begin{bmatrix} -(a_1 + L_{o_1}) & 1 & 0 & \dots & 0 \\ -(a_2 + L_{o_2}) & 0 & 1 & \dots & 0 \\ \vdots & \vdots & \vdots & \ddots & \vdots \\ -(a_{n-1} + L_{o_{n-1}}) & 0 & 0 & \dots & 1 \\ -(a_n + L_{o_n}) & 0 & 0 & \dots & 0 \end{bmatrix} \tag{3.80}$$

which has the characteristic equation

$$|s\mathbf{I} - (\mathbf{\Phi}_o - \mathbf{L}_o\mathbf{H}_o)| = s^n + (a_1 + L_{o_1})s^{n-1} + \ldots + (a_n + L_{o_n}) = 0.$$

Therefore, if a set of desired discrete-time pole locations $\{p_i\}_{i=1}^n$ are selected, then the desired characteristic equation can be computed. The vector $\mathbf{L}_o$ to cause eqn. (3.80) to match the desired characteristic equation can be computed and $\mathbf{L} = \mathbf{U}^{-1}\mathbf{L}_o$.

Example 3.16 *Consider the system described in eqn. (3.4) with $m = 1$, $b = 2$ and $k = 0$. Defining the state vector as $\mathbf{x} = [\dot{p}, p]^\top$, a continuous-time state space model for the system is*

$$\dot{\mathbf{x}}(t) = \begin{bmatrix} -2 & 0 \\ 1 & 0 \end{bmatrix} \mathbf{x}(t) + \begin{bmatrix} 1 \\ 0 \end{bmatrix} f(t) \tag{3.81}$$

$$\mathbf{y}(t) = \begin{bmatrix} 0 & 1 \end{bmatrix} \mathbf{x}(t). \tag{3.82}$$

For a sample period of $T = 0.1$ seconds, the equivalent discrete-time state space model given by eqn. (3.65) is

$$\mathbf{x}_{k+1} = \begin{bmatrix} 0.8187 & 0.0000 \\ 0.0906 & 1.0000 \end{bmatrix} \mathbf{x}_k + \begin{bmatrix} 0.0906 \\ 0.0047 \end{bmatrix} f_k \tag{3.83}$$

$$p_k = \begin{bmatrix} 0.0000 & 1.0000 \end{bmatrix} \mathbf{x}_k. \tag{3.84}$$

The observability matrix is

$$\mathcal{O} = \begin{bmatrix} 0.0000 & 1.0000 \\ 0.0906 & 1.0000 \end{bmatrix}$$

so that $\mathbf{p} = [11.0375, 0.0000]^\top$. The resulting similarity transform uses

$$\mathbf{U} = \begin{bmatrix} 0.0000 & 1.0000 \\ 0.0906 & -0.8187 \end{bmatrix}.$$

With the closed loop estimator poles (i.e. the eigenvalues of $(\mathbf{\Phi} - \mathbf{LH})$ and of $(\mathbf{\Phi}_o - \mathbf{L}_o\mathbf{H}_o)$) specified to be at $0.9 \pm 0.1j$. The observer gain matrices are $\mathbf{L}_o = [0.0187, 0.0013]^\top$ and $\mathbf{L} = [0.1832, 0.0187]^\top$.

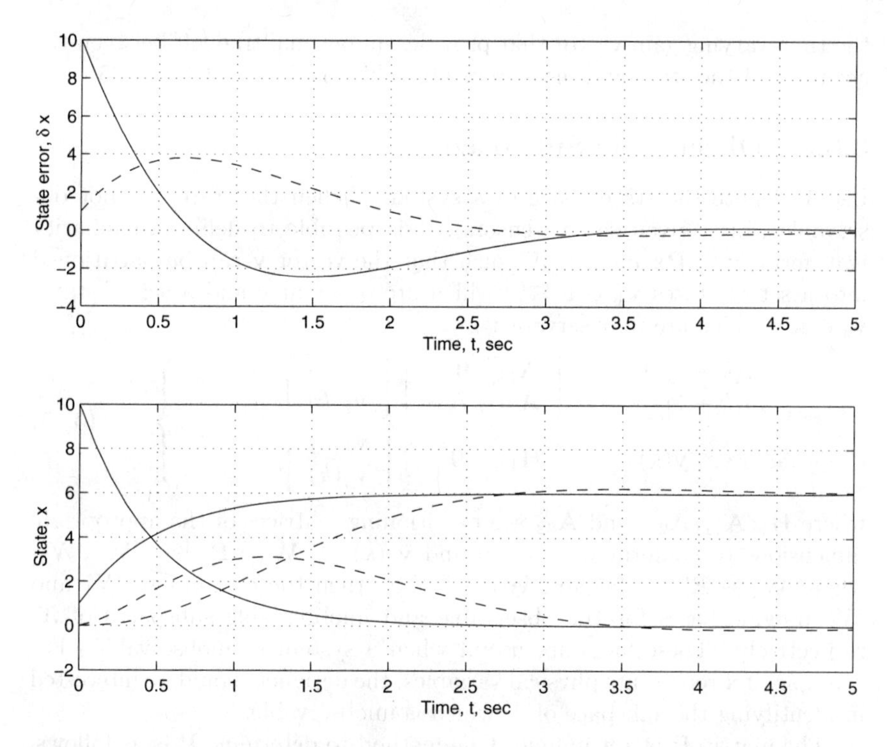

Figure 3.6: State estimator simulation for Example 3.16. Top – State estimation error versus time. The solid line is δx_1 and the dashed line is δx_2. Bottom – State and state estimates versus time. The solid lines are the states and the dashed lines are the estimates.

The convergence of the resulting state estimator is shown in Figure 3.6 for the initial conditions $\mathbf{x} = [10.0, 1.0]^\top$ *and* $\hat{\mathbf{x}} = [0.0, 0.0]^\top$. *The top plot shows the estimation error with* $\mathbf{x}_1$ *as a solid line and* $\mathbf{x}_2$ *as a dashed line. The bottom plot shows state (solid) and the estimated state (dashed). Both plots clearly show the convergence of the estimation error to zero. The plots are shown as continuous lines, but the state estimates are actually only defined at the sampling instants.* △

There are clearly a variety of tradeoffs in the selection of the state estimation gain vector. If $\mathbf{L}$ is small then the convergence will be slow, but measurement noise will have a small effect on the state estimates. Alternatively, when $\mathbf{L}$ is large convergence of the error will be rapid, but measurement noise will have a relatively large effect on the state estimates. When it is possible to propagate estimates of the accuracy of the state estimates and of the measurement accuracy, then it is interesting to consider the time-varying gain vector that provides an optimal tradeoff between the accuracy of the state and measurements, which is done in Chapter 5.

3.6.3 Observable Subspace

Let $\mathcal{X}$ denote the state space of a system. When the system is not observable (for a given set of sensors), it is possible to define a similarity transform, $\mathbf{v} = \mathbf{P}\mathbf{x}$ for $\mathbf{x} \in \mathcal{X}$, such that the vector $\mathbf{v}$ can be partitioned into a set of states $\mathbf{v}_1 \in \mathbb{R}^{n-r_o}$ which are observable and a set of states $\mathbf{v}_2 \in \mathbb{R}^{r_o}$ which are unobservable:

$$\left.\begin{aligned} \begin{bmatrix} \mathbf{v}_1(k+1) \\ \mathbf{v}_2(k+1) \end{bmatrix} &= \begin{bmatrix} \mathbf{A}_{11} & \mathbf{0} \\ \mathbf{A}_{21} & \mathbf{A}_{22} \end{bmatrix} \begin{bmatrix} \mathbf{v}_1(k) \\ \mathbf{v}_2(k) \end{bmatrix} \\ \mathbf{y}(k) &= \begin{bmatrix} \mathbf{H}_1 & \mathbf{0} \end{bmatrix} \begin{bmatrix} \mathbf{v}_1(k) \\ \mathbf{v}_2(k) \end{bmatrix} \end{aligned}\right\} \tag{3.85}$$

where $\mathbf{H}_1, \mathbf{A}_{11}, \mathbf{A}_{21}$, and $\mathbf{A}_{22}$ are partitioning matrices of the appropriate dimensions to conform with $\mathbf{v}_1(k)$ and $\mathbf{v}_2(k)$. If $\mathbf{W} = \mathbf{P}^{-1} = [\mathbf{W}_1, \mathbf{W}_2]$ where $\mathbf{W}_1 \in \mathbb{R}^{n\times(n-r_o)}$ and $\mathbf{W}_2 \in \mathbb{R}^{n\times r_o}$, then the columns of $\mathbf{W}_1$ and $\mathbf{W}_2$ provide bases for the observable and unobservable subspaces of $\mathcal{X}$, respectively. These facts are useful when a system is unobservable. For example, if $\mathbf{x}$ represents physical variables, the designer should be interested in identifying the subspace of $\mathcal{X}$ which is unobservable.

The matrix $\mathbf{P}$ is not unique. One method to determine $\mathbf{P}$ is as follows. Let $r_o = rank(\mathcal{O})$ for the observability matrix $\mathcal{O}$ defined in eqn. (3.73). Select r_o linearly independent rows from $\mathcal{O}$. Call these row vectors $\mathbf{q}_i$ for $i = 1, \ldots, r_o$. Define the remaining $(n - r_o)$ row vectors $\mathbf{q}_i$ for $i = r_o + 1, \ldots, n$ arbitrarily as long as they are linearly independent with each other and with each $\mathbf{q}_i$ for $i = 1, \ldots, r_o$. Then the matrix

$$\mathbf{P} = \left[\begin{array}{c} \mathbf{q}_1 \\ \vdots \\ \mathbf{q}_{r_o} \\ \hline \mathbf{q}_{r_o+1} \\ \vdots \\ \mathbf{q}_n \end{array}\right]$$

is nonsingular and transforms the state space representation for $\mathbf{x}$ into the form of eqns. (3.85).

Example 3.17 *Consider the system described by eqns (3.74–3.75). The similarity transform defined by*

$$\mathbf{P} = \left[\begin{array}{ccc} 0 & 0 & 1 \\ \hline 0 & 1 & 0 \\ 1 & 0 & 0 \end{array}\right]$$

transforms the state space system representation to the form of eqn. (3.85). An observability check shows that only the first component of $\mathbf{v}$ *is observable. Therefore, the second and third columns of*

$$\mathbf{W} = \mathbf{P}^{-1} = \begin{bmatrix} 0 & 0 & 1 \\ 0 & 1 & 0 \\ 1 & 0 & 0 \end{bmatrix}$$

form a basis for the unobservable subspace of $\mathcal{X}$. *The unobservable subspace is*

$$\mathcal{X}_{un} = a \begin{bmatrix} 0 \\ 1 \\ 0 \end{bmatrix} + b \begin{bmatrix} 1 \\ 0 \\ 0 \end{bmatrix}$$

which further clarifies the results previously obtained by showing that any linear combination of position and velocity are unobservable from acceleration measurements. △

Example 3.18 *Consider the system of eqn. (3.105) with the accelerometer and gyro errors represented as constant biases:*

$$\begin{bmatrix} \dot{\phi} \\ \dot{v}_n \\ \dot{\theta} \\ \dot{b}_a \\ \dot{b}_g \end{bmatrix} = \begin{bmatrix} 0 & \frac{1}{R} & 0 & 0 & 0 \\ 0 & 0 & f_d & 1 & 0 \\ 0 & \frac{1}{R} & 0 & 0 & 1 \\ 0 & 0 & 0 & 0 & 0 \\ 0 & 0 & 0 & 0 & 0 \end{bmatrix} \begin{bmatrix} \phi \\ v_n \\ \theta \\ b_a \\ b_g \end{bmatrix}.$$

If this system is position aided with $H = [1, 0, 0, 0, 0]$, *is the state observable. If it is not observable, define the observable subspace.*

The observability matrix

$$\mathcal{O} = \begin{bmatrix} 1 & 0 & 0 & 0 & 0 \\ 0 & \frac{1}{R} & 0 & 0 & 0 \\ 0 & 0 & \frac{f_d}{R} & \frac{1}{R} & 0 \\ 0 & \frac{f_d}{R^2} & 0 & 0 & \frac{f_d}{R} \\ 0 & 0 & \frac{f_d^2}{R^2} & \frac{f_d}{R^2} & 0 \end{bmatrix}$$

has rank 4, so the state is not observable.

The first four rows of $\mathcal{O}$ *are linearly independent, while the fifth row is a multiple of the third row. An orthogonal basis for the first four rows of* $\mathcal{O}$ *is defined by*

$$\begin{aligned} \mathbf{q}_1 &= [1\ 0\ 0\ \ 0\ 0], \\ \mathbf{q}_2 &= [0\ 1\ 0\ \ 0\ 0], \\ \mathbf{q}_3 &= [0\ 0\ f_d\ 1\ 0], \\ \mathbf{q}_4 &= [0\ 0\ 0\ \ 0\ 1]. \end{aligned}$$

The vector $\mathbf{q}_5 = [0 \;\; 0 \;\; -1 \;\; f_d \;\; 0]$ *completes the orthogonal basis for* $\mathbb{R}^5$. *These five vectors define the rows of the matrix* $\mathbf{P}$. *The inverse matrix is*

$$\mathbf{W} = \mathbf{P}^{-1} = \begin{bmatrix} 1 & 0 & 0 & 0 & 0 \\ 0 & 1 & 0 & 0 & 0 \\ 0 & 0 & \frac{f_d}{1+f_d^2} & 0 & \frac{-1}{1+f_d^2} \\ 0 & 0 & \frac{1}{1+f_d^2} & 0 & \frac{f_d}{1+f_d^2} \\ 0 & 0 & 0 & 1 & 0 \end{bmatrix}.$$

Therefore, the fifth column defines the unobservable subspace. As long as f_d *is constant, a position measurement based state estimation scheme cannot discriminate tilt error the error* $\theta = b_a = 0$ *from the error* $\theta = f_d$, $b_a = 1$ *or from any other error such that* $\theta = f_d b_a$. △

Example 3.19 *Let the system of interest be defined by*

$$\begin{bmatrix} x_1(k+1) \\ x_2(k+1) \end{bmatrix} = \begin{bmatrix} 0 & 0 \\ 0 & 0 \end{bmatrix} \begin{bmatrix} x_1(k) \\ x_2(k) \end{bmatrix} \tag{3.86}$$

$$y(k) = x_1(k) + x_2(k) = \begin{bmatrix} 1 & 1 \end{bmatrix} \begin{bmatrix} x_1(k) \\ x_2(k) \end{bmatrix}. \tag{3.87}$$

Define the similarity transform $\mathbf{P} = \left[\begin{array}{cc} 1 & 1 \\ \hline 1 & -1 \end{array}\right]$. *Then,*

$$\begin{bmatrix} v_1(k+1) \\ v_2(k+1) \end{bmatrix} = \begin{bmatrix} 0 & 0 \\ 0 & 0 \end{bmatrix} \begin{bmatrix} v_1(k) \\ v_2(k) \end{bmatrix} \tag{3.88}$$

$$y(k) = \begin{bmatrix} 1 & 0 \end{bmatrix} \begin{bmatrix} v_1(k) \\ v_2(k) \end{bmatrix}. \tag{3.89}$$

By observability analysis, it is straightforward to show that the state v_2 *is unobservable. Since* $\mathbf{W} = \frac{1}{2}\begin{bmatrix} 1 & 1 \\ 1 & -1 \end{bmatrix}$, *a basis for the unobservable portion of* X *is* $\begin{bmatrix} 1 \\ -1 \end{bmatrix}$. *In this example it was straightforward to see that while* $x_1 + x_2$ *was measured, nothing could be concluded about the value of* $x_1 - x_2$. *In more complicated examples it is useful to have a well-defined method for determining a basis for the unobservable portion of* $\mathcal{X}$. △

In the previous example, the subspace of $\mathcal{X}$ spanned by $x_1 - x_2$ was unobservable. This unobservable space is defined by the system dynamics and sensor suite. Once the system and sensor suite are fixed, the designer cannot 'select' the unobservable subspace. If the designer drops either variable in an attempt to estimate the other, the reduced order estimation problem may appear to 'become observable', but the resulting estimate of the retained variable will be biased from the true value of the variable.

Example 3.20 *Continuing from the previous example, assume that the true value of the state vector is* $\mathbf{x} = [1, 1]$. *Also, assume that the designer uses the state estimation model*

$$\hat{x}_1(k+1) = 0, \qquad \hat{y}(k) = \hat{x}_1(k)$$

which a simple observability analysis shows to be observable. Based on the measurement $y = 2$, *the estimator would produce* $\hat{x}_1 = 2$, *which is incorrect and biased from its correct value by the amount* $\mathbf{H}_2\mathbf{x}_2$. △

In subsequent chapters it will be suggested that unobservable variables be combined or dropped in the quest to produce a near optimal estimation scheme with a reasonable amount of computation. When this is done, the new combined variable should be used with caution. In no case should this be interpreted as selection of the unobservable states (subspace). The unobservable subspace can only be affected by changing either the actual system dynamics or the sensor suite.

One method to compute numerically a basis for the observable and unobservable subspaces is to use the singular value decomposition (SVD). Let the SVD of $\mathcal{O}^\top$ be represented as $\mathcal{O}^\top = \mathbf{V\Sigma U}^\top$. The matrix $\mathbf{\Sigma}$ is diagonal with r_o non-zero elements. Let $\mathbf{V}$ be decomposed as $\mathbf{V} = [\mathbf{V}_1, \mathbf{V}_2]$ where $\mathbf{V}_1 \in \mathbb{R}^{n \times r_o}$ and $\mathbf{V}_2 \in \mathbb{R}^{n \times (n-r_o)}$. The columns of $\mathbf{V}_1$ form an orthogonal basis for the range space of $\mathcal{O}^\top$ which is also an orthogonal basis for the rows of $\mathcal{O}$. Therefore, the columns of $\mathbf{V}_1$ serve as a basis for the observable subspace. The columns of $\mathbf{V}_2$ are orthogonal to each other and to the columns of $\mathbf{V}_1$. Therefore, the columns of $\mathbf{V}_2$ serve as a basis for the unobservable subspace.

For exercises related to this topic, see Exercises 3.11 and 4.15. More detailed examples of the definition of unobservable spaces are contained on pages 350, 427, and 449.

3.7 References and Further Reading

There are several good sources for the material from this chapter. At the undergraduate level, two good references are [94, 105]. At the graduate level, two good references are [38, 87]. The main sources for Sections 3.6.2 and 3.6.3 of this chapter were [28] and Sections 8.2.4, 8.3, and 10.1 in [19]. Observability, and its dual concept, controllability are discussed in relation to model reduction in [101].

3.8 Exercises

Exercise 3.1 The purpose of this exercise is to supply the details of the derivation of the transfer function to state space transformation on p. 67.

1. Manipulate eqn. (3.12) to have the form

$$V(s)s^n + d_1V(s)s^{n-1} + \ldots + d_nV(s) = U(s).$$

2. Use inverse Laplace transforms to show that the differential equation relating $u(t)$ to $v(t)$ is

$$v^{(n)} + d_1v^{(n-1)} + \ldots + d_nv = u. \tag{3.90}$$

3. Show that eqn. (3.14) with the state vector defined immediately before eqn. (3.14) is a valid state space representation of eqn. (3.90). To do this, rewrite the derivative of each state variable in terms of only the state and the control. For example, $\dot{x}_1 = \dot{v} = x_2$ which yields the first equation $\dot{x}_1 = x_2$.

Exercise 3.2 Use eqn. (3.51) to show that the transfer function corresponding to the state space model of Example 3.6 on page 69 does in fact match eqn. (3.10) on page 66. To find the inverse matrix required for the computation, use eqn. (B.29).

Exercise 3.3 For the two state space systems represented by eqns. (3.22) and (3.23).

1. Find the state space representation for the parallel connection shown in the left portion of Figure 3.7.

2. Find the state space representation for the feedback connection shown in the right portion of Figure 3.7.

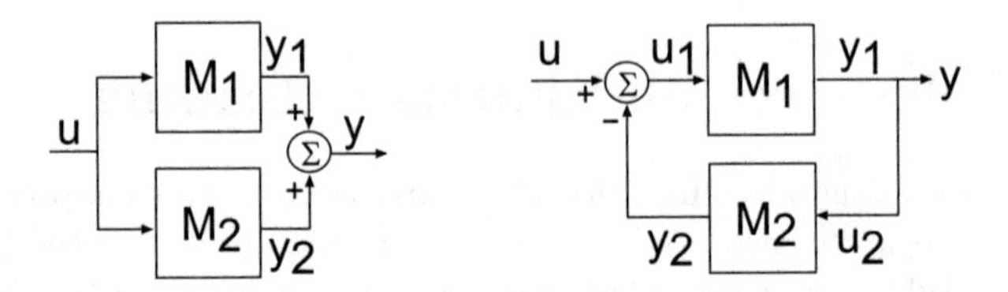

Figure 3.7: Interconnected systems for Exercise 3.3. Left – Parallel interconnection of two state space systems. Right – Feedback interconnection of two state space systems.

Exercise 3.4 Verify the similarity transform in Example 3.10 on page 76. Also use eqn. (3.51) to verify that the input-output transfer function corresponding to eqn. (3.49) still matches eqn. (3.10).

Exercise 3.5 In Example 3.7 on page 70, if the applied force is instead defined by

$$\begin{aligned} \dot{v} &= p - r \\ f &= K_1 v + K_2 (p - r) + K_3 \dot{p}. \end{aligned}$$

where $r(t)$ is an independent input:

1. Find the state space representation for the augmented state vector $\mathbf{z} = [mp, m\dot{p}, v]^\top$ with the output $y = p$.

2. Find the transfer function from r to y.

Exercise 3.6 Use the definition of the fundamental solution and the definition of the state transition matrix to verify the properties in eqns. (3.55–3.59).

Exercise 3.7 A state space representation of the single channel horizontal error dynamics of an INS (see Exercise 3.12) is

$$\begin{bmatrix} \dot{x} \\ \dot{v} \\ \dot{\phi} \end{bmatrix} = \begin{bmatrix} 0 & 1 & 0 \\ 0 & 0 & -g \\ 0 & \frac{1}{R} & 0 \end{bmatrix} \begin{bmatrix} x \\ v \\ \phi \end{bmatrix} + \begin{bmatrix} 0 & 0 \\ 1 & 0 \\ 0 & 1 \end{bmatrix} \begin{bmatrix} \epsilon_a \\ \epsilon_g \end{bmatrix}, \qquad (3.91)$$

where ϵ_a and ϵ_g represent the accelerometer and gyro measurement errors in the navigation frame. Let $g = 9.8\frac{m}{s^2}$ and $R_e = 6 \times 10^6 m$.

1. Treating ϵ_a and ϵ_g as inputs:

   (a) Find the transfer functions from ϵ_a to x.

   (b) Find the transfer functions from ϵ_g to x.

   Discuss the pole locations and the expected characteristics of the response.

2. Find the discrete-time equivalent model assuming that the position x is measured at 0.1 Hz.

3. Is the entire state observable?

4. Find the estimator gain vector that yields discrete-time estimator poles at $p = [0.85 \pm j0.15, 0.75]$.

5. Let $\epsilon_a = \epsilon_g = 0$. A series of noise corrupted measurements and the MATLAB file to produce them is posted on the book's web site. Implement the state estimator and generate plots of the state estimates and the residual measurements.

6. Let $\epsilon_g = 0$. If ϵ_a with dynamic model $\dot{\epsilon}_a = 0$ is augmented to the state vector, defined the new state space model and determine whether the new state vector is observable. If not, define the unobservable subspace.

Exercise 3.8 Consider the state space system

$$\begin{aligned}\dot{\mathbf{v}} &= \begin{bmatrix} -2 & 0 \\ 0 & -3 \end{bmatrix}\mathbf{v} + \begin{bmatrix} -1 \\ 1 \end{bmatrix}u \\ y &= \begin{bmatrix} 1 & 2 \end{bmatrix}\mathbf{v}.\end{aligned}$$

1. Show that the eigenvalues of the $\mathbf{F}$ matrix are -2 and -3.

2. Show that the transfer function from u to y is

$$\frac{Y(s)}{U(s)} = \frac{(s+1)}{(s+3)(s+2)}.$$

   Note that the system poles and eigenvalues are identical.

3. A single transfer function has an infinite number of equivalent state space representations. If $\mathbf{x}$ is a valid state vector and $\mathbf{P}$ is a nonsingular matrix, then $\mathbf{v} = \mathbf{Px}$ is also a valid state vector. This is referred to as a similarity transformation.

   Show that the state space representation for $\mathbf{v}$ of this problem and the state representation for $\mathbf{x}$ of Example 3.11 on p. 77 are similar with $P = \begin{bmatrix} -3 & -1 \\ 2 & 1 \end{bmatrix}$.

4. Find the equivalent discrete-time system assuming that the input $u(t)$ is constant over the time step T.

Exercise 3.9 Show that, for the system described by eqns. (3.74–3.75), the equivalent discrete-time plant is represented by eqns. (3.76–3.77).

Exercise 3.10 Repeat example 3.16 if the desired eigenvalues are 0.85 and 0.90.

Exercise 3.11 Consider the system

$$\dot{\mathbf{x}} = \mathbf{Fx} + \mathbf{G}u \tag{3.92}$$

$$y = \mathbf{Hx} \tag{3.93}$$

where $\mathbf{F} = \begin{bmatrix} 1 & 1 & 0 & 0 & 0 \\ 0 & 1 & 0 & 0 & 0 \\ 0 & 0 & 1 & 0 & 0 \\ 0 & 0 & 0 & 1 & 1 \\ 0 & 0 & 0 & 0 & 1 \end{bmatrix}$ and $\mathbf{H} = [0 \quad 1 \quad 1 \quad 1 \quad 0]$.

1. Is the system observable?

2. If not, determine a basis for the unobservable subspace.

Exercise 3.12 This is a rather long exercise directly related to navigation. It has a few objectives and touches on many of the concepts discussed in Chapter 3. The primary objectives are

1. to step the reader through the navigation system linearization process

2. to derive the error model in eqn. (3.49); and,

3. to introduce the Schuler frequency.

Figure 3.8 depicts a simplified two dimensional navigation system where the Earth is assumed to be spherical and non-rotating, with radius R, and uniform density. The point P is constrained to move in the (p, z) plane. The point P represents the effective location of an inertial measurement unit

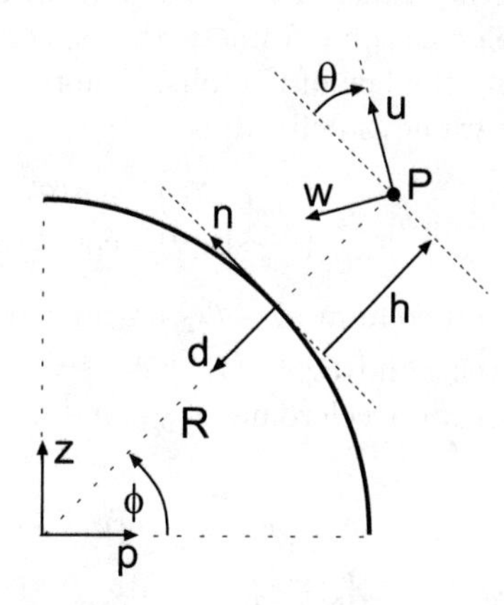

Figure 3.8: Figure for Exercise 3.12.

(IMU) with sensitive axes in the directions indicated by the unit vectors $\mathbf{u}$ and $\mathbf{w}$, which define the platform reference frame. The projection of the point P onto the Earth surface defines the origin of the geographic reference frame. The instantaneous Earth tangent plane at the origin of the geographic frame defines the unit vectors $\mathbf{n}$ and $\mathbf{d}$. The height of the point P above the tangent plane is the altitude h. The gravity vector is $\mathbf{g} = \frac{GM}{(R+h)^2}\mathbf{d}$ which points along the $\mathbf{d}$-axis of the geographic frame. The angles ϕ and θ are defined as positive in the directions indicated in the figure. The $\mathbf{z}$ and $\mathbf{p}$ vectors defined the ECEF reference frame, which for this exercise is identical with the inertial frame of reference.

The kinematics of the IMU at point P are defined by

$$\begin{bmatrix} \dot{\phi} \\ \dot{h} \end{bmatrix} = \begin{bmatrix} \frac{1}{R+h} & 0 \\ 0 & -1 \end{bmatrix} \begin{bmatrix} v_n \\ v_d \end{bmatrix} \tag{3.94}$$

$$\begin{bmatrix} \dot{v}_n \\ \dot{v}_d \end{bmatrix} = \vec{\mathbf{f}}^g + \mathbf{g}^g(h) + \begin{bmatrix} \frac{v_n v_d}{R+h} \\ \frac{-v_n^2}{R+h} \end{bmatrix} \tag{3.95}$$

$$\dot{\theta} = \omega_{gp}^p \tag{3.96}$$

where $\vec{\mathbf{f}}^g = \mathbf{R}_p^g \vec{\mathbf{f}}^p$ and $\vec{\mathbf{f}}^p$ is the specific force vector[1] represented in platform frame, $\omega_{gp}^p = \omega_{ip}^p - \omega_{ig}^p$ with ω_{ip}^p measured by the gyro and $\omega_{ig}^p = -\dot{\phi} = -\frac{v_n}{R+h}$. The gravity vector in navigation frame is

$$\mathbf{g}^g(h) = \begin{bmatrix} 0 \\ \frac{GM}{(R+h)^2} \end{bmatrix}.$$

The vector $[v_n, v_d]$ is the Earth relative velocity of point P represented in the instantaneous tangent plane. The third term in eqn. (3.95) is due to the rotation rate of the geographic frame with respect to the inertial frame and can be derived using the law of Coriolis. The rotation matrix $\mathbf{R}_p^g$ from platform to geographic frame is defined as

$$\mathbf{R}_p^g = [-\theta]_2 = \begin{bmatrix} \cos(\theta) & \sin(\theta) \\ -\sin(\theta) & \cos(\theta) \end{bmatrix}$$

where the second row and column of $[-\theta]_2$ have been removed as they are not relevant to the exercise and $\mathbf{R}_g^p = \left(\mathbf{R}_p^g\right)^\top$.

The IMU consists of an accelerometer $\mathbf{y}_1$ and a gyro y_2 with outputs modeled as

$$\mathbf{y}_1 = \mathbf{a}^p - \mathbf{g}^p - \varepsilon_a^p = \vec{\mathbf{f}}^p - \varepsilon_a^p \tag{3.97}$$

$$y_2 = \omega_{ip}^p - \varepsilon_g \tag{3.98}$$

where $\vec{\mathbf{f}}^p = \mathbf{a}^p - \mathbf{g}^p$ is the specific force vector in platform frame, and ε_a^p and ε_g represent accelerometer and gyro errors, respectively.

A navigation system calculates the navigation state by integration of the following equations

$$\begin{bmatrix} \dot{\hat{\phi}} \\ \dot{\hat{h}} \end{bmatrix} = \begin{bmatrix} \frac{1}{R+\hat{h}} & 0 \\ 0 & -1 \end{bmatrix} \begin{bmatrix} \hat{v}_n \\ \hat{v}_d \end{bmatrix} \tag{3.99}$$

$$\begin{bmatrix} \dot{\hat{v}}_n \\ \dot{\hat{v}}_d \end{bmatrix} = \hat{\mathbf{R}}_p^g (\mathbf{y}_1 + \hat{\varepsilon}_a^p) + \hat{\mathbf{g}}^g(\hat{h}) + \begin{bmatrix} \frac{\hat{v}_n \hat{v}_d}{R+\hat{h}} \\ \frac{-\hat{v}_n^2}{R+\hat{h}} \end{bmatrix} \tag{3.100}$$

[1] The fact that accelerometers measure specific force is discussed in Section 11.1.3.

$$\dot{\hat{\theta}} = \frac{\hat{v}_n}{R+\hat{h}} + y_2 + \hat{\varepsilon}_g \tag{3.101}$$

where $\hat{\boldsymbol{\varepsilon}}_a$ is an estimate of the accelerometer error vector and $\hat{\varepsilon}_g$ is an estimate of the gyro error. Let the state vector for the vehicle kinematics be $\mathbf{x} = [\phi, h, v_n, v_d, \theta]^\top$, the navigation system state vector be $\hat{\mathbf{x}} = [\hat{\phi}, \hat{h}, \hat{v}_n, \hat{v}_d, \hat{\theta}]^\top$, $\mathbf{u}_1 = \left[(\mathbf{f}^p)^\top, \omega_{ip}^p\right]^\top$, and $\mathbf{u}_2^p = \left[\delta\boldsymbol{\varepsilon}_a^\top, \delta\varepsilon_g\right]^\top$ where $\delta\boldsymbol{\varepsilon}_a = \boldsymbol{\varepsilon}_a - \hat{\boldsymbol{\varepsilon}}_a$ and $\delta\varepsilon_g = \varepsilon_g - \hat{\varepsilon}_g$. Finally, define the navigation error state as $\delta\mathbf{x} = \mathbf{x} - \hat{\mathbf{x}}$.

1. List the four frames-of-reference used in this problem and the role of each.

2. Use the Taylor series approximation to $\mathbf{R}_g^p$ to show that

$$\mathbf{R}_p^g = \hat{\mathbf{R}}_p^g + \begin{bmatrix} -\sin\hat{\theta} & \cos\hat{\theta} \\ -\cos\hat{\theta} & -\sin\hat{\theta} \end{bmatrix} \delta\theta.$$

3. Show that, to first order,

$$\hat{\mathbf{R}}_p^g \mathbf{R}_g^p = \begin{bmatrix} 1 & -\delta\theta \\ \delta\theta & 1 \end{bmatrix}.$$

4. Show that for

$$\mathbf{f}(\mathbf{x}, \mathbf{u}_1) = \begin{bmatrix} \frac{x_3}{R+x_2} \\ -x_4 \\ \mathbf{R}_p^g \mathbf{f}^p + \mathbf{g}^g(x_2) + \frac{x_3}{R+x_2} \begin{bmatrix} x_4 \\ -x_3 \end{bmatrix} \\ \frac{x_3}{R+x_2} + \omega_{ip}^p \end{bmatrix}$$

eqns. (3.94–3.96) can be written in the form $\dot{\mathbf{x}} = \mathbf{f}(\mathbf{x}, \mathbf{u}_1)$.

5. Use the fact that $\delta\dot{\mathbf{x}} = \dot{\mathbf{x}} - \dot{\hat{\mathbf{x}}}$, the fact that

$$\mathbf{f}(\mathbf{x}, \mathbf{u}_1) \approx \mathbf{f}(\hat{\mathbf{x}}, \mathbf{u}_1) + \left.\frac{\partial \mathbf{f}}{\partial \mathbf{x}}\right|_{\mathbf{x}=\hat{\mathbf{x}}} \delta\mathbf{x},$$

and the result of Part 3, to show that to first order,

$$\delta\dot{\mathbf{x}} = \begin{bmatrix} 0 & -\frac{\hat{v}_n}{(R+h)^2} & \frac{1}{R+h} & 0 & 0 \\ 0 & 0 & 0 & -1 & 0 \\ 0 & -\frac{v_n v_d}{(R+h)^2} & \frac{v_d}{R+h} & \frac{v_n}{R+h} & \vec{f}_d \\ 0 & \left(\frac{v_n^2}{(R+h)^2} - \frac{2GM}{(R+h)^3}\right) & \frac{-2v_n}{R+h} & 0 & -\vec{f}_n \\ 0 & -\frac{\hat{v}_n}{(R+h)^2} & \frac{1}{R+h} & 0 & 0 \end{bmatrix} \delta\mathbf{x}$$

$$+\begin{bmatrix} 0 & 0 & 0 \\ 0 & 0 & 0 \\ \cos(\theta) & \sin(\theta) & 0 \\ -\sin(\theta) & \cos(\theta) & 0 \\ 0 & 0 & 1 \end{bmatrix} \mathbf{u}_2^p \tag{3.102}$$

where $\vec{f}_n = a_n$ and $\vec{f}_d = a_d - \frac{GM}{(R+h)^2}$ are the north and down components of the specific force vector in the geographic frame, respectively.

6. Let $\mathbf{q} = \mathbf{S}\delta\mathbf{x}$ where $\mathbf{q} = [\delta\phi, \delta v_n, \delta\theta, \delta h, \delta v_d]^\top$. Show that the matrix producing this reordering of the state vector is

$$\mathbf{S} = \begin{bmatrix} 1 & 0 & 0 & 0 & 0 \\ 0 & 0 & 1 & 0 & 0 \\ 0 & 0 & 0 & 0 & 1 \\ 0 & 1 & 0 & 0 & 0 \\ 0 & 0 & 0 & 1 & 0 \end{bmatrix} \quad \text{and that } \mathbf{S}^{-1} = \mathbf{S}^\top.$$

7. Use the similarity transformation matrix $\mathbf{S}$ to show that the dynamic model for $\mathbf{q}$ is

$$\dot{\mathbf{q}} = \left[\begin{array}{ccc|cc} 0 & \frac{1}{R+h} & 0 & -\frac{v_n}{(R+h)^2} & 0 \\ 0 & \frac{v_d}{R+h} & \vec{f}_d & -\frac{v_n v_d}{(R+h)^2} & \frac{v_n}{R+h} \\ 0 & \frac{1}{R+h} & 0 & -\frac{v_n}{(R+\hat{h})^2} & 0 \\ \hline 0 & 0 & 0 & 0 & -1 \\ 0 & \frac{-2v_n}{R+h} & -\vec{f}_n & \left(\frac{\hat{v}_n^2}{(R+h)^2} - \frac{2GM}{(R+h)^3}\right) & 0 \end{array}\right] \mathbf{q}$$

$$+\left[\begin{array}{ccc} 0 & 0 & 0 \\ \cos(\theta) & \sin(\theta) & 0 \\ 0 & 0 & 1 \\ \hline 0 & 0 & 0 \\ -\sin(\theta) & \cos(\theta) & 0 \end{array}\right] \mathbf{u}_2^p \tag{3.103}$$

where the lines are added to separate the vertical state errors δh and δv_d form the horizontal and tilt errors $\delta\phi$, δv_n, and $\delta\theta$.

8. Consider the unforced (i.e., $\mathbf{u}_2^p = \mathbf{0}$), decoupled vertical error dynamics

$$\begin{bmatrix} \delta\dot{h} \\ \delta\dot{v}_d \end{bmatrix} = \begin{bmatrix} 0 & -1 \\ \left(\frac{\hat{v}_n^2}{(R+h)^2} - \frac{2GM}{(R+h)^3}\right) & 0 \end{bmatrix} \begin{bmatrix} \delta h \\ \delta v_d \end{bmatrix}. \tag{3.104}$$

Assume that h small relative to R so that $(R+h)$ can be approximated by R. Assume that v_n is constant so that the stability of the decoupled vertical error dynamics can be checked by eigenvalue analysis. Show that for $v_n^2 < \frac{2GM}{R}$ the system is unstable. If v_n is small enough that the term containing v_n can be neglected, show that the unstable eigenvalue is at $\sqrt{\frac{2g_e}{R}}$ where $g_e = \frac{GM}{R^2}$ is the magnitude of the gravitational acceleration at the Earth surface. Modeling of the gravity error is discussed in greater depth in Section 11.1.4.

9. Assume that the altitude dynamics are stabilized by external means so that δh and δv_d are near zero. Assume also that $h \ll R$ and $v_d \ll R$ so that $(R+h) \approx R$ and $\frac{v_d}{R+h} \approx 0$, then the lateral error dynamics reduce to

$$\begin{bmatrix} \delta\dot{\phi} \\ \delta\dot{v}_n \\ \delta\dot{\theta} \end{bmatrix} = \begin{bmatrix} 0 & \frac{1}{R} & 0 \\ 0 & 0 & \vec{f}_d \\ 0 & \frac{1}{R} & 0 \end{bmatrix} \begin{bmatrix} \delta\phi \\ \delta v_n \\ \delta\theta \end{bmatrix} + \begin{bmatrix} 0 & 0 \\ 1 & 0 \\ 0 & 1 \end{bmatrix} \begin{bmatrix} \delta\varepsilon_a^g \\ \delta\varepsilon_g \end{bmatrix} \quad (3.105)$$

where $\delta\varepsilon_a^g = \begin{bmatrix} \cos(\theta) & \sin(\theta) & 1 \end{bmatrix} \mathbf{u}_2^p$. Show that for $\vec{f}_d = -g_e$ this system has eigenvalues at 0 and $\pm j\omega_s$ where $\omega_s = \sqrt{\frac{g_e}{R}}$ which is referred to as the Schuler frequency. The Schuler period is $T_s = \frac{2\pi}{\omega_s} \approx$ 84 minutes.

If a similarity transform is used to transform the first state from an angular error in radians to a position error (i.e., $\delta n = R\delta\phi$), then the transformed version of eqn. (3.105) is the state space model of eqn. (3.49).

This problem has derived the vertical and horizontal error dynamics under various assumptions. However, the conclusions generalize to more general circumstances (See Section 11.4.). In particular, in the vicinity of the ellipsoidal Earth with its associated gravitational field, the horizontal dynamics are neutrally stable with natural frequency near ω_s and the vertical dynamics are unstable.

Chapter 4

Stochastic Processes

Navigation systems combine uncertain information related to a vehicle to construct an accurate estimate of the vehicle state. The previous chapter discussed deterministic models of dynamic systems that are appropriate for modeling the vehicle dynamics or kinematics. In navigation applications, the information uncertainty does not typically arise from the vehicle dynamics; instead, it arises from noise and imperfections in the sensors. To optimally combine the information from sensed quantities, we will use the vehicle model, quantitative descriptions of uncertainty in the sensed quantities, and models of the relation between the sensed quantities and the vehicle states. Development of the required models and quantitative descriptions is the realm of stochastic process. The objective of this chapter is to introduce the relevant concepts from the field of stochastic processes. This field is quite large; therefore, the chapter must be selective in the information that it presents and will focus on the material required for the implementation of aided navigation systems. It is assumed that the reader already has some background knowledge of statistics and probability theory.

4.1 Basic Stochastic Process Concepts

Knowledge of stochastic processes is required for navigation system design and analysis, yet many students find the topic difficult. This is in part due to the lack of familiarity with the role and utility of stochastic processes in daily life. After introducing a few basic concepts, this section presents a few motivational examples that should be understandable to readers with an engineering background.

Random variables are functions that map random experimental outcomes to real numbers. The idea of a random variable can be more clearly motivated after a few key terms are defined. A random experiment involves a *procedure* and an *observation* of an experimental *outcome.* There is some degree of uncertainty in either the procedure or the observation, which necessitates the experiment. The uncertainty often arises from the extreme complexity of modeling all aspects of the experiments procedure and observation with sufficient detail to determine the output. Consider the example of a coin toss. The procedure involves balancing the coin on the thumbnail, flicking the thumb, allowing the coin to land and settle to a stationary position on a flat surface. The observation involves determining which surface of the coin is on top. Most people have performed this experiment numerous times. Almost nobody would seriously consider trying to determine the initial forces and torques applied by the thumb to the coin, the aerodynamics during flight, or the impact forces and torques upon landing that would be required to predict the result of any given toss. The complexity of such a deterministic prediction motivates the use of a (fair) coin toss as an unbiased tool in many decision making situations.

The utility of probability theory and stochastic processes is that it provides the analyst with the ability to make quantitative statements about events or processes that are too complicated to allow either exact replication of experiments or sufficiently detailed deterministic analysis. For example, even though an analyst might be incapable of predicting the outcome of a single coin toss, by analyzing a sufficiently large number of experiments, the analyst might be able to predict which surface of an unfair coin has a higher likelihood of landing upwards. To be useful, the theory for this analysis should allow quantitative statements concerning the number of observations required to achieve any desired level of confidence in such a prediction. In fact, such analysis is quite commonplace.

The efficacy of random analysis has motivated a mathematically sophisticated theory for probability and probabilistic reasoning. This theory is based on measure theory. Using probability theory, we analyze models of experiments instead of the experiments themselves. Therefore, the analyst must be careful to consider whether the probabilistic model is a sufficiently accurate model of the experiment.

4.1.1 Examples

The topic of stochastic processes is very important to the study of navigation system design as it is the foundation on which we build our ability to quantitatively analyze the propagation of uncertainty through the navigation system as a function of time. A thorough presentation of the related topics of statistics, probability theory, random variables, and stochastic processes is well beyond the scope of this text. Several excellent texts are mentioned at the conclusion of this chapter. Instead, this chapter focuses on topics selected for their relevance to the design and analysis of navigation systems. The following examples are intended to motivate the topics

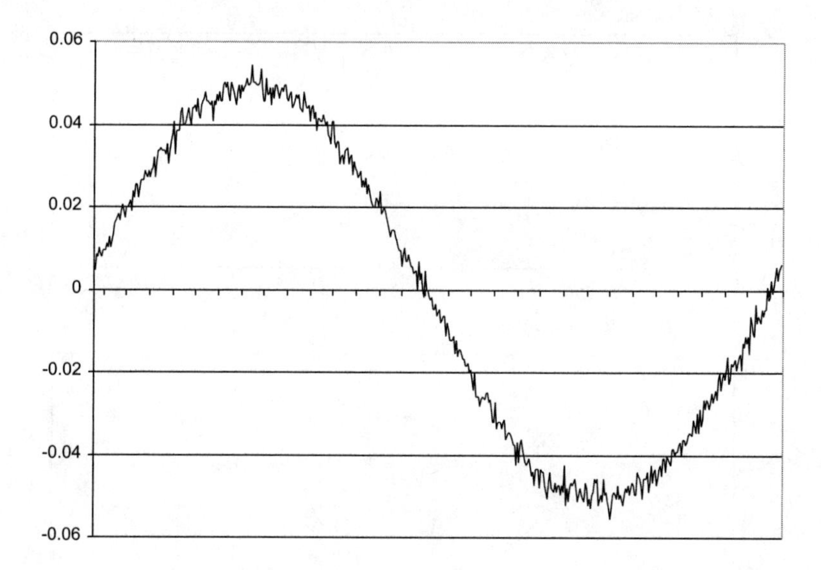

Figure 4.1: Oscilloscope image of a sinusoidal input.

discussed in the balance of this chapter.

Example 4.1 *Most engineers have used an oscilloscope at some point in their careers. An example image from an oscilloscope screen is shown in Figure 4.1. This image is created by directly connecting the output of a sinusoidal source to the oscilloscope input. The image clearly shows a noisy version of a sinusoidal signal. This figure is included to clearly exemplify the concept of measurement noise. The noise is the result of several complex physical processes, e.g., thermal noise in the probes and electronic parts. Quite often it is accurate to describe a measurement in the form*

$$y(t) = Sx(t) + \nu(t),$$

where y represents the measurement, x represents the actual value of some physical property, S is a scale factor, and ν represents measurement noise. Typically, the manufacturer of the instrument is careful to design the instrument such that the scale factor S is constant and accurately known and that the measurement noise ν is not related to the signal x. In addition, the manufacturer will often provide the purchaser with the quantitative description of ν suitable for further analysis. To be able to communicate with sensor manufacturers and other navigation engineers and to be able to perform quantitative analysis, the analyst must be familiar with stochastic process definitions and methods. △

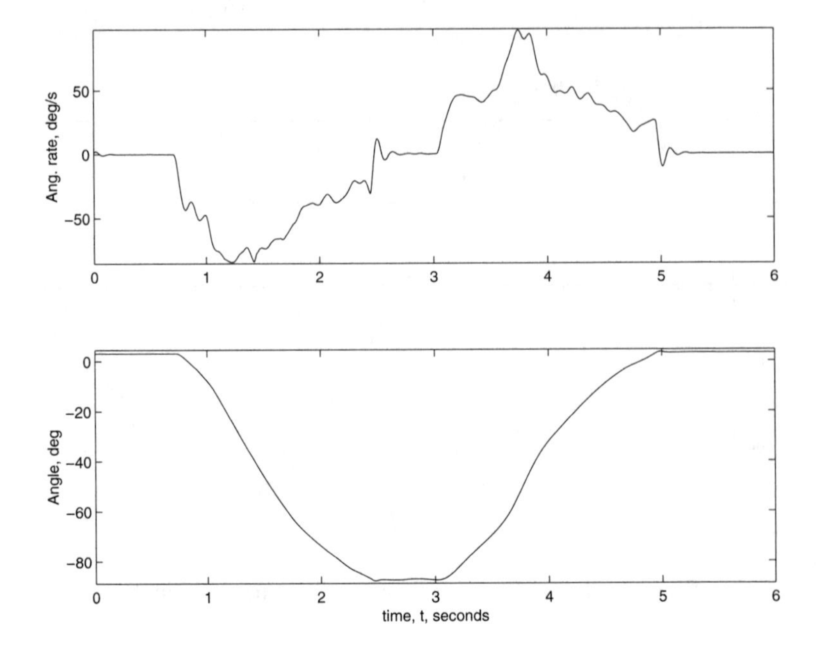

Figure 4.2: Angular rate (top) and integrated angle (bottom) for Example 4.2.

Example 4.2 *This example illustrates the types of processes and signals that occur in navigation systems and some possible questions that are of interest via stochastic analysis methods.*

An angular rate measurement $\tilde{r}(t)$ from a gyro is shown as the top graph of Figure 4.2. The gyro is mounted in such a way that it can only rotate about a single axis that corresponds to the gyro sensitive axis. The analyst would like to understand the nature of the error between the true angle and the angle estimate computed by integration of the gyro output.

To begin, we determine a model for the sensor signal. The model can either be determined by laboratory experiments performed by the analyst or, more typically, provided by the sensor manufacturer. The methodology for estimating the error model is discussed in Section 4.9.2. For this example, assume that the angular rate sensor is accurately modeled as

$$\tilde{r}(t) = r(t) + b_g(t) + \nu(t), \tag{4.1}$$

where r represents the real rotation rate around the sensitive axis of the gyro, b_g represents a very slowly changing component often referred to as a bias, and ν corresponds to measurement error that has a distinct value at each measurement instance. Although we would prefer to integrate $r(t)$ to compute θ, we only have available the signal $\tilde{r}$. Our integration of $\tilde{r}$ yields *the angle estimate $\hat{\theta}$. We will assume that $\hat{\theta}(0) = 0$. A model for the error between θ and $\hat{\theta}$ can be developed as*

$$\hat{\theta}(t) = \int_0^t \tilde{r}(\tau)d\tau \tag{4.2}$$

$$= \int_0^t \big(r(\tau) + b_g(\tau) + \nu(\tau)\big)d\tau \tag{4.3}$$

$$= \theta(t) + \int_0^t (b_g(\tau) + \nu(\tau))\, d\tau \tag{4.4}$$

$$\hat{\theta}(t) - \theta(t) = \int_0^t (b_g(\tau) + \nu(\tau))\, d\tau. \tag{4.5}$$

In this set of equations, eqn. (4.2) shows the actual computation of $\hat{\theta}$ as the integral of the sensor output $\tilde{r}$:

$$\dot{\hat{\theta}}(t) = \tilde{r}(t).$$

Eqns. (4.3–4.5) show analysis. The conclusion from eqn. (4.5) is that the error between the computed estimate $\hat{\theta}(t)$ and the actual angle $\theta(t)$ is the integral of the unknown quantity $b_g(\tau) + \nu(\tau)$ for $\tau \in [0, t]$:

$$\delta\dot{\theta}(t) = b_g(t) + \nu(t) \tag{4.6}$$

where $\delta\theta(t) = \hat{\theta}(t) - \theta(t)$.

For a well designed instrument, the bias and noise terms should be determined completely by events within the instrument (i.e., unaffected by external fields or by the motion of the device). In addition, although the values of b_g and ν may change with time in a nondeterministic manner, the characteristics of the change should be in a form that enables accurate quantitative characterization of the growth of the angle error. The desire to make accurate quantitative statements about characteristics of nondeterministic signals motivates the introduction to and study of stochastic processes.

Returning to Figure 4.2, the integrated gyro signal $\hat{\theta}(t) = \int_0^t \tilde{r}(\tau)d\tau$ is shown as the bottom graph of Figure 4.2. We have returned to this figure to state an important point. During this experiment, the gyro was rotated by -90 deg, *left stationary, rotated by* 90 deg, *and left stationary again. This motion is performed by a human. The erratic nature of the angular rate r is due to the human experimenter and is* not *the object of the stochastic analysis. The objects of the stochastic analysis are the sensor imperfections b_g and ν, and the error $\delta\theta$ that they cause. Although we do not know the true values of θ or $\tilde{\theta}$, based on stochastic analysis we will be able to quantitatively predict the expected magnitude of $\delta\theta$ as a function of time, see Exercises 4.19. This example is further discussed in Exercise 5.7.* △

4.1.2 Plan of Study

The previous discussion and examples have been provided to motivate the material presented in this chapter.

The primary objective for this chapter is to present methods to model and analyze the behavior of the navigation error state as a function of time. The state space error models of such stochastic processes will be generalized versions of eqn. (4.6):

$$\dot{\mathbf{x}}(t) = \mathbf{F}(t)\mathbf{x}(t) + \mathbf{G}(t)\mathbf{w}(t) \qquad \mathbf{y}(t) = \mathbf{H}(t)\mathbf{x}(t) + \mathbf{v}(t) \tag{4.7}$$

where the vector $\mathbf{x}$ represents the error in the navigation state. The vectors $\mathbf{w}$ and $\mathbf{v}$ are random variables determined in part by the navigation instruments. The matrices $\mathbf{F}(t)$ and $\mathbf{H}(t)$ are determined by the design of the navigation system. Such models are discussed in considerable detail in Sections 4.6–4.10. To understand and analyze such models requires knowledge of several additional concepts. Therefore, the chapter first defines several concepts and quantities that have proven to be useful for analysis of stochastic processes. The main topics to be discussed include: definitions of important statistics, definitions of theoretical concepts (e.g., independence, random variables, stationarity), definition of white and colored noise processes, and properties of linear systems with random inputs. Readers already familiar with these topics should skip directly to Section 4.6, while only using Sections 4.2–4.5 for reference.

4.2 Scalar Random Variables

This section presents various fundamental concepts related to probability and random variables that will be required to support subsequent discussions. The discussion will focus on continuous random variables.

4.2.1 Basic Properties

Let w represent a real, scalar random variable. In the following, the notation

$$P\{w < W\}$$

denotes the probability that the random variable w is less than the real number W.

The *distribution function* of the random variable w is defined as the function

$$F_w(W) = P\{w \leq W\}. \tag{4.8}$$

The function $F_w(W)$ is monotonically increasing in W, is continuous from the right with $F_w(-\infty) = 0$ and $F_w(\infty) = 1$. The derivative of the distribution function is the *density function* of the random variable w and will be denoted by

$$p_w(W) = \frac{dF_w}{dW}(W).$$

The density function has the interpretation that $p_w(W)$ represents the probability that the random variable w assumes a value in the differential range $(W, W+dW]$. Because the density is the derivative of a monotone function, $p_w(W) \geq 0$ for any W. Based on the above definitions, for a scalar random variable,

$$P\{W_1 < w \leq W_2\} = F_w(W_2) - F_w(W_1) = \int_{W_1}^{W_2} p(W)dW.$$

Let $g(w)$ be a function of the random variable w. The *expected value* of $g(w)$ is defined by

$$E\langle g(w)\rangle = \int_{-\infty}^{\infty} g(W)p_w(W)dW. \tag{4.9}$$

In particular, the expected value or *mean* of w is

$$\mu_w = E\langle w\rangle = \int_{-\infty}^{\infty} Wp_w(W)dW. \tag{4.10}$$

As a simplified notation, this book will typically use the notation μ_w to indicate the expected value of the random variable w. In cases where the meaning is clear or where double subscripting would otherwise occur, the subscript indicating the random variable may be dropped.

The *variance* of a scalar valued random variable w is defined by

$$var(w) = E\left\langle (w-\mu_w)^2\right\rangle = \int_{-\infty}^{\infty} (W-\mu_w)^2 p(W)dW. \tag{4.11}$$

The variance of a random variable quantifies the variation of the random variable relative to its expected value. If we define $\sigma_w^2 = var(w)$, then σ_w is called the *standard deviation* of w.

The k-th moment of w is defined by

$$E\left\langle w^k\right\rangle = \int_{-\infty}^{\infty} W^k p(W)dW. \tag{4.12}$$

The first moment is the mean value. The variance and second moment are related by eqn. (4.13):

$$\begin{aligned} var(w) &= E\left\langle w^2 - 2w\mu_w + \mu_w^2\right\rangle \\ \sigma_w^2 &= E\left\langle w^2\right\rangle - \mu_w^2. \end{aligned} \tag{4.13}$$

Example 4.3 *Compute the mean, second moment, variance, and standard deviation of the random variable x with the exponential density*

$$p_x(X) = \begin{cases} \alpha \exp(-\alpha X) & \textit{for } X > 0 \\ 0 & \textit{otherwise} \end{cases}$$

where $\alpha > 0$.

The mean is

$$\mu_x = \alpha \int_0^\infty v \exp(-\alpha v) dv = \frac{1}{\alpha}.$$

The second moment is

$$E\left\langle x^2 \right\rangle = \alpha \int_0^\infty v^2 \exp(-\alpha v) dv = \frac{2}{\alpha^2}.$$

The variance is $\sigma_x^2 = \alpha \int_0^\infty (v - \mu_x)^2 \exp(-\alpha v) dv$, but can also be easily calculate from the information above using the eqn. (4.13) as

$$var(x) = \frac{2}{\alpha^2} - \frac{1}{\alpha^2} = \frac{1}{\alpha^2}.$$

Therefore, the standard deviation is $\sigma_x = \frac{1}{\alpha}$. △

4.2.2 Gaussian Distributions

The notation $x \sim N(\mu, \sigma^2)$ is used to indicate that the scalar (or univariate) random variable x has the Gaussian or Normal density function described by

$$p_x(X) = \frac{1}{\sqrt{2\pi}\sigma} \exp\left(-\frac{(X - \mu)^2}{2\sigma^2} \right); \tag{4.14}$$

therefore, the density of a Normal random variable is completely described by two parameters μ and σ. It can be shown that

$$\int_{-\infty}^{\infty} Xp(X)dX = \mu \quad \text{and}$$
$$\int_{-\infty}^{\infty} (X - \mu)^2 p(X)dX = \sigma^2;$$

therefore, the parameters μ and σ^2 are the expected value and variance of the random variable x.

4.2.3 Transformations of Scalar Random Variables

This section presents the method to find the density for the random variable v under the conditions that $v = g(w)$, the function g is invertible, and the density for the random variable w is known.

Due to the assumption that g is monotone, it has an inverse function $f = g^{-1}$. The derivation begins with the basic idea that the probability of the events causing $v \in [V, V + dV]$ and $w \in [W, W + dW]$ for $V = g(W)$ and $V + dV = g(W + dW)$ must be the same whether measured with respect to v or w. Mathematically this is expressed as

$$P\{w \in [W, W + dW]\} = P\{v \in [V, V + dV]\} \tag{4.15}$$

where this formulation assumes that dW is positive. Therefore, $dW = |dW|$ which will be used below. For positive, dW, the quantity dV can still be either positive or negative, depending on the sign of $\frac{dV}{dW} = g'$. The computation of the probabilities in eqn. (4.15) must account for either possible sign:

$$\int_W^{W+dW} p_w(x)dx = \begin{cases} \int_V^{V+dV} p_v(x)dx & \text{if } dV > 0 \\ -\int_V^{V+dV} p_v(x)dx & \text{if } dV < 0. \end{cases}$$

For the second integral, due to dV being negative, the original integral would have been over $[V + dV, V]$ which is equivalent to the integral in the expression above after interchanging the limits of integration and multiplying the integral by -1. The above integral relationships are equivalent to

$$p_w(W)dW = p_v(V)|dV|$$

which can be rewritten as

$$\begin{aligned} p_v(V) &= p_w(f(V)) \left| \frac{dW}{dV} \right| \\ p_v(V) &= p_w(f(V)) \left| f'(V) \right|. \end{aligned} \tag{4.16}$$

Issues related to the transformation of random variables are clarified by the following examples.

Example 4.4 *Find the density for y where $y = ax + b$, $a \neq 0$, and $x \sim N(0, \sigma^2)$.*

In this case, $f(y) = \frac{y-b}{a}$ and $f'(y) = \frac{1}{a}$. Therefore,

$$\begin{aligned} p_y(Y) &= \frac{1}{|a|} p_x \left(\frac{Y - b}{a} \right) \\ &= \frac{1}{\sqrt{2\pi}|a|\sigma} \exp\left(-\frac{(Y - b)^2}{2(a\sigma)^2} \right). \end{aligned}$$

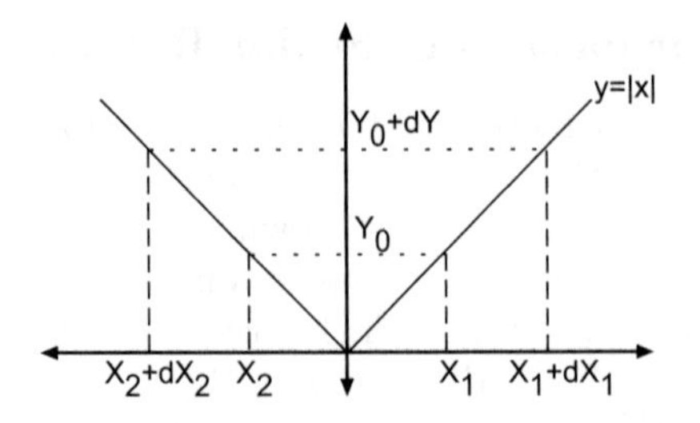

Figure 4.3: Graph of $y = |x|$ with variables defined for Example 4.5.

Note that $P_y(Y)$ has the form of a Gaussian distribution with mean b and variance $(a\sigma)^2$; therefore, y is a Gaussian random variable. In general, all linear operations on Gaussian random variables result in Gaussian random variables. △

Example 4.5 *Find the density for y where $y = |x|$ and $x \sim N(0, \sigma^2)$.*

Because the function g is not invertible, the result of eqn. (4.16) cannot be used. Instead, we have to revert to the main idea of eqn. (4.15) which is that the change of variable must preserve the probability of events. Consider the variables defined in Figure 4.3. For this problem, we have

$$P\{Y_0 \le y \le Y_0 + dY\} = P\{X_2 + dX_2 \le x \le X_2\} + P\{X_1 \le x \le X_1 + dX_1\}$$

where dY represents a positive differential change from Y_0 and

$$\begin{aligned} X_1 &= \ \ Y_0 & \qquad X_1 + dX_1 &= \ \ Y_0 + dY \\ X_2 &= -Y_0 & \qquad X_2 + dX_2 &= -Y_0 - dY. \end{aligned}$$

The above expression translates into the following integral

$$\int_{Y_0}^{Y_0+dY} p_y(v)dv = \int_{X_2+dX_2}^{X_2} p_x(w_2)dw_2 + \int_{X_1}^{X_1+dX_1} p_x(w_1)dw_1. \tag{4.17}$$

Using the changes of variables $y_1 = w_1$ and $y_2 = -w_2$, this integral simplifies as follows

$$\begin{aligned} \int_{Y_0}^{Y_0+dY} p_y(v)dv &= -\int_{Y_0+dY}^{Y_0} p_x(-y_2)dy_2 + \int_{Y_0}^{Y_0+dY} p_x(y_1)dy_1 \\ &= \int_{Y_0}^{Y_0+dY} p_x(-y_2)dy_2 + \int_{Y_0}^{Y_0+dY} p_x(y_1)dy_1 \\ \int_{Y_0}^{Y_0+dY} p_y(v)dv &= \int_{Y_0}^{Y_0+dY} \left(p_x(-v) + p_x(v)\right)dv. \end{aligned}$$

This integral relationship must hold for all nonnegative Y_0 and arbitrarily small dY. In addition, for $x \sim N(0, \sigma^2)$ we have $p_x(v) = p_x(-v)$; therefore,

$$p_y(Y) = \begin{cases} \frac{1}{\sigma}\sqrt{\frac{2}{\pi}} \exp\left(-\frac{Y^2}{2\sigma^2}\right) & \text{for } y \geq 0 \\ 0 & \text{for } y < 0. \end{cases}$$

This density does not have the form of eqn. (4.14); therefore, y is not a Gaussian random variable. In general, nonlinear transformations of Gaussian random variables do not result in Gaussian random variables. △

4.3 Multiple Random Variables

Navigation systems frequently involve multiple random variables. For example, a vector of simultaneous measurements $\tilde{\mathbf{y}}$ might be modeled as

$$\tilde{\mathbf{y}} = \mathbf{y} + \boldsymbol{\nu}$$

where $\mathbf{y}$ represents the signal portion of the measurement and $\boldsymbol{\nu}$ represents a vector of random measurement errors.

In the case of multiple random variables, we are often concerned with how the values of the elements of the random vector relate to each other. When the values are related, then one of the random variables may be useful for estimating the value of the other. Important questions include: how to quantify the interrelation between random variables, how to optimally estimate the value of one random variable when the value of the other is known, and how to quantify the accuracy of the estimate. In this type of analysis, the multivariate density and distribution and the second order statistics referred to as correlation and covariance are important. The discussion of basic properties will focus on two random variables, but the concepts extend directly to higher dimensional vectors.

4.3.1 Basic Properties

Let v and w be random variables. The *joint probability distribution* function of v and w is

$$F_{vw}(V, W) = P\{v \leq V \text{ and } w \leq W\}.$$

The joint distribution has the following properties:

$$\begin{aligned} F_{vw}(V, W) &\in [0, 1] \\ F_{vw}(V, -\infty) &= F_{vw}(-\infty, W) = 0 \\ F_{vw}(\infty, \infty) &= 1 \\ F_{vw}(V, \infty) &= F_v(V) \end{aligned}$$

$$\begin{aligned} F_{vw}(\infty, W) &= F_w(W) \\ F_{vw}(V_1, W) &\le F_{vw}(V_2, W) \text{ for } V_1 < V_2 \\ F_{vw}(V, W_1) &\le F_{vw}(V, W_2) \text{ for } W_1 < W_2. \end{aligned}$$

The last two properties state the the joint distribution is nondecreasing in both arguments.

The *joint probability density* is defined as

$$p_{vw}(V, W) = \frac{\partial^2 F_{vw}}{\partial V \partial W}(V, W). \tag{4.18}$$

The joint density has the following properties:

$$\begin{aligned} p_{vw}(V, W) &\ge 0 \\ F_{vw}(V, W) &= \int_{-\infty}^{V} \int_{-\infty}^{W} p_{vw}(x, y)\; dy\; dx \\ p_v(V) &= \int_{-\infty}^{\infty} p_{vw}(V, y)\; dy \\ p_w(W) &= \int_{-\infty}^{\infty} p_{vw}(x, W)\; dx \\ P\left\{(v, w) \in \mathcal{D} \subset \mathbb{R}^2\right\} &= \iint_{\mathcal{D}} p_{vw}(x, y)\; dx\; dy. \end{aligned}$$

In the above, $p_v(V)$ and $p_w(W)$ are referred to as the marginal density of v and w, respectively. When the meaning is clear from the context, the subscripts may be dropped on either the distribution or the density.

Example 4.6 *Let u and v be random variables with the joint probability density*

$$p_{uv}(U, V) = \begin{cases} \frac{1}{T^2} & (U, V) \in [0, T] \times [0, T] \\ 0 & \textit{otherwise.} \end{cases}$$

Using the properties above, it is straightforward to show that

$$\begin{aligned} p_u(U) &= \begin{cases} \frac{1}{T} & (U) \in [0, T] \\ 0 & \textit{otherwise,} \end{cases} \\ p_v(V) &= \begin{cases} \frac{1}{T} & (V) \in [0, T] \\ 0 & \textit{otherwise,} \end{cases} \\ p_{uv}(U, V) &= p_u(U) p_v(V), \\ F_{uv}(U, V) &= F_u(U) F_v(V). \end{aligned}$$

△

4.3.2 Statistics and Statistical Properties

Two random vector variables $\mathbf{v}$ and $\mathbf{w}$ are *independent* if

$$p_{\mathbf{vw}}(\mathbf{V}, \mathbf{W}) = p_{\mathbf{v}}(\mathbf{V}) p_{\mathbf{w}}(\mathbf{W}). \tag{4.19}$$

When two random variables are independent and have the same marginal densities, they are *independent and identically distributed* (i.i.d.).

Example 4.7 *The random variables u and v in Example 4.6 are independent.* △

There is a result called the *central limit theorem* [31, 107] that states that if

$$\bar{x}_N = A \sum_{i=1}^{N} x_i \tag{4.20}$$

where the x_i are independent random variables, then as N increases the distribution for $\bar{x}_N$ approaches a Gaussian distribution, independent of the distributions for the individual x_i. This rather remarkable result motivates the the importance of Gaussian random variables in applications. Whenever a random effect is the superposition of many small random affects, the superimposed effect can be accurately modeled as a Gaussian random variable.

When two random variables are not independent, it is useful to have metrics to quantify the amount of interdependence. Two important metrics are correlation and covariance.

The *correlation matrix* between two random variables $\mathbf{v}$ and $\mathbf{w}$ is defined by

$$cor(\mathbf{v}, \mathbf{w}) = E\left\langle \mathbf{v}\mathbf{w}^\top \right\rangle = \iint \mathbf{V}\mathbf{W}^\top p(\mathbf{V}, \mathbf{W})\, d\mathbf{V}\, d\mathbf{W}. \tag{4.21}$$

The *covariance matrix* for two vector valued random variables $\mathbf{v}$ and $\mathbf{w}$ is defined by

$$\begin{aligned} cov(\mathbf{v}, \mathbf{w}) &= E\left\langle (\mathbf{v} - \boldsymbol{\mu}_\mathbf{v})(\mathbf{w} - \boldsymbol{\mu}_\mathbf{w})^\top \right\rangle \\ &= \iint (\mathbf{V} - \boldsymbol{\mu}_\mathbf{v})(\mathbf{W} - \boldsymbol{\mu}_\mathbf{w})^\top p(\mathbf{V}, \mathbf{W})\, d\mathbf{V}\, d\mathbf{W}. \end{aligned} \tag{4.22}$$

The correlation coefficient ρ_{vw} is a normalized measure of the correlation between the two scalar random variables v and w that is defined as

$$\rho_{vw} = \frac{cov(v, w)}{\sigma_v \sigma_w}. \tag{4.23}$$

The correlation coefficient always satisfies $-1 \leq \rho_{vw} \leq 1$. When the magnitude of ρ_{vw} is near one, then knowledge of one of the random variables will allow accurate prediction of the other. When $\rho_{vw} = 0$, then v and w are said to be *uncorrelated.* Uncorrelated random variables have $cov(v, w) = 0$ and $cor(v, w) = \mu_v \mu_w$. Independent random variables are uncorrelated, but uncorrelated random variables may or may not be independent. Two vector random variables $\mathbf{v}$ and $\mathbf{w}$ are uncorrelated if $cov(\mathbf{v}, \mathbf{w}) = 0$. Two vector random variables $\mathbf{v}$ and $\mathbf{w}$ are orthogonal if $cor(\mathbf{v}, \mathbf{w}) = 0$.

Example 4.8 *Consider the random variables*

$$\begin{aligned} u &= \cos(\theta) \\ v &= \sin(\theta) \end{aligned}$$

where $\theta \in [-\pi, \pi]$ is a uniform random variable. It is left to the reader to show that

$$E\langle u \rangle = E\langle v \rangle = E\langle uv \rangle = 0.$$

These facts show that u and v are orthogonal and uncorrelated. However, it is also straightforward to show that $u^2 + v^2 = 1$. The fact that the variables are algebraically related shows that u and v are not independent. $\triangle$

Let the matrix $\mathbf{P} = cov(\mathbf{x}, \mathbf{x})$ be the covariance matrix for the vector $\mathbf{x}$. Then by eqn. (4.23), the element in the i-th row and j-th column of $\mathbf{P}$ is

$$\mathbf{P}_{ij} = cov(x_i, x_j) = \begin{cases} \rho_{x_i x_j} \sigma_{x_i} \sigma_{x_j}, & i \neq j \\ \sigma_{x_i}^2, & i = j. \end{cases} \tag{4.24}$$

Therefore, knowledge of the covariance matrix for a random vector allows computation of the variance of each component of the vector and of the correlation coefficients between elements of the vector. This fact is very useful in state estimation applications.

Example 4.9 *An analyst is able to acquire a measurement y that is modeled as*

$$y = a + b + n. \tag{4.25}$$

The measurement is a function of two unknowns a and b and is corrupted by additive measurement noise $n \sim N(0, \sigma_n^2)$. The analyst has no knowledge of the value of a, but based on prior experience the analyst considers a reasonable model for b to be $b \sim N(0, \sigma_b^2)$. The random variables b and n are assumed to be independent. Both σ_b and σ_n are positive.

Based on this model and prior experience, the analyst chooses to estimate the values of a and b as

$$\hat{a} = y \tag{4.26}$$

$$\hat{b} = 0. \tag{4.27}$$

The parameter estimation errors are $\tilde{a} = a - \hat{a}$ and $\tilde{b} = b - \hat{b}$. What are the mean parameter estimation errors, the variance of the parameter estimation errors, and the covariance between the parameter estimation errors?

1. *The mean value of $\tilde{a}$ is $E\langle\tilde{a}\rangle = E\langle -b - n\rangle = 0$.*

2. *The mean value of $\tilde{b}$ is $E\langle\tilde{b}\rangle = E\langle b\rangle - 0 = 0$.*

3. *The covariance of $\tilde{a}$ is*

$$\begin{aligned} E\langle\tilde{a}^2\rangle &= E\langle(b+n)^2\rangle \\ &= E\langle b^2 + 2bn + n^2\rangle = \sigma_b^2 + \sigma_n^2. \end{aligned}$$

4. *The covariance of $\tilde{b}$ is*

$$\begin{aligned} E\langle\tilde{b}^2\rangle &= E\langle(b - \hat{b})^2\rangle \\ &= E\langle b^2\rangle = \sigma_b^2. \end{aligned}$$

5. *The covariance between $\tilde{a}$ and $\tilde{b}$ is*

$$\begin{aligned} E\langle\tilde{a}\tilde{b}\rangle &= E\langle(a - \hat{a})(b - \hat{b})\rangle \\ &= E\langle -(b+n)(b - \hat{b})\rangle \\ &= E\langle -b^2 - nb\rangle \\ &= -\sigma_b^2. \end{aligned}$$

Based on the results of Steps 3—5 and eqn. (4.23), the correlation coefficient between $\tilde{a}$ and $\tilde{b}$ is $\rho_{\tilde{a}\tilde{b}} = \frac{-\sigma_b}{\sqrt{\sigma_b^2+\sigma_n^2}}$.

If we define $\mathbf{P} = cov([\tilde{a}, \tilde{b}]^\top)$, then based on the above analysis

$$\mathbf{P} = \begin{bmatrix} \sigma_b^2 + \sigma_n^2 & -\sigma_b^2 \\ -\sigma_b^2 & \sigma_b^2 \end{bmatrix}$$

which checks with eqn. (4.24). △

Analysis similar to that of Example 4.9 will have utility in Part II for the initialization of navigation systems. Related to this example, the case where **a** is a vector and the case where **y** is a nonlinear function of **a** are considered in Exercise 4.11.

4.3.3 Vector Gaussian Random Variables

For the vector random variable $\mathbf{x} \in \mathbb{R}^n$, the notation $\mathbf{x} \sim N(\boldsymbol{\mu}, \mathbf{P})$ is used to indicate that $\mathbf{x}$ has the multivariate Gaussian or Normal density function described by

$$p_{\mathbf{x}}(\mathbf{X}) = \frac{1}{\sqrt{(2\pi)^n \det(\mathbf{P})}} \exp\left(-\frac{1}{2}(\mathbf{X}-\boldsymbol{\mu})^\top \mathbf{P}^{-1}(\mathbf{X}-\boldsymbol{\mu})\right).$$

Again, the density of the vector Normal random variable is completely described by its expected value $\boldsymbol{\mu} \in \mathbb{R}^n$ and its covariance matrix $\mathbf{P} \in \mathbb{R}^{n\times n}$.

4.3.4 Transformations of Vector Random Variables

If $\mathbf{v}$ and $\mathbf{w}$ are vector random variables in $\mathbb{R}^n$ related by $\mathbf{v} = \mathbf{g}(\mathbf{w})$ where $\mathbf{g}$ is invertible and differentiable with unique inverse $\mathbf{f} = \mathbf{g}^{-1}$, then the formula of eqn. (4.16) extends to

$$p_{\mathbf{v}}(\mathbf{V}) = \frac{1}{|\det(\mathbf{G})|} p_{\mathbf{w}}(\mathbf{f}(\mathbf{V})) \tag{4.28}$$

where $\mathbf{G} = \left.\frac{\partial \mathbf{g}}{\partial \mathbf{W}}\right|_{\mathbf{W}=\mathbf{f}(\mathbf{V})}$ is the Jacobian matrix of $\mathbf{g}$ evaluated at $\mathbf{W} = \mathbf{g}^{-1}(\mathbf{V})$.

Example 4.10 *Find the density of the random variable* $\mathbf{y}$ *where* $\mathbf{y} = \mathbf{A}\mathbf{x}$, $\mathbf{x} \sim N(\boldsymbol{\mu}, \mathbf{P})$, *and* $\mathbf{A}$ *is a nonsingular matrix. Examples such as this are important in state estimation and navigation applications.*

By eqn. (4.28),

$$p_{\mathbf{y}}(\mathbf{Y}) = \frac{1}{|\det(\mathbf{A})|} p_{\mathbf{x}}(\mathbf{A}^{-1}\mathbf{Y})$$

which is equivalent to

$$\frac{1}{\sqrt{(2\pi)^n \det\left(\mathbf{A}\mathbf{P}\mathbf{A}^\top\right)}} \exp\left(-\frac{1}{2}(\mathbf{Y}-\mathbf{A}\boldsymbol{\mu})^\top \left(\mathbf{A}\mathbf{P}\mathbf{A}^\top\right)^{-1} (\mathbf{Y}-\mathbf{A}\boldsymbol{\mu})\right).$$

This shows that $y \sim N\left(\mathbf{A}\boldsymbol{\mu}, \mathbf{A}\mathbf{P}\mathbf{A}^\top\right)$.

This result is considered further in Exercise 4.10. △

As demonstrated in Examples 4.4 and 4.10, affine operations on Gaussian random variables yield Gaussian random variables. Nonlinear functions of Gaussian random variables do not yield Gaussian random variables.

4.4 Stochastic Processes

A stochastic process is a family of random variables indexed by a parameter $t \in \mathbb{R}$ for continuous-time stochastic processes or $k \in Z^+$ for discrete-time stochastic sequences.

For a deterministic signal such as $x(t) = t^2$ the value of x is determined by the value of t. Once t is known, the value of $x(t)$ is known. There is no element of chance. Alternatively, signals such as

- $v(t) = At^2$ where $A \sim N(0, \sigma^2)$ or
- $v(t) = t^2 + w(t)$ where $w(t)$ has density $p_w(W, t)$,

are stochastic processes. In fact, in the above example, $w(t)$ is also a stochastic process. For a stochastic process $v(t)$, the parameter t determines the distribution and statistics of the random variable v, but t does not determine the value of v; instead, given two times t_1 and t_2 we have two distinct random variables $v(t_1)$ and $v(t_2)$. The distributions for $v(t_1)$ and $v(t_2)$ may be the same or distinct, depending on whether or not the stochastic process is stationary.

Example 4.11 *Let $v(t) = t^2 + w(t)$ where $w(t) \sim N(0, \sigma^2)$, the density for v is*

$$p_v(V, t) = \frac{1}{\sqrt{2\pi}\sigma} \exp\left(-\frac{\left(V - t^2\right)^2}{2\sigma^2} \right).$$

At each time, $w(t)$ is a new Gaussian random variable with zero mean and variance σ^2. At each time, $v(t)$ is a new Gaussian random variable with mean t^2 and variance σ^2. △

4.4.1 Statistics and Statistical Properties

The statistical quantities defined previously for random variables become slightly more complicated when applied to stochastic processes, as they may depend on the parameter t or k. For continuous-time processes, the definitions are listed below. Note that these definitions are equally valid whether the parameter t represents a continuous or a discrete-time variable.

The *cross correlation function* between two random processes is defined by

$$\mathbf{R}_{\mathbf{vw}}(t_1, t_2) = cor\left(\mathbf{v}(t_1), \mathbf{w}(t_2)\right) = E\left\langle \mathbf{v}(t_1)\mathbf{w}(t_2)^\top \right\rangle. \tag{4.29}$$

The *autocorrelation function* of a random processes is defined by

$$\mathbf{R}_{\mathbf{v}}(t_1, t_2) = cor\left(\mathbf{v}(t_1), \mathbf{v}(t_2)\right) = E\left\langle \mathbf{v}(t_1)\mathbf{v}(t_2)^\top \right\rangle. \tag{4.30}$$

The autocorrelation function quantifies the similarity of the random process to itself at two different times.

The *cross covariance function* between two random processes is defined by

$$cov\left(\mathbf{v}(t_1), \mathbf{w}(t_2)\right) = E\left\langle \mathbf{v}(t_1)\mathbf{w}(t_2)^\top \right\rangle - E\langle \mathbf{v}(t_1)\rangle E\langle \mathbf{w}(t_2)^\top \rangle. \tag{4.31}$$

The *autocovariance function* of a random processes is defined by

$$cov\left(\mathbf{v}(t_1), \mathbf{v}(t_2)\right) = E\left\langle \mathbf{v}(t_1)\mathbf{v}(t_2)^\top \right\rangle - E\langle \mathbf{v}(t_1)\rangle E\langle \mathbf{v}(t_2)^\top \rangle. \tag{4.32}$$

The above definitions will be critically important in the analysis of subsequent chapters. For example, the Kalman filter will propagate certain error covariance matrices through time as a means to quantify the relative accuracy of various pieces of information that are to be combined.

There are two forms of stationary random processes that will be used in this text.

Definition 4.1 *The random process* $\mathbf{w}(t)$ *is* stationary *if its distribution is independent of time (i.e.,* $p_{\mathbf{w}}(\mathbf{W}, t) = p_{\mathbf{w}}(\mathbf{W})$*).*

If a random process is stationary, then its expected value will be time invariant and both its autocorrelation function and its autocovariance function will only depend on the time difference $\tau = t_1 - t_2$ (i.e., $\mathbf{R}_{\mathbf{v}}(t_1, t_2) = \mathbf{R}_{\mathbf{v}}(\tau)$).

Often this strict sense of stationarity is too restrictive. A relaxed sense of stationarity is as follows.

Definition 4.2 *The random process* $\mathbf{w}(t)$ *is* wide sense stationary *(WSS) if the mean and variance of the process are independent of time.*

Since a Gaussian random process is completely defined by its first two moments, a Gaussian process that is wide sense stationary is also stationary. A wide sense stationary process must have a constant mean, and its correlation and covariance can only depend on the time difference between the occurrence of the two random variables, i.e.

$$cor\left(\mathbf{w}(t_1), \mathbf{w}(t_2)\right) = \mathbf{R}_{\mathbf{ww}}(\tau)$$

where $\tau = t_1 - t_2$.

The *Power Spectral Density* (PSD) of a WSS random process $\mathbf{w}(t)$ is the Fourier transform of the autocorrelation function:

$$\mathbf{S}_{\mathbf{w}}(j\omega) = \int_{-\infty}^{\infty} \mathbf{R}_{\mathbf{w}}(\tau) e^{-j\omega\tau} d\tau. \tag{4.33}$$

For real random variables, the power spectrum can be shown to be real and even.

If the PSD of a WSS random signal is known, then the correlation function can be calculated as the inverse Fourier transform of the PSD,

$$\mathbf{R_w}(\tau) = \frac{1}{2\pi}\int_{-\infty}^{\infty} \mathbf{S_w}(j\omega)e^{j\omega\tau}d\omega. \tag{4.34}$$

In particular, the average power of $\mathbf{w}(t)$ can be calculated as

$$E\langle \mathbf{w}(t)\mathbf{w}(t)^\top \rangle = cor(\mathbf{w}(t), \mathbf{w}(t)) = \mathbf{R_w}(0) = \frac{1}{2\pi}\int_{-\infty}^{\infty} \mathbf{S_w}(j\omega)d\omega. \tag{4.35}$$

Ergodicity is concerned with the question of whether, for a given random process, the temporal averages and ensemble averages yield the same results. For example, given a stochastic process $x(t)$, under what conditions is it reasonable to assume that $\frac{1}{T}\int_0^T x(t)dt$ converges to $E\langle x(t)\rangle$ as T approaches infinity. This may seem to be an esoteric question, but is quite important in applications. In application conditions, it is often quite easy to obtain a time sequence, but may be quite expensive to obtain an ensemble of time sequences. Ergodicity tries to answer the question of whether it is fair to describe the statistics of the ensemble of a stochastic process based on the corresponding temporal statistics computed from the data sequence that is available. Following are two examples of stochastic processes that are not ergodic.

Example 4.12 *Let $x(t) = t + v(t)$ where $v(t) \sim N(0, \sigma^2)$. The ensemble mean is $\mu_x(t) = E\langle x(t)\rangle = t$; however, the limit of the temporal average $\lim_{T\to\infty} \frac{1}{T}\int_0^T x(t)dt$ is not well-defined.* △

Example 4.13 *Let $x(t) = b + v(t)$ where $b \sim N(0, \sigma_b^2)$ and $v(t) \sim N(0, \sigma_v^2)$. Note that b is a random variable that is constant with respect to time. The ensemble mean is $\mu_x(t) = 0$; however, the limit of the temporal average is $\lim_{T\to\infty} \frac{1}{T}\int_0^T x(t)dt = b$.* △

4.4.2 White and Colored Noise

One stochastic process that is particularly useful for modeling purposes is white noise. The adjective 'white' indicates that this particular type of noise has constant power at all frequencies. Any process that does not have equal power per frequency interval will be referred to as *colored*. Because power distribution per frequency interval is the distinguishing factor, for both continuous-time and discrete-time white noise, our discussion will start from the PSD.

4.4.2.1 Continuous-time White Noise

A scalar continuous-time random process $v(t)$ is referred to as a *white noise* process if its PSD is constant:

$$\mathrm{S}_v(j\omega) = \sigma_v^2.$$

Then by eqn. (4.34), the autocorrelation function is

$$\mathrm{R}_v(t) = \sigma_v^2 \delta(\tau) \tag{4.36}$$

where $\delta(\tau)$ is the Dirac delta function: $\delta(\tau) = lim_{\epsilon \to 0}\, \delta_\epsilon(\tau)$ and

$$\delta_\epsilon(\tau) = \begin{cases} \frac{1}{2\epsilon} & \text{if } |\tau| < \epsilon \\ 0 & \text{otherwise} \end{cases}$$

which has units of $\frac{1}{\text{sec}}$. An implication of eqn. (4.36) is that $E\langle v(t)\rangle = 0$. Nevertheless, problem assumptions will often state the fact that $v(t)$ is a zero mean, continuous-time, white noise process with PSD σ_v^2.

The fact that the spectral density is a constant function highlights the fact that continuous-time white noise processes are not realizable. All continuous-time white noise processes have infinite power, as demonstrated by eqn. (4.35):

$$\begin{aligned} \int_{-\infty}^{\infty} v^2(t)dt = R_v(0) &= \frac{1}{2\pi}\int_{-\infty}^{\infty} S_v(j\omega)d\omega \\ &= \frac{1}{2\pi}\int_{-\infty}^{\infty} \sigma_v^2 d\omega = \infty. \end{aligned}$$

Nonetheless, the white noise model is convenient for situations where the noise spectral density is constant over a frequency range significantly larger that the bandwidth of interest in a particular application.

A white noise process can have any probability distribution; however, the Gaussian distribution is often assumed. A key reason is the central limit theorem. If the analyst carefully constructs a model accounting for all known or predictable effects, then the remaining stochastic signals can often be accurately modeled as the output of a linear system that is driven by a Gaussian white noise input. This technique will be discussed in Section 4.5. In this book, the notation $\omega \sim N(0, \sigma^2)$ will sometimes be used to describe a continuous-time, Gaussion, white-noise process ω. In this case, σ^2 is interpreted as the PSD, not the variance.

Reasoning through the units of the symbol σ_v^2 often causes confusion. This can be considered from two perspectives. First, from the PSD perspective, σ_v^2 represents the power per unit frequency interval. Therefore, the units of σ_v^2 are the units of v squared divided by the unit of frequency Hz. Second, from the perspective of correlation, R_v has the units of v

squared and the Dirac delta function $\delta(\tau)$ has units of $Hz = \frac{1}{\text{sec}}$; therefore, σ_v^2 has dimensions corresponding to the square of the units of v times sec. For example, if v is an angular rate in $\frac{\text{deg}}{\text{sec}}$, then σ_v^2 has dimensions $\left(\frac{\text{deg}}{\text{sec}}\right)^2 \text{sec} = \frac{\text{deg}^2}{\text{sec}} = \frac{\text{deg}^2/\text{sec}^2}{Hz}$. The units of σ_v are then either $\frac{\text{deg}}{\sqrt{\text{sec}}}$ or $\frac{\text{deg}/\text{sec}}{\sqrt{Hz}}$.

4.4.2.2 Discrete-time White Noise

A discrete-time random process v_k is a *white noise* process if

$$S_v(j\omega) = \sigma_v^2 \text{ for all } \omega \in [-\pi, \pi].$$

By the definition of the discrete Fourier transform, this implies that

$$cor\,(v_k, v_{k+\kappa}) = R_v(\kappa) = \sigma_v^2 \delta_\kappa \tag{4.37}$$

where δ_κ is the Kronecker delta function

$$\delta_\kappa = \begin{cases} 1 & \text{if } \kappa = 0 \\ 0 & \text{otherwise.} \end{cases}$$

The Kronecker delta function δ_κ is dimensionless. The dimensions of σ_v are the same as the dimensions of v.

Discrete-time white noise is physically realizable. The power is finite, because the integral of the PSD is only over the discrete-frequency range $\omega \in [-\pi, \pi]$.

4.5 Linear Systems with Random Inputs

As will be discussed in Section 4.9 and Chapter 7, navigation systems can often be designed in such a way as to allow performance analysis based on linear systems with stochastic inputs. Therefore, this section considers the statistical properties of the output of a linear system when the system model is known and the input is white.

The results of this section are often used to model error components with significant time correlation via the state augmentation approach. In such situations, linear stochastic systems forced by Gaussian white noise are of special interest. Careful selection of the linear system and white noise statistics can often yield an output stochastic process with statistics matching those of the time correlated error process. Examples of such processes are further discussed in Section 4.6.3.

The output $\mathbf{y}(t)$ of a (deterministic) time-invariant, causal, linear system with (stochastic) input $\mathbf{w}(t)$ is

$$\mathbf{y}(t) = \int_{-\infty}^{t} \mathbf{h}(t-\lambda)\mathbf{w}(\lambda)d\lambda \tag{4.38}$$

where $\mathbf{h}$ represents the impulse response matrix of the linear system. Because $\mathbf{w}(t)$ is a random variable, so is $\mathbf{y}(t)$. The Fourier transform of the impulse response yields the system transfer function

$$\mathbf{H}(j\omega) = \int_{-\infty}^{\infty} \mathbf{h}(t)e^{-j\omega t}dt. \tag{4.39}$$

The *Energy Spectrum* of a time-invariant linear system is

$$|\mathbf{H}(j\omega)|^2 = \mathbf{H}(j\omega)\mathbf{H}(j\omega)^* \tag{4.40}$$

where $\mathbf{H}(j\omega)^*$ represents the complex conjugate transpose of $\mathbf{H}(j\omega)$.

If $\boldsymbol{\mu}_{\mathbf{w}}(t)$ denotes the mean of the system input, then the mean of the system output is

$$\begin{aligned}
\boldsymbol{\mu}_{\mathbf{y}}(t) &= E\langle \mathbf{y}(t)\rangle \\
&= E\left\langle \int_{-\infty}^{t} \mathbf{h}(t-\lambda)\mathbf{w}(\lambda)d\lambda \right\rangle \\
&= \int_{-\infty}^{t} \mathbf{h}(t-\lambda)E\langle \mathbf{w}(\lambda)\rangle d\lambda \\
&= \int_{-\infty}^{t} \mathbf{h}(t-\lambda)\boldsymbol{\mu}_{\mathbf{w}}(\lambda)d\lambda
\end{aligned} \tag{4.41}$$

where the transition from the second to the third line uses the following facts: expectation and time-integration are linear operators and the impulse response $\mathbf{h}$ is deterministic. Eqn. (4.41) shows that the mean of the output process is the convolution of the impulse response with the mean of the input process. In the case where the system is causal and the mean of the input process is constant, the mean of the output reduces to

$$\boldsymbol{\mu}_{\mathbf{y}}(t) = \mathbf{H}(0)\boldsymbol{\mu}_{\mathbf{w}}.$$

The output autocorrelation function is slightly more complicated to derive. Its derivation is addressed in two steps. First, from eqn. (4.29), we find an expression for the cross correlation between $\mathbf{y}$ and $\mathbf{w}$,

$$\mathbf{R}_{\mathbf{wy}}(t_1, t_2) = E\left\langle \mathbf{w}(t_1) \int_{-\infty}^{t_2} \mathbf{w}(\lambda)^\top \mathbf{h}(t_2-\lambda)^\top d\lambda. \right\rangle.$$

Because the linear system is causal, this can be expressed as

$$\begin{aligned}
\mathbf{R}_{\mathbf{wy}}(t_1, t_2) &= E\left\langle \mathbf{w}(t_1) \int_{-\infty}^{\infty} \mathbf{w}(\lambda)^\top \mathbf{h}(t_2-\lambda)^\top d\lambda \right\rangle \\
&= \int_{-\infty}^{\infty} E\left\langle \mathbf{w}(t_1)\mathbf{w}(\lambda)^\top \right\rangle \mathbf{h}(t_2-\lambda)^\top d\lambda
\end{aligned}$$

$$= \int_{-\infty}^{\infty} \mathbf{R_w}(t_1, \lambda)\mathbf{h}(t_2 - \lambda)^\top d\lambda$$
$$= \int_{-\infty}^{\infty} \mathbf{R_w}(t_1, t_2 - \lambda)\mathbf{h}(\lambda)^\top d\lambda. \quad (4.42)$$

Second, we derive the expression for the autocorrelation of $\mathbf{y}$,

$$\begin{aligned}
\mathbf{R_y}(t_1, t_2) &= E\left\langle \int_{-\infty}^{\infty} \mathbf{h}(t_1 - \lambda)\mathbf{w}(\lambda)d\lambda \int_{-\infty}^{\infty} \mathbf{w}(\zeta)^\top \mathbf{h}(t_2 - \zeta)^\top d\zeta \right\rangle \\
&= \int_{-\infty}^{\infty} \mathbf{h}(t_1 - \lambda) \int_{-\infty}^{\infty} E\langle \mathbf{w}(\lambda)\mathbf{w}(\zeta)^\top \rangle \mathbf{h}(t_2 - \zeta)^\top d\zeta d\lambda \\
&= \int_{-\infty}^{\infty} \mathbf{h}(t_1 - \lambda)\mathbf{R_{wy}}(\lambda, t_2)d\lambda \\
&= \int_{-\infty}^{\infty} \mathbf{h}(\lambda)\mathbf{R_{wy}}(t_1 - \lambda, t_2)d\lambda. \quad (4.43)
\end{aligned}$$

If $\mathbf{w}$ is wide sense stationary, then eqns. (4.42-4.43) with $t_1 - t_2 = \tau$ reduce to

$$\mathbf{R_{wy}}(\tau) = \int_{-\infty}^{\infty} \mathbf{R_w}(\tau + \lambda)\mathbf{h}(\lambda)^\top d\lambda \quad (4.44)$$
$$\mathbf{R_y}(\tau) = \int_{-\infty}^{\infty} \mathbf{h}(\lambda)\mathbf{R_{wy}}(\tau - \lambda)d\lambda. \quad (4.45)$$

The Fourier transforms of eqns. (4.44-4.45) provide the cross power spectral density of $\mathbf{y}$ and $\mathbf{w}$ and the power spectral density of $\mathbf{y}$

$$\mathbf{S_{wy}}(j\omega) = \mathbf{S_w}(j\omega)\mathbf{H}(j\omega)^* \quad (4.46)$$
$$\mathbf{S_y}(j\omega) = \mathbf{H}(j\omega)\mathbf{S_{wy}}(j\omega) \quad (4.47)$$
$$= \mathbf{H}(j\omega)\mathbf{S_w}(j\omega)\mathbf{H}(j\omega)^*. \quad (4.48)$$

In the single-input, single-output case where y and w are both scalar random variables, eqn. (4.48) reduces to

$$S_y(j\omega) = |H(j\omega)|^2 S_w(j\omega). \quad (4.49)$$

Eqns. (4.48) and (4.49) are useful for finding a linear system realization $\mathbf{H}(j\omega)$ appropriate for producing a given colored noise process $\mathbf{y}$ from a white noise input $\mathbf{w}$.

Example 4.14 *A linear system with transfer function*

$$H(s) = \frac{\sqrt{2\beta}}{\beta + s}$$

is excited by white noise $w(t)$ with PSD defined by $S_w(j\omega) = \sigma_w^2$, where $\sigma_w, \beta > 0$. The spectral density of the output y is calculated by eqn. (4.49) as

$$\begin{aligned} S_y(j\omega) &= H(j\omega)^* H(j\omega) S_w(j\omega) \\ &= \frac{\sqrt{2\beta}}{\beta - j\omega} \frac{\sqrt{2\beta}}{\beta + j\omega} \sigma_w^2 \\ &= \frac{2\beta\sigma_w^2}{\beta^2 + w^2}. \end{aligned} \quad (4.50)$$

The inverse transform of eqn. (4.50) (which can be found using partial fractions) yields the autocorrelation function for the system output

$$R_y(\tau) = \sigma_w^2 e^{-\beta|\tau|}.$$

The parameter $\frac{1}{\beta}$ is the correlation time of the random variable y. △

The previous example discussed a natural use of stochastic process theory to calculate the PSD and correlation of the output of a linear system output driven by white noise. In at least two cases, the reverse process is of interest:

1. When the output autocorrelation function for a linear system with white noise inputs is known, and it is desired to find the linear system transfer function.

2. When the actual noise driving a process is colored, it may be more convenient for analysis to treat the colored noise as the output of a linear system driven by white noise. Finding the appropriate linear system model is equivalent to solving Item 1.

In either case, an autocorrelation function is given for a WSS stochastic process and the objective is to determine a linear system (represented by a transfer function $G(s)$) and a white noise process (represented by a PSD σ_w^2) such that the output of the linear system driven by the white noise process yields the specified autocorrelation function.

Example 4.15 *In the single channel error model of eqn. (3.49), assume that the additive gyro measurement error ϵ_g is determined to have the autocorrelation function*

$$R_{\epsilon_g}(\tau) = 10^{-4} e^{-10|\tau|} \left(\frac{rad}{s}\right)^2. \quad (4.51)$$

Find a causal linear system with a white noise input process so that the output of the linear system when driven by the white noise process has the correlation function specified in eqn. (4.51). Specify the augmented error state model for the system.

The spectral density of the output process v is

$$\begin{aligned} S_{\epsilon_g}(j\omega) &= \int_{-\infty}^{\infty} 10^{-4} e^{-10|\tau|} e^{-j\omega\tau} d\tau \\ &= \int_{0}^{\infty} 10^{-4} e^{-(10+j\omega)\tau} d\tau + \int_{-\infty}^{0} 10^{-4} e^{(10-j\omega)\tau} d\tau \\ &= \frac{10^{-4}}{10+j\omega} + \frac{10^{-4}}{10-j\omega} = \frac{20 \times 10^{-4}}{100+\omega^2} \\ &= (.01)^2 \left. \frac{\sqrt{20}}{(10+s)} \frac{\sqrt{20}}{(10-s)} \right|_{s=j\omega} . \end{aligned} \tag{4.52}$$

Therefore, by comparison with eqn. (4.49), a suitable linear system gyro error model is

$$G(s) = \frac{\sqrt{20}}{10+s} \tag{4.53}$$

and the PSD of the corresponding white driving noise process w_g is

$$S_{w_g}(j\omega) = 0.01^2 \frac{(rad/s^2)^2}{Hz} \quad \left(i.e., \; R_{w_g}(\tau) = 0.01^2 \delta(\tau) \left(\frac{rad}{s^2} \right)^2 \right).$$

A state space realization of the gyro error transfer function $G(s)$ is

$$\left. \begin{aligned} \dot{b}_g &= -10 b_g + w_g \\ \epsilon_g &= \sqrt{20} b_g . \end{aligned} \right\} \tag{4.54}$$

Appending this state space realization of the gyro error model to the error model of eqn. (3.49) yields

$$\begin{bmatrix} \dot{x} \\ \dot{v} \\ \dot{\phi} \\ \dot{b}_g \end{bmatrix} = \begin{bmatrix} 0 & 1 & 0 & 0 \\ 0 & 0 & -g & 0 \\ 0 & \frac{1}{R} & 0 & \sqrt{20} \\ 0 & 0 & 0 & -10 \end{bmatrix} \begin{bmatrix} x \\ v \\ \phi \\ b_g \end{bmatrix} + \begin{bmatrix} 0 \\ 0 \\ 0 \\ 1 \end{bmatrix} w_g . \tag{4.55}$$

△

Note that the noise process and linear system must be specified jointly, since there are more degrees of freedom than design constraints. Also, the selection of $G(s)$ from the two factors of eqn. (4.52) is not arbitrary. The transfer function $G(s)$ is selected so that its poles and zeros lie in the strict left half of the complex s-plane. This results in a causal, stable linear system model. Such a choice is always possible, when the spectral density is the ratio of polynomials containing only even powers of s. The interested reader is referred to the subject of *spectral factorization*.

When the driving noise is white and Gaussian, the process specified in eqn. (4.54) is an example of a scalar *Gauss-Markov* process. Higher order *vector Gauss-Markov* processes, such as the system of eqn. (4.55), result when the correlation function is more complex. Additional examples of simple Gauss-Markov process that are often useful in navigation system error model are discussed in Section 4.6.3.

4.6 State Models for Stochastic Processes

As motivation for the study of stochastic systems, consider the error dynamics for a single channel of a tangent plane INS, as depicted in Figure 4.4 and the following equation

$$\begin{bmatrix} \dot{x} \\ \dot{v} \\ \dot{\phi} \end{bmatrix} = \begin{bmatrix} 0 & 1 & 0 \\ 0 & 0 & -g \\ 0 & \frac{1}{R} & 0 \end{bmatrix} \begin{bmatrix} x \\ v \\ \phi \end{bmatrix} + \begin{bmatrix} 0 & 0 \\ 1 & 0 \\ 0 & 1 \end{bmatrix} \begin{bmatrix} \varepsilon_a \\ \varepsilon_g \end{bmatrix} \tag{4.56}$$

where $[x, v, \phi]$ represents error in the navigation state. The error model is driven by two sources of error: accelerometer error ϵ_a and gyro error ϵ_g. These errors originate in the sensors, but appear in the error model as process noise. The error at each source may take several forms: white noise, correlated (colored) noise, bias error, scale factor error, etc. After any deterministic error sources have been compensated or modeled, the remaining instrument errors will be modeled as random inputs. The use of random variables to model the error sources is reasonable, since although the analyst may not be able to specify the value of the various error sources at some future time, it is usually possible to specify the dynamic nature and statistics of the error sources as a function of time. It is also reasonable to restrict our attention to linear stochastic systems, since the primary focus of the stochastic analysis in navigation systems will be on linearized error dynamics.

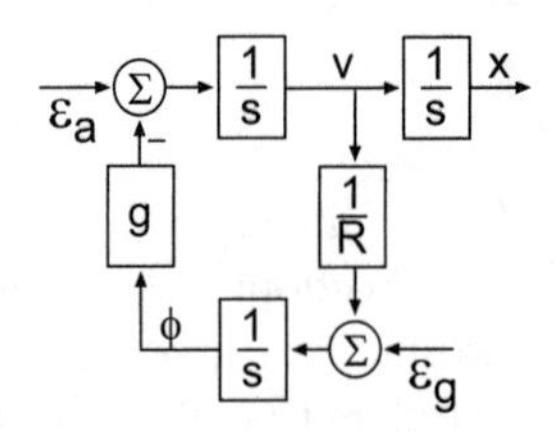

Figure 4.4: Block diagram for single channel INS error model.

The primary objectives for this section are to present: (1) how to account for stochastic sources of error in state space dynamic system modeling; and, (2) how the mean and covariance of system errors propagate in linear stochastic systems. This section presents various results for linear, state space systems with random inputs that will be used throughout the remainder of the book.

4.6.1 Standard Model

The model for a finite-dimensional linear continuous-time system with *stochastic inputs* can be represented as

$$\dot{\mathbf{x}}(t) = \mathbf{F}(t)\mathbf{x}(t) + \mathbf{G}(t)\mathbf{w}(t) \qquad \mathbf{y}(t) = \mathbf{H}(t)\mathbf{x}(t) + \mathbf{v}(t) \tag{4.57}$$

where $\mathbf{w}(t)$ and $\mathbf{v}(t)$ are random variables. The random variable $\mathbf{w}$ is called the *process noise.* The random variable $\mathbf{v}$ is called the *measurement noise.* At each time, $\mathbf{x}(t)$ and $\mathbf{y}(t)$ are also random variables. The designation random variable implies that although the value of the random variable at any time is not completely predictable, the statistics of the random variable may be known. Assuming that the model is constructed so that $\mathbf{v}(t)$ and $\mathbf{w}(t)$ are white, the mean and covariance of the random variables $\mathbf{w}(t)$ and $\mathbf{v}(t)$ will be denoted

$$\mu_{\mathbf{w}}(t) = E\langle\mathbf{w}(t)\rangle = \mathbf{0} \text{ and} \tag{4.58}$$

$$cov\left(\mathbf{w}(t), \mathbf{w}(\tau)\right) = \mathbf{Q}(t)\delta(t-\tau), \tag{4.59}$$

$$\mu_{\mathbf{v}}(t) = E\langle\mathbf{v}(t)\rangle = \mathbf{0} \text{ and} \tag{4.60}$$

$$cov\left(\mathbf{v}(t), \mathbf{v}(\tau)\right) = \mathbf{R}(t)\delta(t-\tau). \tag{4.61}$$

In the analysis that follows, it will often be accurate (and convenient) to assume that the process and measurement noise are independent of the current and previous state

$$cov\left(\mathbf{w}(t), \mathbf{x}(\tau)\right) = \mathbf{0} \text{ for } t \geq \tau \tag{4.62}$$

$$cov\left(\mathbf{v}(t), \mathbf{x}(\tau)\right) = \mathbf{0} \text{ for } t \geq \tau \tag{4.63}$$

and independent of each other

$$cov\left(\mathbf{w}(t), \mathbf{v}(\tau)\right) = \mathbf{0} \text{ for all } t, \tau \geq 0. \tag{4.64}$$

In most applications, the random processes $\mathbf{w}(t)$ and $\mathbf{v}(t)$ will be assumed to be Gaussian.

The linear discrete-time model is

$$\mathbf{x}_{k+1} = \mathbf{\Phi}_k\mathbf{x}_k + \mathbf{w}_k \qquad \mathbf{y}_k = \mathbf{H}_k\mathbf{x}_k + \mathbf{v}_k. \tag{4.65}$$

The mean and covariance of the random variables $\mathbf{w}_k$ and $\mathbf{v}_k$ will be denoted

$$\mu_{\mathbf{w}_k} = E\langle \mathbf{w}_k \rangle = \mathbf{0} \quad \text{and} \tag{4.66}$$
$$cov(\mathbf{w}_k, \mathbf{w}_l) = \mathbf{Qd}_k \delta(k-l), \tag{4.67}$$
$$\mu_{\mathbf{v}_k} = E\langle \mathbf{v}_k \rangle = \mathbf{0} \quad \text{and} \tag{4.68}$$
$$cov(\mathbf{v}_k, \mathbf{v}_l) = \mathbf{R}_k \delta(k-l). \tag{4.69}$$

Assumptions corresponding to eqns. (4.62–4.64) also apply to the discrete-time model.

In the analysis to follow, the time argument of the signals will typically be dropped, to simplify the notation.

4.6.2 Stochastic Systems and State Augmentation

Navigation system error analysis will often result in equations of the form

$$\dot{\mathbf{x}}_n = \mathbf{F}_n(t)\mathbf{x}_n + \mathbf{G}_n(t)\epsilon \tag{4.70}$$
$$\mathbf{y} = \mathbf{H}_n(t)\mathbf{x}_n + \boldsymbol{\alpha} \tag{4.71}$$

where $\boldsymbol{\epsilon}$ and $\boldsymbol{\alpha}$ represent instrumentation error signals and $\mathbf{x}_n$ represents the error in the nominal navigation state. This and subsequent sections of this chapter will present a set of techniques which will allow the error signals $\boldsymbol{\epsilon}$ and $\boldsymbol{\alpha}$ to be modeled as the outputs of linear dynamic systems

$$\dot{\mathbf{x}}_\epsilon = \mathbf{F}_\epsilon \mathbf{x}_\epsilon + \mathbf{G}_\epsilon \boldsymbol{\omega}_\epsilon \tag{4.72}$$
$$\boldsymbol{\epsilon} = \mathbf{H}_\epsilon \mathbf{x}_\epsilon + \boldsymbol{\nu}_\epsilon \tag{4.73}$$

and

$$\dot{\mathbf{x}}_\alpha = \mathbf{F}_\alpha \mathbf{x}_\alpha + \mathbf{G}_\alpha \boldsymbol{\omega}_\alpha \tag{4.74}$$
$$\boldsymbol{\alpha} = \mathbf{H}_\alpha \mathbf{x}_\alpha + \boldsymbol{\nu}_\alpha \tag{4.75}$$

with the noise processes $\boldsymbol{\omega}_\epsilon(t), \boldsymbol{\nu}_\epsilon(t), \boldsymbol{\omega}_\alpha(t)$ and $\boldsymbol{\nu}_\alpha(t)$ being accurately modeled as white noise processes. By the process of *state augmentation*, equations (4.70-4.75) can be combined into the state space error model

$$\begin{aligned}
\dot{\mathbf{x}}_a &= \mathbf{F}_a(t)\mathbf{x}_a(t) + \mathbf{G}_a \boldsymbol{\omega}_a(t) \qquad (4.76)\\
&= \begin{bmatrix} \mathbf{F}_n(t) & \mathbf{G}_n(t)\mathbf{H}_\epsilon & \mathbf{0} \\ \mathbf{0} & \mathbf{F}_\epsilon & \mathbf{0} \\ \mathbf{0} & \mathbf{0} & \mathbf{F}_\alpha \end{bmatrix} \mathbf{x}_a \\
&\quad + \begin{bmatrix} \mathbf{G}_n(t) & \mathbf{0} & \mathbf{0} \\ \mathbf{0} & \mathbf{G}_\epsilon & \mathbf{0} \\ \mathbf{0} & \mathbf{0} & \mathbf{G}_\alpha \end{bmatrix} \begin{bmatrix} \boldsymbol{\nu}_\epsilon \\ \boldsymbol{\omega}_\epsilon \\ \boldsymbol{\omega}_\alpha \end{bmatrix}
\end{aligned}$$

which is in the form of eqn. (4.57) and driven only by white noise processes. In these equations, the augmented state is defined as $\mathbf{x}_a = [\mathbf{x}_n^\top, \mathbf{x}_\epsilon^\top, \mathbf{x}_\alpha^\top]^\top$. The measurements of the augmented system are modeled as

$$\begin{aligned} \mathbf{y} &= \mathbf{H}_a(t)\mathbf{x}_a + \boldsymbol{\nu}_\alpha & (4.77) \\ &= \begin{bmatrix} \mathbf{H}_n(t) & \mathbf{0} & \mathbf{H}_\alpha \end{bmatrix} \mathbf{x}_a + \boldsymbol{\nu}_\alpha & (4.78) \end{aligned}$$

which are corrupted only by additive white noise.

For the state augmented model to be an accurate characterization of the actual system, the state space parameters $(\mathbf{F}_a, \mathbf{G}_a, \mathbf{H}_a)$ corresponding to the appended error models and the statistics of the driving noise processes must be accurately specified. Detailed examples of augmented state models are presented in Sections 4.6.3 and 4.9. Section 4.6.3 also discusses several basic building blocks of the state augmentation process.

4.6.3 Gauss-Markov Processes

For the finite dimensional state space system

$$\dot{\mathbf{x}}(t) = \mathbf{F}(t)\mathbf{x}(t) + \mathbf{G}(t)\mathbf{w}(t) \qquad \mathbf{y}(t) = \mathbf{H}(t)\mathbf{x}(t) + \mathbf{v}(t) \tag{4.79}$$

where $\mathbf{w}(t)$ and $\mathbf{v}(t)$ are stochastic processes, if both $\mathbf{w}$ and $\mathbf{v}$ are Gaussian random processes, then the system is an example of a *Gauss-Markov process.* Since any linear operation performed on a Gaussian random variable results in a Gaussian random variable, the state $\mathbf{x}(t)$ and system output $\mathbf{y}(t)$ will be Gaussian random variables. This is a very beneficial property as the Normal distribution is completely described by two parameters (i.e., the mean and covariance) which are straightforward to propagate through time, as is described in Section 4.6.4 and 4.6.5. The purpose of this section is to present several specific types of Gauss-Markov processes that are useful for error modeling.

4.6.3.1 Random Constants

Some portions of instrumentation error (e.g., scale factor) can be accurately represented as constant (but unknown) random variables. If some portion of the constant error is known, then it can be compensated for and the remaining error can be modeled as an unknown constant. An unknown constant is modeled as

$$\dot{x} = 0, \text{ with } P_x(0) = var(x(0), x(0)). \tag{4.80}$$

This model states that the variable x is not changing and has a known initial variance $P_x(0)$.

Example 4.16 *An acceleration measurement $\tilde{a}$ is assumed to be corrupted by an unknown constant bias b_a:*

$$\tilde{a} = a + b_a.$$

The bias is specified to be zero mean with an initial error variance of $var(b_a) = \sigma_b^2$.

1. *What is an appropriate state space model for the accelerometer output error $\epsilon_a(t)$?*

2. *What is the appropriate state space model when this accelerometer error model is augmented to the tangent plane single channel error model of eqn. (3.91)? Assume that the constant bias is the only form of accelerometer error and that $\epsilon_g = 0$.*

Since the bias is assumed to be constant, an appropriate model is

$$\dot{b}_a(t) = 0 \tag{4.81}$$

with initial condition $b_a(t) = 0$ and $P_{b_a} = \sigma_b^2$. Note that there is no process noise in the bias dynamic equation. Since no other forms of error are being modeled, $\epsilon_a = b_a$. Therefore, the augmented single channel error model becomes

$$\begin{bmatrix} \dot{x} \\ \dot{v} \\ \dot{\phi} \\ \dot{b}_a \end{bmatrix} = \begin{bmatrix} 0 & 1 & 0 & 0 \\ 0 & 0 & -g & 1 \\ 0 & \frac{1}{R} & 0 & 0 \\ 0 & 0 & 0 & 0 \end{bmatrix} \begin{bmatrix} x \\ v \\ \phi \\ b_a \end{bmatrix}. \tag{4.82}$$

Note that for eqn. (4.82), if either a position or velocity measurement were available, an observability analysis will show that the system is not observable. The tilt and accelerometer bias states cannot be distinguished. See Exercise 4.15. △

4.6.3.2 Brownian Motion (Random Walk) Processes

The Gauss-Markov process $x(t)$ defined by

$$x(t) = \int_0^t \omega(q)dq \tag{4.83}$$

with $P_x(0) = 0$ and $R_\omega(t) = \sigma_\omega^2 \delta(t)$ is called a Brownian motion or random walk process. The mean of x can be calculated as

$$\begin{aligned} \mu_x(t) &= E\langle x(t)\rangle = E\left\langle \int_0^t \omega(q)dq \right\rangle \\ &= \int_0^t E\langle w(q)\rangle dq = 0. \end{aligned}$$

By the definition of variance in eqn. (4.22), the covariance function for x can be calculated as

$$\begin{aligned} cov_x(t,\tau) &= E\langle x(t)x(\tau)\rangle = E\left\langle \int_0^t \omega(\lambda)d\lambda \int_0^\tau \omega(\zeta)d\zeta \right\rangle \\ &= \int_0^t \int_0^\tau E\langle \omega(\lambda)\omega(\zeta)\rangle d\lambda d\zeta = \int_0^\tau \int_0^t \sigma_\omega^2 \delta(\lambda-\zeta)d\zeta d\lambda \\ &= \int_0^\tau \sigma_\omega^2 d\lambda, \text{ for } \lambda < t \\ &= \sigma_\omega^2 \min(t,\tau), \end{aligned} \tag{4.84}$$

which yields

$$var_x(t) = \sigma_\omega^2 t. \tag{4.85}$$

In navigation modeling, it is common to integrate the output of a sensor to determine particular navigation quantities. Examples are integrating an accelerometer output to determine velocity or integrating an angular rate to determine an angle, as in Example 4.2. If it is accurate to consider the sensor error as white random noise, then the resulting error equations will result in a random walk model.

As with white noise, the units of the random walk process often cause confusion. Note that if x is measured in degrees, then $var_x(t)$ has units of $\deg^2$. Therefore, the units of σ_ω are $\frac{\deg}{\sqrt{s}} = \frac{\deg/s}{\sqrt{Hz}}$ which makes sense given that σ_ω is the PSD of a white angular rate noise process.

One often quoted measure of sensor accuracy is the random walk parameter. For example, a certain gyro might list its random walk parameter to be $4.0 \deg/\sqrt{hr}$. By eqn. (4.84), the random walk parameter for a sensor quantifies the rate of growth of the integrated (properly compensated) sensor output as a function of time. The $\sqrt{hr}$ in the denominator of the specification reflects that the standard deviation and variance of the random walk variable grow with the square root of time and linearly with time, respectively. When the specification states that the angle random walk parameter is $N\frac{\deg}{\sqrt{hr}}$, then $\sigma_\omega = \frac{N}{60}\frac{\deg/s}{\sqrt{Hz}}$.

For a random walk process, the state space model is

$$\dot{x} = \nu \tag{4.86}$$

where $E\langle x(0)\rangle = 0$, $var(x(0)) = P_x(0) = 0$, and $S_\nu(j\omega) = \sigma_\nu^2$. The transfer function corresponding to eqns. (4.83) and (4.86) is $H(s) = \frac{1}{s}$, so by eqn. (4.49), the PSD of x is

$$S_x(j\omega) = \frac{\sigma_\nu^2}{\omega^2}.$$

Example 4.17 *An accelerometer output is known to be in error by an unknown slowly time-varying bias b_a. Assume that the 'turn-on' bias is* *accurately known and accounted for in the accelerometer calibration; and, that it is accurate to model the residual time-varying bias error as a random walk:*

$$\begin{aligned} \delta b_a(t) &= \int_0^t \omega_b(\tau) d\tau \quad (4.87) \\ var(b(0)) = P_b(0) &= 0 \quad (4.88) \\ var(\omega_b(t), \omega_b(\tau)) &= Q_\omega \delta(t-\tau) \quad (4.89) \end{aligned}$$

where ω_b represents Gaussian white noise and Q_ω is the PSD of ω_b. The parameter Q_ω is specified by the manufacturer.

Appending the random walk bias model to a single channel of the inertial frame INS error model:

$$\begin{bmatrix} \delta\dot{p} \\ \delta\dot{v} \end{bmatrix} = \begin{bmatrix} 0 & 1 \\ 0 & 0 \end{bmatrix} \begin{bmatrix} \delta p \\ \delta v \end{bmatrix} + \begin{bmatrix} 0 \\ 1 \end{bmatrix} \delta b_a$$

yields the following augmented state model

$$\begin{aligned} \begin{bmatrix} \delta\dot{p} \\ \delta\dot{v} \\ \delta\dot{b}_a \end{bmatrix} &= \begin{bmatrix} 0 & 1 & 0 \\ 0 & 0 & 1 \\ 0 & 0 & 0 \end{bmatrix} \begin{bmatrix} \delta p \\ \delta v \\ \delta b_a \end{bmatrix} + \begin{bmatrix} 0 \\ 0 \\ 1 \end{bmatrix} \omega_b(t) \\ \mathbf{P_x}(0) &= \begin{bmatrix} \sigma_{pp}^2 & \sigma_{pv}^2 & 0 \\ \sigma_{vp}^2 & \sigma_{vv}^2 & 0 \\ 0 & 0 & P_b(0) \end{bmatrix}. \end{aligned}$$

This example has made the unrealistic assumption that the turn-on bias is exactly known so that a proper random walk model could be used. In a realistic situation where the initial bias is not perfectly known, a constant plus random walk error model would be identical to the equations shown above with $P_b(0)$ specified to account for the variance in the initial bias estimate. △

4.6.3.3 Scalar Gauss-Markov Process

The scalar Gauss-Markov process refers to the special case of eqn. (4.79) where the state, input, and output are each scalar variables:

$$\left.\begin{aligned} \dot{x} &= -\tfrac{1}{\tau}x + Gw \\ y &= Hx. \end{aligned}\right\} \quad (4.90)$$

where w is Gaussian white noise with PSD denoted by σ^2. In eqn. (4.90), the parameter τ is the *correlation time*. This process has already been used in Examples 4.14 and 4.15.

Eqn. (4.90) is written in its most general form. The PSD of y is

$$S_y(j\omega) = \frac{HG\sigma^2}{\omega^2 + \left(\frac{1}{\tau}\right)^2}$$

which shows that once the correlation time τ is determined, the value of $S_y(j\omega)|_{\omega=0}$ determines the product of the parameters H, G, and σ. Often, one or two of these three parameters is arbitrarily set to one with the remaining parameter values selected to achieve the desired value at $\omega = 0$.

4.6.3.4 Compound Augmented States

The examples of the previous sections each contained a single type of error. In more realistic situations, several forms of error may be present. This section presents an example involving compound error models augmented to simplified INS error equations.

Example 4.18 *The ideal dynamics of a one dimensional INS implemented in inertial space are*

$$\dot{p} = v \tag{4.91}$$
$$\dot{v} = a. \tag{4.92}$$

Two sensors are available. The first sensor measures position. The second sensor measures acceleration.

The position measurement model is

$$\tilde{y}(t) = p(t) + \zeta(t) + \mu(t) \tag{4.93}$$

where $\mu(t)$ represents Gaussian white noise and $\zeta(t)$ is the scalar Gauss-Markov process

$$\dot{\zeta} = -\beta\zeta + \omega_\zeta.$$

The implemented INS is described as

$$\dot{\hat{p}} = \hat{v} \tag{4.94}$$
$$\dot{\hat{v}} = \tilde{a}, \tag{4.95}$$

where the accelerometer model is

$$\tilde{a}(t) = (1 - \delta k)a(t) - \delta b_a(t) - \nu_a(t)$$

where δk is a Gaussian random-constant scale-factor error, δb_a is a constant plus random walk Gaussian bias error, and ν_a is Gaussian white noise. The predicted output at any time is calculated as

$$\hat{y}(t) = \hat{p}(t). \tag{4.96}$$

The differential equations for the error variables are found by subtracting eqns. (4.94-4.95) from eqns. (4.91-4.92)

$$\begin{aligned} \delta\dot{p} &= \delta v \\ \delta\dot{v} &= a(t)\delta k + \delta b_a(t) + \nu_a(t) \end{aligned}$$

with each error term defined as $\delta x = x - \hat{x}$. *The model for the residual measurement* $\delta y = \tilde{y} - \hat{y}$ *is defined by subtracting eqn. (4.96) from eqn. (4.93)*

$$\delta y(t) = \delta p(t) + \zeta(t) + \mu(t).$$

The augmented error state equations are then defined to be

$$\begin{aligned} \begin{bmatrix} \delta\dot{p} \\ \delta\dot{v} \\ \delta\dot{b}_a \\ \delta\dot{k} \\ \dot{\zeta} \end{bmatrix} &= \begin{bmatrix} 0 & 1 & 0 & 0 & 0 \\ 0 & 0 & 1 & a & 0 \\ 0 & 0 & 0 & 0 & 0 \\ 0 & 0 & 0 & 0 & 0 \\ 0 & 0 & 0 & 0 & -\beta \end{bmatrix} \begin{bmatrix} \delta p \\ \delta v \\ \delta b_a \\ \delta k \\ \zeta \end{bmatrix} + \begin{bmatrix} 0 \\ \nu_a \\ \omega_b \\ 0 \\ \omega_\zeta \end{bmatrix} \\ \delta y &= \begin{bmatrix} 1 & 0 & 0 & 0 & 1 \end{bmatrix} \delta\mathbf{x} + \mu. \end{aligned}$$

These error equations have the form of eqn. (4.79) with Gaussian white noise inputs. This model structure is important for the optimal state estimation methods of Chapter 5.

This example has considered a hypothetical one dimensional INS with only two sensors. The state of the original INS was two and three error states were appended. In realistic navigation systems in a three dimensional world with many more sensed quantities, it should be clear that the dimension of the state of the error model can become quite large. △

When designing a system that will be implemented in a real-time application, there is usually a tradeoff required between reasonable cost and computation time and the desire for accurate modeling. Even in the above single axis example, the dimension of the augmented error state vector $\delta\mathbf{x}$ is large (i.e., 5) relative to the dimension of the original state $\mathbf{x} = [p, v]^\top$ of the INS (i.e., 2), and several more error states could be included in the quest for modeling accuracy. In realistic applications, the dimension of the augmented state vector is potentially quite large, often too large for a real-time system. Therefore, the design may result in two models. The most complex model that accounts for all error states considered to be significant is referred to as the *truth model.* A simplified *Design model* may be constructed from the truth model by eliminating or combining certain state variables. The art is to develop a design model small enough to allow its use in practical realtime implementations without paying a significant performance penalty in terms of state estimation accuracy. The methodology for analyzing the performance tradeoffs is discussed in Chapter 6.

4.6.4 Time-propagation of the Mean

From eqn. (4.65), if the mean of the state vector is known at some time k_0 and $E\langle \mathbf{w}_k \rangle = 0$, then the mean can be propagated forward according to

$$\begin{aligned} E\langle \mathbf{x}_{k+1} \rangle &= E\langle \mathbf{\Phi}_k \mathbf{x}_k \rangle + E\langle \mathbf{w}_k \rangle \\ &= \mathbf{\Phi}_k E\langle \mathbf{x}_k \rangle + E\langle \mathbf{w}_k \rangle \\ E\langle \mathbf{x}_{k+1} \rangle &= \mathbf{\Phi}_k E\langle \mathbf{x}_k \rangle. \end{aligned} \tag{4.97}$$

Intuitively, this formula states that since nothing is known a priori about the specific realization of the process noise for a given experiment, the mean of the state is propagated according to the state model:

$$\boldsymbol{\mu}_{\mathbf{x}_{k+1}} = \mathbf{\Phi}_k \boldsymbol{\mu}_{\mathbf{x}_k} \qquad \boldsymbol{\mu}_{\mathbf{y}_k} = \mathbf{H}_k \boldsymbol{\mu}_{\mathbf{x}_k}. \tag{4.98}$$

The equation for $\boldsymbol{\mu}_{\mathbf{y}}$ is found by analysis similar to that for $\boldsymbol{\mu}_{\mathbf{x}_k}$.

4.6.5 Time-propagation of the Variance

The discrete-time error state covariance matrix is defined as

$$\mathbf{P}_k = E\langle (\mathbf{x}_k - \boldsymbol{\mu}_k)(\mathbf{x}_k - \boldsymbol{\mu}_k)^\top \rangle,$$

where $\boldsymbol{\mu}_k = E\langle \mathbf{x}_k \rangle$. The state covariance at time $k+1$

$$\mathbf{P}_{k+1} = E\left\langle (\mathbf{x}_{k+1} - \boldsymbol{\mu}_{k+1})(\mathbf{x}_{k+1} - \boldsymbol{\mu}_{k+1})^\top \right\rangle$$

can be simplified using eqns. (4.65) and (4.98) to determine eqn. (4.99) for propagating the state error covariance through time:

$$\begin{aligned} \mathbf{P}_{k+1} &= E\left\langle (\mathbf{\Phi}_k(\mathbf{x}_k - \boldsymbol{\mu}_k) + \mathbf{w}_k)\left((\mathbf{x}_k - \boldsymbol{\mu}_k)^\top \mathbf{\Phi}_k^\top + \mathbf{w}_k^\top\right)\right\rangle \\ &= E\left\langle \mathbf{\Phi}_k(\mathbf{x}_k - \boldsymbol{\mu}_k)(\mathbf{x}_k - \boldsymbol{\mu}_k)^\top \mathbf{\Phi}_k^\top + \mathbf{w}_k \mathbf{w}_k^\top \right. \\ &\quad \left. + \mathbf{w}_k(\mathbf{x}_k - \boldsymbol{\mu}_k)^\top \mathbf{\Phi}_k^\top + \mathbf{\Phi}_k(\mathbf{x}_k - \boldsymbol{\mu}_k)\mathbf{w}_k^\top \right\rangle \\ &= \mathbf{\Phi}_k E\left\langle (\mathbf{x}_k - \boldsymbol{\mu}_k)(\mathbf{x}_k - \boldsymbol{\mu}_k)^\top \right\rangle \mathbf{\Phi}_k^\top + E\left\langle \mathbf{w}_k \mathbf{w}_k^\top \right\rangle \\ \mathbf{P}_{k+1} &= \mathbf{\Phi}_k \mathbf{P}_k \mathbf{\Phi}_k^\top + \mathbf{Qd}_k. \end{aligned} \tag{4.99}$$

Note that eqns. (4.97) and (4.99) are valid for any zero mean noise process satisfying the assumption of eqn. (4.62). No other assumptions were used in the derivations. If $\mathbf{w}_k$ and $\mathbf{v}_k$ happen to be zero mean Gaussian processes, then $\mathbf{x}_k$ is also a Gaussian stochastic process with mean and variance given by eqns. (4.97) and (4.99).

By limiting arguments (see Chapter 4 in [58]), the covariance propagation for the continuous-time system

$$\dot{\mathbf{x}} = \mathbf{F}\mathbf{x} + \mathbf{G}\boldsymbol{\omega}$$

is described by

$$\dot{\mathbf{P}} = \mathbf{F}\mathbf{P} + \mathbf{P}\mathbf{F}^\top + \mathbf{G}\mathbf{Q}\mathbf{G}^\top. \tag{4.100}$$

Example 4.19 *Consider the scalar Gauss-Markov process with*

$$\dot{x}(t) = -\beta x(t) + \omega(t)$$

with $\beta > 0$ *where the PSD of the white noise process* ω *is* $Q > 0$. *Letting* $var(x) = P$,

$$\begin{aligned} \dot{P} &= -\beta P - P\beta + Q \\ &= -2\beta P + Q. \end{aligned} \tag{4.101}$$

Eqn. (4.101) has the solution

$$P(t) = \frac{Q}{2\beta} + \left(P(0) - \frac{Q}{2\beta}\right) e^{-2\beta t} \tag{4.102}$$

which can be verified by direct substitution. In steady state, $P(\infty) = \frac{Q}{2\beta}$, *which is useful in applications involving scalar Gauss-Markov processes when the steady-state covariance* $P(\infty)$ *and the correlation time* $\frac{1}{\beta}$ *are known and* Q *is to be determined.* △

4.7 Discrete-time Equivalent Models

When the dynamics of the system of interest evolve in continuous time, but analysis and implementation are more convenient in discrete-time, we will require a means for determining a discrete-time model in the form of eqn. (4.65) which is equivalent to eqn. (4.57) at the discrete-time instants $t_k = kT$. Specification of the equivalent discrete-time model requires computation of the discrete-time state transition matrix Φ_k for eqn. (4.65) and the process noise covariance matrix $\mathbf{Qd}$ for eqn. (4.67). These computations are discussed in the following two subsections for time invariant systems. A method to compute $\mathbf{\Phi}$ and $\mathbf{Qd}$ over longer periods of time for which $\mathbf{F}$ or $\mathbf{Q}$ may not be constant is discussed in Section 7.2.5.2.

4.7.1 Calculation of Φ_k from $\mathbf{F}(t)$

For equivalence at the sampling instants when $\mathbf{F}$ is a constant matrix, $\mathbf{\Phi}$ can be determined as in eqn. (3.61):

$$\begin{aligned} \mathbf{\Phi}(t) = \mathbf{\Phi}(t,0) &= e^{\mathbf{F}\,t}, \\ \mathbf{\Phi}(\tau,t) &= e^{\mathbf{F}(\tau - t)}, \text{ and} \\ \mathbf{\Phi}_k &= e^{\mathbf{F}T} \end{aligned} \tag{4.103}$$

where $T = t_k - t_{k-1}$ is the sample period. Methods for computing the matrix exponential are discussed in Section B.12.

Example 4.20 *Assume that for a system of interest,*

$$\mathbf{F} = \begin{bmatrix} \mathbf{0} & \mathbf{F}_{12} & \mathbf{0} \\ \mathbf{0} & \mathbf{0} & \mathbf{F}_{23} \\ \mathbf{0} & \mathbf{0} & \mathbf{F}_{33} \end{bmatrix} \tag{4.104}$$

and the submatrices denoted by $\mathbf{F}_{12}, \mathbf{F}_{23}$, and $\mathbf{F}_{33}$ are constant over the interval $t \in [t_1, t_2]$. Then, $\mathbf{\Phi}(t_2, t_1) = e^{\mathbf{F}T}$ where $T = t_2 - t_1$. Expanding the Taylor series of

$$e^{\mathbf{F}t} = \mathbf{I} + \mathbf{F}t + \frac{1}{2}(\mathbf{F}t)^2 \dots$$

is straightforward, but tedious. The result is

$$\mathbf{\Phi}(t_2, t_1) = \begin{bmatrix} \mathbf{I} & \mathbf{F}_{12}T_2 & \mathbf{F}_{12}\mathbf{F}_{23}\int_{t_1}^{t_2}\int_{t_1}^{t} e^{\mathbf{F}_{33}s}dsdt \\ \mathbf{0} & \mathbf{I} & \mathbf{F}_{23}\int_{t_1}^{t_2} e^{\mathbf{F}_{33}s}ds \\ \mathbf{0} & \mathbf{0} & e^{\mathbf{F}_{33}T_2} \end{bmatrix} \tag{4.105}$$

which is the closed form solution. When $\mathbf{F}_{33}$ can be approximated as zero, the following reduction results

$$\mathbf{\Phi}(t_2, t_1) = \begin{bmatrix} \mathbf{I} & \mathbf{F}_{12}T_2 & \frac{1}{2}\mathbf{F}_{12}\mathbf{F}_{23}T_2^2 \\ \mathbf{0} & \mathbf{I} & \mathbf{F}_{23}T_2 \\ \mathbf{0} & \mathbf{0} & \mathbf{I} \end{bmatrix}. \tag{4.106}$$

Eqn. (4.104) corresponds to the $\mathbf{F}$ matrix for certain INS error models after simplification. △

If the state transition matrix is required for a time interval $[T_{m-1}, T_m]$ of duration long enough that the $\mathbf{F}$ matrix cannot be considered constant, then it may be possible to proceed by subdividing the interval. When the interval can be decomposed into subintervals $T_{m-1} < t_1 < t_2 < \ldots < T_m$, where $\tau = max(t_n - t_{n-1})$ and the $\mathbf{F}$ matrix can be considered constant over intervals of duration less than τ, then by the properties of state transition matrices,

$$\mathbf{\Phi}(t_n, T_{m-1}) = \mathbf{\Phi}(t_n, t_{n-1})\mathbf{\Phi}(t_{n-1}, T_{m-1}) \tag{4.107}$$

where $\mathbf{\Phi}(t_n, t_{n-1})$ is defined as in eqn. (4.103) with $\mathbf{F}$ considered as constant for $t \in [t_n, t_{n-1})$. The transition matrix $\mathbf{\Phi}(t_{n-1}, T_{m-1})$ is defined from previous iterations of eqn. (4.107) where the iteration is initialized at $t = T_{m-1}$ with $\mathbf{\Phi}(T_{m-1}, T_{m-1}) = \mathbf{I}$. The iteration continues for the interval of time propagation to yield $\mathbf{\Phi}(T_m, T_{m-1})$.

4.7.2 Calculation of $\mathbf{Qd}_k$ from $\mathbf{Q}(t)$

For equivalence at the sampling instants, the matrix $\mathbf{Qd}_k$ must account for the integrated effect of $\mathbf{w}(t)$ by the system dynamics over each sampling period. Therefore, by integration of eqn. (4.57) and comparison with eqn. (4.65), $\mathbf{w}_k$ must satisfy

$$\mathbf{x}(t_{k+1}) = e^{\mathbf{F}(t_{k+1}-t_k)}\mathbf{x}(t_k) + \int_{t_k}^{t_{k+1}} e^{\mathbf{F}(t_{k+1}-\lambda)}\mathbf{G}(\lambda)\mathbf{w}(\lambda)d\lambda. \tag{4.108}$$

Comparison with eqn. (4.65) leads to the definition:

$$\mathbf{w}_k = \int_{t_k}^{t_{k+1}} e^{\mathbf{F}(t_{k+1}-\lambda)}\mathbf{G}(\lambda)\mathbf{w}(\lambda)d\lambda. \tag{4.109}$$

Then, with the assumption that $\mathbf{w}(t)$ is a white noise process, we can compute $\mathbf{Qd}_k = cov(\mathbf{w_k})$ as follows:

$$E\left\langle \int_{t_k}^{t_{k+1}} \int_{t_k}^{t_{k+1}} \mathbf{\Phi}(t_{k+1}, s)\mathbf{G}(s)\mathbf{w}(s)\mathbf{w}^\top(\tau)\mathbf{G}^\top(\tau)\mathbf{\Phi}(t_{k+1}, \tau)^\top d\tau ds \right\rangle$$
$$= \int_{t_k}^{t_{k+1}} \int_{t_k}^{t_{k+1}} \mathbf{\Phi}(t_{k+1}, s)\mathbf{G}(s)E\left\langle \mathbf{w}(s)\mathbf{w}^\top(\tau)\right\rangle \mathbf{G}^\top(\tau)\mathbf{\Phi}(t_{k+1}, \tau)^\top d\tau ds$$
$$= \int_{t_k}^{t_{k+1}} \int_{t_k}^{t_{k+1}} \mathbf{\Phi}(t_{k+1}, s)\mathbf{G}(s)\mathbf{Q}(s)\delta(s-\tau)\mathbf{G}^\top(\tau)\mathbf{\Phi}(t_{k+1}, \tau)^\top d\tau ds.$$

Therefore, the solution is

$$\mathbf{Qd}_k = \int_{t_k}^{t_{k+1}} \mathbf{\Phi}(t_{k+1}, s)\mathbf{G}(s)\mathbf{Q}(s)\mathbf{G}^\top(s)\mathbf{\Phi}(t_{k+1}, s)^\top ds. \tag{4.110}$$

If $\mathbf{F}$ and $\mathbf{Q}$ are both time invariant, then $\mathbf{Qd}$ is also time invariant.

A common approximate solution to eqn. (4.110) is

$$\mathbf{Qd} \approx \mathbf{GQG}^\top T \tag{4.111}$$

which is accurate only when the eigenvalues of $\mathbf{F}$ are very small relative to the sampling period T (i.e., $\|\mathbf{F}T\| \ll 1$). This approximation does not account for any of the correlations between the components of the driving noise $\mathbf{w}_k$ that develop over the course of a sampling period due to the integration of the continuous-time driving noise through the state dynamics. The reader should compare the results from Examples 4.21, 4.22, and 4.23.

Example 4.21 *For the double integrator system with*

$$\mathbf{F} = \begin{bmatrix} 0 & 1 \\ 0 & 0 \end{bmatrix} \text{ and } \mathbf{GQG}^\top = \begin{bmatrix} 0.00 & 0.00 \\ 0.00 & 0.01 \end{bmatrix},$$

compute $\mathbf{\Phi}$ *and* $\mathbf{Qd}$ *using eqns. (4.103) and (4.111) for* $T = 1$.

Using the Matlab function 'expm' we obtain $\mathbf{\Phi} = \begin{bmatrix} 1 & 1 \\ 0 & 1 \end{bmatrix}$. *Using eqs. (4.111) we obtain*

$$\mathbf{Qd} \approx \begin{bmatrix} 0 & 0 \\ 0 & 1 \end{bmatrix} \times 10^{-2}. \tag{4.112}$$

△

Given the state of computing power available, there is no reason why more accurate approximations for **Qd** are not used. A few methods for calculating the solution to eqn. (4.110) are described in the following subsections. Each of the following methods assumes that **F** and **Q** are both time invariant. If **F** and **Q** are slowly time varying, then these methods could be used to determine approximate solutions by recalculating $\mathbf{Q}_d$ over each sampling interval.

4.7.2.1 Solution by Matrix Exponentials

It is shown in [133] that the exponential of the $2n \times 2n$ matrix

$$\Xi = \begin{bmatrix} -\mathbf{F} & \mathbf{GQG}^\top \\ 0 & \mathbf{F}^\top \end{bmatrix} T \tag{4.113}$$

is

$$\Upsilon = e^{\Xi} = \begin{bmatrix} -\mathbf{D} & \mathbf{\Phi}^{-1}\mathbf{Qd}_w \\ 0 & \mathbf{\Phi}^\top \end{bmatrix} \tag{4.114}$$

where **D** is a dummy variable representing a portion of the answer that will not be used. Based on the expressions in the second column of eqn. (4.114), $\mathbf{\Phi}$ and $\mathbf{Qd}_w$ are calculated as

$$\begin{aligned} \mathbf{\Phi} &= \Upsilon[(n+1:2n),(n+1):2n]^\top & (4.115) \\ \mathbf{Qd} &= \mathbf{\Phi}\Upsilon[(1:n),(n+1):2n] & (4.116) \end{aligned}$$

where $\Upsilon[(i:j),(k:l)]$ denotes the the sub-matrix of Υ composed of the i through j-th rows and k through l-th columns of matrix Υ.

Example 4.22 *For the simple system defined in Example 4.21, after constructing Ξ and computing its matrix exponential, we have*

$$\Upsilon = \begin{bmatrix} 1 & -1 & -\frac{1}{6}\times 10^{-2} & -\frac{1}{2}\times 10^{-2} \\ 0 & 1 & \frac{1}{2}\times 10^{-2} & 1.0\times 10^{-2} \\ 0 & 0 & 1 & 0 \\ 0 & 0 & 1 & 1 \end{bmatrix}. \tag{4.117}$$

Therefore, using the lower right 2×2 block, we have that $\mathbf{\Phi} = \begin{bmatrix} 1 & 1 \\ 0 & 1 \end{bmatrix}$. *Using the upper right 2×2 block we have that*

$$\mathbf{Qd} = \begin{bmatrix} 1 & 1 \\ 0 & 1 \end{bmatrix} \begin{bmatrix} -\frac{1}{6} & -\frac{1}{2} \\ \frac{1}{2} & 1 \end{bmatrix} \times 10^{-2} = \begin{bmatrix} \frac{1}{3} & \frac{1}{2} \\ \frac{1}{2} & 1 \end{bmatrix} \times 10^{-2}$$

which is significantly different from the result in eqn. (4.112) that was obtained from the approximate solution in eqn. (4.111). △

4.7.2.2 Solution by Taylor Series

If the Taylor series approximation for $\mathbf{\Phi}$

$$\mathbf{\Phi} = e^{\mathbf{F}T} = I + \mathbf{F}T + \frac{1}{2}(\mathbf{F}T)^2 + \frac{1}{3!}(\mathbf{F}T)^3 + \dots \tag{4.118}$$

is substituted into eqn. (4.110), using $k = 0$ to simplify the notation, the result accurate to third order in $\mathbf{F}$ and 4th order in T is

$$\begin{aligned}\mathbf{Qd} \approx\ & \mathbf{Q}T + \left(\mathbf{FQ} + \mathbf{QF}^\top\right)\frac{T^2}{2} + \left(\mathbf{F}^2\mathbf{Q} + 2\mathbf{FQF}^\top + \mathbf{Q}\left(\mathbf{F}^\top\right)^2\right)\frac{T^3}{6} \\ & + \left(\mathbf{F}^3\mathbf{Q} + 3\mathbf{F}^2\mathbf{QF}^\top + 3\mathbf{FQ}\left(\mathbf{F}^\top\right)^2 + \mathbf{Q}\left(\mathbf{F}^\top\right)^3\right)\frac{T^4}{24}.\end{aligned} \tag{4.119}$$

Although the Taylor series approach is approximate for some implementations, eqn. (4.119) is sometimes a convenient means to identify a closed form solution for either $\mathbf{\Phi}$ or $\mathbf{Qd}$. Even an approximate solution in closed form such as eqn. (4.119) is useful in situations where $\mathbf{F}$ is time-dependent and eqns. (4.114–4.116) cannot be solved on-line.

Example 4.23 *For the simple system defined in Example 4.21,* $\mathbf{F}^n = \mathbf{0}$ *for* $n \geq 2$*; therefore, using eqn. (4.118)*

$$\mathbf{\Phi} = \mathbf{I} + \mathbf{F}T = \begin{bmatrix} 1 & 1 \\ 0 & 1 \end{bmatrix}.$$

Simplifying eqn. (4.119) for this specific example gives

$$\mathbf{Qd} = \mathbf{Q}T + \left(\mathbf{FQ} + \mathbf{QF}^\top\right)\frac{T^2}{2} + 2\mathbf{FQF}^\top\frac{T^3}{6}$$

which provides the same result as in Example 4.22. △

4.8 Linear State Estimation

For deterministic systems, Section 3.6 discussed the problem of state estimation. This section considers state estimation for stochastic, linear, discrete-time state space systems

$$\begin{aligned}\mathbf{x}_{k+1} &= \mathbf{\Phi}_k\mathbf{x}_k + \mathbf{\Gamma}_k\mathbf{u}_k + \boldsymbol{\omega}_k \\ \mathbf{y}_k &= \mathbf{H}_k\mathbf{x}_k + \boldsymbol{\nu}_k\end{aligned}$$

where $\mathbf{u}_k$ is a known signal. The standard notation and assumptions stated in Section 4.6.1 apply.

The state estimate is computed according to

$$
\begin{aligned}
\hat{\mathbf{x}}_{k+1}^{-} &= \mathbf{\Phi}_k \hat{\mathbf{x}}_k^{+} + \mathbf{\Gamma}_k \mathbf{u}_k \\
\hat{\mathbf{y}}_k^{-} &= \mathbf{H}_k \hat{\mathbf{x}}_k^{-} \\
\hat{\mathbf{x}}_k^{+} &= \hat{\mathbf{x}}_k^{-} + \mathbf{L}_k \delta \mathbf{y}_k^{-}
\end{aligned}
$$

where $\delta \mathbf{y}_k^{-} = \mathbf{y}_k - \hat{\mathbf{y}}_k^{-}$ is called the measurement residual and $\mathbf{L}_k$ is a designer specified gain vector. The notation $\hat{\mathbf{x}}_k^{-}$ indicates the estimate of $\mathbf{x}$ at time $t = kT$ before correcting the estimate for the information in the measurement $\mathbf{y}_k$ and $\hat{\mathbf{x}}_k^{+}$ denotes the estimate of $\mathbf{x}$ after correcting the estimate for the information in the measurement $\mathbf{y}_k$. Similar interpretations apply to $\mathbf{P}_x^{-}$, $\mathbf{P}_x^{+}$, $\hat{\mathbf{y}}_k^{-}$, and $\mathbf{P}_y^{-}$.

Defining the state estimation error vector at time k, prior and posterior to the measurement correction, as

$$
\begin{aligned}
\delta \mathbf{x}_k^{-} &= \mathbf{x}_k - \hat{\mathbf{x}}_k^{-} && (4.120) \\
\delta \mathbf{x}_k^{+} &= \mathbf{x}_k - \hat{\mathbf{x}}_k^{+}, && (4.121)
\end{aligned}
$$

the state estimation error time propagation, output error, and measurement update equations are

$$
\begin{aligned}
\delta \mathbf{x}_{k+1}^{-} &= \mathbf{\Phi}_k \delta \mathbf{x}_k^{+} + \boldsymbol{\omega}_k && (4.122) \\
\delta y_k^{-} &= \mathbf{H}_k \delta \mathbf{x}_k^{-} + \boldsymbol{\nu}_k && (4.123) \\
\delta \mathbf{x}_k^{+} &= (\mathbf{I} - \mathbf{L}_k \mathbf{H}_k)\, \delta \mathbf{x}_k^{-} - \mathbf{L}_k \boldsymbol{\nu}_k. && (4.124)
\end{aligned}
$$

To determine conditions for stability in the time invariant case, substituting eqn. (4.124) into eqn. (4.122), taking the expected value, and simplifying yields

$$
\boldsymbol{\mu}_{\delta_{k+1}} = \mathbf{\Phi}\,(\mathbf{I} - \mathbf{LH})\, \boldsymbol{\mu}_{\delta_k}
$$

where the notation $\boldsymbol{\mu}_{\delta_k}$ has been used for $E\langle \delta \mathbf{x}_k^{-} \rangle$. Therefore, the stability of the system requires that the eigenvalues of the matrix $\mathbf{\Phi}\,(\mathbf{I} - \mathbf{LH})$ be strictly less that one in magnitude.

From eqn. (4.122), the covariance of the state estimation error prior to the measurement update is given by eqn. (4.99):

$$
\mathbf{P}_x^{-}(k+1) = \mathbf{\Phi}_k \mathbf{P}_x^{+}(k) \mathbf{\Phi}_k^{\top} + \mathbf{Qd}_k. \quad (4.125)
$$

From eqn. (4.123), the covariance matrix for the predicted output error is

$$
\mathbf{P}_y^{-}(k) = \mathbf{H}_k \mathbf{P}_x^{-}(k) \mathbf{H}_k^{\top} + \mathbf{R}_k. \quad (4.126)
$$

From eqn. (4.124), the covariance matrix for the state estimation error posterior to the measurement correction is

$$
\mathbf{P}_x^{+}(k) = (\mathbf{I} - \mathbf{L}_k \mathbf{H}_k)\, \mathbf{P}_x^{-}(k)\, (\mathbf{I} - \mathbf{L}_k \mathbf{H}_k)^{\top} + \mathbf{L}_k \mathbf{R}_k \mathbf{L}_k^{\top}. \quad (4.127)
$$

If the gain sequence $\mathbf{L}_k$ is known, then eqns. (4.125–4.127) can be iterated to quantify the expected accuracy of the state estimation design for the given $\mathbf{L}_k$.

Eqn. (4.127) is true for any state feedback gain $\mathbf{L}_k$; therefore, this equation allows the performance (as measured by state error covariance) of alternative gain sequences to be quantitatively compared. In fact, eqn. (4.127) expresses the covariance of the state estimate immediately after the measurement correction as a function of the estimation gain vector $\mathbf{L}_k$; therefore, a new gain vector could be selected at each time instant to minimize the state estimation error variance. This is the topic of optimal stochastic state estimation which is discussed in Chapter 5.

4.9 Detailed Examples

This section presents a few detailed examples intended to demonstrate the utility of random variables and stochastic processes in relation to navigation applications. Sections 4.9.1 and 4.9.2 focus on frequently asked questions. Sections 4.9.3 and 4.9.4 discuss navigation examples that are relatively simple compared to those in Part II of the book, but follow the same methodology.

4.9.1 System Performance Metrics

Various terms are used to characterize navigation system performance. This section uses results from linear algebra, properties of Gaussian random variables, and the change of variable concepts from Section 4.3.4, to derive the relationships between these alternative performance characterizations.

The discussion makes two basic assumptions:

1. The position error vector $\mathbf{w} = [n, e, d]^\top$ is a Gaussian random variable.

2. The components of the position error are unbiased.

The validity of the assumptions must be considered in the use of the subsequent results.

The Normal (Gaussian) assumption is motivated by the *Central Limit Theorem.* Although the assumption may not be absolutely valid, the large number of small, independent error sources in a navigation system results in error distributions which can be accurately approximated as Gaussian. This assumption is not typically a major source of difficulty.

The unbiased assumption must be considered more carefully. The unbiased assumption can be considered as a distinction between precision and accuracy. Ideally, an analyst would calculate errors as $\delta\mathbf{x} = \mathbf{x} - \hat{\mathbf{x}}$ where

$\mathbf{x}$ denotes the true value of the variable. When $\mathbf{x}$ is unknown, the analyst might instead calculate the error as $\delta\mathbf{x}_c = \hat{\mathbf{x}} - \bar{\mathbf{x}}$ where $\bar{\mathbf{x}}$ denotes the average of the $\hat{\mathbf{x}}$ values[1]. If there is a bias between x and $\bar{x}$, then the second order statistics of $\delta\mathbf{x}_c$ (standard deviation or variance) will indicate the precision of $\hat{\mathbf{x}}$ without indicating its accuracy.

4.9.1.1 Analysis in Three Dimensions

Let $\mathbf{w} = [n, e, d]^\top$ denote the position error variables. Denote the covariance matrix for $\mathbf{w}$ by the symmetric positive definite matrix $\mathbf{P} = E\langle \mathbf{w}\mathbf{w}^\top \rangle$. Given the above assumptions that $\mathbf{w} \sim N(\mathbf{0}, \mathbf{P})$, the probability density is

$$p_{\mathbf{w}}(n, e, d) = \frac{1}{(2\pi)^{3/2} |\mathbf{P}|^{1/2}} \exp\left(-\frac{1}{2} \left(\mathbf{w}^\top \mathbf{P}^{-1} \mathbf{w} \right) \right). \tag{4.128}$$

Because $\mathbf{P}$ is symmetric positive definite, there exists orthonormal matrix $\mathbf{U}$ and positive definite diagonal matrix $\mathbf{\Sigma}$ such that $\mathbf{P} = \mathbf{U}\mathbf{\Sigma}^2\mathbf{U}^\top$. The units of the diagonal elements of $\mathbf{\Sigma} = diag([\sigma_1, \sigma_2, \sigma_3])$ are meters.

To simplify the analysis, we introduce a new random variable $\mathbf{v} = \mathbf{S}\mathbf{w}$ where $\mathbf{S} = \mathbf{\Sigma}^{-1}\mathbf{U}^\top$. The covariance matrix for $\mathbf{v}$ is the identity matrix which is shown as follows:

$$\begin{aligned} E\left\langle \mathbf{v}\mathbf{v}^\top \right\rangle &= E\left\langle \mathbf{S}\mathbf{w}\mathbf{w}^\top\mathbf{S}^\top \right\rangle \\ &= \mathbf{S}\mathbf{P}\mathbf{S}^\top \\ &= \left(\mathbf{\Sigma}^{-1}\mathbf{U}^\top\right)\left(\mathbf{U}\mathbf{\Sigma}^2\mathbf{U}^\top\right)\left(\mathbf{U}\mathbf{\Sigma}^{-1}\right) \\ &= \mathbf{I}. \end{aligned}$$

Therefore, $\mathbf{v} \sim N(0, \mathbf{I})$. The probability density function for $\mathbf{v}$ is

$$p_{\mathbf{v}} = \frac{1}{(2\pi)^{\frac{3}{2}}} \exp\left(-\frac{\mathbf{v}^\top\mathbf{v}}{2} \right).$$

The components of the vector $\mathbf{v}$ are independent and identically distributed with variance equal to one. The components of $\mathbf{v}$ are dimensionless.

Given the previous definitions and a positive scalar c, we have the following relationships

$$Prob\{\|\mathbf{v}\| \leq c\} = Prob\{\mathbf{w}^\top\mathbf{S}^\top\mathbf{S}\mathbf{w} \leq c^2\} = Prob\{\mathbf{w}^\top\mathbf{P}^{-1}\mathbf{w} \leq c^2\}.$$

We will find the value of c such that the sphere in the $\mathbf{v}$ coordinates with radius c contains 50% of the $\mathbf{v}$ samples. In the $\mathbf{w}$ coordinates the region expected to contain 50% of the $\mathbf{w}$ samples will be the ellipsoid

[1] Alternatively, $\bar{\mathbf{x}}$ could be the measurement of $\mathbf{x}$ by a more accurate 'ground truth' system.

$\{\mathbf{w} \in \mathbb{R}^3 \,|\mathbf{w}^\top \mathbf{P}^{-1}\mathbf{w} \leq c^2\}$. Because $\mathbf{w} = \mathbf{U}\boldsymbol{\Sigma}\mathbf{v}$ and $\mathbf{U}$ is an orthonormal matrix, the columns of $\mathbf{U}$ define the principle axes of the ellipsoid and the diagonal elements of $\boldsymbol{\Sigma}$ define the elongation of the ellipsoidal axes. For example, if $\mathbf{U} = [\mathbf{u}_1, \mathbf{u}_2, \mathbf{u}_3]$ where $\mathbf{u}_i$ is the i-th column of $\mathbf{U}$, then a unit vector along the i-th coordinate axis of the $\mathbf{v}$ reference frame transforms to the vector $(\sigma_i \mathbf{u}_i)$ in the $\mathbf{w}$ reference frame.

The probability that $\mathbf{v}$ falls in the region $\|\mathbf{v}\| \leq c$ can be derived as

$$\begin{aligned}
Prob\{\|\mathbf{v}\| \leq c\} &= \int_{\|\mathbf{v}\| \leq c} p_{\mathbf{v}}(\mathbf{v}) d\mathbf{v} \\
&= \frac{1}{(2\pi)^{\frac{3}{2}}} \int_{\|\mathbf{v}\| \leq c} \exp\left(-\frac{\|\mathbf{v}\|}{2}\right) d\mathbf{v} \\
&= \frac{1}{(2\pi)^{\frac{3}{2}}} \int_0^{2\pi} \int_0^{\pi} \int_0^{c} r^2 \exp\left(-\frac{r^2}{2}\right) \sin\phi_1 dr d\phi_1 d\phi_2 \\
&= \sqrt{\frac{2}{\pi}} \int_0^c r^2 \exp\left(-\frac{r^2}{2}\right) dr \\
&= \int_0^c p_\rho(r) dr = Prob\{\rho \leq c\}.
\end{aligned}$$

The random variable $\rho = \|\mathbf{v}\|$ is called a *Maxwell* random variable. The density for a Maxwell random variable is

$$p_\rho(r) = \sqrt{\frac{2}{\pi}} r^2 \exp\left(-\frac{r^2}{2}\right).$$

The value of c such that $Prob\{\|\mathbf{v}\| \leq c\} = 0.5$ is $c = 1.54$. A closed-form, but transcendental, solution to this equation exists [39], but this values was found by table look-up for the cumulative distribution of a Maxwell random variable.

We can introduce an intermediate variable $\mathbf{x}$ such that $\mathbf{v} = \boldsymbol{\Sigma}^{-1}\mathbf{x}$ and $\mathbf{x} = \mathbf{U}^\top \mathbf{w}$. The dimensions of the variable $\mathbf{x}$ are meters. The components of the vector $\mathbf{x}$ are jointly distributed, independent, and Gaussian. In the $\mathbf{x}$ reference frame, for a given value of c, the uncertainty ellipsoid $\mathbf{x}^\top \boldsymbol{\Sigma}^{-2}\mathbf{x} \leq c$ has axes coincident with the $\mathbf{x}$ coordinate axes and

$$Prob\{\mathbf{w}^\top \mathbf{P}^{-1}\mathbf{w} \leq c^2\} = Prob\{\mathbf{x}^\top \boldsymbol{\Sigma}^{-2}\mathbf{x} \leq c^2\}.$$

In the $\mathbf{x}$ coordinate frame the lengths of the ellipsoid principal axes are defined by the diagonal elements of $\boldsymbol{\Sigma}$ with the ellipsoid defined by

$$\frac{x_1^2}{\sigma_1^2} + \frac{x_2^2}{\sigma_2^2} + \frac{x_3^2}{\sigma_3^2} \leq c^2. \tag{4.129}$$

The ellipsoid in $\mathbf{w}$ coordinates is simply a rotation of ellipsoid specified in eqn. (4.129).

Spherical Error Probable (SEP) specifies the radius of the sphere centered on correct location that is expected to contain 50% of the realizations variable $\mathbf{w}$. In the special case where $\sigma = \sigma_1 = \sigma_2 = \sigma_3$, then eqn. (4.129) simplifies to the equation for the sphere

$$\|\mathbf{w}\| = \|\mathbf{x}\| \leq 1.54\sigma.$$

In all other cases, the probability ellipsoid expected to contain 50% of the realizations is defined by eqn. (4.129) with $c = 1.54$. Various approximations exist for estimating the radius of the SEP sphere [39].

4.9.1.2 Analysis in Two Dimensions

The analysis of this section is similar to that of Section 4.9.1.1, but performed in two dimensions. The two dimensions that are of interest are typically the horizontal (e.g., north and east) components. In this section, we will use the vectors $\vec{\mathbf{w}}$, $\vec{\mathbf{v}}$, $\vec{\mathbf{x}}$ which have interpretations parallel to $\mathbf{w}$, $\mathbf{v}$, $\mathbf{x}$ from Section 4.9.1.1. The vector symbol over the symbols indicates that the corresponding variable is an element of $\mathbb{R}^2$.

The joint horizontal position error density in the $\vec{\mathbf{w}}$ coordinates is

$$p_{\vec{\mathbf{w}}} = \frac{1}{2\pi\sqrt{|\vec{\mathbf{P}}|}} exp\left(-\frac{1}{2}\vec{\mathbf{w}}^\top \vec{\mathbf{P}}^{-1}\vec{\mathbf{w}}\right). \tag{4.130}$$

The transformation from the three dimensional distribution of eqn. (4.128) to the two dimensional distribution of eqn. (4.130) is discussed in Exercise 4.10. The joint horizontal position error density in the $\vec{\mathbf{x}}$ coordinates is

$$p_{\vec{\mathbf{x}}} = \frac{1}{2\pi\sigma_1\sigma_2} exp\left(-\left(\frac{x_1^2}{2\sigma_1^2} + \frac{x_2^2}{2\sigma_2^2}\right)\right). \tag{4.131}$$

The joint horizontal position error density in the $\vec{\mathbf{v}}$ coordinates is

$$p_{\vec{\mathbf{v}}} = \frac{1}{2\pi} exp\left(-\frac{\|\vec{\mathbf{v}}\|^2}{2}\right). \tag{4.132}$$

Circular Error Probable (CEP) is the radius of the circle centered on the correct location that contains 50% of the expected horizontal position errors. The *R95* statistic is the radius of the circle expected to contain 95% of the horizontal position errors. The analysis to define both of these statistics is the same. Therefore, we will perform the analysis for probability $\beta \in [0, 1]$. At the conclusion of the analysis, we can compute the results for $\beta = 0.5$ or 0.95.

Let

$$\beta = Prob\{\|\vec{\mathbf{v}}\| \leq c\},$$

then for $\vec{\mathbf{S}}$, $\vec{\mathbf{U}}$, $\vec{\mathbf{\Sigma}}$, defined as in the previous section,

$$\beta = Prob\{\|\vec{\mathbf{v}}\| \leq c\} = Prob\{\vec{\mathbf{w}}^\top \vec{\mathbf{P}}^{-1} \vec{\mathbf{w}} \leq c^2\} = Prob\{\vec{\mathbf{x}}^\top \vec{\mathbf{\Sigma}}^{-2} \vec{\mathbf{x}} \leq c^2\}.$$

The probability that $\vec{\mathbf{v}}$ falls in the circle with radius c can be found, in polar coordinates, as

$$\begin{aligned} \beta = Prob\{\|\vec{\mathbf{v}}\| \leq c\} &= \frac{1}{2\pi} \int_0^c \int_0^{2\pi} \exp\left(\frac{-r^2}{2}\right) r d\theta dr \\ &= \int_0^c r \exp\left(-\frac{r^2}{2}\right) dr \\ &= 1 - \exp\left(-\frac{c^2}{2}\right). \end{aligned} \tag{4.133}$$

The random variable $\vec{\rho} = \|\vec{v}\|$ is a Rayleigh random variable with distribution

$$P_{\vec{\rho}}(r) = 1 - \exp\left(-\frac{r^2}{2}\right)$$

and density

$$p_{\vec{\rho}}(r) = r \exp\left(-\frac{r^2}{2}\right).$$

From eqn. (4.133), the radius such that $P_{\vec{\rho}}(R_{CEP}) = 0.5$ is computed as

$$\begin{aligned} 0.5 &= 1 - \exp\left(-\frac{R_{CEP}^2}{2}\right) \\ R_{CEP} &= \sqrt{2}\sqrt{ln(2)} \\ &\approx 1.1774. \end{aligned}$$

Similarly, the radius such that $P_{\vec{\rho}}(R_{95}) = 0.95$ is calculated as

$$\begin{aligned} 0.95 &= 1 - exp\left(-\frac{R_{95}^2}{2}\right) \\ R_{95} &= \sqrt{2}\sqrt{ln(20)} \\ &\approx 2.4477. \end{aligned}$$

For a given value of c, the ellipse in $\vec{\mathbf{x}}$ coordinates is

$$\frac{x_1^2}{\sigma_1^2} + \frac{x_2^2}{\sigma_2^2} \leq c^2. \tag{4.134}$$

The ellipse in $\vec{\mathbf{w}}$ coordinates is

$$\vec{\mathbf{w}}^\top \vec{\mathbf{P}}^{-1} \vec{\mathbf{w}} \leq c^2. \tag{4.135}$$

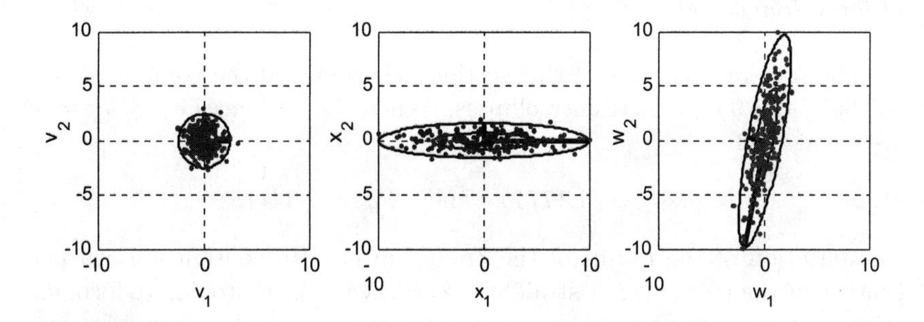

Figure 4.5: Distributions of the $\vec{\mathbf{v}}$, $\vec{\mathbf{x}}$, $\vec{\mathbf{w}}$ variables from Example 4.24 with the 50% and 95% ellipses indicated.

The ellipses defined by eqns. (4.134) and (4.135) each contain the same area $\pi\sigma_1\sigma_2 c^2$ which is distinct from the area enclosed by the circle of radius c in the $\vec{\mathbf{v}}$ coordinates. The ellipse in $\vec{\mathbf{w}}$ coordinates can be drawn numerically by selecting vectors $\vec{\mathbf{v}}$ on the circle of radius c and transforming those vectors according to $\vec{\mathbf{w}} = \vec{\mathbf{U}}\vec{\mathbf{\Sigma}}\vec{\mathbf{v}}$. The locus of points around the circle of radius c in the $\vec{\mathbf{v}}$ coordinates will transform to the ellipse of eqn. (4.135). This procedure is demonstrated in the following example.

Example 4.24 *Consider the situation where*

$$cov(\vec{\mathbf{w}}) = \vec{\mathbf{P}} = \begin{bmatrix} 1 & 3 \\ 3 & 16 \end{bmatrix}$$

which has $\sigma_n = 1m$, $\sigma_e = 4m$, and the correlation coefficient between $\vec{\mathbf{w}}_1$ and $\vec{\mathbf{w}}_2$ is 0.75. The decomposition of $\vec{\mathbf{P}}$, computed by use of the singular value decomposition, is

$$\vec{\mathbf{U}} = \begin{bmatrix} -0.1891 & -0.9820 \\ -0.9820 & 0.1891 \end{bmatrix} \text{ and } \vec{\mathbf{\Sigma}} = \begin{bmatrix} 4.0716 & 0.0000 \\ 0.0000 & 0.6498 \end{bmatrix}.$$

The left graph of Figure 4.5 displays a scatter plot distribution for 200 samples of $\vec{\mathbf{v}}$ where the components of are i.i.d Gaussian (pseudo) random variables with unit variance. The density of $\vec{\mathbf{v}}$ is shown in eqn. (4.132). The graph also shows the circles expected to contain 50% and 95% of the samples. In the experiment shown 97 of the 200 samples (i.e., 48.5%) are within the CEP circle and 191 of 200 (i.e., 95.5%) of the samples are within the R95 circle. Lastly, the figure indicates vectors along the $\vec{\mathbf{v}}$ frame axes.

The center and right graphes of Figure 4.5 displays a scatter plot distribution for the resulting 200 samples of $\vec{\mathbf{x}}$ and $\vec{\mathbf{w}}$, respectively. In each *graph, the 50% and 95% ellipses are indicated as are the transformations of the* $\vec{\mathbf{v}}$ *frame axes.* △

The previous analysis of this section has discussed the computation of the 50% and 95% uncertainty ellipses. When the variances $\sigma_1^2 = \sigma_2^2 = \sigma^2$ are equal then

$$R_{CEP} = 1.1774\sigma \quad \text{and} \quad R_{95} = 2.4477\sigma$$

Computation of the radius of the *circle* expected to contain a given percentage of the points is not straightforward. Various approximate formulae exist. A few examples follow. Let $\bar{\sigma} = max(\sigma_n, \sigma_e)$ and $\underline{\sigma} = min(\sigma_n, \sigma_e)$. Define $\rho = \underline{\sigma}/\bar{\sigma}$. The first formula is

$$R_{CEP} = 0.589(\sigma_n + \sigma_e)$$

which is accurate to 3% for $\rho \in [0.2, 1.0]$. The second formula is

$$R_{CEP} = 0.615\underline{\sigma} + 0.562\bar{\sigma}$$

which is much more accurate than the previous formula for $\rho \in [0.3, 1.0]$. For $\rho \in [0.0, 0.1]$, $R_{CEP} \approx 0.675\bar{\sigma}$.

In addition to the statistics previously discussed, two additional measures are sometimes of interest: drms and 2drms. The *distance root-mean-square (drms)* is the *root-mean-square (RMS)* value of the norm of the horizontal errors. The formula for the drms measure is

$$drms = \sqrt{\sigma_n^2 + \sigma_e^2}.$$

The *2drms* error measure is twice the drms value:

$$2drms = 2(drms).$$

The reason for mentioning the 2drms is that the name is subject to misinterpretation. Note in particular that 2drms is *not* the two-dimensional RMS position error. If the horizontal error distribution is circular (i.e., $\sigma_n = \sigma_e = \sigma$), then

$$\begin{aligned} drms &= \sqrt{2}\sigma \\ 2drms &= 2\sqrt{2}\sigma. \end{aligned}$$

4.9.1.3 Scalar Analysis

Let x represent one component of the position error, the component-wise error density is:

$$p(x) = \frac{1}{\sqrt{2\pi\sigma^2}} exp\left(-\frac{x^2}{2\sigma^2}\right). \tag{4.136}$$

Statistic	RMS	drms	2drms	CEP	R95
Radius	σ	$\sqrt{2}\sigma$	$2\sqrt{2}\sigma$	$\sqrt{2}\sqrt{ln(2)}\sigma$	$\sqrt{2}\sqrt{ln(20)}\sigma$
Probability	.393	.632	.982	.500	.950

Table 4.1: Summary of various horizontal error statistics in $\vec{\mathbf{w}}$ coordinates and their relationships to the error standard deviation σ assuming a circular error distribution (i.e., $\vec{\mathbf{\Sigma}} = \sigma\mathbf{I}$).

Given a set of samples $\{x_i\}_{i=1}^{N}$ of the random variable x, the RMS error is

$$rms_x = \sigma_x = \sqrt{\frac{1}{N}\sum_{i=1}^{N} x_i^2}. \tag{4.137}$$

For large values of N, the RMS value is expected to converge to σ.

From distribution tables for Gaussian random variables (see p. 48 in [107]),

$$Prob\{|x| < 0.674\sigma\} = 50.0\% \qquad Prob\{|x| < \sigma\} = 68.3\%$$
$$Prob\{|x| < 2\sigma\} = 95.5\% \qquad Prob\{|x| < 3\sigma\} = 99.7\%.$$

Therefore, if the altitude h is estimated to be $100m$ with $\sigma = 5m$, then the probability that $h \in [95, 105]m$ is 68%, that $h \in [90, 110]m$ is 95%, etc.

4.9.1.4 Summary Comparison

Table 4.1 summarizes the various horizontal error statistics defined above in relation to the one dimensional error standard deviation σ. The table is useful for converting between error statistics.

Example 4.25 *If an analyst is interested in finding the* R95 *error statistic that is equivalent to a stated* drms *error statistic of 10.0 m, then from Table 4.1:*

$$drms = 10 \quad \rightarrow \quad \sigma = \frac{10}{\sqrt{2}}$$
$$\rightarrow \quad R95 = \sqrt{2ln(20)}\frac{10}{\sqrt{2}}.$$

Therefore, the equivalent R95 *statistic is 17.3 m.* △

Repeating the process of Example 4.25 for each of the other possible combinations of accuracy statistics in two dimensions results in Table 4.2.

	σ	drms	2drms	CEP	R95
σ	1.0	1.4	2.8	1.2	2.4
drms	0.7	1.0	2.0	0.8	1.7
2drms	0.4	0.5	1.0	0.4	0.9
CEP	0.8	1.2	2.4	1.0	2.1
R95	0.4	0.6	1.2	0.5	1.0

Table 4.2: Summary of conversion factors from the statistic listed in the in the leftmost column to the statistic indicated in the top row, assuming a circular error distribution (i.e., $\vec{\mathbf{\Sigma}} = \sigma\mathbf{I}$).

To use this table, find the row corresponding to the given statistic. Multiply this row by the numeric value of that statistic to obtain all the other statistics as indicated in the (top) header row. For example, if the $R95$ statistic is 1.0 m, them the other statistics are

$$\sigma = 0.4m, \ drms = 0.6m, \ 2drms = 1.2m, \text{ and } \ R_{CEP} = 0.5m.$$

4.9.2 Instrument Specifications

Inertial instrument specifications quantify various aspects of sensor performance: range, bandwidth, linearity, random walk, rate random walk, etc. A subset of these parameters specify the expected behavioral characteristics of the stochastic sensor errors. The purpose of this example is to illustrate the relationship between these parameters, the sensor data, and the stochastic error model. The section has two goals:

- to illustrate how the parameters of the instrument error models are determined from a set of instrument data; and
- to clarify how the data from an instrument specification data sheet relates to the type of Gauss-Markov model discussed in Section 4.6.

The example will focus on a gyro error model using a simplified version of the error model described in [9, 10] after setting all the flicker noise components to zero. The methodology is similar for accelerometers [11]. This section contains a very basic description of the error model and model parameter estimation method. Many details of the technical procedures have been excluded.

Assume that the output y of a gyro attached to a non-rotating platform is modeled as

$$\begin{bmatrix} \dot{x}_1 \\ \dot{x}_2 \end{bmatrix} = \begin{bmatrix} 0 & 1 \\ 0 & 0 \end{bmatrix} \begin{bmatrix} x_1 \\ x_2 \end{bmatrix} + \begin{bmatrix} 1 \\ 0 \end{bmatrix} \kappa \tag{4.138}$$

$$y = x_1 + \eta. \tag{4.139}$$

The initial conditions are $x_1(0) \sim N(0, B_0^2)$ and $x_2(0) \sim N(0, R_0^2)$. The stochastic process noise $\kappa(t)$ and measurement noise η are independent, white, and Gaussian with $R_\eta(\tau) = N^2\delta(\tau)$ and $R_\kappa(\tau) = K^2\delta(\tau)$. The nomenclature and units related to the various symbols are shown in Table 4.3.

Symbol	Interpretation	$S_{\tilde{Y}}(j\omega)$	Units
B_0	Random bias, bias, fixed drift	–	$\frac{deg}{hr}$
K	Rate random walk	$\frac{K^2}{\omega^2}$	$\frac{deg/hr}{\sqrt{hr}}$
N	Angular random walk	N^2	$\frac{deg}{\sqrt{hr}}$
R_0	Rate ramp, Random ramp	–	$\frac{deg}{hr^2}$

Table 4.3: Parameter definitions for the gyro error model of eqns. (4.138–4.139). The PSD $S_{\tilde{Y}}(j\omega)$ is defined in eqn. (4.140).

Prior to estimating the model parameters K and N, the effect of the initial conditions B_0 and R_0 must first be removed. For each sequence of available data $Y = \{y(t_k)\}_{k=1}^{M}$, the first step is to remove the trend defined by

$$y_k = b_0 + r_0 t_k.$$

This can be performed by least squares curve fitting as discussed in Section 5.3.2. Representing the data in Y as a column vector, we have that $\mathbf{Y} = \mathbf{A}\boldsymbol{\theta}$ where

$$\mathbf{A} = \begin{bmatrix} 1 & t_1 \\ \vdots & \vdots \\ 1 & t_M \end{bmatrix} \text{ and } \boldsymbol{\theta} = \begin{bmatrix} b_0 \\ r_0 \end{bmatrix}.$$

Because the sampling times are unique, the matrix $\mathbf{A}$ has full row rank and $(\mathbf{A}^\top\mathbf{A})$ is nonsingular. Therefore, the least squares estimate $\hat{\boldsymbol{\theta}} = [\hat{b}_0, \hat{r}_0]^\top$ is computed as (see eqn. (5.26) on p. 176)

$$\hat{\boldsymbol{\theta}} = \left(\mathbf{A}^\top\mathbf{A}\right)^{-1}\mathbf{A}^\top\mathbf{Y}.$$

Given a large set of instruments, with $\hat{\boldsymbol{\theta}}$ estimated for each, the distribution of b_0 and r_0 could be estimated and the variance parameters B_0^2 and R_0^2 can be determined.

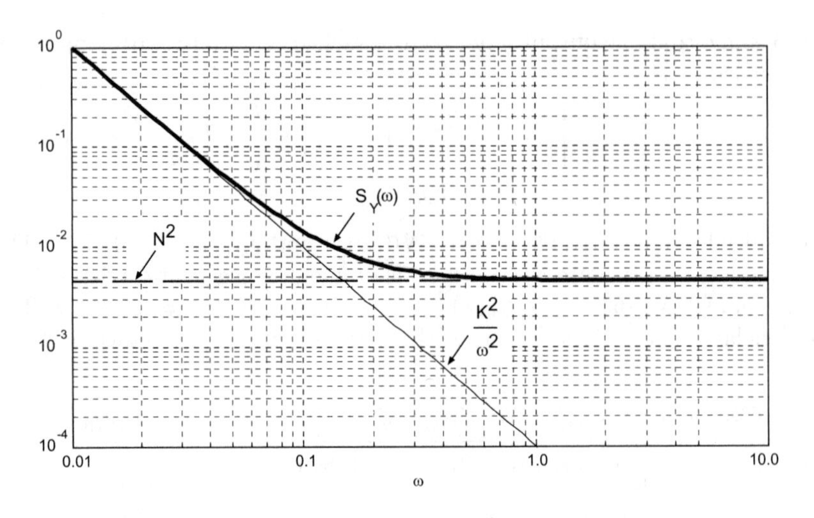

Figure 4.6: Power spectral density of eqn. (4.140).

Given $\hat{\boldsymbol{\theta}}$ and the data set $Y = \{y(t_k)\}_{k=1}^{M}$, the sequence of residual measurements $\tilde{Y} = \{y(t_k) - (\hat{b}_0 + \hat{r}_0 t_k)\}_{k=1}^{M}$ can be formed. This process is referred to as detrending and $\tilde{Y}$ is the detrended data. The covariance sequence $R_{\tilde{Y}}(\tau)$ is computed from $\tilde{Y}$ under the ergodic assumption, and the PSD of $\tilde{Y}$ is computed as the Fourier transform of $R_{\tilde{Y}}(\tau)$. Using the results of Section 4.5, for $B_0 = R_0 = 0$, the transfer function model for the system of eqns. (4.138–4.139) is

$$\tilde{Y}(s) = \begin{bmatrix} \frac{1}{s} & 1 \end{bmatrix} \begin{bmatrix} \kappa(s) \\ \nu(s) \end{bmatrix}.$$

Given that $\eta(t)$ and $\kappa(t)$ are independent, the PSD of $\tilde{y}$ is

$$S_{\tilde{Y}}(j\omega) = N^2 + \frac{K^2}{\omega^2}. \tag{4.140}$$

Therefore, a curve fit in the frequency domain, of the theoretical model for $S_{\tilde{Y}}$ from eqn. (4.140) to the PSD data computed from $\tilde{Y}$ provides the estimates of N and K.

Example 4.26 *Figure 4.6 displays a graph of eqn. (4.140) along with its component parts N^2 and $\frac{K^2}{\omega^2}$. From the graph, N^2 is the approximately* 0.0045*and K^2 is approximately* 1.0×10^{-4}.

△

The instrument specification parameters are sometimes referred to as *Allan variance parameters.* This name refers to the Allan variance method that is an alternative approach from the PSD for processing the data set $\tilde{Y}$ to estimate the parameters of the stochastic error model. The Allan variance method was originally defined as a means to quantify clock (or oscillator) stability.

4.9.3 One Dimensional INS

This example analyzes the growth of uncertainty as measured by error variance in a simple unaided inertial navigation system. The example will illustrate that the analysis methods of this chapter allow quantitative analysis of each contribution to the system error to be considered in isolation.

Consider a one dimensional single accelerometer inertial navigation system (INS) implemented in an inertial frame. The actual system equations are

$$\left.\begin{array}{rcl}\dot{p}(t) & = & v(t) \\ \dot{v}(t) & = & a(t).\end{array}\right\} \tag{4.141}$$

An accelerometer is available that provides the measurement

$$\tilde{u}(t) \quad = \quad a(t) - b(t) - \nu_a(t) \tag{4.142}$$

where the manufacturer specifies that:

- ν_a is Gaussian white noise with PSD equal to $\sigma_{\nu_a}^2 = 2.5 \times 10^{-3} \frac{m^2}{s^3}$;
- b is the random walk process

  $$\dot{b} = \omega_b(t)$$

  where ω_b is Gaussian and white with PSD equal to $\sigma_{\omega_b}^2 = 1.0 \times 10^{-6} \frac{m^2}{s^5}$;
- the distribution of the initial bias is $b(0) \sim N(0, P_{b_0})$.

The plant state vector is $\mathbf{x} = [p, v, b]$.

The implemented navigation equations are

$$\left.\begin{array}{rcl}\dot{\hat{p}}(t) & = & \hat{v}(t) \\ \dot{\hat{v}}(t) & = & \hat{a}(t)\end{array}\right\} \tag{4.143}$$

where $\hat{a}(t) = \tilde{u}(t) + \hat{b}(t)$ and $\hat{b}(t)$ is the best estimate of the accelerometer bias b that is available at time t. These equations are integrated forward in time starting from the expected initial conditions where we assume that $p(0) \sim N(p_0, P_{p_0})$ and $v(0) \sim N(v_0, P_{v_0})$. The navigation system state vector is $\hat{\mathbf{x}} = [\hat{p}, \hat{v}, \hat{b}]$ where the differential equation for $\hat{b}$ is $\dot{\hat{b}} = 0$ because the mean of ω_b is zero.

Therefore, the dynamic equations for the error states $\delta p = p - \hat{p}$, $\delta v = v - \hat{v}$, $\delta b = b - \hat{b}$ are

$$\delta\dot{\mathbf{x}}(t) = \begin{bmatrix} \delta\dot{p}(t) \\ \delta\dot{v}(t) \\ \delta\dot{b}(t) \end{bmatrix} = \underbrace{\begin{bmatrix} 0 & 1 & 0 \\ 0 & 0 & 1 \\ 0 & 0 & 0 \end{bmatrix}}_{\mathbf{F}} \delta\mathbf{x}(t) + \underbrace{\begin{bmatrix} 0 & 0 \\ 1 & 0 \\ 0 & 1 \end{bmatrix}}_{\mathbf{\Gamma}} \begin{bmatrix} v_a(t) \\ \omega_b(t) \end{bmatrix}$$

and using the definition of $\mathbf{Q}$ in eqn. (4.59) in Section 4.6.1

$$\mathbf{\Gamma Q \Gamma}^\top = \begin{bmatrix} 0 & 0 & 0 \\ 0 & \sigma_{v_a}^2 & 0 \\ 0 & 0 & \sigma_{\omega_b}^2 \end{bmatrix} = \begin{bmatrix} 0.0 & 0.0 & 0.0 \\ 0.0 & 2.5\times 10^{-3} & 0.0 \\ 0.0 & 0.0 & 1.0\times 10^{-6} \end{bmatrix}.$$

For any time t, due to the fact that in this example $\mathbf{F}$ and $\mathbf{\Gamma}$ are time invariant, the state error covariance can be found in closed form as

$$\mathbf{P}_x(t) = \mathbf{\Phi}(t)\mathbf{P}_x(0)\mathbf{\Phi}^\top(t) + \int_0^t \mathbf{\Phi}(\tau)\mathbf{\Gamma Q \Gamma}^\top \mathbf{\Phi}^\top(\tau)d\tau. \tag{4.144}$$

Verification that eqn. (4.144) is the solution of eqn. (4.100) is considered in Exercise 4.18. In this example,

$$\mathbf{\Phi}(\tau) = \exp(\mathbf{F}\tau) = \begin{bmatrix} 1 & \tau & \frac{\tau^2}{2} \\ 0 & 1 & \tau \\ 0 & 0 & 1 \end{bmatrix},$$

and the two terms in eqn. (4.144) are

$$\mathbf{\Phi}(t)\mathbf{P}_x(0)\mathbf{\Phi}^\top(t) = \begin{bmatrix} P_{p_0} + P_{v_0}t^2 + P_{b_0}\frac{t^4}{4} & P_{v_0}t + P_{b_0}\frac{t^3}{2} & P_b\frac{t^2}{2} \\ P_{v_0}t + P_{b_0}\frac{t^3}{2} & P_{v_0} + P_{b_0}t^2 & P_{b_0}t \\ P_{b_0}\frac{t^2}{2} & P_{b_0}t & P_{b_0} \end{bmatrix}$$

and

$$\int_0^t \mathbf{\Phi}(\tau)\mathbf{\Gamma Q \Gamma}^\top \mathbf{\Phi}^\top(\tau)d\tau = \begin{bmatrix} \frac{\sigma_{\omega_v}^2 t^3}{3} + \frac{\sigma_{\omega_b}^2 t^5}{20} & \frac{\sigma_{\omega_v}^2 t^2}{2} + \frac{\sigma_{\omega_b}^2 t^4}{8} & \frac{\sigma_{\omega_b}^2 t^3}{6} \\ \frac{\sigma_{\omega_v}^2 t^2}{2} + \frac{\sigma_{\omega_b}^2 t^4}{8} & \sigma_{\omega_v}^2 t + \frac{\sigma_{\omega_b}^2 t^3}{3} & \frac{\sigma_{\omega_b}^2 t^2}{2} \\ \frac{\sigma_{\omega_b}^2 t^3}{6} & \frac{\sigma_{\omega_b}^2 t^2}{2} & \sigma_{\omega_b}^2 t \end{bmatrix}$$

where we have used the fact that $\mathbf{P}(0) = diag(P_{p_0}, P_{v_0}, P_{b_0})$. Therefore, the error variance of each of the three states is described by

$$\begin{aligned} P_p(t) &= \left(P_{p_0} + P_{v_0}t^2 + P_{b_0}\frac{t^4}{4}\right) + \left(\frac{\sigma_{\omega_v}^2 t^3}{3} + \frac{\sigma_{\omega_b}^2 t^5}{20}\right) \\ P_v(t) &= \left(P_{v_0} + P_{b_0}t^2\right) + \left(\sigma_{\omega_v}^2 t + \frac{\sigma_{\omega_b}^2 t^3}{3}\right) \\ P_b(t) &= P_{b_0} + \sigma_{\omega_b}^2 t. \end{aligned}$$

The growth of the error variance has two components. The first component is due to the initial uncertainty in each error state. This component dominates initially. The second component is due to the driving process noise components v_a and ω_b. The driving noise component dominates the growth for large time intervals. The driving noise is determined by the quality of the inertial instruments. Better instruments will yield slower growth of the INS error. However, without some form of aiding measurement, the covariance of the state will increase without bound.

Note that the expression for $\mathbf{P}_x$ in eqn. (4.144) is linear in both $\mathbf{P}_x(0)$ and $\mathbf{Q}$. This fact is used in Chapter 6 to quantitatively compare design tradeoffs.

4.9.4 One Dimensional Position Aided INS

Next, consider the same system as in the previous example, but with a position measurement

$$\tilde{p}(kT) = p(kT) + \nu_p(kT)$$

available at $T = 1$ second intervals. The measurement noise $\nu_p(kT)$ is white and Gaussian with variance $var(\nu_p(kT)) = R_p = 3.0m^2$.

During each interval $t \in ((k-1)T, kT]$, the INS integrates the INS mechanization equations as described in eqn. (4.143). At time $t_k = kT$, the INS position estimate $\hat{p}_k = \hat{p}(t_k)$ is subtracted from the position measurement $\tilde{p}_k = \tilde{p}(t_k)$ to form a position measurement residual

$$\delta y_k = \tilde{p}_k - \hat{p}_k$$

which is equivalent to

$$\delta y_k = \delta p_k + \nu_p(k) = \mathbf{H}\delta\hat{\mathbf{x}}_k + \nu_p(k)$$

where $\mathbf{H} = [1, 0, 0]$. With this $\mathbf{H}$ and the $\mathbf{F}$ from the previous example, the observability matrix is

$$\begin{bmatrix} 1 & 0 & 0 \\ 0 & 1 & 0 \\ 0 & 0 & 1 \end{bmatrix}$$

which has rank 3; therefore, the error state should be observable from the position measurement.

This residual error will be used as the measurement for an error state estimator

$$\left.\begin{aligned} \delta\hat{\mathbf{x}}_{k+1}^- &= \mathbf{\Phi}\delta\hat{\mathbf{x}}_k^+ \\ \delta\hat{\mathbf{x}}_k^+ &= \delta\hat{\mathbf{x}}_k^- - \mathbf{L}_k\delta\mathbf{y}_k \end{aligned}\right\} \tag{4.145}$$

where

$$\mathbf{\Phi} = \begin{bmatrix} 1 & 1 & .5 \\ 0 & 1 & 1 \\ 0 & 0 & 1 \end{bmatrix}.$$

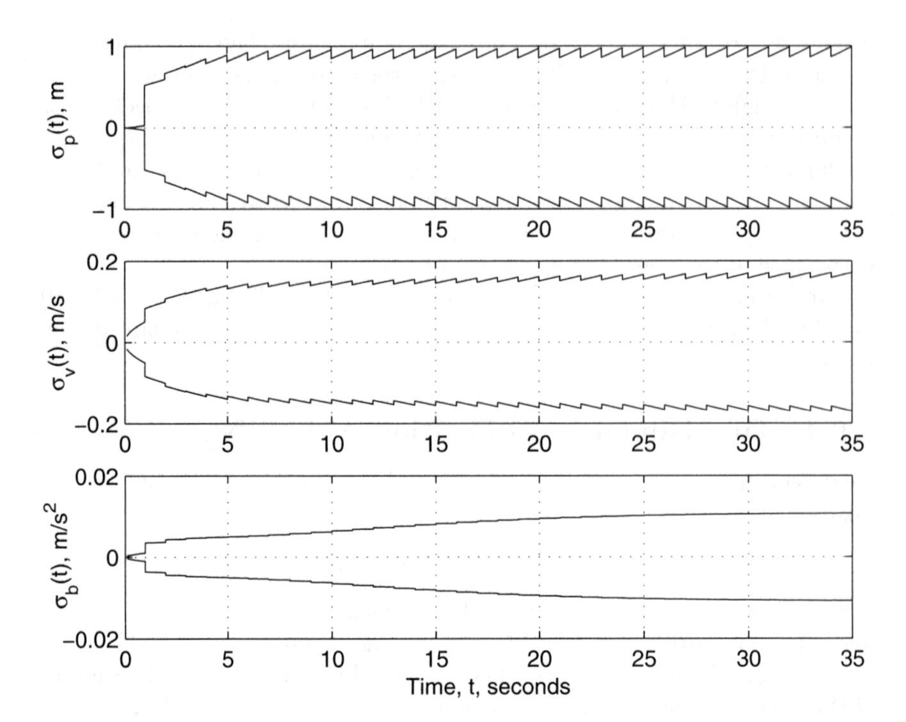

Figure 4.7: Error standard deviation of 1-d inertial frame INS with position measurement aiding.

Either of the methods for determining **Qd** results in

$$\mathbf{Qd} = \begin{bmatrix} 8.33 \times 10^{-4} & 1.25 \times 10^{-3} & 1.67 \times 10^{-7} \\ 1.25 \times 10^{-3} & 2.50 \times 10^{-3} & 5.00 \times 10^{-7} \\ 1.67 \times 10^{-7} & 5.00 \times 10^{-7} & 1.00 \times 10^{-6} \end{bmatrix}.$$

Figure 4.7 displays plots of the error standard deviation for each state as a function of time. In this example, $\mathbf{L} = [0.300, 0.039, 0.002]^\top$ which produces eigenvalues for the discrete-time system near 0.90 ± 0.07, and 0.85. Due to the integration of the noise processes $v_a(t)$ and $\omega_b(t)$, the variance grows between the time instants at which position measurements are available. The position measurement occurs at one second intervals. Typically, the $1.0Hz$ position measurement decreases the error variance of each state estimate, but this is not always the case. Notice for example that at $t = 1.0s$ the standard deviation of all three states increase. Therefore, the gain vector **L** selected for this example is clearly not optimal. At least at this time, the gain vector being zero would have been better. Eventually, by $t = 35s$, an equilibrium condition has been reached where the growth in the error variance between measurements is equal to the decrease in error

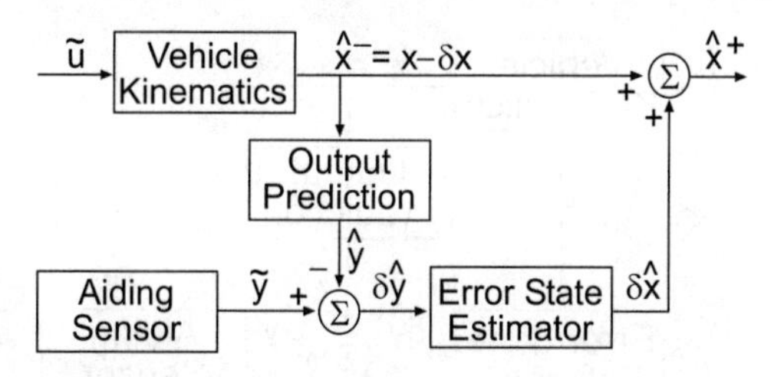

Figure 4.8: Feed-forward complementary filter implementation diagram.

variance due to the measurement update. Although each measurement has an error variance of $3.0m^2$, the filtering of the estimates results in a steady state position error variance less that $1.0m^2$. In comparison with the results in the previous section, the position measurement results in bounded variance for the velocity and accelerometer bias estimation errors.

Notice that there is a distinct difference in the state estimate error variance immediately prior and posterior to the measurement correction. Performance analysis should clearly indicate which value is being presented.

The simulation is for an ideal situation where the initial error variance is identically zero. In a typical situation, the initial error variance may be large, with the measurement updates decreasing the error variance until equilibrium is attained.

4.10 Complementary Filtering

The approach described in Section 4.9.4 and depicted in Figure 4.8 is an example of a feed-forward complementary filter implementation. For this example, the kinematics are given by eqn. (4.141), $\tilde{u}$ represents the acceleration measurement in eqn. (4.142), the output prediction is $\hat{y} = \hat{p}$, and the error estimator is defined in eqns. (4.145) and designed by the choice of $\mathbf{L}$. No '+' or '−' superscript is indicated on the $\delta\hat{\mathbf{x}}$ variable because the error state estimator can compute either variable at any time needed to form the rightmost summation in the figure. The error estimator can also propagate eqns. (4.125) and (4.127) to maintain an estimate of the system accuracy.

The feed-forward complementary filter approach has the navigation system integrating the variable $\tilde{u}$ through the system kinematics to produce $\hat{\mathbf{x}}^-$ and the error state estimator integrating $\delta\mathbf{x}$ between aiding measurements to produce the corrected state estimate denoted as $\hat{\mathbf{x}}^+$ in Figure 4.8.

As an alternative approach, over the sampling interval $t \in [t_{k-1}, t_k)$, starting from the initial condition $\hat{\mathbf{x}}^+(t_{k-1})$ the navigation system could integrate $\tilde{u}(t)$ through the system kinematics to produce $\hat{\mathbf{x}}^-(t_k)$. At time

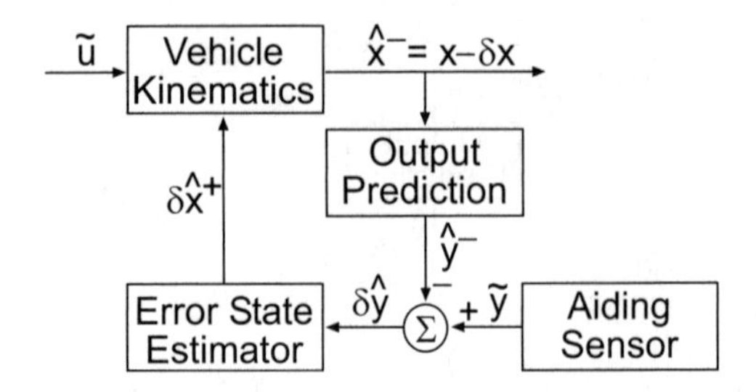

Figure 4.9: Feedback complementary filter implementation diagram.

t_k, the error state estimate

$$\delta\hat{\mathbf{x}}_k^+ = \delta\hat{\mathbf{x}}_k^- - \mathbf{L}_k \delta\mathbf{y}_k$$

is added to the prior state estimate to produce the initial condition for the next interval of integration

$$\hat{\mathbf{x}}^+(t_k) = \hat{\mathbf{x}}^-(t_k) + \delta\hat{\mathbf{x}}_k^+.$$

Because the navigation system state now accounts for the estimated error, the expected value of the error in $\hat{\mathbf{x}}^+(t_k)$ is zero and it is therefore proper to reset

$$\delta\hat{\mathbf{x}}_k^+ = \mathbf{0}.$$

This makes the time update portion of eqn. (4.145) trivial, so that it need not be implemented. This feedback approach is theoretically equivalent to the feed-forward approach. It is depicted in Figure 4.9 where the feedback of $\delta\hat{\mathbf{x}}$ should be read as correcting the initial condition for each period of integration.

4.11 References and Further Reading

This chapter has presented a very brief tour of random variables, stochastic processes, and linear systems with white Gaussian driving noise. The main references for the writing of this chapter are [31, 58, 93, 107]. Additional recent references on the topics of this chapter are [77, 84, 121].

Prerequisites for the chapter were a basic knowledge of probability and statistics. A few good references for this material are [31, 107]. Examples of various probability density functions are graphed on p. 78 in [107]. The main reference for the approach of Section 4.7.2.1 is [133].

This chapter has modeled random processes by state space equations with white noise processes as inputs. There are theoretical difficulties with this approach, due to the fact that the white noise processes are not integrable in the conventional Riemann sense. Fixing these theoretical issues requires fundamental theoretical changes to the concept of integration that are beyond the scope of this text. Interested readers can study the topic of Itô calculus in, for example, [71, 72, 106].

Section 4.9.2 presented a very brief discussion of instrumentation specifications and methods for their estimation. Additional information on these topics is available in the various specification documents, e.g., [9, 10, 11]. The Allan variance method is discussed in [2].

4.12 Exercises

Exercise 4.1 This problem is a computational verification of the central limit theorem. Let the scalar random variables x_i be i.i.d. with density

$$p(X_i) = \begin{cases} 1.0 & |X_i| \leq 0.5 \\ 0.0 & \text{otherwise.} \end{cases}$$

This density is referred to as the uniform density over the range $[-0.5, 0.5]$.

1. For $N = 2$, in MATLAB use the rand function to produce instances of x_i for $i = 1, \ldots, N$. Then compute $\bar{x}_N$ according to eqn. (4.20) with $A = \frac{1}{\sqrt{N}}$. Repeat this experiment $M = 10,000$ times to produce M realizations of $\bar{x}_N$.
   (a) Graph the histogram of the M realizations of $\bar{x}_N$ with the axis limits set to ± 1.0 using about 20 bins.
   (b) Compute the mean μ and the standard deviation σ of the M realizations of $\bar{x}_N$.
   (c) On the same axes as the histogram, graph the Gaussian density with μ and σ determined in the previous step.
2. Repeat the above process for $N = 1, 3$, and 10. Compare the results with the statement of the central limit theorem.

While this problem is not a proof of the central limit theorem, it visually shows that the density of the summed random variables rapidly converges toward a Gaussian density as N increases. An analytic demonstration using circumstances related to this problem is contained in Section 1.13 in [31].

Exercise 4.2 Compute the mean, variance, and second moment of the random variable x with the uniform density over the interval $(0, 1)$:

$$p_x(X) = \begin{cases} 1, & \text{for } X \in (0, 1) \\ 0, & \text{otherwise.} \end{cases}$$

Exercise 4.3 For the uniform density defined in Exercise 4.2, find

1. $P\{x < 0\}$
2. $P\{x < 0.25\}$
3. $P\{x < 0.25 \text{ or } x > 0.50\}$
4. $P\{x < 0.25 \text{ or } x < 0.50\}$
5. $P\{x < a\}$ for $a \in \mathbb{R}$.

Exercise 4.4 Find the density for y where $y = x^3$ and $x \sim N(0, \sigma^2)$.

Exercise 4.5 Find the density for y where $y = x^2$ and

$$p_x(X) = \begin{cases} X \exp(-X), & \text{for } x > 0 \\ 0, & \text{otherwise.} \end{cases}$$

The density for x is a special case of the Gamma or Erlang density.

Exercise 4.6 If $z = x^2$ where $x \sim N(0, \sigma^2)$, use the method of Example 4.5 to show that

$$p_z(Z) = \begin{cases} \frac{1}{\sqrt{2\pi Z}\,\sigma} \exp\left(-\frac{Z}{2\sigma^2}\right) & Z \geq 0, \\ 0, & Z < 0. \end{cases}$$

The random variable z in this exercise is a called a χ-squared random variable with one degree of freedom.

When $z = \sum_{i=1}^{k} x_i^2$ and each $x_i \sim N(0,1)$, then z is a χ-squared random variable with k degrees of freedom. The density is

$$p_z(Z) = \begin{cases} \frac{\left(\frac{1}{2}\right)^{\frac{k}{2}}}{\Gamma\left(\frac{k}{2}\right)} Z^{\frac{k}{2}-1} \exp\left(-\frac{Z}{2}\right), & Z \geq 0 \\ 0, & Z < 0 \end{cases}$$

where $\Gamma(x) = \int_0^\infty t^{x-1} e^{-x} dx$ is an extension of the factorial function to non-integer arguments. It does not have an algebraic closed form expression, but has the property that

$$\Gamma(x+1) = x\Gamma(x) \text{ for } x \neq 0, -1, -2, -3, \ldots.$$

Exercise 4.7 Prove that, for two independent variables, their covariance is zero and their correlation is the product of their means.

Exercise 4.8 Consider the random variables x and y with joint density

$$p_{xy}(X,Y) = \begin{cases} 2, & X \in [0,1] \text{ and } 0 \leq Y \leq X \\ 0, & \text{otherwise.} \end{cases}$$

1. Find the correlation coefficient between x and y.

2. Are these random variables independent?

3. If the objective is to estimate y given knowledge of x, how does the knowledge of x affect the estimate of y? Consider for example the situation where $x = 1$.

Exercise 4.9 Find the density of the random variable $\mathbf{y} = \mathbf{H}\mathbf{x} + \mathbf{b}$, where $\mathbf{x} \sim N(\boldsymbol{\mu}, \mathbf{P})$, and $\mathbf{H}$ is a nonsingular matrix.

Exercise 4.10 Let $\mathbf{y}$ be a random variable with $\mathbf{y} = \mathbf{A}\mathbf{x}$, $\mathbf{x} \sim N(\boldsymbol{\mu}, \mathbf{P})$, $\mathbf{A} \in \mathbb{R}^{m \times n}$ and $\mathbf{A}\mathbf{P}\mathbf{A}^\top$ a nonsingular matrix.

1. Find the expected value $\boldsymbol{\mu}_\mathbf{y}$ of $\mathbf{y}$.

2. Find the covariance matrix $\mathbf{P}_\mathbf{y}$ for $\mathbf{y}$.

3. Use the fact the $\mathbf{y}$ is a linear function of a Gaussian random variable to show that the density $p_\mathbf{y}(\mathbf{Y})$ is the same as in Example 4.10:

$$\frac{1}{\sqrt{(2\pi)^n \det\left(\mathbf{A}\mathbf{P}\mathbf{A}^\top\right)}} \exp\left(-\frac{1}{2}(\mathbf{Y} - \mathbf{A}\boldsymbol{\mu})^\top \left(\mathbf{A}\mathbf{P}\mathbf{A}^\top\right)^{-1} (\mathbf{Y} - \mathbf{A}\boldsymbol{\mu})\right).$$

4. Let $\mathbf{p} = [n, e, d]^\top$ represent the position error vector which is assumed to be $N(\mathbf{0}, \mathbf{P})$. What is the density and covariance matrix for the vector $\vec{\mathbf{p}} = [n, e]^\top$?

Exercise 4.11 Let $\mathbf{y} = \mathbf{f}(\mathbf{a})$ where $\mathbf{a}$ is an unknown deterministic variable. Given a measurement of $\mathbf{y}$ that is modeled as

$$\tilde{\mathbf{y}} = \mathbf{f}(\mathbf{a}) + \mathbf{b} + \mathbf{n} \tag{4.146}$$

where $\mathbf{y}$, $\tilde{\mathbf{y}}$, $\mathbf{a}$, $\mathbf{b}$, $\mathbf{n} \in \Re^3$, and $\mathbf{f}$ is a continuous and invertible function, the analyst chooses to estimate the values of $\mathbf{a}$ and $\mathbf{b}$ as

$$\hat{\mathbf{a}} = \mathbf{g}(\tilde{\mathbf{y}}) \tag{4.147}$$

$$\hat{\mathbf{b}} = \mathbf{0} \tag{4.148}$$

where $\mathbf{g} = \mathbf{f}^{-1}$. Assume that $\mathbf{b} \sim N(0, \mathbf{P}_b)$ and that $\mathbf{n} \sim N(0, \mathbf{P}_n)$ and that $\mathbf{b}$ and $\mathbf{n}$ are independent. The parameter estimation errors are $\delta\mathbf{a} = \mathbf{a} - \hat{\mathbf{a}}$ and $\delta\mathbf{b} = \mathbf{b} - \hat{\mathbf{b}}$.

In the following, use the first-order Taylor series approximation

$$\mathbf{g}(\tilde{\mathbf{y}}) \approx \mathbf{g}(\mathbf{y}) + \mathbf{G}(\mathbf{y})\,(\tilde{\mathbf{y}} - \mathbf{y})$$

where $\mathbf{G} = \frac{\partial \mathbf{g}(\mathbf{y})}{\partial \mathbf{y}}$.

1. Perform analysis similar to that of Example 4.9 to show that:

   (a) The mean parameter estimation errors are

   $$E\langle\delta\mathbf{a}\rangle = \mathbf{0} \text{ and } E\langle\delta\mathbf{b}\rangle = \mathbf{0}.$$

   (b) The variance of the parameter estimation errors are

   $$Var\langle\delta\mathbf{a}\rangle \approx \mathbf{G}(\mathbf{y})(\mathbf{P}_b + \mathbf{P}_n)\mathbf{G}(\mathbf{y})^\top \text{ and } Var\langle\delta\mathbf{b}\rangle = \mathbf{P}_b.$$

   (c) The covariance between the parameter estimation errors is

   $$E\langle\delta\mathbf{a}\delta\mathbf{b}^\top\rangle \approx -\mathbf{G}\mathbf{P}_b.$$

   (d) The covariance matrix for the vector is

   $$\begin{aligned}\mathbf{P} &= cov\langle[\delta\mathbf{a}^\top, \delta\mathbf{b}^\top]^\top\rangle \\ &\approx \begin{bmatrix} \mathbf{G}(\mathbf{P}_b + \mathbf{P}_n)\mathbf{G}^\top & -\mathbf{G}\mathbf{P}_b \\ -\mathbf{P}_b\mathbf{G}^\top & \mathbf{P}_b \end{bmatrix}.\end{aligned}$$

2. If $\boldsymbol{\rho} = \mathbf{A}\mathbf{a}$ and $\hat{\boldsymbol{\rho}} = \mathbf{A}\hat{\mathbf{a}}$:

   (a) Show that the covariance of $\delta\boldsymbol{\rho}$ is

   $$Var\langle\delta\boldsymbol{\rho}\rangle \approx \mathbf{A}\mathbf{G}(\mathbf{P}_b + \mathbf{P}_n)\mathbf{G}^\top\mathbf{A}^\top.$$

   (b) Show that the correlation between $\delta\boldsymbol{\rho}$ and $\delta\mathbf{b}$ is

   $$E\langle\delta\boldsymbol{\rho}\delta\mathbf{b}^\top\rangle \approx -\mathbf{A}\mathbf{G}\mathbf{P}_b.$$

3. Show that the structure of the matrix $\mathbf{P}_x = var(\mathbf{x})$ where $\mathbf{x} = [\boldsymbol{\rho}^\top, \mathbf{b}^\top]^\top$ is

   $$\mathbf{P}_x \approx \begin{bmatrix} \mathbf{A}\mathbf{G}(\mathbf{P}_b + \mathbf{P}_n)\mathbf{G}^\top\mathbf{A}^\top & -\mathbf{A}\mathbf{G}\mathbf{P}_b \\ -\mathbf{P}_b\mathbf{G}^\top\mathbf{A}^\top & \mathbf{P}_b \end{bmatrix}.$$

Exercise 4.12 If $v(t)$ is a continuous-time, white noise process with units of $\frac{m}{s^2}$ and PSD denoted by σ_v^2, what are the units of σ_v?

Exercise 4.13 Repeat the comparison of Examples 4.21, 4.22, and 4.23 for the system having

$$\mathbf{F} = \begin{bmatrix} 0 & 1 \\ 0 & -\lambda \end{bmatrix} \quad \mathbf{G}\mathbf{Q}\mathbf{G}^\top = \begin{bmatrix} 0.00 & 0.00 \\ 0.00 & \sigma^2 \end{bmatrix},$$

with $\lambda = 1$, and $\sigma^2 = 0.01$.

Exercise 4.14 Repeat the comparison of Examples 4.21, 4.22, and 4.23 for the system having

$$\mathbf{F} = \begin{bmatrix} 0 & 1 & 0 \\ 0 & 0 & 1 \\ 0 & 0 & 0 \end{bmatrix}, \quad \mathbf{G} = \begin{bmatrix} 0.00 \\ 0.00 \\ 1.00 \end{bmatrix},$$

and $Q = \sigma^2 = 0.01$.

Exercise 4.15 For the system of eqn. (4.82) with a position measurement $y = x$:

1. Complete an observability analysis to show that the state of the system is not observable.
2. Analyze the observable and unobservable subspaces to determine the linear combination of states that is not observable.

Exercise 4.16 Eqn. (4.125) is derived as in Section 4.6.5. Use similar methods to derive eqns (4.126) and (4.127).

Exercise 4.17 Implement a covariance simulation to reproduce the results of Section 4.9.4.

1. Compute $\mathbf{\Phi}$ and $\mathbf{Qd}$ for $T = 0.1$s.
2. Implement eqn. (4.125) at a 10 Hz update rate. This step will allow the plot of the error standard deviation to show the error growth between measurement updates.
3. For the given value of $\mathbf{L}$, implement eqn. (4.127) at a 1.0 Hz rate. At the measurement instant, store the two values for the covariance. First store the prior value from eqn. (4.125), then store the posterior value from eqn. (4.127).

After duplicating the results from Section 4.9.4:

1. Try alternative initial values for the error covariance matrix $\mathbf{P}$. Do they converge to the same steady state conditions?
2. Experiment with different (stabilizing) gain vectors $\mathbf{L}$ to try to improve the state estimation accuracy.

Exercise 4.18 Verify that eqn. (4.144) is the solution of eqn. (4.100).

1. Differentiate $\mathbf{P}$ from eqn. (4.144), noting that $\dot{\mathbf{\Phi}} = \mathbf{F\Phi}$.
2. Add and subtract the term $\mathbf{\Gamma Q \Gamma}^\top$ to the result of Step 1, then organize the terms that match eqn. (4.100) together. The remaining terms should sum to zero.

Exercise 4.19 For the sensor model of eqn. (4.1) and the estimate of θ formed in eqn. (4.2), let $\nu(t)$ be a Gaussian white noise process with PSD σ_ν^2 where $\sigma_\nu = 0.05\frac{deg/s}{\sqrt{Hz}}$ and let b_g be accurately modeled as

$$\dot{b}_g = \lambda b_g + w$$

where $\lambda = 0.01\frac{1}{s}$ and w is a Gaussian white noise process with PSD σ_w^2 where $\sigma_w = 10^{-4}\frac{deg/s^2}{\sqrt{Hz}}$.

1. Repeat analysis similar to that of Section 4.9.3 to characterize the rate of growth of the error in the estimate of θ.

2. Assume that a measurement of θ is available at a 1 Hz rate and that the measurement is corrupted by Gaussian white noise with standard deviation of 1.0 deg. Design a stable error state estimator and perform analysis similar to that of Section 4.9.4 to characterize the performance of the aided system.

Chapter 5

Optimal State Estimation

The main topic of this chapter is the derivation and study of the topic of optimal state estimation. Sections 3.6 and 4.8 have already introduced the concept of state estimation. In those sections the state estimation gain was represented by $\mathbf{L}$. The gain $\mathbf{L}$ could, for example, be selected by pole placement methods. Given a stochastic model and a stabilizing gain $\mathbf{L}$, the system performance could be evaluated by the methods discussed in Section 4.9.

This chapter presents the state estimation problem from a different perspective: given a stochastic model, can the state estimation gain be selected as a time-varying sequence that optimizes some measure of the performance of the system? The state estimation algorithm that uses this optimal time-varying gain sequence will be referred to as the Kalman filter. The state estimation problem is reviewed in Section 5.1. Using the mean-squared-error of the state covariance sequence as the optimization criteria, we derive the Kalman filter by two distinct methods. Section 5.2 sets up an optimization problem and selects the optimal state estimation gain $\mathbf{K_k}$ at each time step. For an alternative perspective, Section 5.3 starts with the problem of weighted least squares, progresses through recursive weighted least squares, and finally uses time propagation of the state and covariance matrix to derive the Kalman filter gain sequence $\mathbf{K_k}$. Section 5.4 summarizes the derivations of the Kalman filter algorithm and presents different approaches to computing the variance of the state estimation error after a measurement correction. Section 5.5 discusses useful properties of the Kalman filter. Section 5.6 discusses various issues that should be understood for efficient and successful Kalman filter implementations. Section 5.9 discusses two numeric issues important for Kalman filter implementations. Section 5.10 discusses various motivations for and approaches to suboptimal filtering.

Chapter 6 will revisit the topic of state estimation performance analysis.

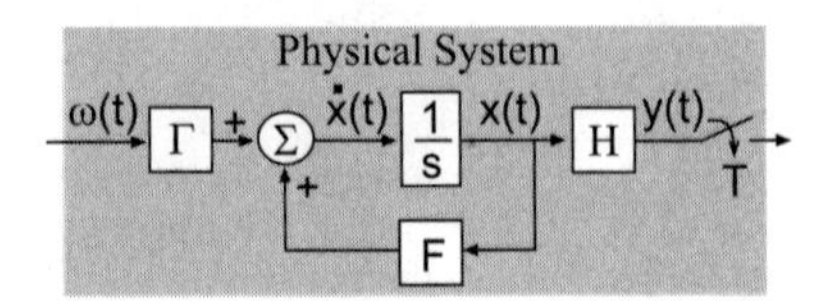

Figure 5.1: Continuous-time system with input $\mathbf{u} = \mathbf{0}$.

The topic of covariance analysis and its use in error budgeting is critical to making informed system level decisions at the design stage.

5.1 State Estimation: Review

This section summarizes the linear stochastic state estimation results from the previous chapters.

For a linear continuous-time system such as that shown in Figure 5.1, to enable state estimation in discrete-time, we find the discrete-time equivalent state-space model of the actual process:

$$\mathbf{x}_k = \mathbf{\Phi}\mathbf{x}_{k-1} + \mathbf{G}\mathbf{u}_{k-1} + \boldsymbol{\omega}_{k-1} \quad (5.1)$$

$$\mathbf{y}_k = \mathbf{H}\mathbf{x}_k + \boldsymbol{\nu}_k \quad (5.2)$$

where $\mathbf{x}_k \in \mathbb{R}^n$, $\mathbf{y}_k \in \mathbb{R}^p$, $\boldsymbol{\nu}_k \in \mathbb{R}^p$, and $\boldsymbol{\omega}_k \in \mathbb{R}^n$ are the stochastic signals representing the state, output, measurement noise and driving noise, respectively; $\mathbf{u} \in \mathbb{R}^m$ is a known input signal. Assuming that the state is observable from the output signal, our objective is to provide an optimal estimate of the state vector. The resulting algorithm is referred to as the Kalman filter. The basic state estimation block diagram is shown in Figure 5.2.

For simplicity of notation, the system is assumed to be time-invariant and the subscript k's on the system parameter matrices have been dropped (e.g., $\mathbf{\Phi}_k = \mathbf{\Phi}$). The assumption of a time-invariant system is not a restriction on the approach. The derivation does go through for and the Kalman filter is often applied to time-varying systems.

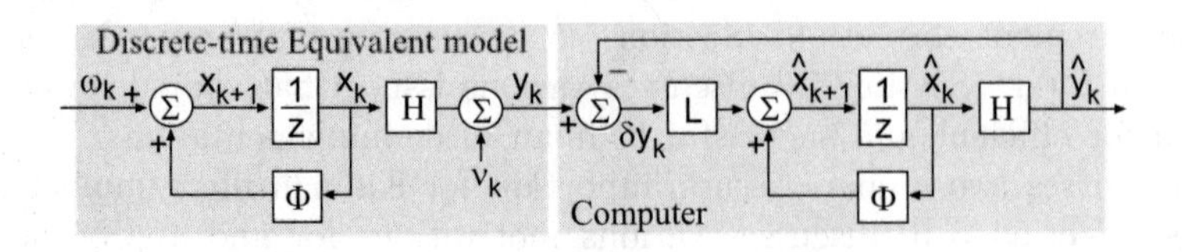

Figure 5.2: Discrete-time state estimator.

In this model, $\boldsymbol{\nu}_k$ and $\boldsymbol{\omega}_k$ are signals determined by the environment. The designer does not know the actual values of these signals, but formulates the problem so that the following assumptions hold:

$$\begin{array}{ll} E\langle\boldsymbol{\nu}_k\rangle = \mathbf{0} & var\,(\boldsymbol{\nu}_k, \boldsymbol{\nu}_l) = \mathbf{R}_k\delta_{kl} \\ E\langle\boldsymbol{\omega}_k\rangle = \mathbf{0} & var\,(\boldsymbol{\omega}_k, \boldsymbol{\omega}_l) = \mathbf{Qd}_k\delta_{kl} \end{array}$$

and the cross-correlation between $\boldsymbol{\nu}_k$ and $\boldsymbol{\omega}_l$ is zero. The symbol δ_{kl} represents the Kronecker delta. In addition, for all $k \geq 0$, $\mathbf{x}_k$ is uncorrelated with $\boldsymbol{\nu}_l$ for all l and $\mathbf{x}_k$ is uncorrelated with $\boldsymbol{\omega}_l$ for $l \geq k$. The matrices $\mathbf{R}_k$ and $\mathbf{Qd}_k$ are assumed to be positive definite and the state is assumed to be controllable from $\boldsymbol{\omega}_k$.

For $k > 0$, assuming that $\hat{\mathbf{x}}_0^+$ and positive definite $\mathbf{P}_0^+$ are known, the state estimate is computed using the algorithm described below. The superscripts $-$ and $+$ denote the estimates prior and posterior to incorporating the measurement, respectively. At time k, given $\hat{\mathbf{x}}_{k-1}^+$, $\mathbf{P}_{k-1}^+$ and $\mathbf{u}_{k-1}$, the prior estimate of the state and output are computed as

$$\begin{aligned} \hat{\mathbf{x}}_k^- &= \boldsymbol{\Phi}\mathbf{x}_{k-1}^+ + \mathbf{G}\mathbf{u}_{k-1} && (5.3) \\ \hat{\mathbf{y}}_k^- &= \mathbf{H}\hat{\mathbf{x}}_k^-. && (5.4) \end{aligned}$$

When the k-th measurement $\mathbf{y}_k$ becomes available, the measurement residual is computed as

$$\delta\mathbf{y}_k^- = \mathbf{y}_k - \hat{\mathbf{y}}_k^-. \tag{5.5}$$

Given the state estimation gain vector $\mathbf{L}_k$, the posterior state estimate is computed as

$$\hat{\mathbf{x}}_k^+ = \hat{\mathbf{x}}_k^- + \mathbf{L}_k\delta\mathbf{y}_k^-. \tag{5.6}$$

Given the above state space models for the state and state estimate, with $\delta\mathbf{x}_k^- = \mathbf{x}_k - \hat{\mathbf{x}}_k^-$ and $\delta\mathbf{x}_k^+ = \mathbf{x}_k - \hat{\mathbf{x}}_k^+$, the previous chapters have derived the following equations for the prior state estimation error, the measurement residual, and the posterior state estimation error:

$$\begin{aligned} \delta\mathbf{x}_{k+1}^- &= \boldsymbol{\Phi}\delta\mathbf{x}_k^+ + \boldsymbol{\omega}_k && (5.7) \\ \delta\mathbf{y}_k^- &= \mathbf{H}\delta\mathbf{x}_k^- + \boldsymbol{\nu}_k && (5.8) \\ \delta\mathbf{x}_k^+ &= (\mathbf{I} - \mathbf{L}_k\mathbf{H}_k)\,\delta\mathbf{x}_k^- - \mathbf{L}_k\boldsymbol{\nu}_k. && (5.9) \end{aligned}$$

With reference to eqn. (4.125), based on eqns. (5.7–5.9), the covariance of the state estimation error prior to the measurement update is given by equation:

$$\mathbf{P}_{k+1}^- = \boldsymbol{\Phi}_k\mathbf{P}_k^+\boldsymbol{\Phi}_k^\top + \mathbf{Qd}_k. \tag{5.10}$$

As in eqn. (4.126), the covariance matrix for the predicted output error is

$$\mathbf{P}_{\mathbf{y}_k}^- = \mathbf{H}_k\mathbf{P}_k^-\mathbf{H}_k^\top + \mathbf{R}_k. \tag{5.11}$$

As in eqn. (4.127), the covariance matrix for the state estimation error posterior to the measurement correction is

$$\mathbf{P}_k^+ = (\mathbf{I} - \mathbf{L}_k\mathbf{H}_k)\,\mathbf{P}_k^-\,(\mathbf{I} - \mathbf{L}_k\mathbf{H}_k)^\top + \mathbf{L}_k\mathbf{R}_k\mathbf{L}_k^\top. \tag{5.12}$$

Given that the state is controllable from $\boldsymbol{\omega}_k$, it can be shown that $\mathbf{P}_k^-$ and $\mathbf{P}_k^+$ are positive definite matrices.

Note that eqns. (5.7–5.9) are not used in the state estimator implementation. These equations are used for analysis to attain eqns. (5.10–5.12) which serve as the basis for the derivation of the optimal state estimation gain in Section 5.2. The optimal state estimator will be implemented using eqns. (5.3–5.6), eqns. (5.10–5.12), and a formula for the optimal state estimation gain.

5.2 Minimum Variance Gain Derivation

Eqn. (5.12) is useful for the selection of the estimator gain $\mathbf{L}_k$, because it allows us to evaluate the covariance matrices $\mathbf{P}_k^+$ that would result from alternative choices of $\mathbf{L}_k$. In fact, the diagonal of $\mathbf{P}_k^+$ contains the variance of each element of the state vector. The scalar function $Tr(\mathbf{P}_k^+) = trace(\mathbf{P}_k^+)$ is the sum of the variances of the individual states (i.e., the mean-squared-error); therefore, for the purposes of optimization it is reasonable to minimize the cost function

$$J(\mathbf{L}_k) = Tr(\mathbf{P}_k^+).$$

5.2.1 Kalman Gain Derivation

Dropping the subscripts and multiplying out expression (5.12) yields

$$\mathbf{P}^+ = \mathbf{P}^- - \mathbf{L}\mathbf{H}\mathbf{P}^- - \mathbf{P}^-\mathbf{H}^\top\mathbf{L}^\top + \mathbf{L}\left(\mathbf{H}\mathbf{P}^-\mathbf{H}^\top + \mathbf{R}\right)\mathbf{L}^\top. \tag{5.13}$$

This expression is a second order matrix polynomial equation in the variable $\mathbf{L}$. Therefore, the mean-squared posterior state estimation error, which can be computed conveniently as $Tr(\mathbf{P}^+)$, is also a function of $\mathbf{L}$. Using the properties of the Tr function, which are reviewed in Section B.2, we can reduce the expression for $Tr(\mathbf{P}^+)$ as follows:

$$\begin{aligned} Tr(\mathbf{P}^+) &= Tr\left[\mathbf{P}^- - \mathbf{L}\mathbf{H}\mathbf{P}^- - \mathbf{P}^-\mathbf{H}^\top\mathbf{L}^\top + \mathbf{L}\left(\mathbf{H}\mathbf{P}^-\mathbf{H}^\top + \mathbf{R}\right)\mathbf{L}^\top\right] \\ &= Tr[\mathbf{P}^-] - 2Tr[\mathbf{L}\mathbf{H}\mathbf{P}^-] + Tr[\mathbf{L}\left(\mathbf{H}\mathbf{P}^-\mathbf{H}^\top + \mathbf{R}\right)\mathbf{L}^\top]. \end{aligned}$$

Selection of $\mathbf{L}$ to minimize $Tr(\mathbf{P}^+)$ requires differentiation of the scalar function $Tr(\mathbf{P}^+)$ with respect to $\mathbf{L}$. Using eqns. (B.55–B.56) for the derivative of Tr with respect to a matrix, results in

$$\frac{d}{d\mathbf{L}}Tr(\mathbf{P}^+) = -2\left(\mathbf{H}\mathbf{P}^-\right)^\top + 2\mathbf{L}\left(\mathbf{H}\mathbf{P}^-\mathbf{H}^\top + \mathbf{R}\right).$$

Setting this expression equal to the zero vector and solving for $\mathbf{L}$ yields the formula for the Kalman gain vector, which will be denoted by $\mathbf{K}$ instead of $\mathbf{L}$:

$$\mathbf{K}_k = \mathbf{P}_k^- \mathbf{H}_k^\top \left(\mathbf{H}_k \mathbf{P}_k^- \mathbf{H}_k^\top + \mathbf{R}_k \right)^{-1} \tag{5.14}$$

where we have used the fact that $\mathbf{P}^-$ is symmetric. Since $\mathbf{P}^-$ and $\mathbf{R}$ are positive definite, the matrix $\left(\mathbf{H}_k \mathbf{P}_k^- \mathbf{H}_k^\top + \mathbf{R}_k \right)$ is also positive definite, which ensures that the inverse in the above formula exist. Since $\left(\mathbf{H}_k \mathbf{P}_k^- \mathbf{H}_k^\top + \mathbf{R}_k \right)$ is also the second derivative of $Tr(\mathbf{P}^+)$ with respect to $\mathbf{L}$, the Kalman gain $\mathbf{K}_k$ yields a minimum value for the mean-squared cost.

5.2.2 Kalman Gain: Posterior Covariance

When the Kalman gain is used, a simplified equation can be derived for the posterior error covariance matrix $\mathbf{P}^+$. The derivation uses the fact that

$$\mathbf{K}_k \left(\mathbf{H}_k \mathbf{P}_k^- \mathbf{H}_k^\top + \mathbf{R}_k \right) = \mathbf{P}_k^- \mathbf{H}_k^\top \tag{5.15}$$

which is easily derived from eqn. (5.14). The derivation proceeds from eqn. (5.13) as follows:

$$\begin{aligned}
\mathbf{P}^+ &= \mathbf{P}^- - \mathbf{KHP}^- + \mathbf{KRK}^\top + \mathbf{KHP}^-\mathbf{H}^\top\mathbf{K}^\top - \mathbf{P}^-\mathbf{H}^\top\mathbf{K}^\top \\
&= [\mathbf{I} - \mathbf{KH}]\,\mathbf{P}^- + \mathbf{K}\left(\mathbf{R} + \mathbf{HP}^-\mathbf{H}^\top\right)\mathbf{K}^\top - \mathbf{P}^-\mathbf{H}^\top\mathbf{K}^\top \\
&= [\mathbf{I} - \mathbf{KH}]\,\mathbf{P}^- + \mathbf{P}^-\mathbf{H}^\top\mathbf{K}^\top - \mathbf{P}^-\mathbf{H}^\top\mathbf{K}^\top \\
\mathbf{P}^+ &= [\mathbf{I} - \mathbf{KH}]\,\mathbf{P}^-.
\end{aligned} \tag{5.16}$$

Note that the final expression in eqn. (5.16) is only valid for the Kalman gain $\mathbf{K}$, while (5.12) is valid for any estimator gain vector $\mathbf{L}$.

5.2.3 Summary

This section has presented a derivation of the Kalman gain formula shown in eqn. (5.14) and has derived a simplified formula for the posterior error covariance matrix that is shown in eqn. (5.16). The Kalman gain is the time-varying gain sequence that minimizes the mean-square state estimation error at each measurement instant. Computation of the Kalman gain sequence requires computation of the prior and posterior error covariance matrices. The prior covariance $\mathbf{P}_k^-$ is used directly in eqn. (5.14). The posterior covariance $\mathbf{P}_k^+$ is used to compute the $\mathbf{P}_{k+1}^-$ for use with the next measurement.

The derivation of this section is short and clearly shows the mean-squared optimality. The following section presents an alternative derivation that obtains the same results. The alternative derivation is considerably longer, but also provides additional insight into the operation of the Kalman filter.

5.3 From WLS to the Kalman Filter

This section considers an alternative derivation of the Kalman filter in a series of steps. Subsection 5.3.1 considers the weighted least squares problem. The batch and recursive solutions to the weighted least squares problem are considered in Subsections 5.3.2 and 5.3.3, respectively. Using the solution to the recursive least squares problem and the state space model, the Kalman filter is derived in Subsection 5.3.4.

5.3.1 Weighted Least Squares (WLS)

In linear weighted least squares estimation, the problem setting is that a set of m noisy measurements $(\tilde{y}_1, \ldots, \tilde{y}_m)$ is available, where each measurement is modeled as

$$\tilde{y}_i = \mathbf{H}_i \mathbf{x} + \nu_i \text{ for } i = 1, \ldots, m. \tag{5.17}$$

The variable $\mathbf{x} \in \mathbb{R}^n$ is assumed to be unknown and constant. For each i, $\mathbf{H}_i^\top \in \mathbb{R}^n$ is assumed to be known, and ν_i is an unknown random variable representing measurement noise. In this derivation, each measurement is assumed to be scalar, but the results do extend to vector measurements. Defining $\boldsymbol{\nu} = [\nu_1, \ldots, \nu_m]^\top$, the vector $\boldsymbol{\nu}$ is assumed to have $E\langle \boldsymbol{\nu} \rangle = \mathbf{0}$ and $E\langle \boldsymbol{\nu}\boldsymbol{\nu}^\top \rangle = \mathbf{R} \in \mathbb{R}^{m \times m}$, where $\mathbf{R}$ is positive definite and diagonal.

The objective is to find an estimate $\hat{\mathbf{x}}$ to minimize the $J_{WLS}(\hat{\mathbf{x}})$ cost function defined as

$$J_{WLS}(\hat{\mathbf{x}}) = \frac{1}{2}(\tilde{\mathbf{Y}} - \mathbf{H}\hat{\mathbf{x}})^\top \mathbf{W} (\tilde{\mathbf{Y}} - \mathbf{H}\hat{\mathbf{x}}) \tag{5.18}$$

where $\mathbf{W} \in \mathbb{R}^{m \times m}$ is a positive definite symmetric matrix,

$$\tilde{\mathbf{Y}} = [\tilde{y}_1, \ldots, \tilde{y}_m]^\top \text{ and } \mathbf{H} = [\mathbf{H}_1^\top, \ldots, \mathbf{H}_m^\top]^\top.$$

The cost function of eqn. (5.18) can be motivated in two contexts. First, in a deterministic sense, it is desirable to minimize some norm of the error between the measurements $\tilde{\mathbf{Y}}$ and the estimated measurements $\hat{\mathbf{Y}} = \mathbf{H}\hat{\mathbf{x}}$. The cost function is a weighted two norm. In a general approach, some measurements may be known to be more accurate than others. The measurement accuracy is characterized by $\mathbf{R}^{-1}$. Therefore, it is natural to select

$$\mathbf{W} = \mathbf{R}^{-1} \tag{5.19}$$

to give the least weighting to the most uncertain measurements. Second, in a probabilistic sense, given the assumption that the measurement noise is Normally distributed, the probability density for $\mathbf{x}$ given the measurements $\tilde{\mathbf{Y}}$ is

$$p(\mathbf{x} : \tilde{\mathbf{Y}}) = A \exp\left\{ -\frac{1}{2}\left(\tilde{\mathbf{Y}} - \mathbf{H}\mathbf{x}\right)^\top \mathbf{R}^{-1} \left(\tilde{\mathbf{Y}} - \mathbf{H}\mathbf{x}\right) \right\}. \tag{5.20}$$

The value of $\mathbf{x}$ that maximizes $p(\mathbf{x} : \tilde{\mathbf{Y}})$ is the *maximum likelihood estimate* (MLE) of $\mathbf{x}$ and coincides with the minimum of J_{WLS} when $\mathbf{W} = \mathbf{R}^{-1}$.

For $m < n$, the problem will be under-determined and an infinite number of solutions will exist. If $m > n$ then the problem may be over-determined, in which case no exact solution will exist. In the latter case, minimization of the objective function in eqn. (5.18) will result in the estimate $\hat{\mathbf{x}}$ that has the minimum weighted 2-norm of $\delta \mathbf{y} = [\delta y_1, \ldots, \delta y_m]^\top$ where $\delta y_i = \tilde{y}_i - \mathbf{H}_i \hat{\mathbf{x}}$.

The following text will build up to the Kalman Filter by discussing sequentially the following set of questions:

1. What is the formula for the WLS estimate of $\mathbf{x}$ (i.e., the minimum of eqn. (5.18))?

2. Given that the WLS estimate $\hat{\mathbf{x}}_m$ has been calculated for a set of m measurements, if an additional measurement y_{m+1} becomes available, can $\hat{\mathbf{x}}_m$ be adjusted to efficiently produce the new WLS estimate $\hat{\mathbf{x}}_{m+1}$?

3. If $\mathbf{x} \in \mathbb{R}^n$ is the state of a dynamic system, which is free to change according to $\mathbf{x}_k = \mathbf{\Phi}_k \mathbf{x}_{k-1} + \mathbf{G}_k \mathbf{u}_{k-1} + \boldsymbol{\omega}_{k-1}$ where $\mathbf{\Phi}_k \in \mathbb{R}^{n \times n}$ is known and $\boldsymbol{\omega}_k \in \mathbb{R}^n$ is a zero mean random vector, how can the estimate of $\mathbf{x}$ at time $k-1$ and the measurement y_k be combined to provide an optimal estimate of $\mathbf{x}$ at time k?

As the above discussion shows, each question is a natural extension of the one before it and as the following sections show, the solutions to these problems are intimately related. The Kalman Filter is the solution to the last problem.

5.3.2 Weighted Least Squares Solution

The WLS estimate $\hat{\mathbf{x}}$ will be found using differential calculus to minimize eqn. (5.18). Eqns. (B.50–B.53) are used in the derivation.

Multiplying out eqn. (5.18) results in

$$J_{WLS} = \frac{1}{2}\left(\tilde{\mathbf{Y}}^\top \mathbf{W} \tilde{\mathbf{Y}} - 2\tilde{\mathbf{Y}}^\top \mathbf{W} \mathbf{H} \hat{\mathbf{x}} + \hat{\mathbf{x}}^\top \mathbf{H}^\top \mathbf{W} \mathbf{H} \hat{\mathbf{x}}\right).$$

Therefore, the extreme points of the objective function are determined by the solution of

$$\frac{\partial J_{WLS}}{\partial \hat{\mathbf{x}}} = -\mathbf{H}^\top \mathbf{W} \tilde{\mathbf{Y}} + \mathbf{H}^\top \mathbf{W} \mathbf{H} \hat{\mathbf{x}} = \mathbf{0} \tag{5.21}$$

which yields

$$\hat{\mathbf{x}} = (\mathbf{H}^\top \mathbf{W} \mathbf{H})^{-1} \mathbf{H}^\top \mathbf{W} \tilde{\mathbf{Y}}. \tag{5.22}$$

The solution of eqn. (5.22) is a unique minimum when $\frac{\partial^2 J_{WLS}}{\partial \mathbf{x}^2} = \mathbf{H}^\top \mathbf{W} \mathbf{H}$ is positive definite. This is the case when $\mathbf{W}$ is positive definite and $\mathbf{H}$ has n linearly independent rows (i.e, $rank(\mathbf{H}) = n$).

For analysis, define

$$\delta \mathbf{x} = \mathbf{x} - \hat{\mathbf{x}}.$$

The following analysis shows that $\delta \mathbf{x}$ is independent of $\mathbf{x}$:

$$\begin{aligned}
\delta \mathbf{x} &= \mathbf{x} - (\mathbf{H}^\top \mathbf{W} \mathbf{H})^{-1} \mathbf{H}^\top \mathbf{W} \tilde{\mathbf{Y}} \\
&= \mathbf{x} - (\mathbf{H}^\top \mathbf{W} \mathbf{H})^{-1} \mathbf{H}^\top \mathbf{W} (\mathbf{H}\mathbf{x} + \boldsymbol{\nu}) \\
&= (\mathbf{I} - (\mathbf{H}^\top \mathbf{W} \mathbf{H})^{-1} \mathbf{H}^\top \mathbf{W} \mathbf{H}) \mathbf{x} - (\mathbf{H}^\top \mathbf{W} \mathbf{H})^{-1} \mathbf{H}^\top \mathbf{W} \boldsymbol{\nu} \\
&= -(\mathbf{H}^\top \mathbf{W} \mathbf{H})^{-1} \mathbf{H}^\top \mathbf{W} \boldsymbol{\nu}.
\end{aligned}$$

Therefore, using the assumed properties of $\boldsymbol{\nu}$,

$$E\{\delta \mathbf{x}\} = \mathbf{0} \quad \text{and} \tag{5.23}$$

$$var\{\delta \mathbf{x}\} = \mathbf{P} = (\mathbf{H}^\top \mathbf{W} \mathbf{H})^{-1} \mathbf{H}^\top \mathbf{W} \mathbf{R} \mathbf{W} \mathbf{H} (\mathbf{H}^\top \mathbf{W} \mathbf{H})^{-1}. \tag{5.24}$$

In the special case where $\mathbf{W} = \mathbf{I}$, the Least Squares (LS) estimate is

$$\hat{\mathbf{x}} = (\mathbf{H}^\top \mathbf{H})^{-1} \mathbf{H}^\top \tilde{\mathbf{Y}} \text{ with} \tag{5.25}$$

$$\mathbf{P} = (\mathbf{H}^\top \mathbf{H})^{-1} \mathbf{H}^\top \mathbf{R} \mathbf{H} (\mathbf{H}^\top \mathbf{H})^{-1}. \tag{5.26}$$

In the special case where $\mathbf{W} = \mathbf{R}^{-1}$, the Maximum Likelihood Estimate (see eqn. (5.20)) is

$$\hat{\mathbf{x}} = \mathbf{P} \mathbf{H}^\top \mathbf{R}^{-1} \tilde{\mathbf{Y}} \tag{5.27}$$

$$\begin{aligned}
\mathbf{P} &= (\mathbf{H}^\top \mathbf{R}^{-1} \mathbf{H})^{-1} \mathbf{H}^\top \mathbf{R}^{-1} \mathbf{R} \mathbf{R}^{-1} \mathbf{H} (\mathbf{H}^\top \mathbf{R}^{-1} \mathbf{H})^{-1} \\
&= (\mathbf{H}^\top \mathbf{R}^{-1} \mathbf{H})^{-1}.
\end{aligned} \tag{5.28}$$

The matrix $\mathbf{P}^{-1} = \mathbf{H}^\top \mathbf{R}^{-1} \mathbf{H}$ is referred to as the *information matrix*. Note that quality of the estimate of $\mathbf{x}$ (i.e., $\mathbf{P}^{-1}$) increases as $\mathbf{R}$ decreases. Note also that $\mathbf{P}$ does not depend on the actual data, but on the statistics of the measurement noise and the matrix $\mathbf{H}$ that relates the vector $\mathbf{x}$ to the noisy measurements.

Table 5.1 estimates the memory and computational requirements to implement eqns. (5.27-5.28). Computational requirements are measured by counting *floating point operations* (FLOP's). The FLOP is a machine independent unit of measurement corresponding to one floating point multiplication and addition. Note that both the computation and memory requirements grow linearly with the number of measurements m.

Example 5.1 *Consider the problem of estimating the value of a scalar constant x from measurements corrupted with i.i.d. zero mean Gaussian white noise. In this problem, each measurement is modeled as*

$$\tilde{y}_k = x + \nu_k$$

Computation	FLOP's	Workspace Memory	Permanent Memory
$\mathbf{d}_1 = \mathbf{H}^\top \mathbf{R}^{-1}$	$m \times n$	$n \times m$	
$\mathbf{d}_2 = \mathbf{d}_1 \mathbf{H}$	$\frac{1}{2}m(n+1)n$	use $\mathbf{P}$	
$\mathbf{P} = (\mathbf{d}_2)^{-1}$	$n^3 + \frac{1}{2}n^2 + \frac{1}{2}n$		$\frac{1}{2}(n+1)n$
$\mathbf{d}_3 = \mathbf{d}_1 \tilde{\mathbf{Y}}$	$n \times m$	use $\mathbf{d}_1$	
$\hat{\mathbf{x}} = \mathbf{P}\mathbf{d}_3$	$n \times n$		n
$\mathbf{H}$			$m \times n$
$\mathbf{R}^{-1}$			m
$\tilde{\mathbf{Y}}$			m
Total	$n^3 + \frac{3}{2}n^2 + \frac{1}{2}n$ $+\frac{1}{2}n^2 m + \frac{5}{2}nm$	nm	$\frac{1}{2}n^2 + n\left(m + \frac{3}{2}\right) + 2m$

Table 5.1: Computational and memory requirements for the weighted least squares algorithm of eqns. (5.27-5.28) assuming $m \geq n$ with $\mathbf{x} \in \mathbb{R}^n$, $\mathbf{H} \in \mathbb{R}^{m \times n}$, $\tilde{\mathbf{Y}} \in \mathbb{R}^m$, and diagonal $\mathbf{R} \in \mathbb{R}^{m \times m}$.

where $\nu_k \sim N(0, \sigma^2)$. Therefore, after m measurements, $\tilde{\mathbf{Y}} = [\tilde{y}_1, \ldots, \tilde{y}_m]^\top \in \mathbb{R}^m$, $\mathbf{H} = [1, \ldots, 1]^\top \in \mathbb{R}^m$ and $\mathbf{R} = \sigma^2 \mathbf{I}_m$. According to eqn. (5.28),

$$P_m = (\mathbf{H}^\top \mathbf{R}^{-1} \mathbf{H})^{-1} = \frac{\sigma^2}{m} \tag{5.29}$$

and according to eqn. (5.27)

$$\hat{x}_m = \mathbf{P}\left(\mathbf{H}^\top \mathbf{R}^{-1} \tilde{\mathbf{Y}}\right) = \frac{\sigma^2}{m}\left(\frac{1}{\sigma^2}\sum_{k=1}^{m} \tilde{y}_k\right)$$

$$\hat{x}_m = \frac{1}{m}\sum_{k=1}^{m} \tilde{y}_k. \tag{5.30}$$

The estimate is the average of the measurements.

In the above formulation, computation of the estimate requires storing all m measurements, m additions, and one division. It is straightforward to manipulate eqn. (5.30) into the recursive form

$$\hat{x}_m = \hat{x}_{m-1} + \frac{1}{m}\left(\tilde{y}_m - \hat{x}_{m-1}\right) \tag{5.31}$$

which only requires the last measurement $\tilde{y}_m$, retention of the last estimate $\hat{x}_{m-1}$, two additions and one division. Derivation of such recursive algorithms is the purpose of the next subsection. △

Example 5.2 *Consider the problem of estimating the deterministic vector* $\mathbf{x}$ *based on observations* $\tilde{\mathbf{Y}}$ *related to* $\mathbf{x}$ *according to*

$$\tilde{\mathbf{Y}} = \mathbf{H}\mathbf{x} + \boldsymbol{\nu}$$

where $\boldsymbol{\nu}$ *is zero mean Gaussian white noise with* $\mathbf{R} = var\{\boldsymbol{\nu}\} = \sigma^2 \mathbf{I}_m$ *with* $\sigma = 0.1$. *At least two criteria are available for judging the quality of the estimate. First, the diagonal of* $\mathbf{P}$ *contains that variance of the estimate of each component of* $\mathbf{x}$. *The variance of the estimation error vector* $\delta\mathbf{x}$ *is quantified by trace*$(\mathbf{P})$. *Second, the residual measurement can be computed as*

$$\delta\mathbf{Y} = \tilde{\mathbf{Y}} - \mathbf{H}\hat{\mathbf{x}}.$$

These two criteria have different properties. The matrix $\mathbf{P}$ *depends only on* $\mathbf{H}$ *and* $\mathbf{R}$, *but not the actual measurements nor the value of* $\mathbf{x}$. *Alternatively,* $\delta\mathbf{Y}$ *is directly affected by the data. The variance of* $\delta\mathbf{Y}$ *and the variance of* $\hat{\mathbf{Y}}$ *are distinct:*

$$var(\delta\mathbf{Y}) = \mathbf{H}\mathbf{P}\mathbf{H}^\top + \mathbf{R} \tag{5.32}$$

while $var(\hat{\mathbf{Y}}) = \mathbf{H}\mathbf{P}\mathbf{H}^\top$.

Consider the set of $m = 3$ *measurements*

$$\tilde{\mathbf{Y}} = [-12.0529, \quad -3.0431, \quad -6.1943]^\top$$

with

$$\mathbf{H} = \begin{bmatrix} -0.6964 & -0.1228 & -0.7071 \\ 0.6330 & -0.7544 & -0.1736 \\ 0.0885 & 0.2432 & -0.9659 \end{bmatrix}.$$

According to eqns. (5.27-5.28),

$$\mathbf{P} = \begin{bmatrix} 0.0160 & 0.0091 & -0.0031 \\ 0.0091 & 0.0208 & -0.0016 \\ -0.0031 & -0.0016 & 0.0075 \end{bmatrix} \quad and \quad \hat{\mathbf{x}} = \begin{bmatrix} 6.8333 \\ 7.7014 \\ 8.9782 \end{bmatrix}.$$

The diagonal of the covariance matrix $\mathbf{P}$ *indicates that the standard deviation of* $\delta\mathbf{x}$ *is* $[0.13, 0.14, 0.09]^\top$. *Since* $m = 3 = dim(\mathbf{x})$ *and* $\mathbf{H}$ *is nonsingular,* $\delta\mathbf{Y} = \mathbf{0}$; *however, this is only a result of the number of measurements being equal to the number of unknowns. It does not imply that* $\delta\mathbf{x} = \mathbf{0}$ *nor that* $\hat{\mathbf{Y}} = \mathbf{Y}$. *Evaluation of* $\mathbf{H}\mathbf{P}\mathbf{H}^\top$ *shows that the standard deviation in the estimation of each component of* $\hat{\mathbf{Y}}$ *is* 0.1. *With three measurements and three unknowns, while an estimate of* $\mathbf{x}$ *is now available, there has been absolutely no reduction in the error in prediction of* $\mathbf{Y}$.

As the number of measurements m *increases,* $\delta\mathbf{Y}$ *will become nonzero, while the variance of both* $\delta\mathbf{x}$ *and* $\delta\mathbf{Y}$ *will tend to decrease. Because* $\delta\mathbf{Y}$ *is directly affected by the data, it can be useful in comparison with the prediction of eqn. (5.32), for detecting anomalies in the data.* △

5.3.3 Recursive Least Squares (RLS)

For the linear measurement model of eqn. (5.17) with the set of measurements $\tilde{\mathbf{Y}} = [\tilde{y}_1, \ldots, \tilde{y}_m]^\top$, using equations (5.27–5.28), it is possible to calculate the estimate $\hat{\mathbf{x}}$ that is the optimal in the WLS, maximum likelihood, and mean-squared senses. The question of this section is how to efficiently produce the new optimal estimate of $\mathbf{x}$ if an additional measurement $\tilde{y}_{m+1}$ becomes available. To be computationally efficient, the estimate $\hat{\mathbf{x}}_{m+1}$ should not start from scratch (i.e., require all elements of $\tilde{\mathbf{Y}}$), but from the best previously available estimate $\hat{\mathbf{x}}_m$. Therefore, based on intuition, the objective is to develop a recursive estimation equation of the form

$$\hat{\mathbf{x}}_{m+1} = \hat{\mathbf{x}}_m + \mathbf{K}(\tilde{y}_{m+1} - \hat{y}_{m+1}) \tag{5.33}$$

where $\hat{y}_{m+1} = \mathbf{H}_{m+1}\hat{\mathbf{x}}_m$ and $\mathbf{K}$ is a vector gain that is to be determined.

To achieve this objective, we start from the prior information. The vector $\hat{\mathbf{x}}_m$ is known with

$$\begin{aligned} E\langle \hat{\mathbf{x}}_m \rangle &= \mathbf{x} \\ var(\hat{\mathbf{x}}_m) &= var(\delta \mathbf{x}_m) = \mathbf{P}_m. \end{aligned}$$

The new measurement is modeled as

$$\tilde{y}_{m+1} = \mathbf{H}_{m+1}\mathbf{x} + \nu_{m+1},$$

where $var(\nu_{m+1}, \nu_j) = R_{m+1,j}\delta_{m+1,j}$ for all j (R_{m+1} will be used as a short hand notation for $R_{m+1,m+1}$ in the following). Therefore, the information available for estimating $\mathbf{x}$ after measurement m+1 is $\hat{\mathbf{x}}_m$ and $\tilde{y}_{m+1}$ which can be organized as

$$\hat{\mathbf{x}}_m = \mathbf{x} - \delta \mathbf{x}_m, \tag{5.34}$$

$$\tilde{y}_{m+1} = \mathbf{H}_{m+1}\mathbf{x} + \nu_{m+1}. \tag{5.35}$$

Constructing a hypothetical set of $m+1$ measurements yields

$$\begin{bmatrix} \hat{\mathbf{x}}_m \\ \tilde{y}_{m+1} \end{bmatrix} = \begin{bmatrix} \mathbf{I} \\ \mathbf{H}_{m+1} \end{bmatrix} \mathbf{x} + \begin{bmatrix} -\delta \mathbf{x}_m \\ \nu_{m+1} \end{bmatrix} \tag{5.36}$$

which has the form of eqn. (5.17). Therefore it can be solved using eqns. (5.27-5.28) by defining the weighting matrix $\mathbf{R}^{-1}$ corresponding to $cov([\delta \mathbf{x}_m^\top, \nu_{m+1}])$.

Eqn. (5.28) results in:

$$\mathbf{P}_{m+1}^{-1} = \begin{bmatrix} \mathbf{I} & \mathbf{H}_{m+1}^\top \end{bmatrix} \begin{bmatrix} \mathbf{P}_m & \mathbf{0} \\ \mathbf{0} & R_{m+1} \end{bmatrix}^{-1} \begin{bmatrix} \mathbf{I} \\ \mathbf{H}_{m+1} \end{bmatrix}$$

$$\mathbf{P}_{m+1}^{-1} = \mathbf{P}_m^{-1} + \mathbf{H}_{m+1}^\top R_{m+1}^{-1} \mathbf{H}_{m+1}. \tag{5.37}$$

Recalling that $\mathbf{P}_m^{-1}$ is the information matrix for $\mathbf{x}_m$, eqn. (5.37) shows the amount that each new measurement increases the information matrix. Note that if all measurements are noisy (i.e., $R_j \neq 0$ for any j), then $\mathbf{H}_{m+1}^\top R_{m+1}^{-1} \mathbf{H}_{m+1}$ is always finite and positive; the information matrix is monotonically increasing; and, the variance of the estimate (i.e, $\mathbf{P}_{m+1}$) monotonically decreases with each new measurement. Multiplying eqn. (5.37) on the right by $\hat{\mathbf{x}}_m$, we obtain

$$\mathbf{P}_{m+1}^{-1} \hat{\mathbf{x}}_m = \mathbf{P}_m^{-1} \hat{\mathbf{x}}_m + \mathbf{H}_{m+1}^\top R_{m+1}^{-1} \hat{y}_{m+1}. \tag{5.38}$$

Multiplying on the left by $\mathbf{P}_{m+1}$ and rearranging yields

$$\mathbf{P}_{m+1} \mathbf{P}_m^{-1} \hat{\mathbf{x}}_m = \hat{\mathbf{x}}_m - \mathbf{P}_{m+1} \mathbf{H}_{m+1}^\top R_{m+1}^{-1} \hat{y}_{m+1} \tag{5.39}$$

which will be needed in the following derivation.

The new estimate is calculated from (5.27) as

$$\begin{aligned}
\hat{\mathbf{x}}_{m+1} &= \mathbf{P}_{m+1} \left[\begin{array}{cc} \mathbf{I} & \mathbf{H}_{m+1}^\top \end{array} \right] \left[\begin{array}{cc} \mathbf{P}_m & \mathbf{0} \\ \mathbf{0} & R_{m+1} \end{array} \right]^{-1} \left[\begin{array}{c} \hat{\mathbf{x}}_m \\ \tilde{y}_{m+1} \end{array} \right] \\
&= \mathbf{P}_{m+1} \left(\mathbf{P}_m^{-1} \hat{\mathbf{x}}_m + \mathbf{H}_{m+1}^\top R_{m+1}^{-1} \tilde{y}_{m+1} \right) \\
&= \hat{\mathbf{x}}_m + \mathbf{P}_{m+1} \mathbf{H}_{m+1}^\top R_{m+1}^{-1} (\tilde{y}_{m+1} - \hat{y}_{m+1}) \\
\hat{\mathbf{x}}_{m+1} &= \hat{\mathbf{x}}_m + \mathbf{K}_{m+1} (\tilde{y}_{m+1} - \hat{y}_{m+1})
\end{aligned} \tag{5.40}$$

where $\mathbf{K}_{m+1} = \mathbf{P}_{m+1} \mathbf{H}_{m+1}^\top R_{m+1}^{-1}$ and eqn. (5.39) was used to obtain the second to the last line of the derivation. Although a linear estimator with the form of eqn. (5.33) was initially stated as an objective, such a linear relationship was never imposed as a constraint. The linear update relationship of eqn. (5.40) is a natural consequence of the problem formulation.

Eqns. (5.37) and (5.40) provide recursive formulas for the estimation of the vector $\mathbf{x}$. With proper initialization, the estimate is exactly the same as that attained by use of a batch approach for m+1 measurements using eqns. (5.27) and (5.28). Table 5.1 shows that, for large m, the memory and computational requirements of the batch algorithm are $O(n^2 m)$ and $O(nm)$, respectively. These requirements are necessary even if the estimate is known for $(m-1)$ measurements prior to the m-th measurement. The computational and memory requirements for incorporating a single additional scalar measurement using the recursive algorithm of eqns. (5.37) and (5.40) are evaluated in Table 5.2. For the recursive algorithm, the memory and computational requirements for incorporating each measurement are determined only by the dimension of the estimated vector.

Eqn. (5.37) computes $\mathbf{P}_{m+1}^{-1}$ where $\mathbf{P}_{m+1}$ is required for the calculation of $\mathbf{K}$; therefore, the algorithm as written requires matrix inversion which is not desirable. Table 5.2 shows that the matrix inversion plays a dominant

Computation	Flops	Workspace Memory	Permanent Memory
$\mathbf{H}_{m+1}$			n
$r = \tilde{y}_{m+1} - \mathbf{H}_{m+1}\hat{\mathbf{x}}_m$	n	1	
$\mathbf{d}_1 = \frac{\mathbf{H}_{m+1}^\top}{R_{m+1}}$	n	n	
$\mathbf{F}_{m+1} = \mathbf{F}_m + \mathbf{d}_1\mathbf{H}_{m+1}$	$\frac{1}{2}(n+1)n$		$\frac{1}{2}(n+1)n$
$\mathbf{P}_{m+1} = \mathbf{F}_{m+1}^{-1}$	$n^3 + \frac{1}{2}n^2 + \frac{1}{2}n$	$\frac{1}{2}(n+1)n$	
$\mathbf{K} = \mathbf{P}_{m+1}\mathbf{d}_1$	n^2	n	
$\hat{\mathbf{x}}_{m+1} = \hat{\mathbf{x}}_m + \mathbf{K}r$	n		n
Total	$n^3 + 2n^2 + 4n$	$\frac{1}{2}n^2 + \frac{5}{2}n + 1$	$\frac{1}{2}n^2 + \frac{5}{2}n$

Table 5.2: Computational and memory requirements for the recursive least squares algorithm of eqns. (5.37) and (5.40) with $\mathbf{x} \in \mathbb{R}^n$, $\mathbf{H}_{m+1} \in \mathbb{R}^{1\times n}$, $\tilde{y} \in \mathbb{R}$, and $R_{m+1} \in \mathbb{R}$. The matrix $\mathbf{F}_{m+1} \in \mathbb{R}^{n\times n}$ represents the information matrix.

role in determining the amount of required computation. Section 5.4 will show that eqns. (5.37) and (5.40) are equivalent to

$$\mathbf{K}_{m+1} = \mathbf{P}_m\mathbf{H}_{m+1}^\top\left(R_{m+1} + \mathbf{H}_{m+1}\mathbf{P}_m\mathbf{H}_{m+1}^\top\right)^{-1} \quad (5.41)$$

$$\hat{\mathbf{x}}_{m+1} = \hat{\mathbf{x}}_m + \mathbf{K}_{m+1}(\tilde{y}_{m+1} - \mathbf{H}_{m+1}\hat{\mathbf{x}}_m) \quad (5.42)$$

$$\mathbf{P}_{m+1} = (\mathbf{I} - \mathbf{K}_{m+1}\mathbf{H}_{m+1})\,\mathbf{P}_m. \quad (5.43)$$

The computational requirements for this set of equations are evaluated in Table 5.3. The workspace and computational requirements are significantly reduced relative to the previous algorithms.

In addition to requiring smaller amounts of memory and computation, the recursive least squares algorithm provides iterative estimates immediately following the measurement time. This has the potential to provide estimates with reduced delay over an approach which waits to accumulate a fixed sized batch of m samples before calculating an estimate.

Example 5.3 *This example reconsiders the problem stated in Example 5.1 using the algorithm of eqns. (5.41–5.43).*

For this problem, $n = 1$, $H_{m+1} = 1$ and $R_{m+1} = \sigma^2$. Using the first measurement and eqns. (5.29–5.30) to initialize the state and covariance, we have $\hat{x}_1 = \tilde{y}_1$ and $P_1 = \sigma^2$. The recursion then starts at $m = 2$:

$$K_{m+1} = \frac{P_m}{\sigma^2 + P_m}$$

$$P_{m+1} = (1 - K_{m+1})\,P_m$$

Computation	Flops	Workspace Memory	Permanent Memory
$\mathbf{H}_{m+1}$			n
$r = \tilde{y}_{m+1} - \mathbf{H}_{m+1}\hat{\mathbf{x}}_m$	n	1	
$\mathbf{d}_1 = \mathbf{H}_{m+1}\mathbf{P}_m$	n^2	n	
$d_2 = R_{m+1} + \mathbf{d}_1\mathbf{H}_{m+1}^\top$	n	1	
$\mathbf{K} = \frac{\mathbf{d}_1^\top}{d_2}$	n	n	
$\hat{\mathbf{x}}_{m+1} = \hat{\mathbf{x}}_m + \mathbf{K}r$	n		n
$\mathbf{P}_{m+1} = \mathbf{P}_m - \mathbf{K}\mathbf{d}_1$	$\frac{1}{2}(n+1)n$		$\frac{1}{2}(n+1)n$
Total	$\frac{3}{2}n^2 + \frac{9}{2}n$	$2n+2$	$\frac{1}{2}n^2 + \frac{5}{2}n$

Table 5.3: Computational and memory requirements for the recursive least squares algorithm of eqns. (5.41–5.43) with $\mathbf{x} \in \mathbb{R}^n$, $\mathbf{H}_{m+1} \in \mathbb{R}^{1\times n}$, $\tilde{y} \in \mathbb{R}$, and $R_{m+1} \in \mathbb{R}$.

$$= \left(1 - \frac{P_m}{\sigma^2 + P_m}\right) P_m$$

$$= \left(\frac{\sigma^2}{\sigma^2 + P_m}\right) P_m = K_{m+1}\sigma^2.$$

Table 5.4 evaluates K_m and P_m versus m. The covariance of the m-th estimate is $P_m = \frac{\sigma^2}{m}$ which matches the result of the batch computation in eqn. (5.29). Also, eqn. (5.42) becomes

$$\hat{x}_m = \hat{x}_{m-1} + \frac{1}{m}(\tilde{y}_m - \hat{x}_{m-1})$$

which matches eqn. (5.31). △

	1	2	3	m
K_m		$\frac{1}{2}$	$\frac{1}{3}$	$\frac{1}{m}$
P_m	σ^2	$\frac{1}{2}\sigma^2$	$\frac{1}{3}\sigma^2$	$\frac{1}{m}\sigma^2$

Table 5.4: Evaluation of K_m and P_m versus m for Example 5.3.

The previous example demonstrates that with proper initialization of the state estimate and the error covariance, the recursive and batch solutions yield the same results.

The following subsection will extend the present analysis to the derivation of the Kalman filter for applications where $\mathbf{x}$ is nonconstant and the dynamic model for $\mathbf{x}$ is given by eqn. (5.1).

5.3.4 Kalman Filtering

The last extension to be considered in this section concerns the estimation of the state of a dynamic system described by a linear ordinary difference equation,

$$\mathbf{x}_k = \mathbf{\Phi}_{k-1}\mathbf{x}_{k-1} + \mathbf{G}_{k-1}\mathbf{u}_{k-1} + \boldsymbol{\omega}_{k-1} \tag{5.44}$$

when only a noisy linear combination of the system states

$$y_k = \mathbf{H}_k\mathbf{x}_k + \nu_k \tag{5.45}$$

can be measured. All quantities in the above equations are defined in Section 5.1. For the problem formulation, we assume that

$$E\langle \boldsymbol{\omega}_k \boldsymbol{\omega}_j^\top \rangle = \mathbf{Qd}_k \delta_{jk} \text{ and } E\langle \nu_k \nu_j^\top \rangle = R_k \delta_{jk} \tag{5.46}$$

are known. At $t = kT$, the objective is to produce the optimal estimate of $\mathbf{x}_k$ which is corrected for the measurement y_k. We assume that, based on the previous sequence of measurements $\{y_i\}_{i=0}^{k-1}$, we have an unbiased estimate $\hat{\mathbf{x}}_{k-1}^+ = E\langle \mathbf{x}_{k-1} \rangle$ with known variance

$$\mathbf{P}_{k-1}^+ = var\left\{ \left(\hat{\mathbf{x}}_{k-1}^+\right)\left(\hat{\mathbf{x}}_{k-1}^+\right)^\top \right\} = var\left\{ \left(\delta\mathbf{x}_{k-1}^+\right)\left(\delta\mathbf{x}_{k-1}^+\right)^T \right\}.$$

This objective will be achieved in two steps.

First, consider the objective of estimating $\mathbf{x}_k$ if there were no measurement available at time k. This estimate is denoted as $\hat{\mathbf{x}}_k^-$ where the superscript indicates that this is the estimate prior to using the new information available from the measurement. Taking the expectation of both sides of eqn. (5.44) results in:

$$\begin{aligned} E\langle \mathbf{x}_k \rangle &= \mathbf{\Phi}_{k-1}E\langle \mathbf{x}_{k-1} \rangle + \mathbf{G}_{k-1}\mathbf{u}_{k-1} \\ &= \mathbf{\Phi}_{k-1}\hat{\mathbf{x}}_{k-1}^+ + \mathbf{G}_{k-1}\mathbf{u}_{k-1}. \end{aligned}$$

Therefore, the unbiased state estimate at $t = kT$ prior to incorporating the measurement is

$$\hat{\mathbf{x}}_k^- = \mathbf{\Phi}_{k-1}\hat{\mathbf{x}}_{k-1}^+ + \mathbf{G}_{k-1}\mathbf{u}_{k-1}. \tag{5.47}$$

To determine the error variance of this prior estimate, we first difference eqns. (5.44) and (5.47) to produce the state space model for $\delta\mathbf{x}_k^- = \mathbf{x}_k - \hat{\mathbf{x}}_k^-$

$$\delta\mathbf{x}_k^- = \mathbf{\Phi}_{k-1}\delta\mathbf{x}_{k-1}^+ + \boldsymbol{\omega}_{k-1}, \tag{5.48}$$

then proceed with the analysis similar to that described in Section 4.6.5:

$$\begin{aligned} \mathbf{P}_k^- &= E\langle (\delta\mathbf{x}_k^-)(\delta\mathbf{x}_k^-)^T \rangle \\ &= \mathbf{\Phi}_{k-1}\mathbf{P}_{k-1}^+\mathbf{\Phi}_{k-1}^\top + \mathbf{Qd}_{k-1}. \end{aligned} \tag{5.49}$$

Second, at time $t = kT$ (i.e., prior to incorporating the new measurement), treating $\hat{\mathbf{x}}_k^-$ as a measurement with measurement error $\delta\mathbf{x}_k^-$, we have two sets of information about the value of $\mathbf{x}_k$:

$$\begin{aligned} \hat{\mathbf{x}}_k^- &= \mathbf{x}_k - \delta\mathbf{x}_k^-, \quad \text{where } var(\delta\mathbf{x}_k^-) = \mathbf{P}_k^-, \text{ and} \\ \tilde{y}_k &= \mathbf{H}_k\mathbf{x}_k + \nu_k, \text{ where } var(\nu_k) = R_k \end{aligned}$$

which is exactly the same form as eqns. (5.34–5.35) in the RLS derivation. Therefore, the optimal estimate $\hat{\mathbf{x}}_k^+$ of $\mathbf{x}_k$ corrected for the measurement $\tilde{y}_k$ is given by eqns. (5.37) and (5.40). After changing the notation, the covariance and state measurement update equations are

$$\left(\mathbf{P}_k^+\right)^{-1} = (\mathbf{P}_k^-)^{-1} + \mathbf{H}_k^\top R_k^{-1}\mathbf{H}_k \tag{5.50}$$

$$\hat{\mathbf{x}}_k^+ = \hat{\mathbf{x}}_k^- + \mathbf{K}_k(\tilde{y}_k - \mathbf{H}_k\hat{\mathbf{x}}_k^-) \tag{5.51}$$

where $\mathbf{K}_k = \mathbf{P}_k^+\mathbf{H}_k^\top R_k^{-1}$.

Therefore, the Kalman filter update is a two step process, the time update by eqns. (5.47) and (5.49) and the measurement correction by eqns. (5.50) and (5.51). These equations propagate both the estimate and its error variance. The error variance is used to adjust the state estimation gain for each new measurement to optimally weight the new measurement information relative to all past information as represented by $\hat{\mathbf{x}}$ and $\mathbf{P}$.

5.4 Kalman Filter Derivation Summary

For the system described by

$$\mathbf{x}_k = \mathbf{\Phi}_{k-1}\mathbf{x}_{k-1} + \mathbf{G}_{k-1}\mathbf{u}_{k-1} + \boldsymbol{\omega}_{k-1} \tag{5.52}$$

$$\tilde{\mathbf{y}}_k = \mathbf{H}_k\mathbf{x}_k + \boldsymbol{\nu}_k. \tag{5.53}$$

Section 5.3.4 derived the following version of the Kalman filter:

$$\hat{\mathbf{x}}_k^- = \mathbf{\Phi}_{k-1}\hat{\mathbf{x}}_{k-1}^+ + \mathbf{G}_{k-1}\mathbf{u}_{k-1} \tag{5.54}$$

$$\hat{\mathbf{y}}_k = \mathbf{H}_k\hat{\mathbf{x}}_k^- \tag{5.55}$$

$$\mathbf{P}_k^- = \mathbf{\Phi}_{k-1}\mathbf{P}_{k-1}^+\mathbf{\Phi}_{k-1}^\top + \mathbf{Qd}_{k-1} \tag{5.56}$$

$$\left(\mathbf{P}_k^+\right)^{-1} = (\mathbf{P}_k^-)^{-1} + \mathbf{H}_k^\top \mathbf{R}_k^{-1}\mathbf{H}_k \tag{5.57}$$

$$\mathbf{K}_k = \mathbf{P}_k^+\mathbf{H}_k^\top\mathbf{R}_k^{-1} \tag{5.58}$$

$$\mathbf{r}_k = \tilde{\mathbf{y}}_k - \hat{\mathbf{y}}_k \tag{5.59}$$

$$\hat{\mathbf{x}}_k^+ = \hat{\mathbf{x}}_k^- + \mathbf{K}_k r_k. \tag{5.60}$$

Eqns. (5.54-5.56) are referred to as the Kalman filter time update equations. Eqns. (5.57-5.60) are referred to as the Kalman filter measurement update equations. Eqn. (5.59) defines the Kalman filter measurement residual. Analysis of the measurement residual can provide information about the validity of sensor information or the validity of the system and measurement models.

Section 5.2 also presented a derivation of the Kalman filter. The time update equations were the same as eqns. (5.54–5.56); however, the measurement update equations for $\mathbf{P}_k^+$ and $\mathbf{K}_k$ were

$$\mathbf{K}_k = \mathbf{P}_k^- \mathbf{H}_k^\top \left(\mathbf{H}_k \mathbf{P}_k^- \mathbf{H}_k^\top + \mathbf{R}_k\right)^{-1} \tag{5.61}$$

$$\mathbf{P}_k^+ = \left[\mathbf{I} - \mathbf{K}_k \mathbf{H}_k\right] \mathbf{P}_k^-. \tag{5.62}$$

The purpose of the next subsection is to show that the formulation of eqns. (5.57-5.58) is equivalent to the formulation of eqns. (5.61-5.62). Then Subsection 5.4.2 will derive two additional forms of the covariance update equation. The various implementations are summarized in Table 5.5. Additional alternative implementations, including square-root implementations, are presented in [60].

It is important that the reader remember that although the Kalman filter can be implemented by a variety of techniques, the techniques are equivalent theoretically. However, some of the implementation techniques require less computation while others have better numeric properties. It is up to the designer to select the most appropriate form for a given application.

For example, the form of the measurement update in eqns. (5.57-5.58) is often not convenient for real-time implementations due to the required matrix inversion. The alternative formulation of eqns. (5.61-5.62) also involves a matrix inversion; however, the inverted matrix has dimensions corresponding to the number of measurements instead of the number of states. Since the number of measurements is often significantly less than the number of states, the update of eqns. (5.61-5.62) is often preferred. In fact, when the problem can be formulated with only scalar measurements, as discussed in Section 5.6.1, then the matrix inversion in eqn. (5.61) reduces to scalar division. Such issues are discussed in Section 5.6.

Table 5.5 summarizes the discrete-time Kalman filter equations. The Kalman filter could be implemented by one of (at least) four techniques. All the approaches would use the same time update (eqns. (5.54-5.56)). For the measurement covariance update, four methods have been presented. The user must select one approach. In comparing the various covariance measurement update equations, eqn. (5.62) is the simplest formula and requires the least computation. Although eqn. (5.57) results naturally from the derivation presented in Section 5.3.4, it is not as efficient to implement as the alternative, equivalent solutions. Eqns. (5.68) and (5.69) involve only symmetric operations. Of the two, eqn. (5.69) is more numerically stable. Note that eqn. (5.69) is the same as eqn. (5.12) which is valid for

Initialization	$\hat{\mathbf{x}}_0^- = E\langle \mathbf{x}_0 \rangle$ $\mathbf{P}_0^- = var(\mathbf{x}_0^-)$
Gain Calculation	$\mathbf{K}_k = \mathbf{P}_k^- \mathbf{H}_k^\top \left(\mathbf{R}_k + \mathbf{H}_k \mathbf{P}_k^- \mathbf{H}_k^\top\right)^{-1}$
Measurement Update	$\hat{\mathbf{x}}_k^+ = \hat{\mathbf{x}}_k^- + \mathbf{K}_k \left(\tilde{\mathbf{y}}_k - \hat{\mathbf{y}}_k\right)$
Covariance Update (choose one)	$\mathbf{P}_k^+ = [\mathbf{I} - \mathbf{K}_k \mathbf{H}_k]\mathbf{P}_k^-$ $\mathbf{P}_k^+ = [\mathbf{I} - \mathbf{K}_k \mathbf{H}_k]\mathbf{P}_k^- [\mathbf{I} - \mathbf{K}_k \mathbf{H}_k]^\top + \mathbf{K}_k \mathbf{R} \mathbf{K}_k^\top$ $\mathbf{P}_k^+ = \mathbf{P}_k^- - \mathbf{K}_k \left(\mathbf{R}_k + \mathbf{H}_k \mathbf{P}_k^- \mathbf{H}_k^\top\right) \mathbf{K}_k^\top$ $(\mathbf{P}_k^+)^{-1} = (\mathbf{P}_k^-)^{-1} + \mathbf{H}_k^\top \mathbf{R}_k^{-1} \mathbf{H}_k$
Time Propagation	$\hat{\mathbf{x}}_{k+1}^- = \mathbf{\Phi}_k \hat{\mathbf{x}}_k^+ + \mathbf{G}_k \mathbf{u}_k$ $\mathbf{P}_{k+1}^- = \mathbf{\Phi}_k \mathbf{P}_k^+ \mathbf{\Phi}_k^\top + \mathbf{Qd}_k$

Table 5.5: Discrete-time Kalman filter equations. Computation of the matrices $\mathbf{\Phi}_k$ and $\mathbf{Qd}_k$ is discussed in Section 4.7.1.

any stabilizing state feedback gain vector $\mathbf{L}$. The approaches described in eqns. (5.57), (5.62), and (5.68) apply only for the Kalman filter gain $\mathbf{K}_k$.

5.4.1 Equivalent Measurement Updates

To show the equivalence of eqns. (5.57-5.58) to eqns. (5.61-5.62), we will apply the *Matrix Inversion Lemma* (see Appendix B) to eqn. (5.57) with $\mathbf{A} = \mathbf{P}^-$, $\mathbf{B} = \mathbf{P}^- \mathbf{H}^\top$, $\mathbf{C} = -\left(\mathbf{R} + \mathbf{H}\mathbf{P}^-\mathbf{H}^\top\right)^{-1}$, and $\mathbf{D} = \mathbf{H}\mathbf{P}^-$. The result is

$$
\begin{aligned}
(\mathbf{P}^+)^{-1} &= \left[\mathbf{P}^- - \mathbf{P}^-\mathbf{H}^\top \left(\mathbf{R} + \mathbf{H}\mathbf{P}^-\mathbf{H}^\top\right)^{-1} \mathbf{H}\mathbf{P}^-\right]^{-1} \\
\mathbf{P}^+ &= \mathbf{P}^- - \mathbf{P}^-\mathbf{H}^\top \left(\mathbf{R} + \mathbf{H}\mathbf{P}^-\mathbf{H}^\top\right)^{-1} \mathbf{H}\mathbf{P}^- && (5.63) \\
\mathbf{P}^+ &= \mathbf{P}^- \left[\mathbf{I} - \mathbf{H}^\top \left(\mathbf{R} + \mathbf{H}\mathbf{P}^-\mathbf{H}^\top\right)^{-1} \mathbf{H}\mathbf{P}^-\right] && (5.64) \\
\mathbf{P}^+ &= \left[\mathbf{I} - \mathbf{P}^-\mathbf{H}^\top \left(\mathbf{R} + \mathbf{H}\mathbf{P}^-\mathbf{H}^\top\right)^{-1} \mathbf{H}\right] \mathbf{P}^-. && (5.65)
\end{aligned}
$$

Substituting eqn. (5.64) into eqn. (5.58) yields

$$
\begin{aligned}
\mathbf{K} &= \mathbf{P}^+\mathbf{H}^\top\mathbf{R}^{-1} \\
&= \mathbf{P}^-\mathbf{H}^\top \left[\mathbf{I} - \left(\mathbf{R} + \mathbf{H}\mathbf{P}^-\mathbf{H}^\top\right)^{-1} \mathbf{H}\mathbf{P}^-\mathbf{H}^\top\right] \mathbf{R}^{-1} \\
&= \mathbf{P}^-\mathbf{H}^\top \left(\mathbf{R} + \mathbf{H}\mathbf{P}^-\mathbf{H}^\top\right)^{-1} \left[\mathbf{R} + \mathbf{H}\mathbf{P}^-\mathbf{H}^\top - \mathbf{H}\mathbf{P}^-\mathbf{H}^\top\right] \mathbf{R}^{-1} \\
&= \mathbf{P}^-\mathbf{H}^\top \left(\mathbf{R} + \mathbf{H}\mathbf{P}^-\mathbf{H}^\top\right)^{-1}. && (5.66)
\end{aligned}
$$

Substituting eqn. (5.66) into eqn. (5.65), simplifies the latter to

$$\mathbf{P}^+ = [\mathbf{I} - \mathbf{KH}]\,\mathbf{P}^- \tag{5.67}$$

where eqns. (5.66–5.67) correspond to eqns. (5.61-5.62).

5.4.2 Equivalent Covariance Measurement Updates

An alternative use of eqns. (5.66) and (5.63) is to show that

$$\mathbf{P}^+ = \mathbf{P}^- - \mathbf{K}\left(\mathbf{R} + \mathbf{HP}^-\mathbf{H}^\top\right)\mathbf{K}^\top. \tag{5.68}$$

Lastly, eqn. (5.67) can be manipulated as shown below:

$$\begin{aligned}
\mathbf{P}^+ &= [\mathbf{I} - \mathbf{KH}]\,\mathbf{P}^- + \mathbf{P}^-\mathbf{H}^\top\mathbf{K}^\top - \mathbf{P}^-\mathbf{H}^\top\mathbf{K}^\top \\
&= [\mathbf{I} - \mathbf{KH}]\,\mathbf{P}^- + \mathbf{K}\left(\mathbf{R} + \mathbf{HP}^-\mathbf{H}^\top\right)\mathbf{K}^\top - \mathbf{P}^-\mathbf{H}^\top\mathbf{K}^\top \\
&= \mathbf{P}^- - \mathbf{KHP}^- + \mathbf{KRK}^\top + \mathbf{KHP}^-\mathbf{H}^\top\mathbf{K}^\top - \mathbf{P}^-\mathbf{H}^\top\mathbf{K}^\top \\
&= [\mathbf{I} - \mathbf{KH}]\,\mathbf{P}^-\,[\mathbf{I} - \mathbf{KH}]^\top + \mathbf{KRK}^\top.
\end{aligned} \tag{5.69}$$

where eqn. (5.66) was used in the transition from the first to second line. Eqn. (5.69) is referred to as the *Joseph form* of the covariance propagation equations [34].

5.4.3 Kalman Filter Examples

This section presents a few Kalman filter examples.

The following example is useful in various applications including the estimation of the GPS ionospheric error using data from a two frequency receiver as is discussed in Section 8.6 and carrier smoothing of code measurements as is discussed in Section 8.7.

Example 5.4 *Consider the scenario where at each measurement epoch two measurements are available*

$$y_k = x_k + n_k \tag{5.70}$$

$$z_k = x_k + v_k + B \tag{5.71}$$

where $n_k \sim N(0, \sigma^2)$, $v_k \sim N(0, (\mu\sigma)^2)$ with $\mu \ll 1$, B is a constant, and x_k can change arbitrarily from one epoch to another, and all the variables in the righthand sides of the above equations are mutually uncorrelated. The objective is to estimate x_k.

One solution using a complementary filter is illustrated in Figure 5.3. The residual measurement $r_k = z_k - y_k$ is equivalent to

$$r_k = B + w_k$$

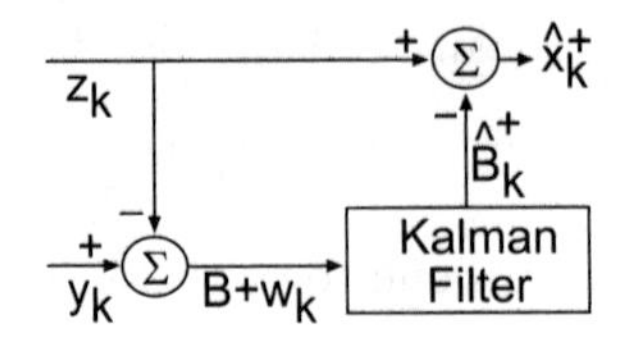

Figure 5.3: Measurement bias complementary filter for Example 5.4.

where $w_k = v_k - n_k$ and $w_k \sim N(0, (1+\mu^2)\sigma^2)$. Assuming that there is no prior information about B, this is the problem of estimating a constant from independent noise corrupted measurements of the constant that was discussed in Example 5.3. The solution is

$$\hat{B}_k = \hat{B}_{k-1} + \frac{1}{k}(\tilde{r}_k - \hat{B}_{k-1})$$

starting at $k = 1$ with $B_0 = 0$. Note that the value of B_0 is immaterial. The estimation error variance for $\hat{B}_k$ is $P_B(k) = (1+\mu^2)\frac{\sigma^2}{k}$. Given this estimate of B, using eqn. (5.71), the value of x_k can be computed as

$$\hat{x}_k = z_k - \hat{B}_k,$$

where the error variance for $\hat{x}_k$ is $P_x(k) = \left((1-\mu^2)\frac{1}{k} + \mu^2\right)\sigma^2$. Initially, the variance of $\hat{x}$ is σ^2 as the number of measurements k increases, the variance of $\hat{x}$ decreases toward $\mu^2\sigma^2$.

This example has relevance to GPS applications for estimation of ionospheric delay and for carrier smoothing of the pseudorange observable. △

This example is further investigated in Exercise 5.11. That exercise considers the case where the measurement error process n_k is not white.

Example 5.5 *Section 4.9.4 presented a one dimension aided INS example using a state estimation gain vector designed via the pole placement approach. That example is continued here using a Kalman filter. The state space description and noise covariance matrices are specified in Section 4.9.4. In the example, the initial error covariance matrix $\mathbf{P}^-(0)$ is assumed to be a zero matrix.*

The performance of the estimator is indicated by error standard deviation plots are shown in the left column of Figure 5.4 where $var(\hat{p}) = \sigma_p^2$, $var(\hat{v}) = \sigma_v^2$, and $var(\hat{b}) = \sigma_b^2$. The portion of this curve that appears as a wide band in the left column of figures is due to the covariance growth between position measurements followed by the covariance decrease at the measurement instant. If the time axis was magnified, this growth and decrease would be clearly evident as it was in Figure 4.7. The resulting time-varying Kalman filter gains are plotted in the right column. The Kalman

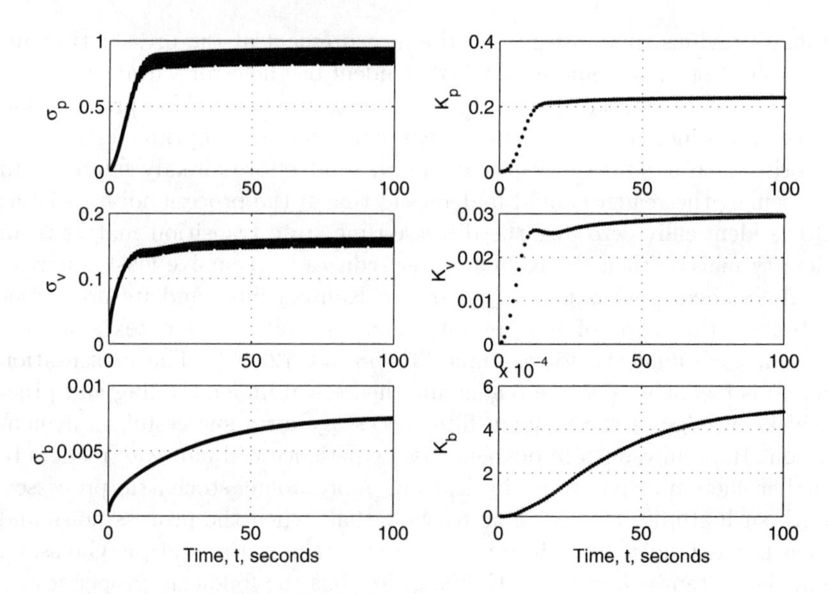

Figure 5.4: Kalman filter estimation performance for the simplified navigation system of Example 5.5. Left – Standard deviation of estimation error. Right – Kalman filter gain sequence.

gains are plotted as one dot at each sampling instant to clearly indicate that they are only defined at the sampling instants. Due to the assumption of initial perfect knowledge (i.e., $\mathbf{P}^{-}(0) = \mathbf{0}$*), the Kalman filter gains are initially small, but grow with time to optimally tradeoff the noisy measurement with the increasingly noisy state estimates. In comparison with Figure 4.7, each of the Kalman filter state error variance terms is smaller than the corresponding error variance terms of the pole placement observer, as expected of the optimal design.* △

5.5 Kalman Filter Properties

Section 5.2 derived the Kalman filter in the stochastic process framework as an unbiased, minimum variance, linear stochastic estimator. The results of that derivation approach can yield valuable insight into the Kalman filter and its performance.

Section 5.3 derived the Kalman filter as an extension of the weighted least squares approach from linear sets of algebraic equations to linear systems of ordinary difference equations in the presence of additive process driving noise. In this context, the Kalman filter is derived as an optimal combination (that turns out to be linear) of the time propagated estimate from a previous time instant and the measurement at the present time instant. The optimal combination is dependent on the error variance of both the prior state estimate and the current measurement. This presentation method was included to present the Kalman filter, as an optimal estimation algorithm, in a context with which many readers are already familiar. In particular, the reader should understand that if the process noise variance $\mathbf{Qd}$ is identically zero and the discrete-time state transition matrix is an identity matrix, then the Kalman filter reduces to recursive least squares.

A detailed stochastic analysis of the Kalman filter and its properties is beyond the scope of the present book. Several excellent texts already present such analysis, for example [31, 58, 60, 72, 93]. The presentation herein is meant to give the reader the theoretical understanding and practical knowledge of the Kalman filter necessary for a successful implementation. Hopefully, such implementation experience will motivate readers to further their understanding by learning more about stochastic processes. With such studies it is possible to show that, when the process noise and measurement noise are white and Gaussian, the initial state is Gaussian, and the system is linear, the Kalman filter has the following properties:

1. The Kalman filter estimate is *unbiased.*

2. The Kalman filter estimate is the *maximum likelihood estimate.* It is also the minimum mean-squared error estimate. Under the given assumptions, the filter state is Gaussian. By propagating the mean and error variance of the state, the Kalman filter is propagating the distribution of the state conditioned on all available measurements. Since the conditional distribution is Gaussian, the mean, mode, and median are identical.

3. The conditional mean (hence, the Kalman filter estimate) is the minimum of any positive definite quadratic cost function.

4. The conditional mean (hence, the Kalman filter estimate) is the minimum of almost any reasonable non-decreasing function of the estimation error [31].

5. The Kalman filter, although a linear algorithm, is the optimal (linear or nonlinear) state estimation algorithm for a linear system.

6. The Kalman filter residual state error $\mathbf{x}_k - \hat{\mathbf{x}}_k$ at time step k is orthogonal (stochastically) to all previous measurements $\mathbf{y}_k$:

$$E\langle[\mathbf{x}_k - \hat{\mathbf{x}}_k]\mathbf{y}_i\rangle = 0, \quad i = 1, \ldots, k.$$

7. The measurement residual sequence $\mathbf{r}_k = \mathbf{y}_k - \hat{\mathbf{y}}_k$ is white with variance $\left(\mathbf{R}_k + \mathbf{H}_k\mathbf{P}_k^-\mathbf{H}_k^\top\right)$ as given by eqn. (5.11).

8. Given that the state is controllable from the driving noise and is observable from the measurements, the Kalman filter estimation error dynamics are asymptotically stable. Therefore, for an observable system, the effects of the initial conditions ($\mathbf{x}_0^-$ and $\mathbf{P}_0^-$) decay away and do not affect the solution as $k \rightarrow \infty$.

Since the above properties rely on three assumptions, it is natural to consider the reasonableness of these assumptions. Although most physical systems are in fact non-linear, the linearity assumption can be locally applied when the system non-linearities are linearizable and the distance from the linearizing trajectory is small. These conditions are usually valid in navigation problems, especially when aiding information such as GPS is available. The white noise assumption is also valid, since colored driving noise can be modeled by augmenting a linear system with white driving noise to the system model. The Gaussian assumption is valid for most driving noise sources as expected based on the *Central Limit Theorem* [107]. Even when an application involves non-Gaussian noise, it is typical to proceed as if the noise source were Gaussian with appropriately defined first and second moments. In such cases, although a better nonlinear estimator may exist, the Kalman filter will provide the minimum variance, *linear*, unbiased state estimate.

5.6 Implementation Issues

This section discusses several issues related to the real-time implementation of Kalman filters.

5.6.1 Scalar Measurement Processing

The Kalman filter algorithm as presented in Table 5.5 is formulated to process a vector of m simultaneous measurements. The portion of the measurement update that requires the most computing operations (i.e., FLOP's) is the covariance update and gain vector calculation. For example, the standard algorithm

$$\mathbf{K} = \mathbf{P}^-\mathbf{H}^\top\left(\mathbf{R}+\mathbf{H}\mathbf{P}^-\mathbf{H}^\top\right)^{-1} \tag{5.72}$$

$$\hat{\mathbf{x}}^+ = \hat{\mathbf{x}}^- + \mathbf{K}(\tilde{\mathbf{y}} - \mathbf{H}\hat{\mathbf{x}}^-) \tag{5.73}$$

$$\mathbf{P}^+ = (\mathbf{I} - \mathbf{K}\mathbf{H})\mathbf{P}^- \tag{5.74}$$

can be programmed to require $\left(\frac{3}{2}n^2m + \frac{3}{2}nm^2 + nm + m^3 + \frac{1}{2}m^2 + \frac{1}{2}m\right)$ FLOP's. Alternatively, when $\mathbf{R}$ is a diagonal matrix, the measurements can be equivalently treated as m sequential measurements with a zero-width time-interval between measurements, which results in significant computational savings.

At time $t = kT$, define

$$\left.\begin{array}{ll} \mathbf{P}_1 = \mathbf{P}^-(k) & \hat{\mathbf{x}}_1 = \hat{\mathbf{x}}^-(k) \\ \mathbf{H} = \begin{bmatrix} \mathbf{H}_1 \\ \vdots \\ \mathbf{H}_m \end{bmatrix} & \mathbf{R} = \begin{bmatrix} R_1 & \cdots & 0 \\ \vdots & & \vdots \\ 0 & \cdots & R_m \end{bmatrix}. \end{array}\right\} \tag{5.75}$$

Then, the equivalent scalar measurement processing algorithm is, for $i = 1$ to m,

$$\mathbf{K}_i = \frac{\mathbf{P}_i\mathbf{H}_i^\top}{R_i + \mathbf{H}_i\mathbf{P}_i\mathbf{H}_i^\top} \tag{5.76}$$

$$\hat{\mathbf{x}}_{i+1} = \hat{\mathbf{x}}_i + \mathbf{K}_i(\tilde{y}_i - \mathbf{H}_i\hat{\mathbf{x}}_i) \tag{5.77}$$

$$\mathbf{P}_{i+1} = (\mathbf{I} - \mathbf{K}_i\mathbf{H}_i)\,\mathbf{P}_i, \tag{5.78}$$

with the state and error covariance matrix posterior to the set of measurements defined by

$$\hat{\mathbf{x}}^+(k) = \hat{\mathbf{x}}_{m+1} \tag{5.79}$$

$$\mathbf{P}^+(k) = \mathbf{P}_{m+1}. \tag{5.80}$$

The total number of computations for the m scalar measurement updates is $m\left(\frac{3}{2}n^2 + \frac{5}{2}n\right)$ plus m scalar divisions. Thus it can be seen that m scalar updates are computationally cheaper for all m.

At the completion of the m scalar measurements $\hat{\mathbf{x}}^+(k)$ and $\mathbf{P}^+(k)$ will be identical to the values that would have been computed by the corresponding vector measurement update. The state Kalman gain vectors $\mathbf{K}_i$ corresponding to the scalar updates are *not* equal to the columns of the state feedback gain matrix that would result from the corresponding vector update. This is due to the different ordering of the updates affecting the error covariance matrix $\mathbf{P}_i$ at the intermediate steps during the scalar updates.

Example 5.6 *To illustrate the differences between the vector and scalar measurement updates, consider one measurement update for a second-order system having*

$$\mathbf{P}^-(k) = \begin{bmatrix} 100 & 0 \\ 0 & 100 \end{bmatrix}, \quad \hat{\mathbf{x}}^-(k) = \begin{bmatrix} 0 \\ 0 \end{bmatrix},$$

$$\mathbf{H} = \begin{bmatrix} 1.0 & 0.0 \\ 0.0 & 1.0 \\ 0.7 & 0.3 \\ 0.5 & 0.5 \end{bmatrix}, \quad \tilde{\mathbf{y}} = \begin{bmatrix} 10.24 \\ 21.20 \\ 13.91 \\ 14.84 \end{bmatrix}, \quad \textit{and } \mathbf{R} = \mathbf{I}_4.$$

For the vector update, by eqn. (5.72), the Kalman gain is

$$\mathbf{K}(k) = \begin{bmatrix} 0.6276 & -0.2139 & 0.3752 & 0.2069 \\ -0.2139 & 0.8136 & 0.0944 & 0.2999 \end{bmatrix}$$

which results by eqn. (5.73-5.74) in the state estimate

$$\hat{\mathbf{x}}^+(k) = \begin{bmatrix} 10.1823 \\ 20.8216 \end{bmatrix} \text{ with covariance } \mathbf{P}^+(k) = \begin{bmatrix} 0.6276 & -0.2139 \\ -0.2139 & 0.8136 \end{bmatrix}.$$

For the set of $m = 4$ scalar updates, by eqn. (5.76) the gain sequence is

m	*1*	*2*	*3*	*4*
$\hat{\mathbf{K}}_i$	0.9901 0.0000	0.0000 0.9901	0.4403 0.1887	0.2069 0.2999

The sequence of state estimates from eqn. (5.77) is

m	*1*	*2*	*3*	*4*	*5*
$\hat{\mathbf{x}}_i$	0 0	10.1386 0.0000	10.1386 20.9901	10.3658 21.0874	10.1823 20.8216

By eqn. (5.78) the second, fourth, and final error covariance matrices are

$$\mathbf{P}_2 = \begin{bmatrix} 0.9901 & 0.0000 \\ 0.0000 & 100.0000 \end{bmatrix}, \quad \mathbf{P}_4 = \begin{bmatrix} 0.6850 & -0.1308 \\ -0.1308 & 0.9341 \end{bmatrix}, \text{ and}$$

$$\mathbf{P}^+(k) = \mathbf{P}_5 = \begin{bmatrix} 0.6276 & -0.2139 \\ -0.2139 & 0.8136 \end{bmatrix}.$$

Note that the final state estimate and covariance matrix from the sequence of scalar updates is identical to the result from the vector update; however, the Kalman gains are distinct for the two implementations. △

5.6.2 Correlated Measurements

The scalar measurement processing algorithm is only valid in situations where the measurement noise corrupting the set of scalar measurements is uncorrelated (i.e., $\mathbf{R}$ is a diagonal matrix). In situations where this assumption is not satisfied, the actual measurements can be transformed into an equivalent set of measurements $\mathbf{z}(k)$ with uncorrelated measurement noise. The transformed set of measurements can then be used in the scalar processing algorithm.

With $\tilde{\mathbf{y}} = \mathbf{H}\mathbf{x} + \boldsymbol{\nu}$, let

$$E\langle \boldsymbol{\nu}\boldsymbol{\nu}^\top \rangle = \mathbf{R} = \mathbf{U}\mathbf{D}\mathbf{U}^\top,$$

where $\mathbf{U}$ is an orthonormal matrix (i.e., $\mathbf{U}^\top\mathbf{U} = \mathbf{I}$). The UDU decomposition is possible for any symmetric matrix, see Section B.11. If we choose $\mathbf{z}$ to be defined as

$$\mathbf{z} = \mathbf{U}^\top\mathbf{y}.$$

Then the measurement $\tilde{\mathbf{z}}$ satisfies

$$\begin{aligned}\tilde{\mathbf{z}} &= \mathbf{U}^\top\tilde{\mathbf{y}}\\ &= \mathbf{U}^\top\mathbf{H}\mathbf{x} + \mathbf{U}^\top\boldsymbol{\nu}\\ &= \bar{\mathbf{H}}\mathbf{x} + \bar{\boldsymbol{\nu}}.\end{aligned}$$

The equivalent additive noise $\bar{\boldsymbol{\nu}}$ has variance

$$E\langle\bar{\boldsymbol{\nu}}\bar{\boldsymbol{\nu}}^\top\rangle = E\langle\mathbf{U}^\top\boldsymbol{\nu}\boldsymbol{\nu}^\top\mathbf{U}\rangle = \mathbf{U}^\top\mathbf{U}\mathbf{D}\mathbf{U}^\top\mathbf{U} = \mathbf{D}.$$

Therefore, the scalar processing algorithm can be used with the measurement $\mathbf{z}$ having measurement noise with variance described by the diagonal matrix $\mathbf{D}$, and related to the state by the observation matrix $\bar{\mathbf{H}}$.

5.6.3 Bad or Missing Data

In many applications, a measurement expected at the k-th sample time may occasionally be missing. Such instances have no effect on the Kalman filter algorithm, as long as the missing measurement is 'ignored' correctly. When a measurement is missing, the corresponding measurement update (both state and covariance) must not occur. Alternative, the missing measurement can be processed using an arbitrary value for $\mathbf{y}_k$ as long as the Kalman gain vector is set to zero, which corresponds to $\mathbf{R}_k$ being infinite. In addition, the state estimate and error covariance must be propagated through time to the time of the next expected measurement.

Even when a measurement is available, it is sometimes desirable to reject certain measurements as faulty or erroneous. The Kalman filter provides a mechanism for detecting candidate measurements for exclusion. Let the state error covariance be $\mathbf{P}^-$ before the incorporation of the scalar measurement $\tilde{y} = \mathbf{H}\mathbf{x} + \nu$. Then the measurement residual $r = \tilde{y} - \hat{y}$ is a scalar Gaussian random variable with zero mean and variance

$$E\langle rr\rangle = \mathbf{H}\mathbf{P}^-\mathbf{H}^\top + R$$

where $R = var(\nu)$. The designer can select a threshold λ such that

$$Prob\left\{r^2 > \lambda\left(\mathbf{H}\mathbf{P}^-\mathbf{H}^\top + R\right)\right\} = \mu$$

where $\mu \in (0,1)$ is usually quite small. This probability can be evaluated using the probability density for chi-squared random variables which is defined on p. 164. When the condition

$$r^2 > \lambda(\mathbf{H}\mathbf{P}^-\mathbf{H}^\top + R) \tag{5.81}$$

is satisfied, the measurement is declared invalid and the corresponding measurement update (state and covariance) is skipped. Repeated violations of the test condition can be used to declare the sensor invalid.

Use of such conditions to remove measurements deemed to be invalid can be problematic. If the state error $\delta x = \mathbf{x} - \hat{\mathbf{x}}$ is for some reason larger than that predicted by $\mathbf{P}^-$, then the above logic may remove all measurements. In realistic applications, the conditions for discarding measurements is usually considerably more descriptive that indicated by eqn. (5.81).

In addition, when the condition $r^2 > \lambda(\mathbf{HP}^-\mathbf{H}^\top + R)$ occurs, it is often useful to log an assortment of descriptive data (e.g., y_k, $\hat{y}_k$, $\hat{\mathbf{x}}$, R_k, $\mathbf{H}_k$, and $\mathbf{H}_k\mathbf{P}_k^-\mathbf{H}_k^\top$). The data logged at this condition may allow detection of modeling errors or debugging of software errors.

5.7 Implementation Sequence

The discrete-time Kalman filter equations are summarized in Table 5.5. Because of the fact that the Kalman gain and covariance update equations are not affected by the measurement data, the sequence of implementation of the equations can be manipulated to decrease the latency between arrival of the measurement and computation of the corrected state.

After finishing with the k-th measurement, but prior to the arrival of the measurement $(k+1)$, the following quantities can be precomputed

$$\hat{\mathbf{x}}_{k+1}^- = \mathbf{\Phi}_k\hat{\mathbf{x}}_k^+ + \mathbf{G}_k\mathbf{u}_k \tag{5.82}$$

$$\mathbf{P}_{k+1}^- = \mathbf{\Phi}_k\mathbf{P}_k^+\mathbf{\Phi}_k^\top + \mathbf{Qd}_k \tag{5.83}$$

$$\mathbf{K}_{k+1} = \mathbf{P}_{k+1}^-\mathbf{H}_{k+1}^\top\left(\mathbf{R}_{k+1} + \mathbf{H}_{k+1}\mathbf{P}_{k+1}^-\mathbf{H}_{k+1}^\top\right)^{-1}$$

$$\mathbf{P}_{k+1}^+ = \mathbf{P}_{k+1}^- - \mathbf{K}_{k+1}\mathbf{H}_{k+1}\mathbf{P}_{k+1}^-.$$

After the arrival of the measurement $\tilde{\mathbf{y}}_{k+1}$ at time t_{k+1}, the state estimate is computed

$$\hat{\mathbf{x}}_{k+1}^+ = \hat{\mathbf{x}}_{k+1}^- + \mathbf{K}_{k+1}\left(\tilde{\mathbf{y}}_{k+1} - \hat{\mathbf{y}}_{k+1}\right).$$

This reorganization of the KF equations requires that $\mathbf{\Phi}_k$, $\mathbf{G}_k$, $\mathbf{Qd}_k$, $\mathbf{R}_{k+1}$, and $\mathbf{H}_{k+1}$ all be independent of the data so that they can be precomputed. This is possible, for example, for time-invariant systems.

5.8 Asynchronous Measurements

Asynchronous do not occur at a periodic rate. It is straightforward to handle asynchronous measurements within the Kalman filter framework. Note that the Kalman filter derivation did not assume the measurements to be periodic. The Kalman filter does require propagation of the state estimate and its covariance from one measurement time instant to the next according to eqns. (5.82–5.83). Therefore, the Kalman filter can accommodate asynchronous measurements so long as the matrices $\mathbf{\Phi}_k$ and $\mathbf{Qd}_k$ can be computed. Because the time period between measurements is not constant, for asynchronous measurements, the matrices $\mathbf{\Phi}_k$ and $\mathbf{Qd}_k$ are usually not constant.

5.9 Numeric Issues

One of the implicit assumptions in the derivation of the Kalman filter algorithm was that the gains and the arithmetic necessary to calculate them would take place using infinite precision real valued numbers. Computer implementations of the Kalman filter use a finite accuracy subset of the rational numbers. Although the precision of each computation may be quite high (especially in floating point applications), the number of calculations involved in the Kalman filter algorithm can allow the cumulative effect of the machine arithmetic to affect the filter performance. This issue becomes increasingly important in applications with either a long duration or a high state dimension. Both factors result in a greater number of calculations.

If the filter state is observable from the measurements and controllable from the process driving noise, then the effects of finite machine precision are usually minor. Therefore, it is common to add a small amount of process driving noise even to states which are theoretically considered to be constants.

Two common results of finite machine precision are loss of symmetry or loss of positive definiteness of the covariance matrix. These two topics are discussed in the following subsections.

5.9.1 Covariance Matrix Symmetry

The error covariance matrix $\mathbf{P} = E\langle \delta\mathbf{x}\delta\mathbf{x}^\top \rangle$ is by its definition symmetric. However, numeric errors can result in the numeric representation of $\hat{\mathbf{P}}$ becoming non-symmetric. Four techniques are common for maintaining the symmetry of $\hat{\mathbf{P}}$:

1. Use a form of the Kalman filter measurement update, such as the *Joseph form*, that is symmetric.

2. Resymmetrize the covariance matrices at regular intervals using the equation
$$\hat{\mathbf{P}} = \frac{1}{2}\left(\hat{\mathbf{P}} + \hat{\mathbf{P}}^\top\right).$$

3. Because $\hat{\mathbf{P}}$ should be symmetric, its upper and lower triangular portions are redundant. Instead of computing both, only calculate the main diagonal and upper triangular portion of $\hat{\mathbf{P}}$, then (if necessary) copy the upper portion appropriately into the lower triangular portion. This approach also substantially decreases the memory and computational requirements.

4. Use a UD factorized or square root implementation of the Kalman Filter as discussed in Subsection 5.9.2.

5.9.2 Covariance Matrix Positive Definiteness

The error covariance matrix $\mathbf{P} = E\langle \delta\mathbf{x}\delta\mathbf{x}^\top \rangle$ is by its definition positive semi-definite. Computed values of $\mathbf{P}$ can lose the positive semi-definite property. When this occurs, the Kalman filter gains for the corresponding states have the wrong sign, and the state may temporarily diverge. Even if the sign eventually corrects itself, subsequent performance will suffer since the covariance matrix is no longer accurate. In addition, the interim divergence can be arbitrarily bad.

Any symmetric matrix, in particular $\mathbf{P}$, can be factorized as

$$\mathbf{P} = \mathbf{U}\mathbf{D}\mathbf{U}^\top.$$

Special purpose algorithms have been derived which propagate the factors $\mathbf{U}$ and $\mathbf{D}$ instead of $\mathbf{P}$ itself. The factors contain the same information as $\mathbf{P}$, the factorized algorithms can be shown to be theoretically equivalent to the Kalman filter, and the algorithms automatically maintain both the positive definiteness and symmetry of the covariance matrix. The main drawback of square root algorithms are that they are more complex to program. Also, some formulations require more computation.

More general square root implementations of the Kalman filter have also been derived. Various algorithms are presented and compared in for example [60].

5.10 Suboptimal Filtering

When the "cost function" involves only the mean square estimation error and the system is linear, the Kalman filter provides the optimal estimation algorithm. In many applications, the actual system is nonlinear. Suboptimal filtering methods for nonlinear systems are discussed in Section 5.10.5. In addition, more factors may have to be considered in the overall filter optimization than the mean squared error alone. For example, the complete Kalman filter may have memory or computational requirements beyond those feasible for a particular project. In this case, the system engineer will attempt to find an implementable suboptimal filter that achieves performance as close as possible to the benchmark optimal Kalman Filter algorithm. Dual state covariance analysis, which is discussed in Chapter 6, is the most common method used to analyze the relative performance of suboptimal filters. A few techniques for generating sub-optimal filters within the Kalman filter framework are discussed in the following subsections.

5.10.1 Deleting States

The best model available to represent the actual system, referred to as the 'truth' model, will usually be high dimensional. In INS applications, the number of states can be near 100. GPS error states could add on the order of 6 states per satellite. Since the number of FLOP's required to implement a Kalman filter is on the order of n^2m, where n is the state dimension and m is the measurement dimension, the question arises of whether some of the states can be removed or combined in the implemented filter. For each hypothesized filter model, Monte Carlo or covariance studies can be performed to determine the performance of the designed filter relative to the 'truth' model. Several rules have been developed through analysis and experience to guide the designer in the choice of states to combine or eliminate.

5.10.1.1 Broadband Noise

Consider a driving noise that is modeled as a Gauss-Markov process. If the process PSD is flat at low frequencies and rolls off at frequencies sufficiently higher than the bandwidth of the process it excites, then the corresponding process model can be replaced with white driving noise with a PSD matching that of the Gauss-Markov process at low frequencies.

Example 5.7 *Consider the scalar Gauss-Markov process $y(t)$ described by*

$$\dot{x} = -\beta x + \sqrt{2\beta G}\nu \tag{5.84}$$

$$y = x \tag{5.85}$$

where β and $\mathbf{G}$ are constants and ν is a Gaussian white noise process with PSD equal to σ_ν^2. According to eqn. (4.49) the PSD of y is defined by

$$PSD_y(\omega) = \frac{2\beta G}{\beta^2 + \omega^2}\sigma_\nu^2. \tag{5.86}$$

If y is driving a low pass system with bandwidth much less than β rad/s, then the driving noise can be replaced by a Gaussian white noise process

$\tilde{y} = \xi$ *with PSD equal to* σ_ξ^2. *To be equivalent at low frequencies, it is necessary that*

$$\sigma_\xi^2 = \frac{2G}{\beta}\sigma_\nu^2.$$

△

In general, if a wideband Gauss-Markov noise process is driving a narrow band process, it may be possible to find a lower state dimension driving noise process that closely matches the PSD of the actual driving noise on the frequency range of interest.

5.10.1.2 Unobservable States

When two states have the same dynamics and affect the output and other states in identical manners, they can be combined.

Example 5.8 *A sensor bias b is modeled as a random walk plus a random constant:*

$$\begin{aligned}\mathbf{x}_{k+1} &= \begin{bmatrix} 1 & 0 \\ 0 & 1 \end{bmatrix}\mathbf{x}_k + \begin{bmatrix} 0 \\ 1 \end{bmatrix}\omega_k \\ b_k &= \begin{bmatrix} 1 & 1 \end{bmatrix}\mathbf{x}_k\end{aligned}$$

with the initial variance of the random constant defined as

$$\mathbf{P}_0 = \begin{bmatrix} P_c(0) & 0 \\ 0 & 0 \end{bmatrix}.$$

Without any loss in performance, the sensor error model can be represented with the following one-state model:

$$\begin{aligned}x_{k+1} &= x_k + \omega_k \\ b_k &= x_k\end{aligned}$$

with initial variance of the state defined as $P_0 = P_c(0)$. △

In Example 5.8, the two states were not separately observable. In general, unobservable (or weakly observable) states can be considered for removal or combination with other states.

5.10.2 Schmidt-Kalman Filtering

When states are removed from the full model to produce an implementable sub-optimal filter, the filter equations of the implemented filter must somehow account for the increased modeling error. This can be accomplished by ad-hoc techniques such as increasing $\mathbf{Q}$ and $\mathbf{R}$, but that approach can be problematic, see Example 6.1. A rigorous approach is presented below.

Consider a system described by

$$\begin{bmatrix} \mathbf{x}_1 \\ \mathbf{x}_2 \end{bmatrix}_{k+1} = \begin{bmatrix} \mathbf{\Phi}_{11} & \mathbf{0} \\ \mathbf{0} & \mathbf{\Phi}_{22} \end{bmatrix} \begin{bmatrix} \mathbf{x}_1 \\ \mathbf{x}_2 \end{bmatrix}_k + \begin{bmatrix} \boldsymbol{\omega}_1 \\ \boldsymbol{\omega}_2 \end{bmatrix}_k \tag{5.87}$$

$$\mathbf{y}_k = \begin{bmatrix} \mathbf{H}_1 & \mathbf{H}_2 \end{bmatrix} \begin{bmatrix} \mathbf{x}_1 \\ \mathbf{x}_2 \end{bmatrix}_k + \boldsymbol{\nu}_k \tag{5.88}$$

where $\boldsymbol{\nu}$, $\boldsymbol{\omega}_1$, and $\boldsymbol{\omega}_2$ are mutually uncorrelated, Gaussian, white noise processes. The full state vector $\mathbf{x}$ has been partitioned such that $\mathbf{x}_2$ contains those state vector elements to be eliminated and $\mathbf{x}_1$ contains those state vector elements that the implemented filter will retain. The measurement update equations are

$$\begin{bmatrix} \mathbf{x}_1^+ \\ \mathbf{x}_2^+ \end{bmatrix}_k = \begin{bmatrix} \mathbf{x}_1^- \\ \mathbf{x}_2^- \end{bmatrix}_k + \begin{bmatrix} \mathbf{K}_1 \\ \mathbf{K}_2 \end{bmatrix}_k (\tilde{\mathbf{y}}_k - \hat{\mathbf{y}}_k). \tag{5.89}$$

By the standard Kalman filter equations, the optimal filter gains can be computed as

$$\begin{aligned} \begin{bmatrix} \mathbf{K}_1 \\ \mathbf{K}_2 \end{bmatrix}_k &= \begin{bmatrix} \mathbf{P}_{11}^-\mathbf{H}_1^\top + \mathbf{P}_{12}^-\mathbf{H}_2^\top \\ \mathbf{P}_{21}^-\mathbf{H}_1^\top + \mathbf{P}_{22}^-\mathbf{H}_2^\top \end{bmatrix} (\mathbf{H}_1\mathbf{P}_{11}^-\mathbf{H}_1^\top + \mathbf{H}_2\mathbf{P}_{21}^-\mathbf{H}_1^\top \\ &\quad + \mathbf{H}_1\mathbf{P}_{12}^-\mathbf{H}_2^\top + \mathbf{H}_2\mathbf{P}_{22}^-\mathbf{H}_2^\top + \mathbf{R})^{-1}. \end{aligned}$$

The decision not to estimate $\mathbf{x}_2$ is equivalent to assuming that $\mathbf{x}_2(0) = \mathbf{0}$ and defining $\mathbf{K}_2(k) = \mathbf{0}$. Therefore, the implemented filter should propagate the covariance matrices for the suboptimal gain vector

$$\mathbf{L}_k = \begin{bmatrix} \mathbf{K}_1 \\ \mathbf{0} \end{bmatrix}_k.$$

Due to the decoupling of the $\mathbf{x}_1$ and $\mathbf{x}_2$ states in the time propagation equations, the time propagation of the covariance is straightforward:

$$\begin{aligned} \mathbf{P}_{11}^- &= \mathbf{\Phi}_{11}\mathbf{P}_{11}^+\mathbf{\Phi}_{11}^\top + \mathbf{Q}_1 \\ \mathbf{P}_{12}^- &= \mathbf{\Phi}_{11}\mathbf{P}_{12}^+\mathbf{\Phi}_{22}^\top \\ \mathbf{P}_{22}^- &= \mathbf{\Phi}_{22}\mathbf{P}_{22}^+\mathbf{\Phi}_{22}^\top + \mathbf{Q}_2. \end{aligned}$$

Substituting the suboptimal gain vector into the covariance measurement update of eqn. (5.12) yields:

$$\begin{aligned} \mathbf{P}_{11}^+ &= (\mathbf{I} - \mathbf{K}_1\mathbf{H}_1)\mathbf{P}_{11}^-(\mathbf{I} - \mathbf{K}_1\mathbf{H}_1)^\top - \mathbf{K}_1\mathbf{H}_2\mathbf{P}_{21}^-(\mathbf{I} - \mathbf{K}_1\mathbf{H}_1)^\top \\ &\quad - (\mathbf{I} - \mathbf{K}_1\mathbf{H}_1)\mathbf{P}_{12}^-\mathbf{H}_2^\top\mathbf{K}_1^\top + \mathbf{K}_1\mathbf{H}_2\mathbf{P}_{22}^-\mathbf{H}_2^\top\mathbf{K}_1^\top + \mathbf{K}_1 R\mathbf{K}_1^\top \\ \mathbf{P}_{12}^+ &= (\mathbf{I} - \mathbf{K}_1\mathbf{H}_1)\mathbf{P}_{12}^- - \mathbf{K}_1\mathbf{H}_2\mathbf{P}_{22}^- \\ \mathbf{P}_{22}^+ &= \mathbf{P}_{22}^-. \end{aligned}$$

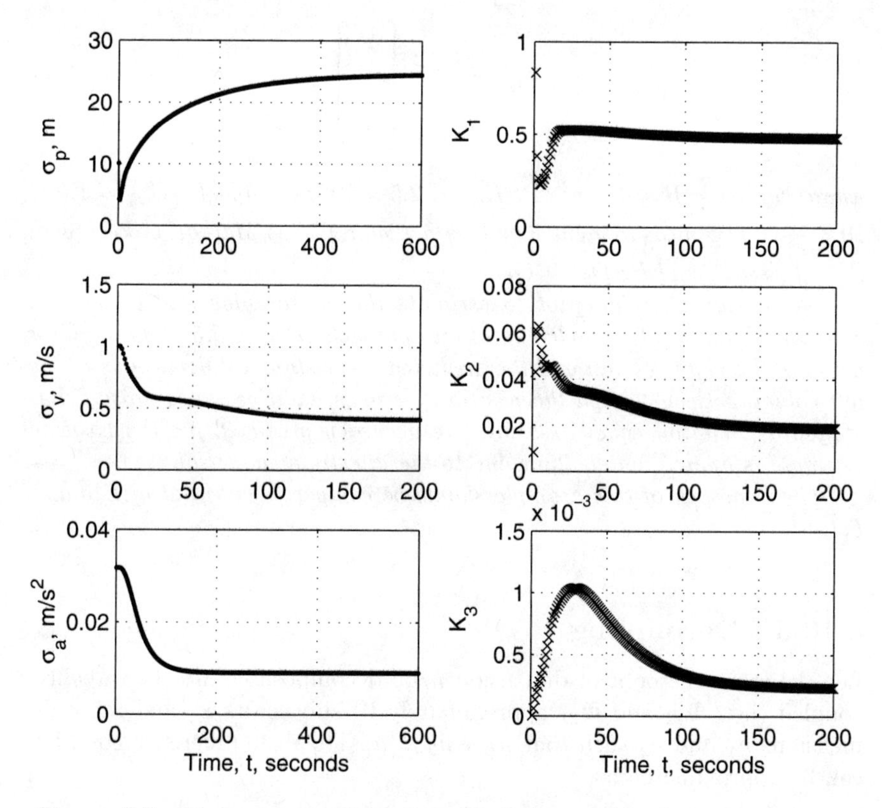

Figure 5.5: Performance of the Schmidt-Kalman Filter from Example 5.9. The left column of plots shows the error standard deviation of each estimate. The right column of plots shows the filter gains.

The Schmidt-Kalman filter eliminates the calculations that would be required in an optimal implementation to propagate $\mathbf{P}_{22}$; however, the filter correctly accounts for the uncertainty of the $\mathbf{x}_2$ states in the calculation of the gain vector $\mathbf{K}_1$.

Example 5.9 *Consider the one dimension inertial frame INS aided by a biased position measurement. The error equations for the INS are*

$$\begin{bmatrix} \dot{p} \\ \dot{v} \\ \dot{a} \\ \dot{b} \end{bmatrix} = \begin{bmatrix} 0 & 1 & 0 & 0 \\ 0 & 0 & 1 & 0 \\ 0 & 0 & 0 & 0 \\ 0 & 0 & 0 & -\beta \end{bmatrix} \begin{bmatrix} p \\ v \\ a \\ b \end{bmatrix} + \begin{bmatrix} 0 \\ \omega_v \\ \omega_a \\ \omega_b \end{bmatrix} \tag{5.90}$$

$$y = \begin{bmatrix} 1 & 0 & 0 & 1 \end{bmatrix} \begin{bmatrix} p \\ v \\ a \\ b \end{bmatrix} + \nu \tag{5.91}$$

where $var(\nu) = R = 10\ m^2$, $PSD_{\omega_v} = 2.5 \times 10^{-3} \frac{m^2}{s^3}$ *and* $PSD_{\omega_a} = 1.0 \times 10^{-6} \frac{m^2}{s^5}$. *The measurement bias* b *is a scalar Gauss-Markov process with* $\beta = \frac{1}{300}\ sec^{-1}$ *and* $PSD_{\omega_b} = Q_b = 4\frac{m^2}{s}$.

The designer decides not to estimate the sensor bias and to use the Schmidt-Kalman filter implementation approach. Figure 5.5 plots the filter gains in the right column and the predicted error standard deviations in the left column. Even though the position sensor's additive white noise ν *has standard deviation equal to 3.2m, the accuracy predicted for the position estimate is approximately 25m due to the unestimated sensor bias.*

The approach of this example should be compared with that of Example 6.1. △

5.10.3 Decoupling

Let the state vector $\mathbf{x}$ of dimension n be decomposable into two weakly-coupled (i.e., $\mathbf{\Phi}_{12}$ and $\mathbf{\Phi}_{21}$ approximately $\mathbf{0}$) subvectors $\mathbf{x}_1$ and $\mathbf{x}_2$ of dimensions n_1 and n_2 such that $n_1 + n_2 = n$. Then the difference equation can be approximated as

$$\mathbf{x}_{k+1} = \begin{bmatrix} \mathbf{x}_1 \\ \mathbf{x}_2 \end{bmatrix}_{k+1} = \begin{bmatrix} \mathbf{\Phi}_{11} & \mathbf{0} \\ \mathbf{0} & \mathbf{\Phi}_{22} \end{bmatrix} \begin{bmatrix} \mathbf{x}_1 \\ \mathbf{x}_2 \end{bmatrix}_k + \begin{bmatrix} \omega_1 \\ \omega_2 \end{bmatrix}_k$$

where subscript k's have been dropped in the right-hand side expressions. In this case, the error covariance can be time propagated according to

$$\begin{bmatrix} \mathbf{P}_{11} & \mathbf{P}_{12} \\ \mathbf{P}_{21} & \mathbf{P}_{22} \end{bmatrix}_{k+1} = \begin{bmatrix} \mathbf{\Phi}_{11}\mathbf{P}_{11}\mathbf{\Phi}_{11}^\top & \mathbf{\Phi}_{11}\mathbf{P}_{12}\mathbf{\Phi}_{22}^\top \\ \mathbf{\Phi}_{22}\mathbf{P}_{21}\mathbf{\Phi}_{11}^\top & \mathbf{\Phi}_{22}\mathbf{P}_{22}\mathbf{\Phi}_{22}^\top \end{bmatrix} + \begin{bmatrix} \mathbf{Qd}_1 & \mathbf{0} \\ \mathbf{0} & \mathbf{Qd}_2 \end{bmatrix}.$$

The decoupled equations require $\frac{3}{2}n_1^3 + \frac{1}{2}n_1^2 + \frac{3}{2}n_2^3 + \frac{1}{2}n_2^2 + n_1^2 n_2 + n_1 n_2^2$ FLOP's. The full equation requires $\frac{3}{2}n^3 + \frac{1}{2}n^2$ FLOP's. Therefore, the decoupled equations require $\frac{7}{2}\left(n_1^2 n_2 + n_1 n_2^2\right) + n_1 n_2$ fewer FLOP's per time iteration than the full covariance propagation equations.

Similar special purpose covariance equations can be derived for the case where only one of the coupling terms is non-zero.

5.10.4 Off-line Gain Calculation

The covariance update and Kalman gain calculations do not depend on the measurement data; therefore, they could be computed off-line. Since these calculations represent the greatest portion of the filter computations, this approach can greatly decrease the on-line computational requirements, possibly at the expense of greater memory requirements. The memory requirements depend on which of three approaches is used when storing the Kalman gains. The Kalman gains can be stored for each measurement epoch, curve fit, or stored as steady-state values.

Pre-calculation of the Kalman gains will always involve at least the risk of RMS estimation error performance loss. In the case where only the steady-state gains are stored, the initial transient response of the implemented system will deteriorate as will the settling time. Even in the case where the entire gain sequence is stored, the absence of expected measurements will result in discrepancies between the actual and expected error covariance matrices. The effect of such events should be thoroughly analyzed prior to committing to a given approach.

Example 5.10 *Consider using the steady-state Kalman filter gains to estimate the state of the system defined as:*

$$\begin{aligned} \begin{bmatrix} \dot{\mathbf{x}}_1 \\ \dot{\mathbf{x}}_2 \end{bmatrix} &= \begin{bmatrix} 0.0 & 1.0 \\ -1.0 & -0.2 \end{bmatrix} \begin{bmatrix} \mathbf{x}_1 \\ \mathbf{x}_2 \end{bmatrix} + \begin{bmatrix} 0.0 \\ 1.0 \end{bmatrix} \omega \\ y_k &= \begin{bmatrix} 1.0 & 0.0 \end{bmatrix} \begin{bmatrix} \mathbf{x}_1 \\ \mathbf{x}_2 \end{bmatrix}_k + \nu_k, \ \textit{with } \mathbf{P}_0 = \begin{bmatrix} 16.0 & 0.0 \\ 0.0 & 1.0 \end{bmatrix}, \end{aligned}$$

$PSD_\omega = 1.0$ and $var(\nu_k) = 10$. The measurements are available at a 1.0 Hz rate. The performance of the steady-state filter (dashed-dotted) relative to the performance of the Kalman filter with time-varying gains (solid) is displayed in Figure 5.6. The two estimators have the same performance until the instant of the first measurement at $t = 1.0$ s. The time-varying Kalman filter makes a much larger correction initially due to the large initial error covariance. During the initial transient, the relative performance difference is large with $\sigma_1^2 + \sigma_2^2$ always smaller for the Kalman filter than for the steady state filter. Since both filters are stable and both filters use the same gains in steady-state, the steady-state performance of the steady state filter is identical to that of the full Kalman filter implementation. △

Note that the steady state filter approach will fail to apply directly if either the Kalman gains are zero in steady state or do not approach constant values as $t \to \infty$.

5.10.5 Nonlinear Filtering

The previous sections have concentrated on the design and analysis of optimal (and suboptimal) estimation algorithms for linear systems. In many

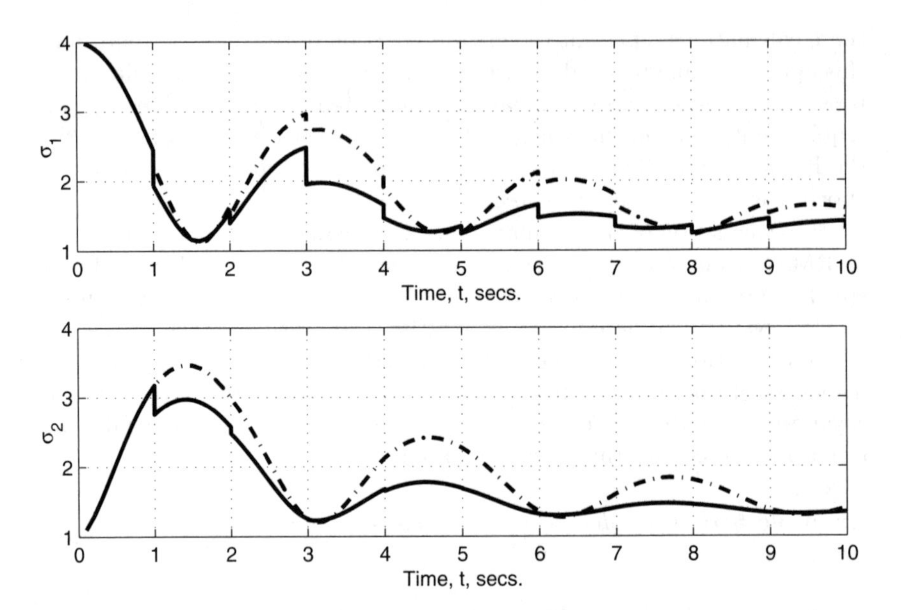

Figure 5.6: Standard deviation of the estimation errors of steady state filter (dashed-dotted) relative to time varying Kalman filter (solid). The error variance of $\hat{x}_1$ is σ_1^2. The error variance of $\hat{x}_2$ is σ_2^2.

navigation applications either the system dynamic equations or the measurement equations are not linear. For example, the GPS range measurement

$$r = \sqrt{(x_s - x_r)^2 + (y_s - y_r)^2 + (z_s - z_r)^2} \tag{5.92}$$

is nonlinear in the receiver antenna position (x_r, y_r, z_r). It is therefore necessary to consider extensions of the linear optimal estimation equations to nonlinear systems. This is the objective of the present section. The section mainly considers continuous-time systems with discrete measurements, as they are most pertinent to navigation applications. The presentation herein is limited to two of the traditionally most common nonlinear filtering techniques referred to as the Linearized Kalman filter and the Extended Kalman filter. Other nonlinear optimal estimation approaches such as particle filters or unscented Kalman filters are discussed for example in [121]. The results of this section will be used extensively in the applications in Part II of this text.

Consider the nonlinear system with dynamics and measurement described by

$$\dot{\mathbf{x}}(t) = \mathbf{f}(\mathbf{x}(t), \mathbf{u}(t), t) + \boldsymbol{\omega}(t) \tag{5.93}$$

$$\mathbf{y}(t) = \mathbf{h}(\mathbf{x}(t), t) + \boldsymbol{\nu}(t). \tag{5.94}$$

Although this model is restrictive in assuming additive uncorrelated white noise processes $\boldsymbol{\omega}$ and $\boldsymbol{\nu}$, the model is typical of the systems found in navigation applications. The signal $\mathbf{u}(t)$ is a known signal and $\mathbf{f}$ and $\mathbf{h}$ are known smooth functions.

Define a reference trajectory $\bar{\mathbf{x}}$ satisfying

$$\dot{\bar{\mathbf{x}}}(t) = \mathbf{f}(\bar{\mathbf{x}}(t), \mathbf{u}(t), t) \tag{5.95}$$

$$\bar{\mathbf{y}}(t) = \mathbf{h}(\bar{\mathbf{x}}(t), t) \tag{5.96}$$

where $\mathbf{x}(0) \approx \bar{\mathbf{x}}(0)$ and $\|\mathbf{x}(t) - \bar{\mathbf{x}}(t)\|$ is expected to be small for the range of t of interest. Two common methods for the specification of $\bar{\mathbf{x}}$ will be discussed in the sequel.

Let

$$\mathbf{x} = \bar{\mathbf{x}} + \delta\mathbf{x}$$

then, dropping the time arguments of all signals to simplify notation, we obtain the following intermediate result:

$$\begin{aligned} \dot{\mathbf{x}} = \dot{\bar{\mathbf{x}}} + \delta\dot{\mathbf{x}} &= \mathbf{f}(\bar{\mathbf{x}} + \delta\mathbf{x}, \mathbf{u}, t) + \boldsymbol{\omega} \\ \dot{\bar{\mathbf{x}}} + \delta\dot{\mathbf{x}} &= \mathbf{f}(\bar{\mathbf{x}}, \mathbf{u}, t) + \left.\frac{\partial \mathbf{f}}{\partial \mathbf{x}}\right|_{\mathbf{x}=\bar{\mathbf{x}}} \delta\mathbf{x} + h.o.t.'s + \boldsymbol{\omega}. \end{aligned} \tag{5.97}$$

Using eqn. (5.95) and eqn. (5.97), while dropping terms higher than first order, results in the linearized dynamic equation for the trajectory perturbation error state

$$\delta\dot{\mathbf{x}} = \mathbf{F}(\bar{\mathbf{x}}, \mathbf{u}, t)\delta\mathbf{x} + \boldsymbol{\omega} \text{ where} \tag{5.98}$$

$$\mathbf{F}(\bar{\mathbf{x}}, \mathbf{u}, t) = \left.\frac{\partial \mathbf{f}}{\partial \mathbf{x}}\right|_{\mathbf{x}=\bar{\mathbf{x}}}. \tag{5.99}$$

The measurement equation can also be expanded as

$$\begin{aligned} \mathbf{y} &= \mathbf{h}(\bar{\mathbf{x}} + \delta\mathbf{x}, t) + \boldsymbol{\nu} \\ &= \mathbf{h}(\bar{\mathbf{x}}, t) + \left.\frac{\partial \mathbf{h}}{\partial \mathbf{x}}\right|_{\mathbf{x}=\bar{\mathbf{x}}} \delta\mathbf{x} + h.o.t.'s + \boldsymbol{\nu}. \end{aligned} \tag{5.100}$$

Using the definition of $\bar{\mathbf{y}}$ in eqn. (5.96), the linearized residual output signal is defined by

$$\mathbf{z} = \mathbf{y} - \bar{\mathbf{y}} = \mathbf{H}(\bar{\mathbf{x}}, t)\delta\mathbf{x} + \boldsymbol{\nu} \text{ where} \tag{5.101}$$

$$\mathbf{H}(\bar{\mathbf{x}}, t) = \left.\frac{\partial \mathbf{h}}{\partial \mathbf{x}}\right|_{\mathbf{x}=\bar{\mathbf{x}}}. \tag{5.102}$$

With the linearized error dynamics defined in eqns. (5.98) and (5.101), an approximation to the optimal nonlinear estimator can be defined using the linear Kalman Filter.

Initialization	$\hat{\mathbf{x}}_0^- = \bar{\mathbf{x}}_0$ $\mathbf{P}_0^- = var(\delta \mathbf{x}_0^-)$
Measurement Update	$\mathbf{K}_k = \mathbf{P}_k^- \mathbf{H}_k^\top \left(\mathbf{R}_k + \mathbf{H}_k \mathbf{P}_k^- \mathbf{H}_k^\top\right)^{-1}$ $\mathbf{z}_k = \mathbf{y}_k - \bar{\mathbf{y}}_k$ $\delta\hat{\mathbf{x}}_k^+ = \delta\hat{\mathbf{x}}_k^- + \mathbf{K}_k \left(\mathbf{z}_k - \mathbf{H}_k \delta\hat{\mathbf{x}}_k^-\right)$ $\mathbf{P}_k^+ = [\mathbf{I} - \mathbf{K}_k \mathbf{H}_k] \mathbf{P}_k^-$
Time Propagation	$\delta\hat{\mathbf{x}}_{k+1}^- = \mathbf{\Phi}_k \delta\hat{\mathbf{x}}_k^+$ $\mathbf{P}_{k+1}^- = \mathbf{\Phi}_k \mathbf{P}_k^+ \mathbf{\Phi}_k^\top + \mathbf{Qd}_k$
Definitions	$\dot{\bar{\mathbf{x}}}(t) = \mathbf{f}(\bar{\mathbf{x}}(t), u(t), t)$ $\bar{\mathbf{y}}(t) = \mathbf{h}(\bar{\mathbf{x}}(t), t)$ $\delta\mathbf{x} = \mathbf{x} - \bar{\mathbf{x}}$ $\hat{\mathbf{x}} = \bar{\mathbf{x}} + \delta\mathbf{x}$ $\mathbf{F}(\bar{\mathbf{x}}(t), \mathbf{u}(t), t) = \frac{\partial \mathbf{f}}{\partial \mathbf{x}}\Big\vert_{\mathbf{x}=\bar{\mathbf{x}}}$ $\mathbf{H}_k = \mathbf{H}(\bar{\mathbf{x}}(t_k), t_k) = \frac{\partial \mathbf{h}}{\partial \mathbf{x}}\Big\vert_{\mathbf{x}(t_k)=\bar{\mathbf{x}}(t_k)}$

Table 5.6: Approximate nonlinear optimal estimation equations. Computation of the matrices $\mathbf{\Phi}_k$ and $\mathbf{Qd}_k$ is discussed in Section 4.7.1.

The implementation equations are accumulated for ease of reference in Table 5.6. With reference to Table 5.6, a few points are worth specific mention:

- The estimator in Table 5.6 is designed based entirely on the error state and the error state dynamics.
  - The measurement update works with residual measurements, not measurements. Note that $\mathbf{z}_k$ is the measurement minus the prediction of the measurement based on the reference trajectory.
  - The estimator determines the error state, not the state. Because the reference trajectory is known and the error state is estimated, the best estimate of the state can be constructed (outside the estimator) as

$$\hat{\mathbf{x}}(t) = \bar{\mathbf{x}}(t) + \delta\hat{\mathbf{x}}(t).$$

  - Due to the actual dynamics being nonlinear, the matrix $\mathbf{P}$ is only an approximation to the actual error covariance matrix. Therefore, Monte Carlo simulations should be used in addition to covariance analysis to ensure that the results of the covariance analysis are accurate.

This approach is analogous to the idea of a complementary filter and is further discussed in Sections 5.10.5.3 and 7.3.

- The linearized system is typically time-varying, see eqns. (5.99) and (5.102).

- The matrix $\mathbf{\Phi}_k$ is the state transition matrix defined in eqn. (3.54) corresponding to $\mathbf{F}(\bar{\mathbf{x}}(t), \mathbf{u}(t), t)$ for $t \in [t_k, t_{k+1})$. Both $\mathbf{\Phi}_k$ and $\mathbf{Qd}_k$ will therefore be time varying. A method for computing $\mathbf{\Phi}_k$ and $\mathbf{Qd}_k$ online is discussed in Section 7.2.5.2.

Several examples and applications are discussed in subsequent chapters. In navigation applications, the $\mathbf{F}$ matrix is typically not asymptotically stable. Therefore, without aiding, the trajectory perturbation typically diverges with time. This is exemplified in Sections 4.9.3 and 4.9.4 where the divergence between aiding measurements is clearly shown. For the linearized error dynamics to remain accurate, either the time interval of interest should be short or some aiding mechanism should be used to ensure that $\|\mathbf{x}(t) - \bar{\mathbf{x}}(t)\|$ remains small.

5.10.5.1 Linearized Kalman Filter

In certain applications, orbiting satellites or interplanetary travel, the vehicle is expected to closely follow a predetermined trajectory $\bar{\mathbf{x}}(t)$ that is computed in advance.

In such applications, where $\bar{\mathbf{x}}(t)$ is a predetermined trajectory, the approximate optimal estimation equations are referred to as a *Linearized Kalman Filter.* In this case, the on-line calculations can be significantly simplified since the on-line measurements do not affect the calculation of $\mathbf{F}, \Phi, \mathbf{H}, \mathbf{P}$, $\mathbf{Qd}$, or $\mathbf{K}$. Table 5.7 organizes the calculations of Table 5.6 into on-line and off-line calculations. The estimate of the filter can be interpreted as the error between the actual and prespecified nominal trajectory. A control law can be designed to attempt to drive this error toward zero.

5.10.5.2 Extended Kalman Filter

In many applications, human driven vehicles or autonomous vehicles with data-reactive missions, the vehicle trajectory cannot be accurately predicted during the design of the navigation system. See Part II for several applications of this approach.

In such applications, the nominal trajectory is defined to be equal to the estimated trajectory, $\bar{\mathbf{x}}(t) = \hat{\mathbf{x}}(t)$. In this case, the state estimation gain vector $\mathbf{K}$ (which is a function of $\bar{\mathbf{x}}$) will be a function of the stochastic process $\hat{\mathbf{x}}(t)$. Therefore, the gain vector is a stochastic process. As long as the state is observable from the measurements and the measurements are

Off-line Calculations	$\dot{\bar{\mathbf{x}}}(t) = \mathbf{f}(\bar{\mathbf{x}}(t), u(t), t)$ $\bar{\mathbf{y}}_k = \mathbf{h}(\bar{\mathbf{x}}_k, k)$ $\mathbf{F}(\bar{\mathbf{x}}(t), \mathbf{u}(t), t) = \frac{\partial \mathbf{f}}{\partial \mathbf{x}}\Big\|_{\mathbf{x}=\bar{\mathbf{x}}}$ $\mathbf{H}_k = \mathbf{H}(\bar{\mathbf{x}}(t_k), t_k) = \frac{\partial \mathbf{h}}{\partial \mathbf{x}}\Big\|_{\mathbf{x}(t_k)=\bar{\mathbf{x}}(t_k)}$ $\mathbf{K}_k = \mathbf{P}_k^- \mathbf{H}_k^\top \left(\mathbf{R}_k + \mathbf{H}_k \mathbf{P}_k^- \mathbf{H}_k^\top\right)^{-1}$ $\mathbf{P}_k^+ = [\mathbf{I} - \mathbf{K}_k \mathbf{H}_k]\mathbf{P}_k^-$ $\mathbf{P}_{k+1}^- = \mathbf{\Phi}_k \mathbf{P}_k^+ \mathbf{\Phi}_k^\top + \mathbf{Qd}_k$
Initialization	$\hat{\mathbf{x}}_0^- = \bar{\mathbf{x}}_0$ $\mathbf{P}_0^- = var(\delta \mathbf{x}_0^-)$
Measurement Update	$\mathbf{z}_k = \mathbf{y}_k - \bar{\mathbf{y}}_k$ $\delta\hat{\mathbf{x}}_k^+ = \delta\hat{\mathbf{x}}_k^- + \mathbf{K}_k \left(z_k - \mathbf{H}_k \delta\hat{\mathbf{x}}_k^-\right)$
Time Propagation	$\delta\hat{\mathbf{x}}_{k+1}^- = \mathbf{\Phi}_k \delta\hat{\mathbf{x}}_k^+$

Table 5.7: Linearized Kalman filter equations. Computation of the matrices $\mathbf{\Phi}_k$ and $\mathbf{Qd}_k$ is discussed in Section 4.7.1.

sufficiently accurate, then 1) the state estimate should be near the actual state; 2) the linearization should be accurate; and, 3) the performance should be good. If however, the estimate is far from the actual state, the linearization will be inaccurate and the estimate may diverge rapidly. The possibility of using a poor state estimate in the filter gain calculations makes the Extended Kalman Filter (EKF) riskier than the linearized Kalman filter. Alternatively, the fact that the estimated state is typically (under good observability conditions) closer to the actual state than a predefined nominal trajectory usually allows the extended Kalman filter to produce better performance than the linearized Kalman filter.

The implementation equations for the EKF are summarized in Table 5.8. The algorithm can be manipulated to correctly work with either total states (i.e., $\hat{\mathbf{x}}(t)$) or error states (i.e., $\delta\hat{\mathbf{x}}(t)$), as discussed further in Section 5.10.5.3.

5.10.5.3 Complementary Filtering and the EKF

In the implementation of the EKF, the solution to

$$\dot{\hat{\mathbf{x}}}^-(t) = \mathbf{f}(\hat{\mathbf{x}}^-(t), \mathbf{u}(t), t), \text{ for } t \in [t_k, t_{k+1}) \text{ with } \hat{\mathbf{x}}^-(t_k) = \hat{\mathbf{x}}_k^+$$

is required between sampling instants. This book will primarily focus on navigation systems implemented using a complementary filter structure,

Initialization	$\hat{\mathbf{x}}_0^- = \bar{\mathbf{x}}_0$ $\mathbf{P}_0^- = var(\delta \mathbf{x}_0^-)$
Measurement Update	$\mathbf{K}_k = \mathbf{P}_k^- \mathbf{H}_k^\top \left(\mathbf{R}_k + \mathbf{H}_k \mathbf{P}_k^- \mathbf{H}_k^\top\right)^{-1}$ $\mathbf{z}_k = \mathbf{y}_k - \bar{\mathbf{y}}_k$ $\hat{\mathbf{x}}_k^+ = \hat{\mathbf{x}}_k^- + \mathbf{K}_k \mathbf{z}_k$ $\mathbf{P}_k^+ = [\mathbf{I} - \mathbf{K}_k \mathbf{H}_k] \mathbf{P}_k^-$
Time Propagation	$\dot{\hat{\mathbf{x}}}^-(t) = \mathbf{f}(\hat{\mathbf{x}}^-(t), \mathbf{u}(t), t)$, for $t \in [t_k, t_{k+1})$ with $\hat{\mathbf{x}}^-(t_k) = \hat{\mathbf{x}}_k^+$ $\mathbf{P}_{k+1}^- = \mathbf{\Phi}_k \mathbf{P}_k^+ \mathbf{\Phi}_k^\top + \mathbf{Qd}_k$
Definitions	$\mathbf{F}(\bar{\mathbf{x}}(t), \mathbf{u}(t), t) = \frac{\partial \mathbf{f}}{\partial \mathbf{x}}\Big\|_{\mathbf{x}=\bar{\mathbf{x}}}$ $\bar{\mathbf{y}}_k = \bar{\mathbf{y}}(t_k) = \mathbf{h}(\hat{\mathbf{x}}^-(t_k), t_k)$ $\mathbf{H}_k = \mathbf{H}(\bar{\mathbf{x}}(t_k), t_k) = \frac{\partial \mathbf{h}}{\partial \mathbf{x}}\Big\|_{\mathbf{x}(t_k)=\bar{\mathbf{x}}(t_k)}$

Table 5.8: Extended Kalman filter in feedback form. Computation of the matrices $\mathbf{\Phi}_k$ and $\mathbf{Qd}_k$ is discussed in Section 4.7.1.

see Chapter 7 and Part II. Feedforward and feedback complementary filter block diagrams are shown in Figures 5.7 and 5.8, respectively. In the complementary filter approach, a set of high sample rate sensors provide measurements of the signals $\mathbf{u}$. The navigation system will integrate $\mathbf{u}$ using the vehicle kinematic equations denoted by $\mathbf{f}(\hat{\mathbf{x}}^-, \tilde{\mathbf{u}}, t)$ to provide the necessary reference trajectory. At appropriate sampling instants, the navigation system will provide an estimate of the value of the aiding sensor information $\mathbf{y}$. The residual measurement, $\mathbf{z} = \tilde{\mathbf{y}} - \hat{\mathbf{y}}$, contains measurement noise and information about the state error $\delta\mathbf{x}$. The purpose of the EKF is to remove the measurement noise and provide an accurate estimate of the state error vector.

The complementary filter structure is desirable because it manipulates the available data so that the Kalman filter design assumptions are satisfied (to first order). The measurements $\mathbf{z} = \mathbf{H}\delta\mathbf{x} + \boldsymbol{\nu}$ that drive the EKF are a linear combination of two random process: $\delta\mathbf{x}$ and $\boldsymbol{\nu}$. The state vector of the EKF, $\delta\mathbf{x}$, over short time intervals can be accurately modeled by a linear state space model driven by white noise. The issues stated in this section will be further clarified in Chapter 7.

The reader should carefully compare the EKF algorithms summarized in Tables 5.6 and 5.8. The feedforward approach of Table 5.6 clearly shows that the EKF is designed using the linearized error dynamic equations and estimates the error state vector $\delta\mathbf{x}$. Both the measurement update and time propagation of the error state vector are accounted for explicitly. For

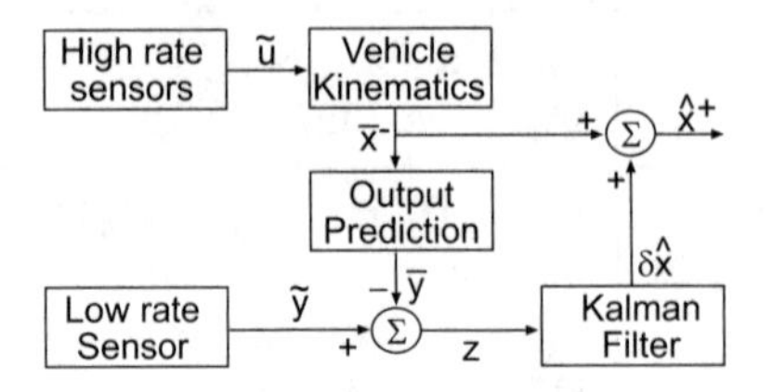

Figure 5.7: Feedforward complementary filter implementation of the algorithm in Table 5.6.

the feedback algorithm summarized in Table 5.8, the error state is absent; however, the equations for $\mathbf{K}_k$, $\mathbf{P}_k^+$ and $\mathbf{P}_{k+1}^-$ are exactly the same. These issues often cause confusion and therefore deserve further comment.

The algorithm in Table 5.6 is illustrated in Figure 5.7. Integration of the vehicle kinematics provides the reference trajectory $\bar{\mathbf{x}}$ which allows calculation of the output $\bar{\mathbf{y}}$. The EKF estimates and propagates the error state which is added to the navigation system output to produce the total navigation state estimate.

The algorithm in Table 5.8 is illustrated in Figure 5.8. Again, integration of the vehicle kinematics provides the EKF with the nominal trajectory and output. At initialization $\hat{\mathbf{x}}_0 = E\langle \mathbf{x}(0)\rangle$; therefore, $\delta\mathbf{x}^+(0) = \mathbf{0}$ and $\mathbf{P}_0^+ = var(\delta\mathbf{x}^+(0)) = var(\mathbf{x}(0))$. Assuming that the first measurement occurs for $k = 1$, the error state and its covariance at time t_1 are computed as

$$\begin{aligned} \delta\mathbf{x}_1^- &= \mathbf{\Phi}_0 \delta\mathbf{x}^+(0) = \mathbf{0} \\ \mathbf{P}_1^- &= \mathbf{\Phi}_0 \mathbf{P}_0^+ \mathbf{\Phi}_0^\top + \mathbf{Qd}_0. \end{aligned}$$

Because of the fact that $\delta\mathbf{x}_1^- = \mathbf{0}$, the measurement update

$$\delta\hat{\mathbf{x}}_1^+ = \delta\hat{\mathbf{x}}_1^- + \mathbf{K}_1 \left(\mathbf{z}_1 - \mathbf{H}_1 \delta\hat{\mathbf{x}}_1^-\right)$$

reduces to

$$\delta\hat{\mathbf{x}}_1^+ = \mathbf{K}_1 \mathbf{z}_1 \tag{5.103}$$

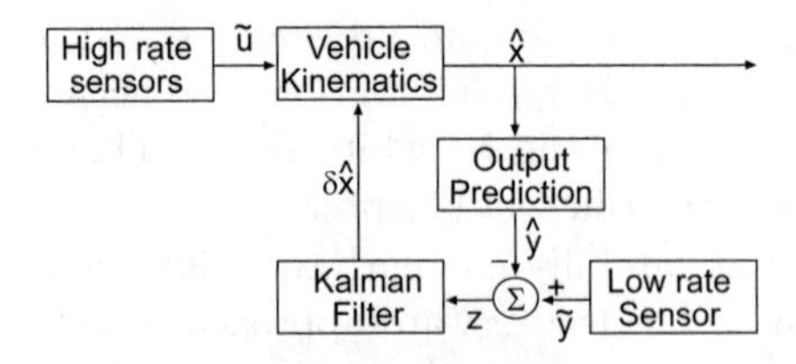

Figure 5.8: Complementary filter feedback implementation of the algorithm in Table 5.8.

with $\mathbf{K}_1$ and $\mathbf{P}_1^+$ computed as shown in Table 5.8. Given $\hat{\mathbf{x}}_1^-$ and $\delta\hat{\mathbf{x}}_1^+$, the best estimate of the total state after the measurement is

$$\hat{\mathbf{x}}_1^+ = \hat{\mathbf{x}}_1^- + \delta\hat{\mathbf{x}}_1^+ = \hat{\mathbf{x}}_1^- + \mathbf{K}_1\mathbf{z}_1 \tag{5.104}$$

which is equal to the expected value of $\mathbf{x}(t_1)$ after accounting for the measurement information at time t_1, which is denoted as $\mathbf{x}_1^+$. Because the state correction of eqn. (5.104) results in $\hat{\mathbf{x}}^+(t_1) = E\langle\mathbf{x}_1^+\rangle$ then it is now true that

$$\delta\hat{\mathbf{x}}_1^+ = \hat{\mathbf{x}}_1^+ - E\langle\mathbf{x}_1^+\rangle = \mathbf{0}. \tag{5.105}$$

The software implementation of the algorithm must make the assignment $\delta\hat{\mathbf{x}}_1^+ \doteq \mathbf{0}$ immediately after implementing eqn. (5.104). Failure to reset $\delta\hat{\mathbf{x}}_1^+$ to zero after correcting the total state estimate as in eqn. (5.104) would result in double accounting for $\delta\hat{\mathbf{x}}_1^+$ in the subsequent measurement correction. The result of eqn. (5.104) serves as the initial condition for the next period of integration. Because $\delta\hat{\mathbf{x}}_1^+ = \mathbf{0}$, we will have that $\delta\mathbf{x}_2^- = \mathbf{0}$ and the measurement correction at t_2 will still reduce as in eqn. (5.103). The above procedure can be replicate at each measurement instant. Note that it is not proper to state that "the error state is not propagated through time." It is correct to state that the error state time propagation is trivial and is therefore not implemented.

A detailed example of the use of the error state to correct the total state is presented in Section 12.6.

5.11 References and Further Reading

There are many excellent books on Kalman filter theory and design. The main references for the presentation of this chapter were [31, 34, 58, 60, 93]. The complementary filter and its relation to the Kalman filter is described well in [30, 31]. Alternative nonlinear estimation approaches, including the unscented and particle filters, are presented in a unified notation similar to that used herein in [121].

5.12 Exercises

Exercise 5.1 In eqn. (5.22) on p. 175, let $\mathbf{S} = \mathbf{H}(\mathbf{H}^\top\mathbf{W}\mathbf{H})^{-1}\mathbf{H}^\top\mathbf{W}$.

1. If $\mathbf{W} = \mathbf{R}^{-1} = \sigma^{-2}\mathbf{I}$ show that $\mathbf{S} = \mathbf{H}(\mathbf{H}^\top\mathbf{H})^{-1}\mathbf{H}^\top$.

2. For $\mathbf{W}$ symmetric and positive definite, use direct multiplication to show that $\mathbf{S}$ is idempotent.

3. Show that $(\mathbf{I} - \mathbf{S})$ is idempotent.

Exercise 5.2 If $\tilde{\mathbf{Y}} \in \mathbb{R}^{N \times 1}$, $\mathbf{H} \in \mathbb{R}^{N \times n}$, and $\mathbf{x} \in \mathbb{R}^{n \times 1}$, the solution to the least squares problem from eqn. (5.25) is

$$\hat{\mathbf{x}} = (\mathbf{H}^\top \mathbf{H})^{-1} \mathbf{H}^\top \tilde{\mathbf{Y}}. \tag{5.106}$$

The solution to this equation can be problematic when $\mathbf{H}$ is poorly conditioned or nearly singular.

Let the singular value decomposition (see p. 469) of $\mathbf{H}$ be

$$\mathbf{H} = [\mathbf{U}_1, \mathbf{U}_2] \begin{bmatrix} \mathbf{\Sigma}_1 & \mathbf{0} \\ \mathbf{0} & \mathbf{\Sigma}_2 \end{bmatrix} \begin{bmatrix} \mathbf{V}_1^\top \\ \mathbf{V}_2^\top \end{bmatrix}$$

where $\mathbf{U}_1 \in \mathbb{R}^{N \times r}$, $\mathbf{\Sigma}_1 \in \mathbb{R}^{r \times r}$ is nonsingular and diagonal, and $\mathbf{V}_1 \in \mathbb{R}^{r \times n}$. The parameter r is the rank of the matrix $\mathbf{H}$ and can be selected to exclude small singular values (i.e., $\mathbf{\Sigma}_2 \approx \mathbf{0}$).

Representing $\mathbf{H}$ as

$$\mathbf{H} = \mathbf{U}_1 \mathbf{\Sigma}_1 \mathbf{V}_1^\top,$$

show that the solution to the least squares problem can be computed as

$$\hat{\mathbf{x}} = \mathbf{V}_1 \mathbf{\Sigma}_1^{-1} \mathbf{U}_1^\top \tilde{\mathbf{Y}}. \tag{5.107}$$

Exercise 5.3 Use eqns. (5.66) and (5.63) to derive eqn. (5.68).

Exercise 5.4 Assume that two Kalman filters provide two normally distributed and unbiased estimates of $\mathbf{x} \in \mathbb{R}^n$. Denote the estimates as $\hat{\mathbf{x}}_1$ and $\hat{\mathbf{x}}_2$. The covariance of $\hat{\mathbf{x}}_1$ is $\mathbf{P}_1$. The covariance of $\hat{\mathbf{x}}_2$ is $\mathbf{P}_2$. The estimates $\hat{\mathbf{x}}_1$ and $\hat{\mathbf{x}}_2$ are independent random variables.

Construct a hypothetical set of $2n$ measurements:

$$\begin{bmatrix} \hat{\mathbf{x}}_1 \\ \hat{\mathbf{x}}_2 \end{bmatrix} = \begin{bmatrix} \mathbf{I} \\ \mathbf{I} \end{bmatrix} \mathbf{x} + \begin{bmatrix} \mathbf{n}_1 \\ \mathbf{n}_2 \end{bmatrix} \tag{5.108}$$

where $\mathbf{n}_1 \sim N(\mathbf{0}, \mathbf{P}_1)$ and $\mathbf{n}_2 \sim N(\mathbf{0}, \mathbf{P}_2)$.

1. Use a method similar to that in Section 5.3.3 to show that the optimal (minimum mean squared error) combined estimate can be computed by either of the following equations:

$$\hat{\mathbf{x}}_c = \hat{\mathbf{x}}_1 + \left(\mathbf{P}_1^{-1} + \mathbf{P}_2^{-1}\right)^{-1} \mathbf{P}_2^{-1} (\hat{\mathbf{x}}_2 - \hat{\mathbf{x}}_1)$$
$$\hat{\mathbf{x}}_c = \hat{\mathbf{x}}_2 + \left(\mathbf{P}_1^{-1} + \mathbf{P}_2^{-1}\right)^{-1} \mathbf{P}_1^{-1} (\hat{\mathbf{x}}_1 - \hat{\mathbf{x}}_2).$$

2. Show that either equation in part 1 can be reduced to

$$\hat{\mathbf{x}}_c = \left(\mathbf{P}_1^{-1} + \mathbf{P}_2^{-1}\right)^{-1} \left(\mathbf{P}_1^{-1} \hat{\mathbf{x}}_1 + \mathbf{P}_2^{-1} \hat{\mathbf{x}}_2\right).$$

3. In the special case that $n = 1$, show that either equation reduces to

$$\hat{x}_c = \frac{\mathbf{P}_2}{\mathbf{P}_1 + \mathbf{P}_2}\hat{\mathbf{x}}_1 + \frac{\mathbf{P}_1}{\mathbf{P}_1 + \mathbf{P}_2}\hat{\mathbf{x}}_2.$$

4. Discuss the physical intuition that supports this result. Consider the limiting cases where $\mathbf{P}_1$ is near 0 or ∞ and $\mathbf{P}_2$ is a finite nonzero constant.

Exercise 5.5 For the system described as

$$\begin{aligned} \dot{\mathbf{x}} &= \begin{bmatrix} 0 & 1 \\ 0 & 0 \end{bmatrix} \mathbf{x} + \begin{bmatrix} 0 \\ 1 \end{bmatrix} n_x \\ y &= \begin{bmatrix} 1 & 0 \end{bmatrix} \mathbf{x} + n_y \end{aligned}$$

with Gaussian white noise processes $n_x \sim N(0, 0.01)$ and $n_y \sim N(0, 1)$, complete the following.

1. Calculate $\mathbf{\Phi}$ and $\mathbf{Qd}$ for the discrete-time equivalent model for sample time T.

2. Assume that $\mathbf{x}(0) \sim N(0, \mathbf{0})$ and the measurement y is available at 1.0 Hz.

   (a) Implement the Kalman filter gain and covariance propagation equations.

   (b) Plot the Kalman gains and the diagonal of the covariance matrix versus time.

3. Assume that $\mathbf{x}(0) \sim N(0, 100\mathbf{I})$. Repeat the two steps of item 2.

4. Compare the results of the two simulations for t near zero and for large t. Discuss both the Kalman gain and the diagonal of the covariance matrix.

This problem is further discussed in Exercise 6.3.

Exercise 5.6 This problem returns to the application considered in Section 1.1.1. As derived in Section 1.1.1, the model for the error state vector is

$$\left.\begin{aligned} \delta\dot{p} &= \delta v, \\ \delta\dot{v} &= \delta b + \omega_1, \\ \delta\dot{b} &= -\lambda_b \delta b + \omega_b. \end{aligned}\right\}$$

Assume that $\lambda_b = \frac{1}{600s}$ and that the PSD of ω_b is $\frac{0.01}{300}\frac{(m/s^2)^2}{Hz}$. The variance of acceleration measurement is $\left(0.05\frac{m}{s^2}\right)^2$, The aiding sensor residual measurement model is

$$\delta y = \delta p + \nu_p$$

where $\nu_p \sim N(0, \sigma_p^2)$, $\sigma_p = 1m$, and the measurement is available at $1Hz$.

Assuming that the system is initially (nominally) stationary at an unknown location, design a Kalman filter by specifying the stochastic model parameters and initial conditions: $\hat{\mathbf{x}}(0)$, $\mathbf{P}(0)$, $\mathbf{F}$, $\mathbf{Q}$, $\mathbf{\Gamma}$, $\mathbf{H}$, $\mathbf{\Phi}$, $\mathbf{Qd}$, and $\mathbf{R}$.

See also Exercise 7.5.

Exercise 5.7 In Example 4.2, the angle error model was

$$\delta\dot{\theta}(t) = b_g(t) + \nu(t). \tag{5.109}$$

Assume that b_g is a Gauss-Markov process with

$$R_b(\tau) = \sigma_b^2 e^{-\lambda_b|\tau|}.$$

that $\nu \sim N(0, \sigma_g^2)$ and that an angle measurement is available at $0.5Hz$ that is modeled as $\tilde{\theta} = \theta + \nu_\theta$ where $\nu_\theta \sim N(0, (1.0°)^2)$. Be careful with units.

Design a Kalman filter by specifying the stochastic model parameters and initial conditions: $\hat{\mathbf{x}}(0)$, $\mathbf{P}(0)$, $\mathbf{F}$, $\mathbf{Q}$, $\mathbf{\Gamma}$, $\mathbf{H}$, $\mathbf{\Phi}$, $\mathbf{Qd}$, and $\mathbf{R}$.

See also Exercise 7.1.

Exercise 5.8 Assume that

$$y(t) = x(t) + b(t) + n(t)$$

where $n(t) \sim N(0, \sigma_n^2)$ is white and x and b are stationary, independent random processes with correlation functions

$$\begin{aligned} R_x(\tau) &= \sigma_x^2 e^{-\lambda_x|\tau|} \\ R_b(\tau) &= \sigma_b^2 e^{-\lambda_b|\tau|}. \end{aligned}$$

If the state vector is $\mathbf{z} = [x, b]^\top$ and measurements are available at 0.5 Hz, specify a set of state space model parameters ($\mathbf{F}$, $\mathbf{Q}$, $\mathbf{\Gamma}$, $\mathbf{H}$, $\mathbf{\Phi}$, $\mathbf{Qd}$, and $\mathbf{R}$) and initial conditions ($\hat{\mathbf{x}}(0)$, $\mathbf{P}(0)$) for the Kalman filter design.

Exercise 5.9 For the system described in Exercise 3.7 on p. 97, assume that the variance of the 0.1 Hz position measurement is $10m$, and that $\epsilon_a \sim N(0, \sigma_a)$ with $\sigma_a = 0.01\frac{m/s^2}{\sqrt{Hz}}$ and $\epsilon_g \sim N(0, \sigma_g)$ with $\sigma_g = 2.0 \times 10^{-5}\frac{rad/s}{\sqrt{Hz}}$.

1. Design a Kalman filter by specifying the discrete-time stochastic model parameters and initial conditions: $\hat{\mathbf{x}}(0)$, $\mathbf{P}(0)$, $\mathbf{F}$, $\mathbf{Q}$, $\mathbf{\Gamma}$, $\mathbf{H}$, $\mathbf{\Phi}$, $\mathbf{Qd}$, and $\mathbf{R}$.

2. Implement the Kalman filter (time and measurement) covariance propagation equations. Plot the diagonal of the covariance matrix versus time.

3. Implement the covariance propagation equations for Exercise 3.7. Use the observer gain from that exercise. Be careful to use the correct form of covariance propagation for the measurement update. Plot the diagonal of the covariance matrix versus time.

4. Compare the performance of the two estimators.

Exercise 5.10 For the system described in Exercises 3.7 and 5.9, assume that the variance of the 0.1 Hz position measurement is $10m$ and that $\epsilon_g \sim N(0, \sigma_g)$ with $\sigma_g = 2.0 \times 10^{-5} \frac{rad/s}{\sqrt{Hz}}$. Also, assume that $\epsilon_a = b_a + \nu_a$ where b_a is a random constant with initial covariance $\left(0.001 \frac{m}{s^2}\right)^2$ and $\nu_a \sim N(0, \sigma_a)$ with $\sigma_a = 0.01 \frac{m/s}{\sqrt{Hz}}$.

1. Design a Kalman filter by specifying the discrete-time stochastic model parameters and initial conditions: $\hat{\mathbf{x}}(0)$, $\mathbf{P}(0)$, $\mathbf{F}$, $\mathbf{Q}$, $\mathbf{\Gamma}$, $\mathbf{H}$, $\mathbf{\Phi}$, $\mathbf{Qd}$, and $\mathbf{R}$.

2. Implement the Kalman filter (time and measurement) covariance propagation equations. Plot the diagonal of the covariance matrix versus time. Discuss this plot and why it has the form that it does.

Exercise 5.11 This exercise extends the analysis of Example 5.4 which relates to integer ambiguity resolution in carrier phase GPS processing.

Considered the scenario where at each measurement epoch ($T = 1s$) two measurements are available

$$y_k = x_k + n_k \tag{5.110}$$

$$z_k = x_k + v_k + B\lambda \tag{5.111}$$

where $\lambda = 0.2$; v_k is a zero mean, white, Gaussian process with variance $\sigma_z^2 = (0.01)^2$; and x_k can change arbitrarily from one epoch to another. The constant B is known to be an integer, but for this exercise we will treat it as a real number. We are interested in determining the time required for the standard deviation of the estimate of B to be less 0.1 (i.e., the time until we can round $\hat{B}$ to the nearest integer and expect it to have a high probability of being correct). The variable n_k is a scalar, Gauss-Markov process modeled by

$$n_{k+1} = an_k + w_k \tag{5.112}$$

where w_k is a zero mean, white Gaussian process with variance σ_w^2. The parameter $a = exp(-1/\beta)$ where β is the correlation time in seconds. The positive parameters σ_w and a are related by $\sigma_w^2 = 1 - a^2$.

1. Denote the variance of n_k as $P_n(k) = var(n_k)$. Confirm that propagation of the variance P_n, using the model given by eqn. (5.112), will result in $P_n = 1.0$ in steady-state.

2. Design a (complementary form) Kalman filter for the system with state $\mathbf{x}_k = [B, n_k]^\top$ and measurement $r_k = z_k - y_k$.

   - Show that the state transition model for the state has

   $$\mathbf{\Phi} = \begin{bmatrix} 1 & 0 \\ 0 & a \end{bmatrix}, \qquad \mathbf{Qd} = \begin{bmatrix} 0 & 0 \\ 0 & \sigma_w^2 \end{bmatrix},$$

   $\mathbf{H} = [1, -1]$ and $R = \sigma_z^2$.

   - At $k = 1$, assume that $\mathbf{x}_k = \mathbf{0}$ and that

   $$\mathbf{P}_1^- = \begin{bmatrix} \sigma_0^2 & 0 \\ 0 & 1 \end{bmatrix}$$

   with σ_0 near infinity. Multiply out eqn. (5.61), then take the limit as σ_0 approaches infinity to show that $\mathbf{K}_1 = [1, 0]^\top$.

   - Implement the Kalman filter to compute $\mathbf{K}_k$, $\mathbf{P}_k^+$, and $\mathbf{P}_k^-$. For $\beta = 60s$, run the simulation and answer the following questions. How many iterations are require for the standard deviation of the estimate of B to be less that 0.1. In steady-state, what is the Kalman gain $\mathbf{K}$?

3. Refer back to Example 5.4 where the correlation of n_k was ignored. For that example, answer the following questions. How many iterations are require for the standard deviation of the estimate of B to be less that 0.1. In steady-state, what is the Kalman gain $\mathbf{K}$?

Exercise 5.12 For the system describe in Exercise 5.11, design a Schmidt-Kalman filter that estimates B, but not n_k.

Chapter 6

Performance Analysis

This chapter presents methods to evaluate the performance of a state estimator. The state estimator is assumed to produce an unbiased estimate (i.e. $E\langle \hat{\mathbf{x}}(t) \rangle = E\langle \mathbf{x}(t) \rangle$) and performance is quantified by the state estimate error covariance matrix.

Chapter 5 assumed that the estimator design model was an exact model of the actual system. This assumption is optimistic. In the design process, as discussed in Section 5.10, a series of design tradeoffs may be considered that result in the design model being a reduced order approximation to the *truth model* (i.e., best available linear model for the system). Due to such tradeoffs and the fact that any mathematical model is only an approximation to the actual physical system, the estimator design model will not exactly match the real system. Sections 6.1 and 6.2 are concerned with methods to study estimator performance under these more realistic conditions. Section 6.3 presents methods that allow the designer to consider each source of system error separately to determine its contribution to the overall *error budget*.

6.1 Covariance Analysis

Consider a system described by

$$\left. \begin{array}{rcl} \dot{\mathbf{x}}(t) &=& \mathbf{F}(t)\mathbf{x}(t) + \mathbf{G}(t)\boldsymbol{\omega}(t) \\ \mathbf{y}(t) &=& \mathbf{H}(t)\mathbf{x}(t) + \boldsymbol{\nu}(t) \end{array} \right\} \tag{6.1}$$

which generates the discrete-time equivalent "truth model":

$$\left. \begin{array}{rcl} \mathbf{x}_{k+1} &=& \boldsymbol{\Phi}_k \mathbf{x}_k + \boldsymbol{\omega}_d(k) \\ \mathbf{y}_k &=& \mathbf{H}_k \mathbf{x}_k + \boldsymbol{\nu}_k. \end{array} \right\} \tag{6.2}$$

The truth model is a high-fidelity model of the system. It may be too complex for use in actual applications. In such cases, a reduced-order design model may be used for the application. The continuous-time design model is

$$\left.\begin{array}{rcl}\dot{\hat{\mathbf{x}}}(t) & = & \hat{\mathbf{F}}(t)\hat{\mathbf{x}}(t)+\hat{\mathbf{G}}(t)\hat{\boldsymbol{\omega}}(t)\\ \hat{\mathbf{y}}(t) & = & \hat{\mathbf{H}}(t)\hat{\mathbf{x}}(t)+\hat{\boldsymbol{\nu}}(t)\end{array}\right\}\qquad(6.3)$$

which generates the discrete-time equivalent "design model:"

$$\hat{\mathbf{x}}_{k+1} = \hat{\boldsymbol{\Phi}}\hat{\mathbf{x}}_k+\hat{\boldsymbol{\omega}}_d(k)\qquad(6.4)$$

$$\hat{\mathbf{y}}_k = \hat{\mathbf{H}}_k\hat{\mathbf{x}}_k+\hat{\boldsymbol{\nu}}_k.\qquad(6.5)$$

From the design model, the Kalman filter design equations are

$$\hat{\mathbf{P}}_{k+1}^- = \hat{\boldsymbol{\Phi}}_k\hat{\mathbf{P}}_k\hat{\boldsymbol{\Phi}}_k^\top+\hat{\mathbf{Q}}\mathbf{d}_k\qquad(6.6)$$

$$\hat{\mathbf{K}}_k = \hat{\mathbf{P}}_k^-\hat{\mathbf{H}}_k^\top\left(\hat{\mathbf{R}}_k+\hat{\mathbf{H}}_k\hat{\mathbf{P}}_k^-\hat{\mathbf{H}}_k^\top\right)^{-1}\qquad(6.7)$$

$$\hat{\mathbf{P}}_k^+ = \left[\mathbf{I}-\hat{\mathbf{K}}_k\hat{\mathbf{H}}_k\right]\hat{\mathbf{P}}_k^-.\qquad(6.8)$$

The objective of covariance analysis is to evaluate the performance of the state estimated using the design model when using data obtained from the truth model. The state definitions for $\mathbf{x}$ and $\hat{\mathbf{x}}$ are distinct with it usually being the case that $dim(\mathbf{x})$ larger than $dim(\hat{\mathbf{x}})$. Therefore, we define a matrix $\mathbf{V}$ to select the appropriate linear combination of the actual system states to correspond with the estimator states. This section is then concerned with calculation of the error variance of

$$\mathbf{z} = (\mathbf{V}\mathbf{x}-\hat{\mathbf{x}}).$$

To analyze the performance of the coupled system, the joint state space representation for the system and the implemented estimator will be required:

$$\begin{bmatrix}\mathbf{x}_{k+1}^-\\ \hat{\mathbf{x}}_{k+1}^-\end{bmatrix} = \begin{bmatrix}\boldsymbol{\Phi} & \mathbf{0}\\ \mathbf{0} & \hat{\boldsymbol{\Phi}}\end{bmatrix}_k\begin{bmatrix}\mathbf{x}_k^+\\ \hat{\mathbf{x}}_k^+\end{bmatrix}+\begin{bmatrix}\mathbf{I}\\ 0\end{bmatrix}\omega_d(k)\qquad(6.9)$$

$$\mathbf{z}_k = \begin{bmatrix}\mathbf{V} & -\mathbf{I}\end{bmatrix}\begin{bmatrix}\mathbf{x}_k\\ \hat{\mathbf{x}}_k\end{bmatrix}\qquad(6.10)$$

$$y_k^- = [\mathbf{H}_k,\mathbf{0}]\begin{bmatrix}\mathbf{x}_k^-\\ \hat{\mathbf{x}}_k^-\end{bmatrix}\qquad(6.11)$$

$$\hat{y}_k^- = \left[\mathbf{0},\hat{\mathbf{H}}_k\right]\begin{bmatrix}\mathbf{x}_k^-\\ \hat{\mathbf{x}}_k^-\end{bmatrix}\qquad(6.12)$$

$$\mathbf{K}_k = \begin{bmatrix}\mathbf{0}_k\\ \hat{\mathbf{K}}_k\end{bmatrix}\qquad(6.13)$$

$$\begin{bmatrix}\mathbf{x}_k^+\\ \hat{\mathbf{x}}_k^+\end{bmatrix} = \begin{bmatrix}\mathbf{x}_k^-\\ \hat{\mathbf{x}}_k^-\end{bmatrix}+\mathbf{K}_k\left(y_k^- - \hat{y}_k^-\right).\qquad(6.14)$$

This expression represents the actual time propagation of the real system and the estimate. Since the estimate is calculated exactly[1], no process driving noise is represented in the corresponding rows of eqn. (6.9). Also, the zeros concatenated into the estimation gain vector account for the fact that the estimator corrections do not affect the state of the actual system.

For this coupled system, the covariance propagates between sampling times according to

$$\mathbf{P}_{11}^{-}(k+1) = \mathbf{\Phi}\mathbf{P}_{11}^{+}\mathbf{\Phi}^{\top} + \mathbf{Qd}^{\top} \tag{6.15}$$

$$\mathbf{P}_{12}^{-}(k+1) = \mathbf{\Phi}\mathbf{P}_{12}^{+}\hat{\mathbf{\Phi}}^{\top} \tag{6.16}$$

$$\mathbf{P}_{22}^{-}(k+1) = \hat{\mathbf{\Phi}}\mathbf{P}_{22}^{+}\hat{\mathbf{\Phi}}^{\top} \tag{6.17}$$

where $\mathbf{P}$ is partitioned as $\mathbf{P}_{11} = cov(\mathbf{x}, \mathbf{x})$, $\mathbf{P}_{21}^{\top} = \mathbf{P}_{12} = cov(\mathbf{x}, \hat{\mathbf{x}})$, $\mathbf{P}_{22} = cov(\hat{\mathbf{x}}, \hat{\mathbf{x}})$.

The measurement update is computed by eqn. (6.14). Based on eqn. (6.14), the covariance of the state after the measurement update is

$$\mathbf{P}^{+}(k) = \left(\mathbf{I} + \mathbf{K}_k\bar{\mathbf{H}}_k\right)\mathbf{P}^{-}(k)\left(\mathbf{I} + \mathbf{K}_k\bar{\mathbf{H}}_k\right)^{\top} + \mathbf{K}_k\mathbf{R}_k\mathbf{K}_k^{\top}. \tag{6.18}$$

Noting that $\bar{\mathbf{H}} = [\mathbf{H}, -\hat{\mathbf{H}}]$, eqn. (6.18) reduces to the following set of equations,

$$\mathbf{P}_{11}^{+}(k) = \mathbf{P}_{11}^{-} \tag{6.19}$$

$$\mathbf{P}_{12}^{+}(k) = \mathbf{P}_{12}^{-}\left(\mathbf{I} - \hat{\mathbf{H}}^{\top}\hat{\mathbf{K}}^{\top}\right) + \mathbf{P}_{11}^{-}\mathbf{H}^{\top}\hat{\mathbf{K}}^{\top} \tag{6.20}$$

$$\begin{aligned}\mathbf{P}_{22}^{+}(k) = {} & \left(\mathbf{I} - \hat{\mathbf{K}}\hat{\mathbf{H}}\right)\mathbf{P}_{22}^{-}\left(\mathbf{I} - \hat{\mathbf{H}}^{\top}\hat{\mathbf{K}}^{\top}\right) + \hat{\mathbf{K}}\mathbf{H}\mathbf{P}_{11}^{-}\mathbf{H}^{\top}\hat{\mathbf{K}}^{\top} + \hat{\mathbf{K}}\mathbf{R}\hat{\mathbf{K}}^{\top} \\ & + \hat{\mathbf{K}}\mathbf{H}\mathbf{P}_{12}^{-}\left(\mathbf{I} - \hat{\mathbf{H}}^{\top}\hat{\mathbf{K}}^{\top}\right) + \left(\mathbf{I} - \hat{\mathbf{K}}\hat{\mathbf{H}}\right)\mathbf{P}_{21}^{-}\mathbf{H}^{\top}\hat{\mathbf{K}}^{\top}.\end{aligned} \tag{6.21}$$

The error variance of the variable of interest, $\mathbf{z}$, is defined at any instant by

$$cov(\mathbf{z}) = \mathbf{V}\mathbf{P}_{11}\mathbf{V}^{\top} + \mathbf{P}_{22} - \mathbf{V}\mathbf{P}_{12} - \mathbf{P}_{21}\mathbf{V}^{\top}. \tag{6.22}$$

Note that both the time and measurement covariance update equations are dependent on the characteristics of the measurement data (i.e. update rate, $\mathbf{Qd}$, and $\mathbf{R}_k$), but independent of the actual measurement data. Therefore, these equations can be precomputed off-line during the design phase. There are two reasons for being interested in such precomputation. First, the covariance propagation accounts for the majority of filter computation. In certain applications where the on-line computational capabilities are limited, it may be necessary to determine and use either curves fit to the

[1] Quantization error can be modeled by a second additive noise source affecting the estimate, but is dropped in this analysis for convenience. In floating point processors, the effect is small.

Kalman gains or the steady-state Kalman gains. Although this will lighten the on-line computational load, performance may suffer. For example, when steady-state gains are used, performance will suffer during transient conditions such as start-up. Second, the covariance equations allow prediction of the state estimation accuracy. Therefore, it is often useful to perform covariance analysis in simulation prior to hardware construction to analyze implementation tradeoffs or ensure satisfaction of system specifications.

To summarize, assuming that $\mathbf{P}^+(0)$ is known, starting with $k = 0$, the steps of the covariance analysis are as follows for $t \in [t_k, t_{k+1})$ and $k \geq 0$:

1. Specify the "truth model:" $\mathbf{F}$, $\mathbf{G}$, and $\mathbf{Q}$.

2. Specify the "design model:" $\hat{\mathbf{F}}$, $\hat{\mathbf{G}}$, and $\hat{\mathbf{Q}}$.

3. Compute the discrete-time equivalent design model parameters: $\hat{\mathbf{\Phi}}_k$ and $\mathbf{Qd}_k$.

4. Use eqn. (6.6) to propagate the error covariance of the design model from t_k to t_{k+1}.

5. At time t_{k+1}, increment k by one and compute $\mathbf{H}$ and $\hat{\mathbf{H}}$.

6. Compute and store the Kalman filter gain sequence $\hat{\mathbf{K}}_k$ using eqn. (6.7) and compute the posterior design model error covariance using eqn. (6.8).

7. Use the matrix parameters from the previous steps to process eqns. (6.15–6.22).

Iterate the above process over the time duration of interest or until steady-state is achieved.

Example 6.1 demonstrates the covariance analysis process.

Example 6.1 *Consider the one dimension inertial frame INS aided by a biased position measurement. The error equations for the INS are*

$$\begin{bmatrix} \dot{p} \\ \dot{v} \\ \dot{a} \\ \dot{b} \end{bmatrix} = \begin{bmatrix} 0 & 1 & 0 & 0 \\ 0 & 0 & 1 & 0 \\ 0 & 0 & 0 & 0 \\ 0 & 0 & 0 & -\beta \end{bmatrix} \begin{bmatrix} p \\ v \\ a \\ b \end{bmatrix} + \begin{bmatrix} 0 \\ \omega_v \\ \omega_a \\ \omega_b \end{bmatrix} \tag{6.23}$$

$$y = \begin{bmatrix} 1 & 0 & 0 & 1 \end{bmatrix} \begin{bmatrix} p \\ v \\ a \\ b \end{bmatrix} + \nu \tag{6.24}$$

where $var(\nu) = R = 10\ m^2$, $PSD_{\omega_v} = 2.5 \times 10^{-3} \frac{m^2}{s^3}$ *and* $PSD_{\omega_a} = 1.0 \times 10^{-6} \frac{m^2}{s^5}$. *The navigation mechanization equations are integrated in* *discrete-time using the small time step determined by the IMU to compute the navigation state. The error state dynamics are modeled in continuous-time. The position aiding measurement denoted by y is available at a $1.0Hz$ rate.*

The measurement bias b is a scalar Gauss-Markov process with $\beta = \frac{1}{300}$ sec^{-1} and $PSD_{\omega_b} = Q_b = 4\frac{m^2}{s}$. For a scalar Gauss-Markov process with these parameters,

$$\begin{aligned} \dot{P}_b &= -2\beta P_b + Q_b \\ P_b(t) &= e^{-2\beta t} P_b(0) + \left(1 - e^{-2\beta t}\right) \frac{Q_b}{2\beta} \\ P_b(\infty) &= \frac{Q_b}{2\beta}. \end{aligned}$$

Therefore, the steady-state bias error covariance for b would be $600m^2$.

Covariance analysis can be used to determine how a filter which ignores the measurement bias would actually perform. In this case,

$$\Phi = \begin{bmatrix} 1 & T & \frac{T^2}{2} & 0 \\ 0 & 1 & T & 0 \\ 0 & 0 & 1 & 0 \\ 0 & 0 & 0 & \rho \end{bmatrix} \qquad \hat{\Phi} = \begin{bmatrix} 1 & T & \frac{T^2}{2} \\ 0 & 1 & T \\ 0 & 0 & 1 \end{bmatrix} \tag{6.25}$$

where $\rho = e^{-\beta T}$. For both of the covariance analysis simulations described below, $T = 1.0$ s and $\mathbf{P}(0)$ is a diagonal matrix with elements 100.0, 1.0, 0.001, and 625.

Figure 6.1 displays the actual performance (bold dotted line) and internal filter prediction (narrow solid line) of performance for the case where $\hat{R} = 10$ m^2. Therefore, except for the neglected measurement bias, the model of the implemented filter is correct. Based on the design model, the filter expects an accurate, unbiased measurement of position. Therefore, the filter gains are relatively high. The design model covariance sequence of eqn. (6.8) predicts that the position error standard deviation in steady-state will be less than $1.5m$, while the covariance analysis predicts a steady-state position error standard deviation great than $24m$. The design neglecting the bias is overly optimistic.

Figure 6.2 displays the actual performance (bold dotted line) and internal filter prediction (narrow solid line) of performance for the case where $\hat{R} = 625m^2$. For this simulation, the implemented filter does not model the bias; instead, the design model increases $\hat{R}$ to a value meant *to indicate that the error in each measurement has a standard deviation of $25m$. Based on this design model, the filter expects a measurement of position with a relatively large amount of white noise; however, if the noise was white, as modeled, then suitable filtering of the measurement should produce an accurate estimate of position. Therefore, the filter gains are lower than the*

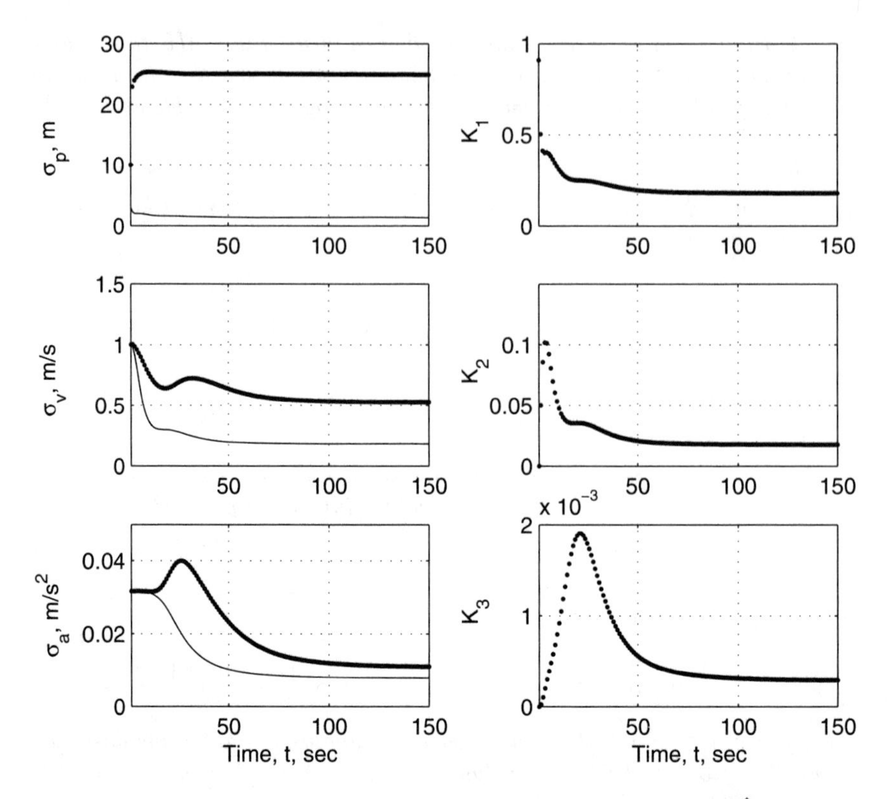

Figure 6.1: Results of the covariance analysis of Example 6.1 for $\hat{\mathbf{R}} = 10$ m^2. The standard deviation of each component of the estimated state is plotted versus time in the left column. Actual performance is indicated by the bold dotted curve. The performance predicted by the implemented filter is the narrow solid curve. The right column plots the Kalman filter gains versus time.

previous case. The filter trusts the filter state, which represents a time average of all past measurements more than the measurement. The performance prediction by the filter is still optimistic, $\sigma_p = 7m$ predicted in steady-state while $\sigma_p = 25m$ is achieved, because the model does not properly account for the time correlation of the position measurement error. The filter is not averaging white noise corrupted measurements, but measurements with a highly correlated error. The actual (correlated error) will not average out over any reasonable time span.

A better method for addressing the sensor bias of this example without including it as an augmented state is discussed in Section 5.10.2. △

In the special case where the state definitions are the same and it is reasonable to assume that $\mathbf{\Phi} = \hat{\mathbf{\Phi}}$ and $\mathbf{H} = \hat{\mathbf{H}}$ with $\mathbf{V} = \mathbf{I}$, the error state

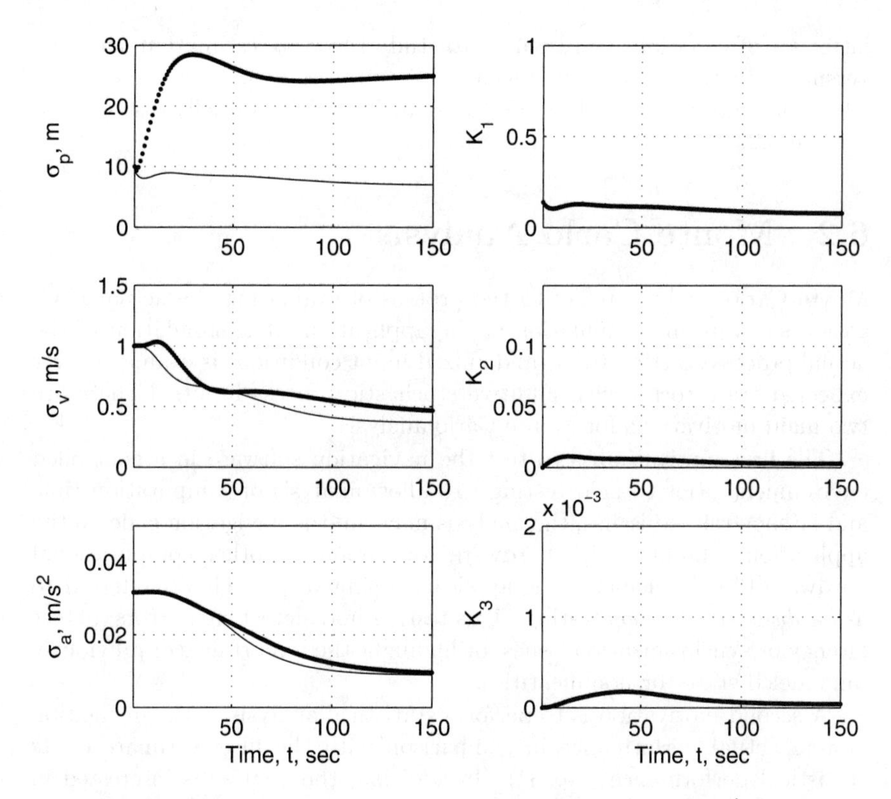

Figure 6.2: Results of the covariance analysis of Example 6.1 for $\hat{\mathbf{R}} = 625\ \text{m}^2$. The standard deviation of each component of the estimated state is plotted versus time in the left column. Actual performance is indicated by the bold dotted curve. The performance predicted by the implemented filter is the solid narrow curve. The right column plots the Kalman filter gains versus time.

$\tilde{\mathbf{x}} = (\mathbf{x} - \hat{\mathbf{x}})$ can be defined and the covariance matrix propagation equations for the error state simplify considerably to

$$\mathbf{M}_{k+1}^{-} = \mathbf{\Phi}\mathbf{M}_k^{+}\mathbf{\Phi}^{\top} + \mathbf{Qd}_k \tag{6.26}$$

$$\mathbf{M}_k^{+} = \left(\mathbf{I} - \hat{\mathbf{K}}_k\mathbf{H}_k\right)\mathbf{M}_k^{-}\left(\mathbf{I} - \mathbf{H}_k^{\top}\hat{\mathbf{K}}_k^{\top}\right) + \hat{\mathbf{K}}_k\mathbf{R}_k\hat{\mathbf{K}}_k^{\top} \tag{6.27}$$

where $\mathbf{M} = cov(\tilde{\mathbf{x}}, \tilde{\mathbf{x}})$. These equations will be used to support the discussion of Section 6.3. In that section, to simplify the notation, all subscript k's in eqn. (6.27) will be dropped.

To test the robustness of the filter design to model assumptions, the covariance analysis may be repeated for various instances of the design model assumptions. For example, because the values of $\hat{\mathbf{Q}}\mathbf{d}_k$ and $\hat{\mathbf{R}}_k$ in eqns. (6.6-6.8) and $\mathbf{Qd}_k$ and $\mathbf{R}_k$ in eqns. (6.26-6.27) can be distinct, the latter set of equations can be used to study the state estimation accuracy versus the variance of a component of the process or measurement noise. This leads into the idea of Monte Carlo analysis that is briefly discussed in Section 6.2.

6.2 Monte Carlo Analysis

Monte Carlo analysis refers to the process of evaluating the actual navigation software in a simulation of the application. The simulation of the actual process starting from randomized input conditions is evaluated over expected trajectories with additive stochastic process inputs. There are two main motivations for Monte Carlo analysis.

The first motivation is to test the navigation software in a controlled environment prior to field testing. To allow analysis of computation time and latency related issues, the analysis may run the navigation code on the application computer with hardware connections to other computational hardware that implements the application simulation. This is referred to as *hardware-in-the-loop* testing. This testing may detect sign errors, detect latency or synchronization issues, or highlight the importance of previously unmodeled states or nonlinearities.

A second motivation is to perform statistical analysis of the navigation system actual performance in comparison with the filter estimate of its statistical performance (i.e., $\hat{\mathbf{P}}$). In addition, the analyst is interested in determining whether the mean estimation error is unbiased. For statistically valid results, numerous simulations calculating the filter estimation error will be required. The error statistics will have to be calculated over the ensemble of simulations. The accuracy of the statistical results determined by either Monte Carlo or covariance analysis techniques is dependent on the accuracy of the "reality" or "truth" model that is assumed.

6.3 Error Budgeting

In the design of a state estimation system, there is usually a conflicting set of desires. Better performance may be achieved by either using more or better sensors, or by appending more error states to the model; however, either approach may increase the system cost. Covariance analysis techniques provide a systematic approach to evaluating alternative system implementations. When performing a covariance analysis as described in Section 6.1, the result is the covariance of the output error due to *all* error sources. For an analysis of system design tradeoffs, it is usually more useful to be able to calculate the output error variance due to *each* source independently. The output error variance due to each error source allows the designer to easily determine the dominant error sources and decide whether it is feasible to reduce the error effects (through either better hardware or more detailed modeling) to the point where the design specifications are achieved. This process is referred to as error budgeting.

Either eqns. (6.15-6.21) or eqns. (6.26-6.27) can be used for error budgeting analysis. Note that when $\mathbf{K}$ is fixed, both sets of equations are linear in $\mathbf{P}_0, \mathbf{Qd}$, and $\mathbf{R}$ (or $\mathbf{M}_0, \mathbf{Qd}$, and $\mathbf{R}$). For brevity, this discussion only considers eqns. (6.26-6.27) with $\mathbf{Q}$ and $\mathbf{R}$ being time-invariant. Since this set of equations is linear in each of the driving noise terms, the principal of superposition can be applied.

Let $\mathbf{M}_0 = \sum_{i=1}^{l_m} \mathbf{M}_i(0)$, $\mathbf{Q} = \sum_{i=1}^{l_q} \mathbf{Q}_i$, and $\mathbf{R} = \sum_{i=1}^{l_r} \mathbf{R}_i$. The symbols l_m, l_q, and l_r denote the number of independent error or noise terms in the matrices $\mathbf{M}_0, \mathbf{Q}$, and $\mathbf{R}$, respectively, and each of the matrices $\mathbf{M}_i(0)$, $\mathbf{Q}_i$, and $\mathbf{R}_i$ are symmetric and positive definite. For simplicity of notation, let each of these matrices be diagonal. The design eqns. (6.5-6.8) are first simulated once, with all noise sources turned on, to determine the gain sequence $\mathbf{K}_k$. This gain sequence must be saved and used in each of the subsequent component simulations of the error budget analysis. The components of $\mathbf{P}_k$ that are of interest should be saved at the corresponding times of interest, for later analysis. For the subsequent $l = l_m + l_q + l_r$ simulations of eqns. (6.26-6.27), the gain vector sequence is considered as given.

For each independent component of $\mathbf{M}_0$ (i.e., $l = 1, \ldots, l_m$), we calculate

$$\mathbf{M}_l^-(k+1) = \mathbf{\Phi}\mathbf{M}_l^+(k)\mathbf{\Phi}^\top \tag{6.28}$$

$$\mathbf{M}_l^+ = \left(\mathbf{I} - \hat{\mathbf{K}}\mathbf{H}\right)\mathbf{M}_l^-\left(\mathbf{I} - \mathbf{H}^\top\hat{\mathbf{K}}^\top\right) \tag{6.29}$$

with $\mathbf{M}_l(0)$ being the l-th independent component of $\mathbf{M}(0)$ as previously defined. This set of simulations produces the error covariance due to each independent source of initial condition uncertainty in the absence of process and measurement noise.

For each independent component of $\mathbf{Qd}$ (i.e., $l = (l_m+1), \ldots, (l_m+l_q)$), we calculate

$$\mathbf{M}_l^-(k+1) = \mathbf{\Phi}\mathbf{M}_l^+(k)\mathbf{\Phi}^\top + \mathbf{Qd}_i \tag{6.30}$$

$$\mathbf{M}_l^+ = \left(\mathbf{I} - \hat{\mathbf{K}}\mathbf{H}\right)\mathbf{M}_l^-\left(\mathbf{I} - \mathbf{H}^\top\hat{\mathbf{K}}^\top\right) \tag{6.31}$$

with $\mathbf{M}_l(0)$ being a zero matrix and $\mathbf{Qd}_i$ being the discrete-time equivalent of $\mathbf{Q}_i$, where $i = l - l_m$. This set of simulations produces the error covariance due to each independent source of process driving noise in the absence of initial condition errors and measurement noise.

For each independent component of $\mathbf{R}$ (i.e., $l = (l_m + l_q + 1), \ldots, (l_m + l_q + l_r)$), we calculate

$$\mathbf{M}_l^-(k+1) = \mathbf{\Phi}\mathbf{M}_l^+(k)\mathbf{\Phi}^\top \tag{6.32}$$

$$\mathbf{M}_l^+ = \left(\mathbf{I} - \hat{\mathbf{K}}\mathbf{H}\right)\mathbf{M}_l^-\left(\mathbf{I} - \mathbf{H}^\top\hat{\mathbf{K}}^\top\right) + \hat{\mathbf{K}}\mathbf{R}_i\hat{\mathbf{K}}^\top \quad (6.33)$$

with $\mathbf{M}_l(0)$ being a zero matrix and $i = l - l_m - l_q$. This set of simulations produces the error covariance due to each independent source of measurement noise in the absence of initial condition errors and process driving noise.

At the completion of the $(l_m + l_q + l_r)$-th covariance simulation, each independent source of system error has been accounted for exactly once. By the principle of superposition, for each k the total error covariance from all sources $\mathbf{P}_k$ is computed as

$$\mathbf{P}_k = \sum_{l=1}^{(l_m+l_q+l_r)} \mathbf{M}_l(k).$$

Usually it is some subset of the diagonal elements of the $\mathbf{P}_k$ matrices that are of interest. It is straightforward to tabulate the appropriate elements of $\mathbf{P}_k$ versus the dominant error terms (i.e., $\mathbf{M}_0, \mathbf{Qd}, \mathbf{R}$) to determine the appropriate amount that each error source must be reduced to meet a given specification. This analysis also indicates states that are unobservable and therefore candidates for either combination or deletion.

Example 6.2 *Consider the one dimensional INS system described in Example 4.9.4 on page 159. Assume that a position measurement is available at a 1 Hz rate for the first 100 seconds of a 200-second application and that the position measurement is corrupted by white measurement noise with variance of 1.0 m^2.*

A 200 second simulation of eqns. (6.26-6.27) and the Kalman filter gain calculation equation results in the position error covariance illustrated in Figure 6.3. During the first 100 seconds when the position aiding is available, the position covariance converges to approximately $0.3m^2$ ($\sigma_p = 0.55m$). For $t \in [100, 200]$, while the aiding signal is unavailable, the position error covariance increases to approximately 3200 m^2 ($\sigma_p = 56.5m$).

In this simple example, we will consider three independent error sources as performance limiting factors. If the present performance is not satisfactory, it is necessary to further analyze the contribution of each of the error sources to the total error covariance so that educated decisions can be made concerning which of the system components should be improved to have the greatest impact on the total system performance.

Figure 6.4 displays the results of performing separate covariance simulations for each of the following cases:

1. Total Error. *This simulation produces the total error depicted in Figure 6.3 and produces the Kalman gain sequence that will be used for the following three simulations. For this simulation, $R = 1.0m^2$, $\mathbf{Q}$ is*

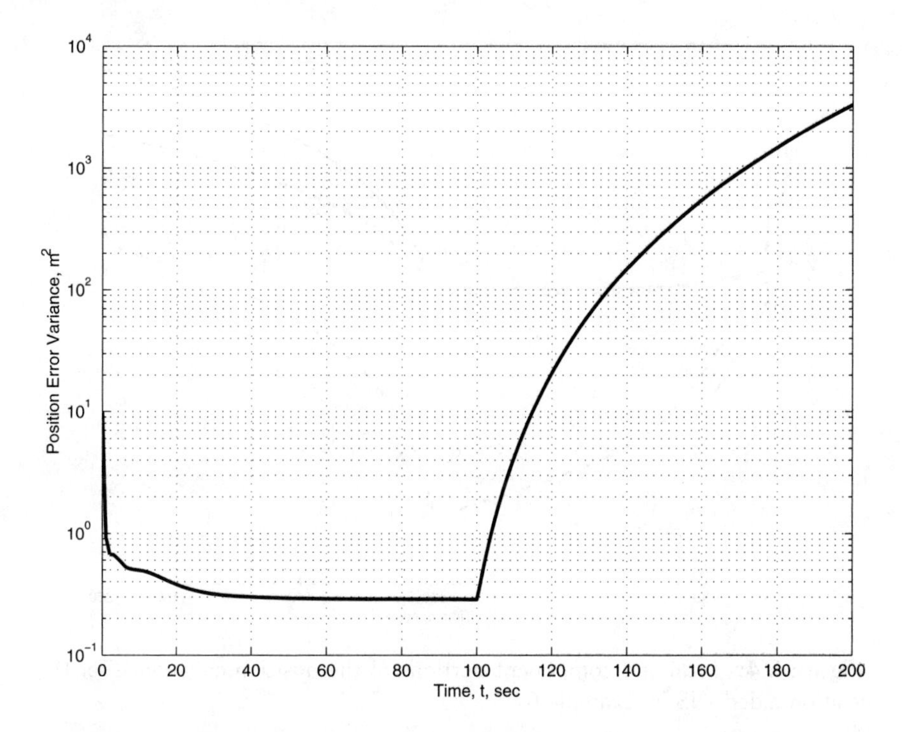

Figure 6.3: Position covariance for the position aided INS in Example 6.2.

a diagonal matrix of the following elements $[0.0, 2.5\times10^{-3}, 1.0\times10^{-6}]$, *and* $\mathbf{M}_0$ *is a diagonal matrix with elements* $[10.0, 1.0, 0.01]$. *The position covariance sequence resulting from this simulation is shown in Figure 6.3 and Figure 6.4 as the solid line.*

2. Uncertain Initial Conditions. *This simulation uses the Kalman gain vector of simulation 1, with* R *and* $\mathbf{Q}$ *set equal to appropriately dimensioned zero matrices. The* $\mathbf{M}_0$ *matrix assumes its original value as defined in Simulation 1. Therefore, this simulation shows the effect of the uncertain initial conditions if both the INS and position sensors were perfect. The results of the simulation are shown in Figure 6.4 as the dashed-dotted line.*

3. Position Measurement Noise. *This simulation uses the Kalman gain vector of simulation 1. The* $\mathbf{M}_0$ *and* $\mathbf{Q}$ *matrices were set equal to appropriately dimensioned zero matrices. The* R *matrix assumes its original value as defined in Simulation 1. Therefore, this simulation shows the effect of the position measurement noise if there were no uncertainty in the initial conditions or INS sensors. The results of*

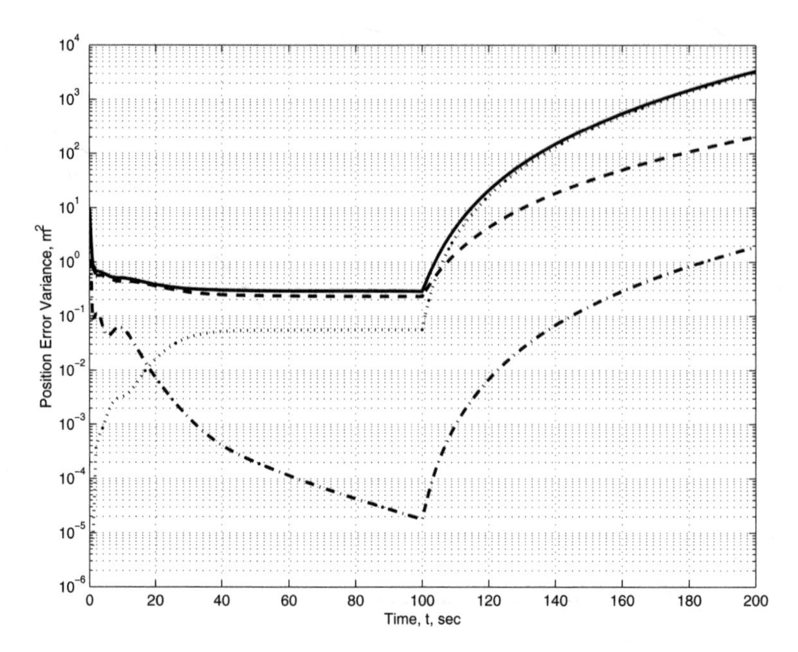

Figure 6.4: Total and component portions of the position covariance for the position aided INS in Example 6.2.

the simulation are shown in Figure 6.4 as the dashed line.

4. INS Sensor Error. *This simulation uses the Kalman gain vector of simulation 1. The* $\mathbf{M}_0$ *and* R *matrices were set equal to appropriately dimensioned zero matrices. The* $\mathbf{Q}$ *matrix assumed its original value as defined in Simulation 1. Therefore, this simulation shows the effect of the INS sensor noise and bias drift if there were no uncertainty in the initial conditions or position sensor. The results of the simulation are shown in Figure 6.4 as the dotted line. The individual components of* $\mathbf{Q}$ *were not analyzed separately.*

As in any correct application of the principle of superposition, each (error) source is accounted for exactly once and the results corresponding to each source individually sum to the total result with all sources forcing the system at the same time. Figure 6.4 illustrates this fact.

Figure 6.4 shows that during the first 100 seconds, the system has essentially reached steady state. The error due to the initial condition $\mathbf{M}(0)$ *is asymptotically approaching zero as expected in a stable system. The main factor limiting performance during the initial 100 seconds is the additive noise on the position measurement.*

For the time interval $t \in (100, 200]$ *seconds, no position measurements*

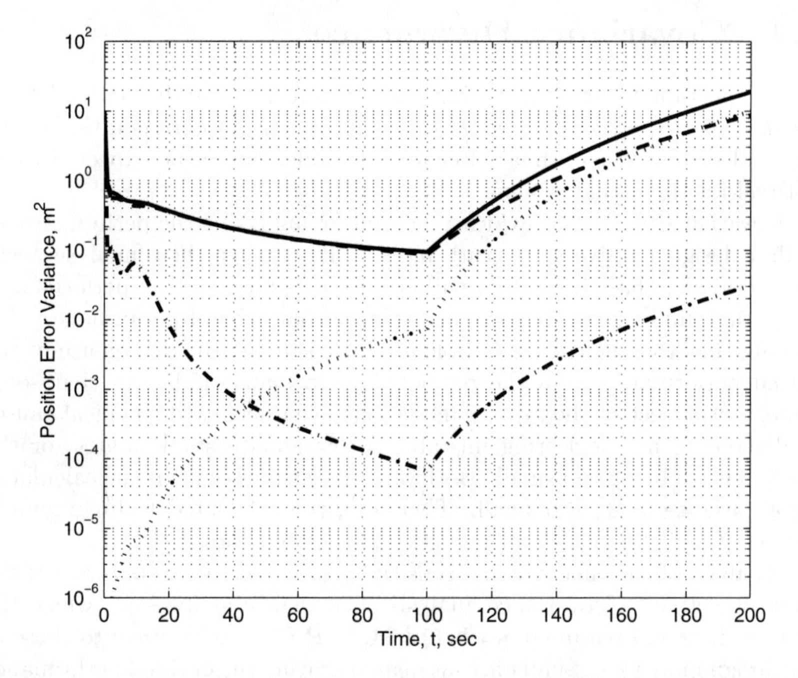

Figure 6.5: Total and component portions of the position covariance for the position aided (improved) INS in Example 6.2.

are available. On this interval, the position error covariance due to all three sources increases with time. Within 10 seconds of losing the position measurement, the INS errors (i.e., $\mathbf{Q}$*) dominate the error growth. Therefore, if better position accuracy is required for* $t \in (110, 200]$*, the system performance can be improved considerably by investing in better inertial sensors.*

If the series of four simulations is repeated assuming that the nominal value of the diagonal values of $\mathbf{Q}$ *are* $[0.0, 5.0 \times 10^{-6}, 2.0 \times 10^{-9}]$*, then the error budget simulations change to those shown in Figure 6.5. The design values for* $\mathbf{Q}$ *were selected to make the INS and position sensor portions of the error covariance at* $t = 200s$ *be approximately equal. In an actual analysis, each component of* $\mathbf{Q}$ *could be analyzed separately to determine which characteristic of the original system was driving the original performance.*

Note that the Kalman gains used in these simulations to produce Figure 6.5 are distinct from those used to produce Figure 6.4. The change in the system design results in different system characteristics in the first simulation of each figure, which yields the gains for the remaining simulations.
△

6.4 Covariance Divergence

It is important to emphasize that the Kalman filter is the optimal filter *for the modeled process.* If the model is an inaccurate representation of the physical system, the optimal filter for the model cannot be expected to be optimal for the actual system.

Covariance analysis provides a method for analyzing the performance of a filter design relative to a 'truth model'. Even when these analysis techniques predict that the system will achieve a desired level of performance, the actual implementation may not achieve that level of performance. Assuming that the algorithm is implemented correctly, the divergence of the actual performance from the predicted performance indicates that some important aspect of the system model has been neglected. Typical causes of divergence are neglecting unstable or marginally stable states, or the process noise being too small. Both of these effects result in the calculated error variance being too small. Therefore, the calculated Kalman gain is too small and the measurements are not weighted enough.

Analysis of the effects of and sensitivity to modeling errors is one of the main objectives of covariance analysis. Covariance analysis for which the system state and parameters $\{\mathbf{\Phi}, \mathbf{Qd}, \mathbf{H}, \mathbf{R}, \mathbf{P}(0)\}$ are identical to those of the design model are useful only for demonstrating the optimal performance that can be achieved under perfect modeling conditions. A good analyst will follow this optimal analysis with a study to determine a set of filter parameters that results in near optimal performance while being robust to expected ranges of model uncertainty.

State estimators designed to be robust to model error and noise parameter uncertainty in the system model are called robust state estimators. Such methods are reviewed in, for example, [121].

Example 6.3 *To illustrate various ways that the actual and predicted filter performance might not match, consider the example of estimating the state of the random walk plus random constant system described by*

$$\dot{x}(t) = \omega(t) \tag{6.34}$$

with $P(0) = var(x(0)) = 10$ *and* $Q(t) = var(\omega(t)) = \sigma^2$ *from discrete measurements occurring at two second intervals:*

$$y(t_k) = x(t_k) + \nu(t_k)$$

where $t_k = 2.0k$ *seconds and* $R = var(\nu) = 5.0$. *Therefore, the Kalman filter is designed using* $\hat{H} = 1$, $\hat{\Phi} = 1$, $\hat{P}(0) = 10$ *and* $\hat{R} = 5$. *Assume for the purpose of this example that the value of* σ^2 *is not accurately known by the analyst at the design stage. For the purposes of the example, the actual value will be taken as* $Q = 0.5$.

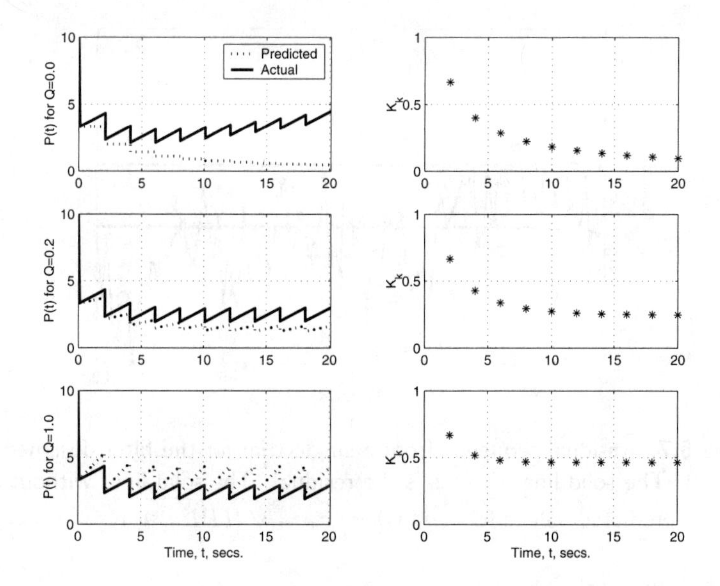

Figure 6.6: Various forms of performance mismatch. Top - Divergence. Middle - Optimistic performance prediction. Bottom - 'Pessimistic' or conservative design.

Each row of plots in Figure 6.6 portrays the actual and predicted covariance, and the Kalman gain for an assumed value of $\hat{Q}$. In the first row, the analyst incorrectly models the state as a random constant (i.e., $\hat{Q} = 0$). Both the Kalman filter gain K_k and predicted covariance (dotted line) are approaching zero. However, the actual error covariance (solid line) is diverging and will grow without bound. This is due to the uncertainty in the actual state growing, while the (erroneously) computed variance of the estimated state approaches zero. The second row corresponds to an optimistic design. The design value of the driving noise spectral density is too small (i.e., $\hat{Q} = 0.2$), resulting in the Kalman filter predicting better performance than is actually achieved. The third row corresponds to a pessimistic design in which the analyst has selected too large a value for the driving noise spectral density. The Kalman filter gain is larger than the optimal gain, and the filter achieves better performance than it predicts.

When the filter design is known to not perfectly match the plant, the pessimistic approach corresponds to the preferred situation. If the analyst designs a pessimistic filter and achieves the design specification, then the actual system will also achieve the design specification.

Figure 6.7 shows data from a single trial run corresponding to the filter design shown in the top row of Figure 6.6 (i.e., $Q_h = 0.0$). In this case, the measurement residual sequence is clearly not a white noise sequence with

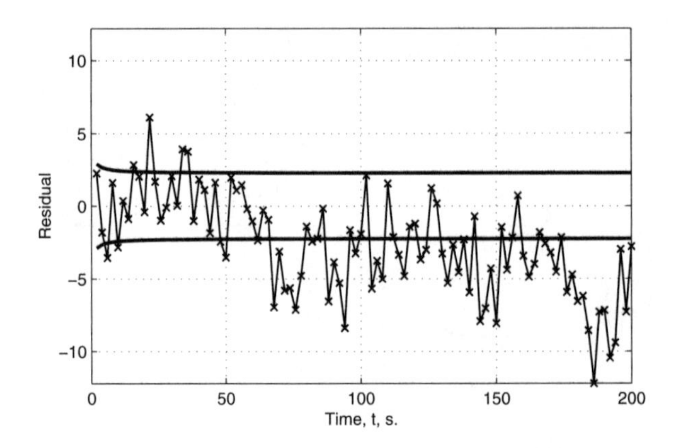

Figure 6.7: Residual sequence from a single trial for the filter designed with $Q_h = 0.0$. The solid line with x's is the residual. The solid lines without x's is the filters predicted values for $\pm\sigma_r$ where $\sigma_r = \sqrt{\hat{H}\hat{P}\hat{H}^\top + \hat{R}}$.

the variance predicted by $\hat{P}$. Analysis of residual sequences is one of the best methods to detect model deficiencies. In doing so, the analyst should keep in mind that $\hat{H}_k\hat{P}_k\hat{H}_k^\top + \hat{R}_k$ is predicting the variance of the residual r_k over an ensemble of experiments. Therefore, for example, the residual sequence of Figure 6.7 does indicate a model deficiency, but by itself does not indicate that the residual is biased. If the experiment was repeated N times, with the residual sequence from the i-th trial run denoted by r_{k_i}, and all residual sequences tended to drift in the same direction, then the analyst might reconsider the system model to find the cause of and model the bias. Alternatively, if the average of the residual sequences $\bar{r}_k = \sum_{i=1}^{N} r_{k_i}$ is essentially zero mean, but the ensemble of residual sequences is not well-modeled by the $\hat{P}$, then the analyst must consider possible alternative explanations.

To illustrate the nature of a residual sequence for a filter designed with a correct model, Figure 6.8 shows the residual for the filter designed with $Q_h = 0.5$. The residual is essentially zero mean, has no noticeable trends, and has steady-state variance that closely matches the $\hat{P}$. △

6.5 References and Further Reading

The main references for the presentation of this chapter were [58, 93].

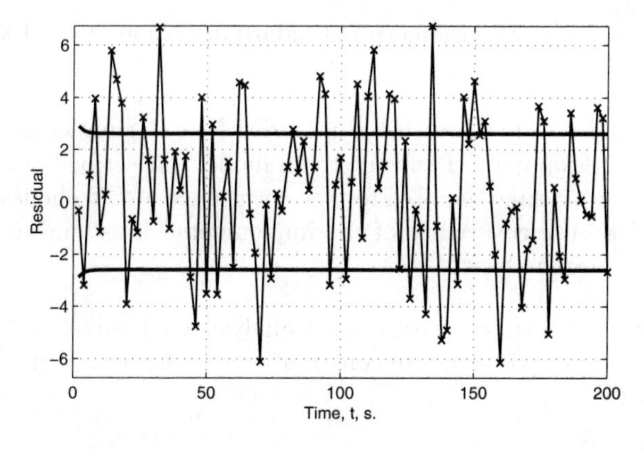

Figure 6.8: Residual sequence from a single trial for the filter designed with $Q_h = 0.5$. The solid line with x's is the residual. The solid lines without x's is the filters predicted values for $\pm\sigma_r$ where $\sigma_r = \sqrt{\hat{H}\hat{P}\hat{H}^\top + \hat{R}}$.

6.6 Exercise

Exercise 6.1 This example is intended to motivate the importance of having an accurate design model.

Let the actual system dynamics (i.e., eqn. (6.1)) be

$$\begin{aligned} \dot{\mathbf{x}} &= \begin{bmatrix} 0 & 1 \\ -\omega_n^2 & 0 \end{bmatrix} \mathbf{x} + \begin{bmatrix} 0 \\ 1 \end{bmatrix} n_x \\ y &= \begin{bmatrix} 1 & 0 \end{bmatrix} \mathbf{x} + n_y \end{aligned}$$

where $n_x \sim N(0, \sigma_x^2)$, $n_y \sim N(0, \sigma_y^2)$, and $\mathbf{x}(0) \sim N(0, \mathbf{I})$. Assume that the estimator design model is based on the assumption that ω_n is zero, so that eqn. (6.3) is

$$\begin{aligned} \dot{\hat{\mathbf{x}}} &= \begin{bmatrix} 0 & 1 \\ 0 & 0 \end{bmatrix} \hat{\mathbf{x}} + \begin{bmatrix} 0 \\ 1 \end{bmatrix} n_x \\ \hat{y} &= \begin{bmatrix} 1 & 0 \end{bmatrix} \hat{\mathbf{x}} + n_y. \end{aligned}$$

The aiding measurements y are available at 1.0 Hz.

Perform a covariance analysis using $\sigma_x = 0.01$ and $\sigma_y = 1.00$.

1. For $\omega_n = 0$ (i.e., the design model is correct), note the approximate time at which steady-state performance is achieved and the steady-state prior and posterior state estimation error standard deviation.

2. Repeat the simulation for values of $\omega_n \in [0, 0.1]$. Let the simulation run for approximately 1000 s. Is a steady-state performance achieved?

Plot the state estimation error standard deviation at $t = 1000s$ versus ω_n.

Exercise 6.2 Perform a covariance analysis where the implemented filter is described as in Example 5.4 and the truth model is described as in Exercise 5.11. Compare the time required for the covariance of B to be less than 0.1 as predicted by the covariance of the implemented filter and as predicted by the covariance analysis.

Exercise 6.3 Perform an error budget analysis for items 2 and 3 of Exercise 5.5. Determine the separate contributions to the overall error from n_x, n_y, and $var\left(\mathbf{x}(0)\right)$.

Chapter 7

Navigation System Design

Chapter 1 used the rather simple example of point motion along a line to motivate study of the various topics that have been presented in the previous chapters. That chapter also raises various questions and topics of study, most of which could not be considered at that point without knowledge of the intervening material. This chapter revisits that simple example with discussion of how the material in Chapters 2–6 is useful in the analysis and study of aided navigation applications. The main objective of this chapter is to summarize the methodology outlined in Section 1.1.2 in the context of the theoretical tools discussed in Part I of this text. This methodology and set of theoretical tools forms the basis for the more realistic applications that are considered in Part II.

7.1 Methodology Summary

The approach summary is as follows. We derive a kinematic model

$$\begin{aligned} \dot{\mathbf{x}} &= \mathbf{f}(\mathbf{x}, \mathbf{u}) \\ \mathbf{y} &= \mathbf{h}(\mathbf{x}) \end{aligned}$$

and have knowledge of $E\langle \mathbf{x}(t_{k-1}) \rangle$ and $var(\mathbf{x}(t_{k-1}))$. We also have measurements

$$\begin{aligned} \tilde{\mathbf{u}} &= \mathbf{u} + \mathbf{n}_u \\ \tilde{\mathbf{y}} &= \mathbf{y} + \mathbf{n}_y, \end{aligned}$$

where $\tilde{\mathbf{u}}$ is available at the rate f_1 and $\tilde{\mathbf{y}}$ is available at the rate f_2 with $f_1 \gg f_2$. The state estimate is produced by integrating

$$\dot{\hat{\mathbf{x}}} = \mathbf{f}(\hat{\mathbf{x}}, \tilde{\mathbf{u}})$$

over $t \in (t_{k-1}, t_k]$ with the initial condition $\hat{\mathbf{x}}(t_{k-1}) = E\langle \mathbf{x}(t_{k-1})\rangle$. This integration is performed numerically with a time increment $dt = \frac{1}{f_1}$. The symbol t_k represents the next time at which the measurement of $\mathbf{y}$ is available. We assume that $Nf_2 = f_1$ so that after N integration time steps $t = t_k^-$.

At time t_k, the extended Kalman filter is used to estimate the error state vector using the information in the residual measurement $\mathbf{z}_k = \tilde{\mathbf{y}}_k - \hat{\mathbf{y}}_k$ where $\hat{\mathbf{y}}_k = \mathbf{h}(\hat{\mathbf{x}}(t_k))$. The EKF approach requires the initial error covariance $var(\delta\mathbf{x}(t_{k-1})) = \mathbf{P}_{k-1} = var(\mathbf{x}(t_{k-1}))$ and the linearized dynamics of the error state

$$\delta\mathbf{x}_k = \boldsymbol{\Phi}_{k-1}\delta\mathbf{x}_{k-1} + \boldsymbol{\omega}_{k-1} \tag{7.1}$$

$$\mathbf{z}_k = \mathbf{H}_k\delta\mathbf{x}_k + \boldsymbol{\nu}_k. \tag{7.2}$$

The EKF algorithm is summarized in Table 5.8 on p. 209.

As discussed in Section 5.10.5.3, the EKF is formulated using a complementary filter framework. The main assumptions of the Kalman filter derivation were that the system model was linear, driven by Gaussian white noise processes, with measurements corrupted by Gaussian white noise processes. The complementary filter approach ensures that these main assumptions are satisfied because the Kalman filter is designed to estimate the error state vector. The Kalman filter derivation, properties, implementation issues, and examples are presented in Chapter 5.

High performance navigation systems require accurate deterministic and stochastic modeling. Deterministic modeling requires careful analysis of the relationships between quantities represented in distinct reference frames to determine the proper system model as represented by the functions $\mathbf{f}$ and $\mathbf{h}$. There are a few issues that motivate the need for stochastic modeling. First, the initial conditions are not exactly known, but are conveniently represented by a mean value and an error covariance matrix. Second, the sensors for $\mathbf{u}$ and $\mathbf{y}$ are not perfect, but are affected by various corrupting factors. The imperfections in sensing $\mathbf{u}$ may be addressed by additive white process noise or by state augmentation. The imperfections in sensing $\mathbf{y}$ may be modeled by white measurement noise or by state augmentation (with additional process noise contributions). Through linearization and the state augmentation process, the designer is able to construct a dynamic error model that is linear with white Gaussian measurement and process noise vectors. Reference frames, deterministic modeling, and stochastic modeling are discussed in Chapters 2, 3, and 4, respectively.

The expected system performance and the performance under various design configurations can be analyzed by the methods in Chapter 6.

7.2 Methodology: Detailed Example

Sections 1.1.2 and 7.1 outlined the aided navigation system design and analysis methodology. This section contains a short subsection that performs each step indicated in Sections 1.1.2 and 7.1 using the tools from Chapters 2–6. The intent is to provide a rather complete presentation of the methodology, subject to design assumptions, for a very simple application. This is the same methodology and presentation format that will be used in each of the aided navigation applications considered in Part II.

The navigation problem that is of interest in this subsection is stated as follows. A point is free to translate along a line with position p, velocity v, and acceleration a. Two navigation sensors are available: an accelerometer and a position sensor.

The accelerometer is sampled at the high rate of $f_1 = 100Hz$ and is accurately modeled as

$$u = a - \alpha a - b_u - \eta_1 \tag{7.3}$$

where η_1 is a Gaussian, white noise process with PSD equal to $\sigma_1^2 = \left(10^{-3}\right)^2 \frac{\left(m/s^2\right)^2}{Hz}$. The symbol α represents a constant scale factor error with $\alpha(0) \sim N(0, (0.01)^2)$. The accelerometer output signal u contains a bias b_u that is modeled as a constant plus random walk process

$$\dot{b}_u = \omega_1 \tag{7.4}$$

where $E\langle b_u(0)\rangle = \mu_{b_u}$, $var(b_u(0)) = P_{b_u}(0) = 0.01^2$. The process noise ω_1 is white, Gaussian, and uncorrelated with p, v, a, and η_1. The PSD of ω_1 is $\sigma_{b_u}^2 = (0.01)^2 \frac{\left(m/s^3\right)^2}{Hz}$.

The aiding sensor provides information about the position. At discrete-time instants, with a sampled rate of $f_2 = 1.0$ Hz, the sensor output is accurately modeled as

$$y_k = p_k + b_y(k) + \eta_2(k). \tag{7.5}$$

The discrete-time noise samples $\eta_2(k)$ are zero mean and Gaussian with variance $\sigma_2^2 = (1.0m)^2$. The sensor y contains a bias b_y that is modeled as

$$\dot{b}_y = -\lambda_y b_y + \omega_2 \tag{7.6}$$

where $\lambda_y \approx \frac{1}{60}$, $E\langle b_y(0)\rangle = \mu_{b_y}$, $var(b_y(0)) = P_{b_y}(0) = 3^2\ m^2$ and ω_2 is white, Gaussian and uncorrelated with p, v, a, η_1, η_2, and ω_1. The PSD of ω_2 is $\sigma_{b_y}^2 = \left(\frac{3}{\sqrt{30}}\right)^2 \frac{(m/s)^2}{Hz}$.

Assuming that the above information is accurate, the objective is to design a system that accurately estimates p, v, and a at the frequency

f_1. The navigation system should maintain the position accuracy to better than 5.0 m while position aiding information is available and maintain an accuracy of better than 10 m for at least 50 seconds after loss of the position sensor information.

7.2.1 Augmented Kinematic Model

The system kinematics are described as

$$\dot{p} = v \tag{7.7}$$

$$\dot{v} = a. \tag{7.8}$$

To these equations, we must augment the accelerometer and position sensor calibration parameters. The augmented system kinematics have the form

$$\dot{\mathbf{x}} = \mathbf{f}(\mathbf{x}, a, \boldsymbol{\omega}) \tag{7.9}$$

where $\mathbf{x} = [p, v, b_u, b_y, \alpha]^\top$, $\boldsymbol{\omega} = [\omega_1, \omega_2, \eta_1]^\top$ and

$$\mathbf{f}(\mathbf{x}, a, \boldsymbol{\omega}) = \begin{bmatrix} v \\ a \\ \omega_1 \\ -\lambda_y b_y + \omega_2 \\ 0 \end{bmatrix} = \begin{bmatrix} x_2 \\ a \\ \omega_1 \\ -\lambda_y x_4 + \omega_2 \\ 0 \end{bmatrix}. \tag{7.10}$$

The initial value of the augmented state is

$$\mathbf{x}_0 = [E\langle p(0)\rangle, E\langle v(0)\rangle, \mu_{b_u}, \mu_{b_y}, 0]^\top$$

with covariance

$$\mathbf{P}_{\mathbf{x}}(0) = diag\left([\,P_p(0), P_v(0), (0.01)^2, 3^2, (0.01)^2\,]^\top\right).$$

7.2.2 Navigation Mechanization Equations

The navigation mechanization equations are

$$\dot{\hat{\mathbf{x}}} = \mathbf{f}(\hat{\mathbf{x}}, \hat{a}, \mathbf{0}) \tag{7.11}$$

where $\hat{\mathbf{x}} = [\hat{p}, \hat{v}, \hat{b}_u, \hat{b}_y, \hat{\alpha}]^\top$, acceleration is estimated as $\hat{a} = \left(\frac{u+\hat{b}_u}{1-\hat{\alpha}}\right)$, and

$$\mathbf{f}(\hat{\mathbf{x}}, \hat{a}, \mathbf{0}) = \begin{bmatrix} \hat{v} \\ \hat{a} \\ 0 \\ -\lambda_y \hat{b}_y \\ 0 \end{bmatrix} = \begin{bmatrix} \hat{x}_2 \\ \left(\frac{u+\hat{x}_3}{1-\hat{x}_5}\right) \\ 0 \\ -\lambda_y \hat{x}_4 \\ 0 \end{bmatrix}. \tag{7.12}$$

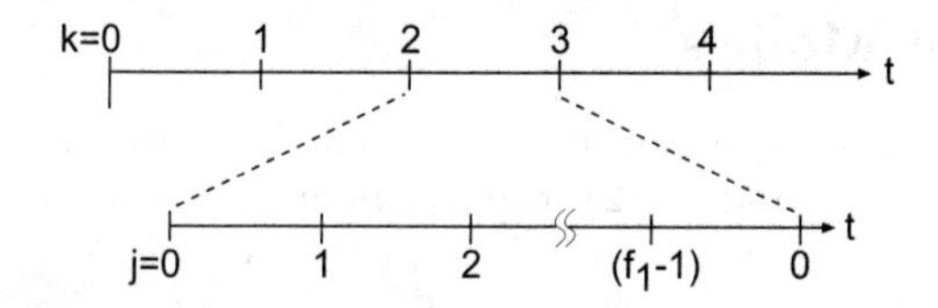

Figure 7.1: Timeline illustration of the relationship between the fast time-index j used for state integration and the slow time-index k used for Kalman filter position aiding for the case that $f_2 = 1Hz$.

In the general mathematical formulation given above, the mechanization equations may not look quite as familiar as they do when written out in the component-wise discrete-time equivalent form

$$\left.\begin{aligned} \hat{p}_{j+1} &= \hat{p}_j + \hat{v}_j dt + \tfrac{1}{2}\hat{a}_j dt^2 \\ \hat{v}_{j+1} &= \hat{v}_j + \hat{a}_j dt \\ \hat{b}_{u_{j+1}} &= \hat{b}_{u_j} \\ \hat{b}_{y_{j+1}} &= \beta \hat{b}_{y_j} \\ \hat{\alpha}_{j+1} &= \hat{\alpha}_j \end{aligned}\right\} \tag{7.13}$$

where $\beta = exp(-\lambda_y dt)$ and $dt = \frac{1}{f_1}$. At least the first two equations should look very familiar from elementary physics. These equations are iterated at the f_1 rate. The (fast) time index $j = [0, f_1 - 1]$ is used instead of the (slow) time index $k = 0, 1, 2, \ldots$ to make clear the distinction. The relation between the fast and slow time indices is illustrated in Figure 7.1. At the time of each position measurement update: prior to the update j should be $(f_1 - 1)$, after the measurement update j is reset to zero, and the result of the measurement update serves as the initial condition (i.e., at $j = 0$) for the next interval of integration.

7.2.3 Sensor Models

For this sample problem, all the sensor models are stated in the problem statement. Given the estimated state $\hat{\mathbf{x}}$, the aiding sensor output is predicted at discrete-time instants as

$$\hat{y}_k = \hat{p}_k + \hat{b}_y(k) \tag{7.14}$$

where $T = 1.0$ since the aiding measurements are obtained at a 1.0 Hz sample rate.

7.2.4 Error Models

Defining the error state vector as $\delta \mathbf{x} = \mathbf{x} - \hat{\mathbf{x}}$, by linearizing the difference between eqns. (7.10) and (7.12), the dynamic equation for the state error vector is

$$\delta \dot{\mathbf{x}} = \begin{bmatrix} 0 & 1 & 0 & 0 & 0 \\ 0 & 0 & 1 & 0 & u \\ 0 & 0 & 0 & 0 & 0 \\ 0 & 0 & 0 & -\lambda_y & 0 \\ 0 & 0 & 0 & 0 & 0 \end{bmatrix} \delta \mathbf{x} + \begin{bmatrix} 0 & 0 & 0 \\ 0 & 0 & 1 \\ 1 & 0 & 0 \\ 0 & 1 & 0 \\ 0 & 0 & 0 \end{bmatrix} \boldsymbol{\omega}. \tag{7.15}$$

In the derivation of this model, in the standard form of eqn. (4.57), the elements F_{23}, F_{25}, and G_{23} warrant additional discussion. These terms are a linearization of

$$a - \left(\frac{u + \hat{x}_3}{1 - \hat{x}_5} \right)$$

evaluated for $\mathbf{x}$ near $\hat{\mathbf{x}}$. Using the fact that for $|\delta x| \ll 1$, $\frac{1}{1-\delta x} \approx (1 + \delta x)$ the linearization proceeds as follows

$$\begin{aligned} a - \left(\frac{u + \hat{x}_3}{1 - \hat{x}_5} \right) &\approx a - (u + \hat{x}_5 u + \hat{x}_3 + \hat{x}_3 \hat{x}_5) \\ &= a - (a - x_5 a - x_3 - \eta_1 + \hat{x}_5 u + \hat{x}_3 + \hat{x}_3 \hat{x}_5) \\ &\approx \delta x_3 + u \delta x_5 + \eta_1 \end{aligned} \tag{7.16}$$

where $u\delta x_5 = x_5 a - \hat{x}_3 \hat{x}_5 - \hat{x}_5 u$ to first order. Eqn. (7.16) results in the second row of the $\mathbf{F}$ and $\mathbf{G}$ matrices.

The aiding output error model is the linearization of the difference between eqns. (7.5) and (7.14). This linearization yields

$$\begin{aligned} \delta y_k &= \delta x_1(k) + \delta x_4(k) + \eta_2(k) \\ &= \mathbf{H} \delta \mathbf{x}_k + \eta_2(k) \end{aligned} \tag{7.17}$$

where $\mathbf{H} = [1, 0, 0, 1, 0]$.

7.2.5 State Estimator Design

Each subsection that follows describes one aspect of the estimator design.

7.2.5.1 Observability Analysis

To initiate the estimator design, we perform an analysis of eqns. (7.15) and (7.17) to determine the conditions under which the error state is observable.

Using the results of Section 3.6.1, the observability matrix is

$$\mathcal{O} = \begin{bmatrix} 1 & 0 & 0 & 1 & 0 \\ 0 & 1 & 0 & -\lambda_y & 0 \\ 0 & 0 & 1 & \lambda_y^2 & u \\ 0 & 0 & 0 & -\lambda_y^3 & 0 \\ 0 & 0 & 0 & \lambda_y^4 & 0 \end{bmatrix}.$$

The first four states are observable because $\lambda_y > 0$. The accelerometer scale factor error α is only observable over time intervals during which $u(t) \neq 0$. Because the application is expected to satisfy this condition frequently, we expect to be able to estimate the error state vector. The accuracy of the estimate cannot be assessed without further analysis. Kalman filter design and subsequent performance analysis will require the discrete-time equivalent model.

Note that formally this is an analysis of the observability of the *error* state $\delta\mathbf{x}$, not the state $\mathbf{x}$ or the estimated state $\hat{\mathbf{x}}$. In fact, $\hat{\mathbf{x}}$ is computed by the navigation system and is known; however, the fact that $\hat{\mathbf{x}}$ is known does not mean that it is an accurate representation of $\mathbf{x}$. Because $\delta\mathbf{x} = \mathbf{x} - \hat{\mathbf{x}}$, the observability of $\delta\mathbf{x}$ implies that we can maintain the accuracy of $\hat{\mathbf{x}}$ as we can estimate $\delta\mathbf{x}$ and then correct $\hat{\mathbf{x}}$ using the assignment $\hat{\mathbf{x}} + \delta\mathbf{x}$. When $\delta\mathbf{x}$ is not observable it is possible for the unobservable portion of $\delta\mathbf{x}$ to become large, which implies that the corresponding 'unobservable' portion of $\hat{\mathbf{x}}$ is not accurate.

7.2.5.2 Computation of $\mathbf{\Phi}$ and Qd

Section 7.2.2 described the integration of the navigation state estimate with the time increment $dt = \frac{1}{f_1}$ determined by the instrument that measures u. As the navigation state is integrated, error will develop between the computed state and the actual state, due for example to imperfections in the measurement of u. It is the job of the Kalman filter to use the aiding measurements at the rate f_2 to estimate and characterize the accuracy of the state estimation error.

The Kalman filter implementation will require the matrices $\mathbf{\Phi}_k$ and $\mathbf{Qd}_k$ for the linear discrete-time model

$$\delta\mathbf{x}_{k+1} = \mathbf{\Phi}_k \delta\mathbf{x}_k + \mathbf{w}_k \tag{7.18}$$

that is equivalent to eqn. (7.15) at the sampling instants $t_k = kT$ where $T = \frac{1}{f_2}$. The error model matrix parameters $\mathbf{\Phi}_k$ and $\mathbf{Qd}_k$ are used by the Kalman filter to propagate the error covariance matrix from t_k to t_{k+1}. When the matrix $\mathbf{F}$ is not constant, as in this example, it may be necessary to subdivide the time interval $t \in [t_k, t_{k+1}]$ into smaller subintervals of length τ such that it is reasonable to consider $\mathbf{F}$ as constant on each interval of length τ where $N\tau = T$. On each interval $[t_k, t_{k+1}]$, define $N > 1$ and select $(N-1)$ uniformly space time instants satisfying

$$t_{k-1} = \tau_0 < \tau_1 < \ldots < \tau_N = t_k.$$

By the method of Section 4.7.2.2, the discrete-time state transition matrix for an interval $t \in [\tau_{i-1}, \tau_i]$ of length $\tau = (\tau_i - \tau_{i-1})$ is

$$\mathbf{\Phi}(\tau_i, \tau_{i-1}) = \begin{bmatrix} 1 & \tau & \frac{1}{2}\tau^2 & 0 & \frac{u}{2}\tau^2 \\ 0 & 1 & \tau & 0 & u\tau \\ 0 & 0 & 1 & 0 & 0 \\ 0 & 0 & 0 & e^{-\lambda_y \tau} & 0 \\ 0 & 0 & 0 & 0 & 1 \end{bmatrix} \tag{7.19}$$

where τ should be selected small enough so that $u(t)$ can be considered constant over each interval of length τ. Over each one second interval, using the results of Section 3.5.3, the $\mathbf{\Phi}$ matrix is accumulated as

$$\mathbf{\Phi}(\tau_{i+1}, \tau_0) = \mathbf{\Phi}(\tau_{i+1}, \tau_i)\mathbf{\Phi}(\tau_i, \tau_0) \tag{7.20}$$

with $\mathbf{\Phi}_{k-1} = \mathbf{\Phi}(\tau_N, \tau_0)$ and $\mathbf{\Phi}(\tau_0, \tau_0) = \mathbf{I}$.

The $\mathbf{Qd}$ matrix can be computed by various methods, as discussed in Section 4.7.2. Due to $\mathbf{F}$ being a function of the time varying signal $u(t)$, we choose to compute $\mathbf{Qd}$ by numeric integration of eqn. (4.110). Over each one second interval, the $\mathbf{Qd}_{k-1}$ matrix is accumulated according to

$$\mathbf{Q}(\tau_i, \tau_0) = \mathbf{\Phi}(\tau_i, \tau_{i-1})\mathbf{Q}(\tau_{i-1}, \tau_0)\mathbf{\Phi}^\top(\tau_i, \tau_{i-1}) + \mathbf{G}\mathbf{Q}_{\boldsymbol{\omega}}\mathbf{G}^\top\tau \tag{7.21}$$

for $i = 1, \ldots, N$ with

$$\mathbf{Qd}_{k-1} = \mathbf{Q}(\tau_N, \tau_0)$$

and $\mathbf{Q}(\tau_0, \tau_0) = \mathbf{0}$, where

$$\mathbf{Q}_{\boldsymbol{\omega}} = diag\left(\left[\sigma_{b_u}^2, \sigma_{b_y}^2, \sigma_1^2\right]\right)$$

is the PSD matrix for the continuous-time process noise vector $\boldsymbol{\omega}$.

Eqn. (7.21) is not obvious. It is derived as follows:

$$\begin{aligned} \mathbf{Q}(\tau_i, \tau_0) &= \int_{\tau_0}^{\tau_i} \mathbf{\Phi}(\tau_i, s)\mathbf{G}(s)\mathbf{Q}(s)\mathbf{G}^\top(s)\mathbf{\Phi}^\top(\tau_i, s)ds \\ &= \sum_{j=0}^{i-1} \int_{\tau_j}^{\tau_{j+1}} \mathbf{\Phi}(\tau_i, s)\mathbf{G}\mathbf{Q}\mathbf{G}^\top\mathbf{\Phi}^\top(\tau_i, s)ds. \end{aligned}$$

At this point, we note that $\mathbf{\Phi}(\tau_i, s) = \mathbf{\Phi}(\tau_i, \tau_{j+1})\mathbf{\Phi}(\tau_{j+1}, s)$ and to simplify notation we define the matrix $\mathbf{A}_j = \int_{\tau_j}^{\tau_{j+1}} \mathbf{\Phi}(\tau_{j+1}, s)\mathbf{GQG}^\top\mathbf{\Phi}^\top(\tau_{j+1}, s)ds$. With this definition,

$$\begin{aligned}
\mathbf{Q}(\tau_1, \tau_0) &= \mathbf{A}_0 \\
\mathbf{Q}(\tau_2, \tau_0) &= \mathbf{\Phi}(\tau_2, \tau_1)\mathbf{A}_0\mathbf{\Phi}^\top(\tau_2, \tau_1) + \mathbf{A}_1 \\
\mathbf{Q}(\tau_3, \tau_0) &= \mathbf{\Phi}(\tau_3, \tau_1)\mathbf{A}_0\mathbf{\Phi}^\top(\tau_3, \tau_1) + \mathbf{\Phi}(\tau_3, \tau_2)\mathbf{A}_1\mathbf{\Phi}^\top(\tau_3, \tau_2) + \mathbf{A}_2 \\
&= \mathbf{\Phi}(\tau_3, \tau_2)\left(\mathbf{\Phi}(\tau_2, \tau_1)\mathbf{A}_0\mathbf{\Phi}^\top(\tau_2, \tau_1) + \mathbf{A}_1\right)\mathbf{\Phi}^\top(\tau_3, \tau_2) + \mathbf{A}_2 \\
&= \mathbf{\Phi}(\tau_3, \tau_2)\mathbf{Q}(\tau_2, \tau_0)\mathbf{\Phi}^\top(\tau_3, \tau_2) + \mathbf{A}_2.
\end{aligned}$$

For the i-th iteration the corresponding formula is

$$\mathbf{Q}(\tau_i, \tau_0) = \mathbf{\Phi}(\tau_i, \tau_{i-1})\mathbf{Q}(\tau_{i-1}, \tau_0)\mathbf{\Phi}^\top(\tau_i, \tau_{i-1}) + \mathbf{A}_{i-1}.$$

Eqn. (7.21) results when $\mathbf{A}_j$ is approximated to first order in τ as

$$\mathbf{A}_j = \mathbf{G}_j\mathbf{Q}_j\mathbf{G}_j^\top\tau$$

where $\mathbf{G}_j$ and $\mathbf{Q}_j$ are the constant values of $\mathbf{G}(t)$ and $\mathbf{Q}(t)$ for $t \in [\tau_j, \tau_{j+1}]$.

The above discussion has presented one means of accumulating $\mathbf{\Phi}$ and $\mathbf{Qd}$ over a given time period to allow propagation of $\mathbf{P}_k$ at the f_2 rate. In the above approach, N is a design parameter that allows a tradeoff between computational load and performance. The accumulation of $\mathbf{Qd}$ in eqn. (7.21) has the same structure as the time propagation of $\mathbf{P}_k$. One possibility is to propagate the error covariance matrix $\mathbf{P}_k$ at the rate f_1, which is equivalent to choosing $N = f_1$, and replaces the accumulation of $\mathbf{Qd}$ with the high rate computation of $\mathbf{P}_k$.

7.2.5.3 Time Propagation

This subsection formulates the equations for propagation of the state estimate and error covariance matrix between position corrections. The parenthesis notation '$x(t)$' is used to designate variables that are evolving in continuous time. The subscript notation 'x_k' is used to denote variables at discrete instants of time.

The EKF is implemented as follows, starting with $\hat{\mathbf{x}}_0^+ = \mathbf{x}_0$ and $\mathbf{P}_0^+ = \mathbf{P}_\mathbf{x}(0)$. Assuming that the first position measurement occurs at $t = kT = 1s$, for $k \geq 1$ the state estimate is updated by integration of eqn. (7.12) for $t \in [(k-1)T, kT]$ from the initial condition $\hat{\mathbf{x}}((k-1)T) = \hat{\mathbf{x}}_{k-1}^+$ to produce $\hat{\mathbf{x}}_k^- = \hat{\mathbf{x}}(kT)$. One implementation of the numeric integration is given in eqn. (7.13).

Given $\mathbf{\Phi}_{k-1}$ and $\mathbf{Qd}_{k-1}$, the covariance time propagation can be performed as derived in eqn. (5.49)

$$\mathbf{P}_k^- = \mathbf{\Phi}_{k-1}\mathbf{P}_{k-1}^+\mathbf{\Phi}_{k-1}^\top + \mathbf{Qd}_{k-1}. \tag{7.22}$$

7.2.5.4 EKF Measurement Correction

This section is concerned with correcting the state estimate and error covariance matrix $\mathbf{P}$ for any new information that may be available from the position measurement at the discrete-time instants $t = kT = t_k$. The corrected information serves as the initial conditions for the next period of time integration.

At any time $t = kT$, for which a position measurement is available, as summarized in Table 5.5, the Kalman gain vector is defined as

$$\mathbf{K}_k = \mathbf{P}_k^- \mathbf{H}^\top \left(\mathbf{H}\mathbf{P}_k^- \mathbf{H}^\top + R\right)^{-1}$$

where $R = \sigma_2^2$. Given the Kalman gain vector, the measurement correction for the state and covariance are computed as

$$\begin{aligned}\hat{\mathbf{x}}_k^+ &= \hat{\mathbf{x}}_k^- + \mathbf{K}_k \left(y_k - \hat{y}_k^-\right) \\ \mathbf{P}_k^+ &= [\mathbf{I} - \mathbf{K}_k \mathbf{H}_k]\, \mathbf{P}_k^- .\end{aligned}$$

where $\hat{y}_k^- = \mathbf{H}\hat{\mathbf{x}}_k^-$. At this point, both the state estimate and the error covariance matrix are appropriately defined to serve as the initial conditions for the next period of integration, as defined in Section 7.2.5.3.

At any time $t = kT$, if for some reason, no position measurement is available, then it is straightforward to define

$$\begin{aligned}\hat{\mathbf{x}}_k^+ &= \hat{\mathbf{x}}_k^- \\ \mathbf{P}_k^+ &= \mathbf{P}_k^- .\end{aligned}$$

These equations correctly account for the fact that no new information is available. Neither the state estimate nor the error covariance matrix are changed, but the initial conditions for the next period of integration, as defined in Section 7.2.5.3, are appropriately defined.

7.2.6 Covariance Analysis

This section uses the covariance analysis methods of Chapter 6 to analyze the performance of the system relative to the specification:

- $|\delta p| < 5.0m$ with position aiding;
- $|\delta p| < 10.0m$ for $50s$ following the loss of position aiding.

Prior to starting any numeric analysis, it is useful to have some physical intuition into the required performance, to use as a reasonableness metric for the analysis results. Based on the above specifications it is reasonable to expect that velocity accuracy on the order of $0.1\frac{m}{s}$ will be required to meet the second portion of the spec.

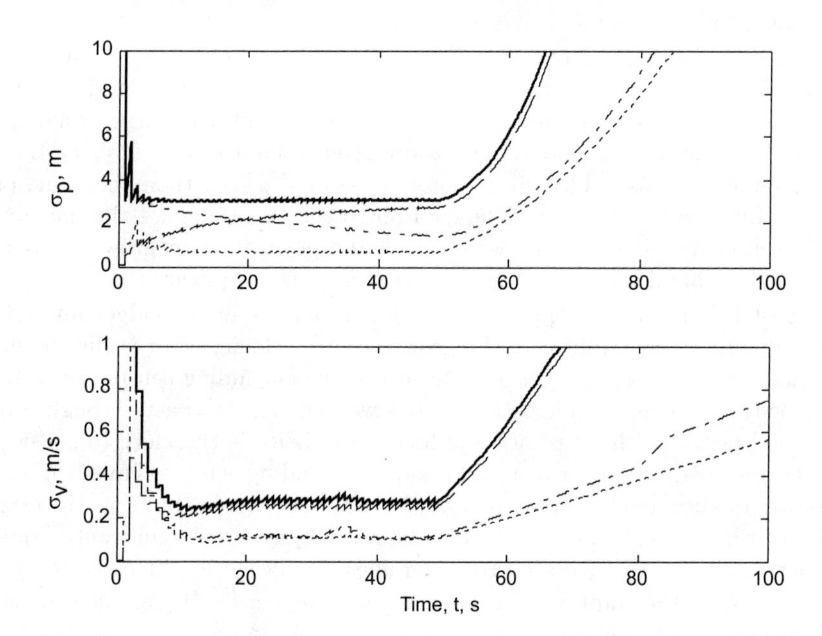

Figure 7.2: Covariance analysis results for the system of Section 7.1. The top set of graphs show results for the standard deviation of position σ_p. The bottom set of graphs show results for the standard deviation of velocity σ_v. The wide solid line is the total error due to initial conditions, $\mathbf{Q}$, and R. The dotted line is the error due to the driving noise terms represented by $\mathbf{Q}$. The dashed line is the error due to the measurement noise represented by R. The dashed-dotted line is the error due to initial conditions represented by $\mathbf{P}_0$.

A straightforward covariance analysis for the system described above with the variance and PSD parameters as summarized in Table 7.1 shows that the first specification is safely achieved, but that the system fails to meet the second specification. The result of that analysis is the wide solid curve in each of the graphs of Figures 7.2 and 7.3. Position aiding is available for $t \in [0, 50]s$, but not for $t > 50s$. At approximately 65 seconds, the second specification is violated.

The analysis requires definition of a specific acceleration profile because the system matrix $\mathbf{F}$ depends on the acceleration. For the analysis herein, it was assumed that the system accelerated forward at $3\frac{m}{s^2}$ for $t \in [30, 35]s$ and backward at $3\frac{m}{s^2}$ for $t \in [80, 85]s$. Such analysis should be repeated for a variety of acceleration profiles to ensure that the level of performance indicated is not particular to the assumed profile. The effect of the scale factor error is slightly evident in the plot of the standard deviation of the velocity for $t \in [30, 35]s$.

Due to the failure to meet the specification, an error budget analysis (see Section 6.3) is performed to determine whether the problem is fixable. The result of the error budget analysis is shown by the three remaining curves in Figures 7.2 and 7.3. If at any given time, the values of the three narrow curves are added in a root-sum-squared sense, then the value on the wide curve will result. The caption to the figure defines the meaning of each curve. For $t > 50s$ we see that the position error grows approximately parabolically due to the velocity error that appears to be growing almost linearly with slope near $\frac{0.05m/s}{s}$. From the error budget analysis, the dominant contributor to the growth of the velocity error is the the position sensor noise variance R. Given the current aiding sensor suite, the velocity and bias cannot be calibrated well enough to coast through a 50 second period without position aiding. Our choice is therefore to consider a better position sensor or to add additional aiding sensors. Investing in a better quality accelerometer is not a high priority at this point as the error budget analysis shows that the driving noise $\mathbf{Q}$ makes a significantly lower contribution to the overall error than does the measurement noise R.

Consider the addition of a velocity sensor with a 1.0 Hz sample rate and a measurement modeled as

$$y_v = v + \nu_3$$

where ν_3 is Gaussian white noise with standard deviation $\sigma_3 = 0.6\frac{m}{s}$. For convenience, of presentation, the position and velocity measurements are assumed to both occur simultaneously. The velocity sensor is completely independent of the position sensor and will be available even when the position measurement is unavailable. Therefore, for $t \in [0, 50]s$ at 1.0 Hz we can now make both position and velocity corrections. In this case,

$$\mathbf{H} = \begin{bmatrix} 1 & 0 & 0 & 1 & 0 \\ 0 & 1 & 0 & 0 & 0 \end{bmatrix} \text{ and } \mathbf{R} = \begin{bmatrix} 1.00 & 0.00 \\ 0.00 & 0.36 \end{bmatrix}.$$

For $t > 50s$ at 1.0 Hz we will only make velocity corrections. In this case,

$$\mathbf{H} = \begin{bmatrix} 0 & 1 & 0 & 0 & 0 \end{bmatrix} \text{ and } R = 0.36.$$

σ_1	σ_2	σ_{b_u}	σ_{b_u}	P_p	P_v	P_{b_u}	P_{b_y}	P_α
0.001	1	0.01	$\frac{3}{\sqrt{30}}$	100	25	0.01^2	3^2	0.01^2

Table 7.1: Summary of stochastic error model parameters and initial conditions.

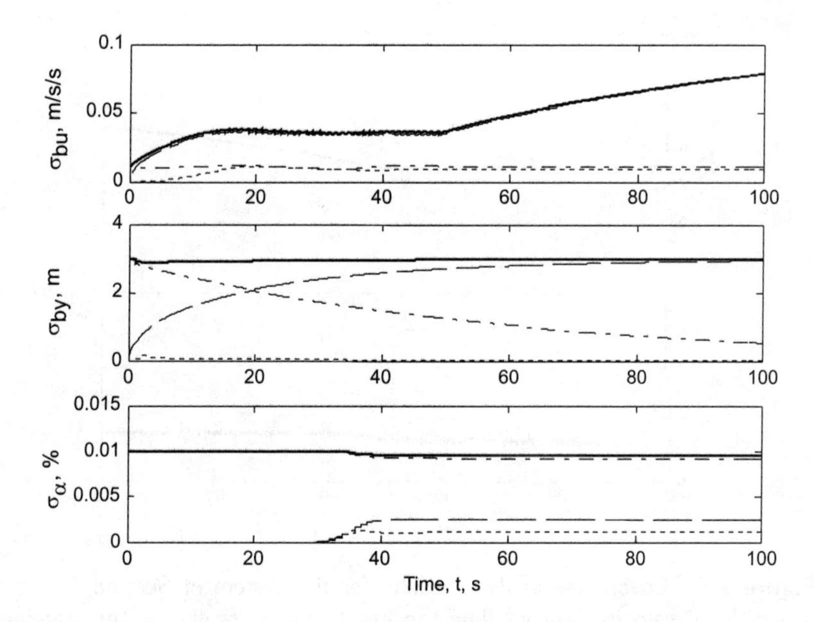

Figure 7.3: Covariance analysis results for the system of Section 7.1. The top set of graphs show results for the standard deviation of accelerometer bias, σ_{b_u}. The middle set of graphs show results for the standard deviation of position sensor time correlated error, σ_{b_y}. The bottom set of graphs show results for the standard deviation of scale factor, σ_{α}. The wide solid line is the total error due to initial conditions, $\mathbf{Q}$, and R. The dotted line is the error due only to the driving noise terms represented by $\mathbf{Q}$. The dashed line is the error due only to the measurement noise represented by R. The dashed-dotted line is the error due only to initial conditions represented by $\mathbf{P}_0$.

The covariance analysis results for this sensor suite are shown in Figures 7.4 and 7.5. With the additional velocity sensor, the position and velocity estimation accuracy is changed only slightly for $t \in [0, 50]$; however, for $t > 50$ the improvement is clear. The velocity is bounded at approximately $30cm/s$ and position error is growing slow enough that the second specification is now safely achieved. Additional analysis of this example is considered in Exercise 7.3.

7.3 Complementary Filtering

The design approach of Section 7.1 is a form of complementary filtering. The basic idea of and motivation for complementary filtering is discussed

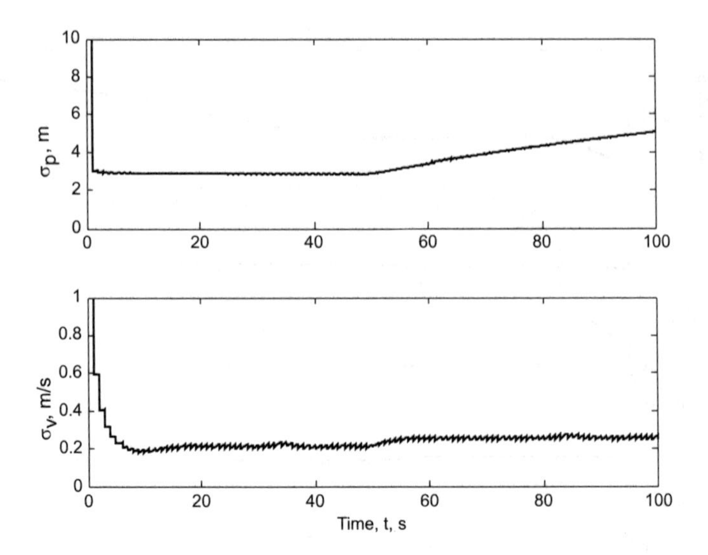

Figure 7.4: Covariance analysis results for the system of Section 7.1 with an additional velocity sensor. The top graph shows results for the standard deviation of position σ_p. The bottom graph shows results for the standard deviation of velocity σ_v.

in this section.

7.3.1 Frequency Domain Approach

The designer has two sensors available

$$\begin{aligned} y_1(t) &= s(t) + n_1(t) \\ y_2(t) &= s(t) + n_2(t) \end{aligned}$$

where the measurement noise terms n_1 and n_2 are stationary random processes with known power spectral densities. The signal portion $s(t)$ may be deterministic or nonstationary; however, a model for $s(t)$ is not available.

A general approach to the problem is illustrated in Figure 7.6. The output of the filtering scheme of Figure 7.6 is [1]

$$\hat{s}(t) = G_1(s)[y_1(t)] + G_2(s)[y_2(t)]. \tag{7.23}$$

The objective would be to design the filters $G_1(s)$ and $G_2(s)$ to optimally estimate the signal $s(t)$ from the noisy measurements $y_1(t)$ and $y_2(t)$. In

[1]The notation $z(t) = G_1(s)[y_1(t)]$ should be read as $z(t)$ is the signal output by the filter with transfer function $G_1(s)$ given the input $y_1(t)$.

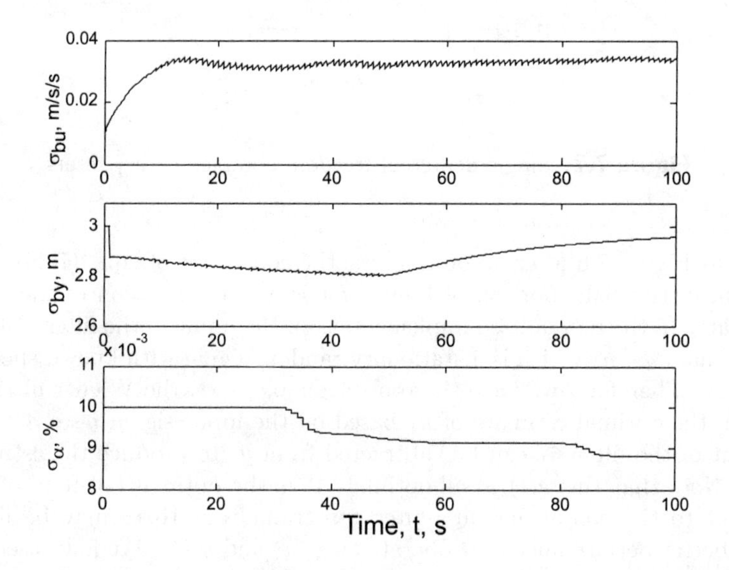

Figure 7.5: Covariance analysis results for the system of Section 7.1 with an additional velocity sensor. The top graph shows results for the standard deviation of accelerometer bias, σ_{b_u}. The middle graph shows results for the standard deviation of position sensor time correlated error, σ_{b_y}. The bottom graph shows results for the standard deviation of scale factor, σ_α.

the frequency domain, if the spectral characteristics of $s(t)$ were known and stationary, this would correspond to the Wiener filter problem [31].

In the frequency domain, eqn. (7.23) is equivalent to

$$\hat{S}(s) = (G_1(s) + G_2(s))\, S(s) + G_1(s)N_1(s) + G_2(s)N_2(s). \tag{7.24}$$

Therefore, to pass the signal $s(t)$ without distortion, we may wish to impose the constraint that

$$G_1(s) + G_2(s) = 1, \text{ for all } s. \tag{7.25}$$

This results in the complementary filter approach to the problem as illustrated in Figure 7.7. The system in the left portion of Figure 7.7 is the

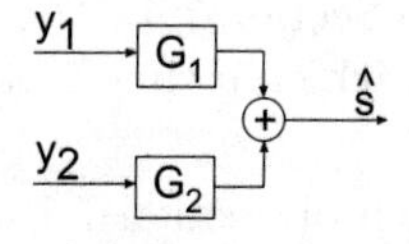

Figure 7.6: Two degree of freedom filtering application.

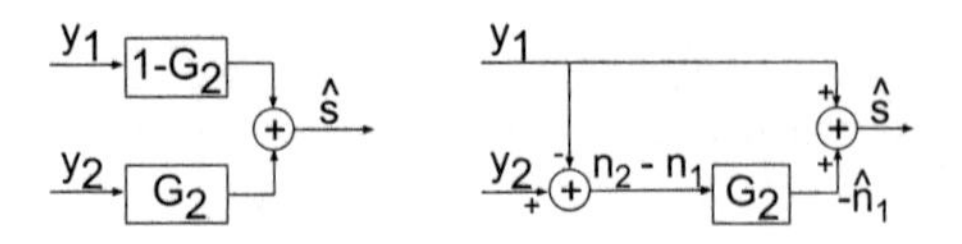

Figure 7.7: Single degree of freedom complementary filters.

same as Figure 7.6 after accounting for the constraint of eqn. (7.25). The system in the right portion of Figure 7.7 is a reorganization of the block diagram. In the reorganized implementation, the input to the filter $G_2(s)$ is the signal $n_2 - n_1$ which is a stationary random signal with known spectral content. Therefore, we have the tools available (e.g., the Wiener filter) to design the optimal estimate of n_1 based on the input signal $n_2 - n_1$. The output of the filter $\hat{n}_1$ can be subtracted from y_1 to produce the estimate $\hat{s}(t)$. Note that this $\hat{s}(t)$ is suboptimal. It is the optimal estimate of $s(t)$ subject to the complementary filter constraint, but there may be filters with better performance that operate on $y_1(t)$ and $y_2(t)$. We have used the complementary filter constraint to convert the original problem into a new problem for which our available design and analysis methods are appropriate. The Wiener filter design approach was not applicable to the original problem because no stochastic model was available for the signal s.

The complementary filter approach is especially useful in applications where the sensors have complementary spectral characteristics. For example, if n_1 has its spectral content predominantly in the low frequency range while n_2 has its spectral content predominantly in the high frequency range, then the filter $G_2(s)$ would be some form of low pass filter.

7.3.2 Kalman Filter Approach

The previous section presented the complementary filter from a frequency domain perspective. Because the Kalman filter is a time varying filter, the transfer function perspective is not directly applicable; however, the complementary filter idea does extend to Kalman filter applications as discussed in Section 5.10.5.3 and the ideas related to frequency separation between the various error and noise sources is still important.

The direct application of the Kalman filter to the navigation problem of Section 7.1 is considered in Section 7.4. The problematic issue in Section 7.4 will be the signal $a(t)$ for which a time invariant stochastic model does not exist.

Alternatively, the approach of Section 7.1 addresses this issue by using the Kalman filter in a complementary filter format that eliminates $a(t)$ from the estimation problem. Note that the Kalman filter is designed for the error dynamics of eqn. (7.15) aided by the residual measurement

$$\begin{aligned}\delta y_k &= y_k - \hat{y}_k \\ &= \delta p_k + \delta b_y(k) + \eta_2 \\ &= \mathbf{H}\delta\mathbf{x} + \eta_2\end{aligned}$$

as shown in eqn. (7.17). Therefore, for the linearized system, the assumptions of the Kalman filter derivation are satisfied: the state space model is known; it is driven by mutually uncorrelated, white, random noise processes with known power spectral densities; and, the measurement noise satisfies similar assumptions. The approach of Section 7.1 is represented in block diagram form in Figure 5.8.

Note that the output $\hat{y}$ in eqn. (7.14) is the results of an integrative process. It can be represented as

$$\hat{y}_k = (p_k + b_y(k)) - (\delta p_k + \delta b_y(k))$$

where the first term is the signal and the second term is the noise. As the output of an integrative process, the error term $(\delta p_k + \delta b_y(k))$ is predominantly low frequency and a state space model for its time variation is available. The model is given by eqn. (7.15) which is linear with white noise inputs. The output y of eqn. (7.5) can be expressed as

$$y_k = (p_k + b_y(k)) + \eta_2(k)$$

where η_2 is a white noise process. Due to the spectral separation provided by these two measurements, the complementary filter approach works well.

7.4 An Alternative Approach

The previous sections of this chapter have presented and discussed the approach that is the primary focus of this text. This section presents an alternative *total state approach* that is often pursued instead of the approach described above. This section is included, not because it is a better approach, but because it describes the approach that students often try first. The approaches have many similarities and some important distinctions. The two approaches will be compared in Section 7.5.

Define the navigation system state vector as $\hat{x} = [\hat{p}, \hat{v}, \hat{a}, \hat{b}_u, \hat{b}_y, \hat{\alpha}]^\top$ with the same initial conditions as stated in Section 7.1. Note that this state definition includes an estimate of the acceleration. In this approach the Kalman filter measurement equations are propagated for each acceleration measurement.

7.4.1 Total State: Kinematic Model

In this formulation, the system augmented kinematics are described as

$$\dot{\mathbf{x}} = \bar{\mathbf{f}}(\mathbf{x}, \boldsymbol{\omega}) \tag{7.26}$$

where $\boldsymbol{\omega} = [\omega_1, \omega_2, \omega_3]$

$$\bar{\mathbf{f}}(\mathbf{x}, \boldsymbol{\omega}) = \begin{bmatrix} 0 & 1 & 0 & 0 & 0 & 0 \\ 0 & 0 & 1 & 0 & 0 & 0 \\ 0 & 0 & 0 & 0 & 0 & 0 \\ 0 & 0 & 0 & 0 & 0 & 0 \\ 0 & 0 & 0 & 0 & -\lambda_y & 0 \\ 0 & 0 & 0 & 0 & 0 & 0 \end{bmatrix} \mathbf{x} + \begin{bmatrix} 0 & 0 & 0 \\ 0 & 0 & 0 \\ 0 & 0 & 1 \\ 1 & 0 & 0 \\ 0 & 1 & 0 \\ 0 & 0 & 0 \end{bmatrix} \boldsymbol{\omega}. \tag{7.27}$$

In this model, ω_3 is a fictitious driving noise.

For the acceleration state, the above model has assumed that

$$\dot{a} = \omega_3.$$

For simplicity of the description, this discussion will assume that ω_3 is a Gaussian white noise process with PSD of σ_3^2. Many other possible acceleration models could be hypothesized. In fact, this is one of the drawbacks of this approach. The model is attempting to force a time invariant Markov model on the acceleration signal when in fact the acceleration process is not even a stationary random signal. Typically, the parameter σ_3 of the acceleration model will be selected so that the $[\mathbf{P}]_{33}$ element is large, to cause the Kalman filter to use the acceleration measurement to correct the acceleration 'state.' Note that the selection of the acceleration model is a *hypothesis*. It does not have the same rigorous support as does the modeling of the stationary sensor error processes that were discussed in Section 4.9.2 and used in the approach of Section 7.1. As a result, the acceleration model may require significant tuning to achieve satisfactory performance.

7.4.2 Total State: Time Update

The navigation mechanization equations are

$$\dot{\hat{\mathbf{x}}} = \bar{\mathbf{f}}(\hat{\mathbf{x}}, \mathbf{0}) \tag{7.28}$$

where $\hat{\mathbf{x}} = [\hat{p}, \hat{v}, \hat{a}, \hat{b}_u, \hat{b}_y, \hat{\alpha}]^\top$ and

$$\bar{\mathbf{f}}(\hat{\mathbf{x}}, \boldsymbol{\omega}) = \begin{bmatrix} 0 & 1 & 0 & 0 & 0 & 0 \\ 0 & 0 & 1 & 0 & 0 & 0 \\ 0 & 0 & 0 & 0 & 0 & 0 \\ 0 & 0 & 0 & 0 & 0 & 0 \\ 0 & 0 & 0 & 0 & -\lambda_y & 0 \\ 0 & 0 & 0 & 0 & 0 & 0 \end{bmatrix} \hat{\mathbf{x}}. \tag{7.29}$$

The navigation system state vector is propagated between measurements using the following equations

$$\left.\begin{aligned}\hat{p}^-_{k+1} &= \hat{p}^+_k + \hat{v}^+_k\,dt + \tfrac{1}{2}\hat{a}^+_k\,dt^2\\ \hat{v}^-_{k+1} &= \hat{v}^+_k + \hat{a}^+_k\,dt\\ \hat{a}^-_{k+1} &= \hat{a}^+_k\\ \hat{b}^-_{u_{k+1}} &= \hat{b}^+_{u_k}\\ \hat{b}^-_{y_{k+1}} &= \beta\hat{b}^+_{y_k}\\ \hat{\alpha}^-_{k+1} &= \hat{\alpha}^+_k.\end{aligned}\right\}\tag{7.30}$$

The superscript + and − notations are incorporated at this point because a measurement update will be performed following each acceleration measurement. These equations are iterated at the rate f_1 (i.e., $dt = \frac{1}{f_1}$).

Because the Kalman updates will occur at the f_1 rate, the covariance matrix **P** must also be updated at the rate f_1, which is a major disadvantage. The **P** matrix is propagated between measurement as

$$\mathbf{P}^-_{k+1} = \mathbf{\Phi}\mathbf{P}^+_k\mathbf{\Phi}^\top + \mathbf{Qd}\tag{7.31}$$

where $\mathbf{\Phi} \in \mathbb{R}^{6\times6}$ and $\mathbf{Qd} \in \mathbb{R}^{6\times6}$ are constant matrices found by the method in Section 4.7.2.1 with $T = \frac{1}{f_1}$, with **F** and **G** as defined in eqn. (7.27), and $\mathbf{Q} = diag\left([\sigma^2_{b_u}, \sigma^2_{b_y}, \sigma^2_3]\right)$. Alternatively, $\mathbf{\Phi}$ can be extracted from eqn. (7.30).

7.4.3 Total State: Measurement Update

At the time of a (vector) measurement, the Kalman gain is computed as

$$\mathbf{K}_k = \mathbf{P}^-_k\mathbf{H}^\top\left(\mathbf{H}\mathbf{P}^-_k\mathbf{H}^\top + \mathbf{R}\right)^{-1}.$$

Given the Kalman gain vector, the measurement correction for the state and **P** matrix are computed as

$$\begin{aligned}\hat{\mathbf{x}}^+_k &= \hat{\mathbf{x}}^-_k + \mathbf{K}_k\left(\mathbf{y}_k - \hat{\mathbf{y}}^-_k\right)\\ \mathbf{P}^+_k &= [\mathbf{I} - \mathbf{K}_k\mathbf{H}_k]\,\mathbf{P}^-_k.\end{aligned}$$

The total state approach would include $f_1 = 100$ acceleration and $f_2 = 1$ position measurement updates per second. The acceleration update occurs alone $(f_1 - f_2) = 99$ times per second. For these instances, the Kalman gain and update are computed using

$$\mathbf{H} = \left[\begin{array}{cccccc} 0 & 0 & 1 & -1 & 0 & -\hat{a}\end{array}\right] \quad \text{and} \quad \mathbf{R} = R = \left[\begin{array}{c}\sigma^2_1\end{array}\right].$$

The accelerometer and position measurements occur simultaneously $f_2 = 1$ times per second. For these instances, the Kalman gain and update are computed using

$$\mathbf{H} = \begin{bmatrix} 1 & 0 & 0 & 0 & 1 & 0 \\ 0 & 0 & 1 & -1 & 0 & -\hat{a} \end{bmatrix} \quad \text{and} \quad \mathbf{R} = \begin{bmatrix} \sigma_2^2 & 0 \\ 0 & \sigma_1^2 \end{bmatrix}.$$

7.5 Approach Comparison

Even for this simple application, both approaches implement an EKF due to the nonlinearity imposed by the scale factor error. In the approach of Section 7.1 the nonlinearity is induced by the accelerometer scale factor compensation in the time update. In the approach of Section 7.4 the nonlinearity appears in the accelerometer portion of the measurement update. In more realistic applications, the kinematics typically involve nonlinearities, so an EKF will be required in either approach.

In the approach of Section 7.1, there are no free 'tuning' parameters. All the parameter values are determined from the characteristics of the instruments. Based on the results of observability, covariance, or simulation analysis, the designer may choose to combine or remove states or to tune the parameters determined from the sensor characteristics; however, there is a rigorous basis for the initial settings and subsequent tuning. The basic design model is a Markov process driven by stationary white noise processes and therefore satisfies the derivation assumptions for the Kalman filter. The approach of Section 7.4 starts with a model hypothesis for the acceleration state. In most applications, there is no time invariant Markov model for this state because the signal characteristics are determined by the end user and usually cannot be reliably predicted during the system design phase. The design does involve free tuning parameters and does not satisfy the basic modeling assumptions of the Kalman filter.

Because both approaches implement EKF's the $\mathbf{P}$ matrix is not, strictly speaking, the error covariance matrix in either case. However, as long as $\delta\mathbf{x}$ is kept small, in the approach of Section 7.1 the linearization error is small, and $\mathbf{P}$ is typically a reasonable representation of the accuracy of $\delta\mathbf{x}$. In the case of the approach of Section 7.4 the matrix $\mathbf{P}$ cannot be expected to be a reasonable characterization of the accuracy of $\delta\mathbf{x}$ unless the assumed model for the acceleration turns out to be accurate.

Similarly, the EKF measurement update in Section 7.1 is the optimal linear correction. The EKF measurement update in Section 7.4 is only optimal if the assumed model for the acceleration turns out to be accurate.

The computational load of aided navigation systems is typically dominated by the time update of the $\mathbf{P}$ matrix. For the two algorithms, the computational load for the $\mathbf{P}$ matrix time and measurement propagation over each second of operation are compared in the Table 7.2. Eqn. (7.21)

	Section 7.1	Section 7.4
State dimension	n_e	n_t
Time Update	$\mathcal{O}(n_e^3)\left(\frac{1}{\tau}+f_2\right)$	$\mathcal{O}(n_t^3)f_1$
Measurement Update	$\mathcal{O}(n_e^3)f_2$	$\mathcal{O}(n_t^3)f_1$

Table 7.2: General comparison of computational load for the navigation systems of Sections 7.1 and 7.4. Typically, $n_e < n_t$ and $f_2 << f_1$.

	Section 7.1	Section 7.4
State dimension	5	6
Time Update	$\mathcal{O}(125)\left(\frac{1}{\tau}+1\right)$	$\mathcal{O}(21,600)$
Measurement Update	$\mathcal{O}(125)$	$\mathcal{O}(21,600)$

Table 7.3: Specific comparison of computational load for the navigation systems of Sections 7.1 and 7.4 $f_2 = 1$ and $f_1 = 100$.

is updated at the rate $\frac{1}{\tau}$. In cases where the systems **F** and **G** matrices are constant, τ can be set equal to $\frac{1}{f_2}$.

Using the numerical values from the specific example discussed in this chapter, the computational loading is displayed in Table 7.3. Assuming that the vehicle control bandwidth is approximately 1.0 Hz, then a reasonable value of τ would be 10 Hz. For $\tau = 0.1s$, the approach of Section 7.1 requires on the order of 1500 FLOPS per second while the approach of Section 7.4 requires on the order of 43000 FLOPS per second. Many examples show similar significant differences in computational loading.

7.6 A Caution

Specification of the initial error covariance matrix **P** is often a cause of difficulties. Being careless in the definition of **P**, especially the portions of **P** related to the attitude errors, can have serious detrimental effects on the performance of the system. It is often best to use the sensor readings during a short period at the start of operation to initialize the state vector. Based on the statistics of the sensor measurements and the initialization period duration, the error covariance matrix **P** can be specified reasonably. See for example Section 10.5.5.

In particular, the initial error covariance should not be initialized as

$\mathbf{P} = \sigma^2\mathbf{I}$ with σ being a large positive number. Although this setting is often suggested in linear estimation problems, it can be problematic in nonlinear estimation problems. Such a setting provides no guidance to the estimation algorithm as too which components of the state are erroneous. Poor initialization of the portion of $\mathbf{P}$ relating to the attitude variables, which have strong nonlinearities, is particularly problematic. An initial value of the error covariance on the order of $(\pi/2)^2$ for an attitude error means that the direction of motion is completely unknown. Therefore a relatively small position error could result in large changes to the attitude.

7.7 References and Further Reading

The main references for the discussion of the complementary filter are [30, 31]. According to [31], the principle of the complementary filter originates in [141] even though the term *complementary filter* appears first in [3]. Exercise 7.4 considers an simplified aircraft instrumented landing system (ILS). The ILS was the topic of [3, 141] and was considered in Example 4.6 in [31]. The ILS model and parameters have been altered to fit the context of this text. The Wiener filter problem is discussed in [31, 139].

7.8 Exercises

Exercise 7.1 Consider the following two sensed quantities:

$$\begin{aligned} u &= \omega + b + n_1 \\ y &= \theta + n_2 \end{aligned}$$

are available at $100Hz$, where $n_1 \sim N(0, \sigma_1^2)$, $n_2 \sim N(0, \sigma_2^2)$, and b is a scalar Gauss-Markov process with $R_b(\tau) = \sigma_3^2 e^{-\lambda|\tau|}$:

$$\dot{b} = -\lambda b + \sqrt{2\lambda} n_3$$

with $n_3 \sim N(0, \sigma_3^2)$, and $\lambda > 0$. The symbol θ represents an angle and ω represents the angular rate. The kinematic relationship between θ and ω is $\dot{\theta} = \omega$, as in Example 4.2.

1. Define the mechanization equations by which u would be used to compute $\hat{\theta}$.

2. Define the error state to be $\delta\mathbf{x} = [\delta\theta, \delta b]^\top$. Using the complementary filter approach, define the error state dynamic and measurement equations.

3. Assume that $\sigma_1 = 0.005\frac{rad}{s}$, $\sigma_2 = 1.0°$, $\sigma_3 = 0.0005\frac{rad/s^2}{\sqrt{Hz}}$, $\lambda = \frac{1}{300}sec^{-1}$, and that the angle sensor is sampled at $0.5Hz$.

(a) Design a Kalman filter by specifying the stochastic model parameters and initial conditions: $\hat{\mathbf{x}}(0)$, $\mathbf{P}(0)$, $\mathbf{F}$, $\mathbf{Q}$, $\mathbf{\Gamma}$, $\mathbf{H}$, $\mathbf{\Phi}$, $\mathbf{Qd}$, and $\mathbf{R}$.

(b) Angular rate and angle measurement data are posted on the website for the book.

i. Implement the equations that integrate u to compute $\hat{\theta}$.

ii. Implement the Kalman filter that uses y to estimate $\delta\theta$. Use this $\delta\theta$ estimate to correct $\hat{\theta}$.

Plot the residual measurements and the standard deviation of the residual measurements at the measurement time instants.

(c) Compare the actual and predicted performance of the system based on the analysis of the residuals.

Exercise 7.2 For the system defined in Exercise 7.1, the error dynamic equations are time invariant. The error state is observable from the residual measurement and controllable from the process noise. A stable steady-state filter exists. The steady-state gain $\bar{\mathbf{K}}$ can be found by a variety of methods, one of which is running the covariance simulation until steady-state is approximately achieved.

1. Find the steady-state gain $\bar{\mathbf{K}}$ of the Kalman filter.

2. Perform covariance analysis for the time-varying filter of Exercise 7.1 and for the steady-state filter. Discuss the performance (transient and steady-state) and computational tradeoffs.

Exercise 7.3 For the system of Section 7.2.6:

1. Complete an observability analysis of the error state $\delta\mathbf{x}$ for the sensor suite being used for $t > 50s$. Relate the results of this observability analysis to the results shown in Figures 7.4–7.5.

2. For $t \in [0, 100]$, complete an error budget analysis similar to that in Figures 7.2 and 7.3 for the case of the additional velocity sensor.

Exercise 7.4 This exercise discusses a simplified aircraft instrument landing system (ILS). The objective of this exercise is to estimate the lateral deviation d and its rate of change $\dot{d}$. The aircraft is assumed to be flying at a known constant speed $v = 200km/hr$.

The ILS is assumed to supply to the aircraft a signal

$$\tilde{d}_k = d_k + \eta_k + n_k \tag{7.32}$$

at 1.0 Hz where n is Gaussian white noise with variance $\sigma_n^2 = (10m)^2$ and

$$\dot{\eta}(t) = -\lambda\eta(t) + \sqrt{2\lambda}\sigma_1 w_1(t) \tag{7.33}$$

where w_1 is Gaussian white noise with PSD $\sigma_1^2 = 10\frac{(m/s)^2}{Hz}$, and $\lambda = \frac{1}{2}\frac{rad}{s}$.

The rate of change of $d(t)$ is $\dot{d} = v\psi$ where ψ is the difference between the aircraft and runway heading. Because the heading of the runway is constant and known, ψ can be estimated from the yaw rate gyro as

$$\dot{\hat{\psi}}(t) = \tilde{r}(t) \tag{7.34}$$

where $\tilde{r}(t) = r(t) + w_2(t)$ and r represents the yaw rate, w_2 is Gaussian white noise with PSD $\sigma_2^2 = 9 \times 10^{-8}\frac{(rad/s)^2}{Hz}$.

Define $\mathbf{x} = [d, \eta, \psi]^\top$ and $\boldsymbol{\omega} = [w_1, w_2]^\top$.

1. Define the kinematic equations for the state vector.

2. Define the state model for the system with the output $\hat{d}$.

3. Show that the state space model for the error state vector is

$$\begin{aligned}\delta\dot{\mathbf{x}} &= \begin{bmatrix} 0 & 0 & v \\ 0 & -\lambda & 0 \\ 0 & 0 & 0 \end{bmatrix}\delta\mathbf{x} + \begin{bmatrix} 0 & 0 \\ \sqrt{2\lambda}\sigma_1 & 0 \\ 0 & -1 \end{bmatrix}\boldsymbol{\omega} \\ \delta d_k &= \begin{bmatrix} 1 & 1 & 0 \end{bmatrix}\delta\mathbf{x}_k + n_k\end{aligned}$$

   and that the PSD matrix for $\boldsymbol{\omega}$ is

$$\mathbf{Q} = \begin{bmatrix} 1 & 0 \\ 0 & \sigma_2^2 \end{bmatrix}.$$

4. Analyze the observability of the system.

5. Predict the estimation accuracy for both $\hat{d}$ and $\dot{\hat{d}}$.

6. The system model is time invariant. Find the steady state Kalman gain $\bar{\mathbf{K}}$ from the previous step. Analyze the performance and computational savings that could be achieved by using the $\bar{\mathbf{K}}$ instead of $\mathbf{K}_k$ in the filter implementation.

Exercise 7.5 This problem returns to the application considered in Section 1.1.1. Exercise 5.6 derived the parameters of the Kalman filter design. Accelerometer and position aiding measurements are posted on the website for the book.

1. Implement the equations to integrate the (corrected) accelerometer measurements to produce an estimate of the navigation state vector.

2. Implement the Kalman filter error estimator and use the estimated error state to correct the navigation state vector.

3. Check the system design by analyzing the behavior of the residual aiding measurements relative to their expected behavior.

Part II

Application

Part II Overview

This portion of the book applies the theory from Part I in a few navigation applications. A detailed discussion of the GPS system is presented in Chapter 8. GPS is a very commonly used aiding system in navigation applications such as encoder-based dead-reckoning (see Chapter 9) or inertial navigation (see Chapter 11). There are several other aiding sensors available (e.g., alternative GNSS, pseudolites, radar, lidar, imagery, and acoustic rangers). In fact, there are too many to present them all herein. Instead, the objective herein is to present the methodology in sufficient detail, using GPS as an example aiding system, that the readers will be able to adapt the approach to the specific aiding sensors appropriate for their applications.

Chapters 9 discusses GPS aided encoder-based dead-reckoning. This sensor combination is used in a variety of commercial applications and student projects involving land vehicles. This aided navigation application is presented first as students often consider its model and mathematical analysis to be easier than the inertial applications in the subsequent chapters.

Each of the aided navigation chapters (i.e., Chapters 9–12) will follow a similar method of presentation: basic discussion of the sensor technology, derivation of kinematic equations, presentation of the corresponding navigation mechanization equations, derivation of the navigation error state dynamic equations, discussion of state augmentation, derivation of the model of the residual aiding signal measurement model, and discussion of performance related issues. The similar method of presentation in each chapter is intended to highlight the methodology so that readers can easily extend the necessarily frugal analysis herein to their own circumstances and applications.

Chapter 10 considers the estimation of the attitude and heading of a vehicle. This is the only chapter in the book to discuss and utilize quaternions. The application can be easily built and tested in a laboratory environment and makes a very good first design and implementation project. Completion of the project ensures that the quaternion algorithms are implemented correctly and those algorithms are then available for other applications.

Chapters 11 and 12 consider strapdown inertial navigation systems. Chapter 11 discusses standard material on the topic of inertial navigation. GPS aided inertial navigation is one of the most common aided navigation applications. Chapter 12 considers an inertial navigation system aided by Doppler velocity and acoustic ranging measurements for an underwater vehicle. This application is included as it allows discussion of the modeling of measurements involving delayed states (i.e., the measurement $y(t)$ is a function of $\mathbf{x}(t)$ and $\mathbf{x}(t-\tau)$ for some $\tau > 0$).

Chapter 8

Global Positioning System

The purpose of this chapter is to introduce the Global Positioning System (GPS), model equations, and system characteristics in the notation that will be utilized in the following chapters. The chapter presents the essential technical information necessary for the interested reader to understand and utilize GPS in real-time navigation systems.

This chapter discusses three GPS positioning methods: standard GPS, differential GPS (DGPS), and carrier phase DGPS. For each method, the expected accuracy is discussed. Throughout this chapter, GPS solution techniques are presented in the context of realtime point positioning. Each time sample is considered independently (i.e., no filtering). This presentation approach enables straightforward discussion of GPS position accuracy and solution methods. Clear benefits are obtainable by filtering the GPS data. Use of GPS as an aiding signal is discussed in Chapters 9 and 11. This chapter assumes that certain necessary quantities (e.g., satellite positions) can be calculated. These calculations are described in Appendix C.

The outline of this chapter is as follows. Section 8.1 provides a high-level view of the GPS system and an overview of the signal characteristics. Section 8.2 discusses the GPS pseudorange observable, related notation, and a method for determining the receiver antenna location from a set of simultaneous pseudorange measurements. Section 8.3 discusses the basic measurements of a receiver and the formation of the pseudorange, carrier phase, and Doppler observables from the basic measurements. Sections 8.2–8.3 will have introduced and used notation to describe the various GPS measurement errors. The various error sources are characterized in Section 8.4. The effect of GPS measurement errors on position estimation accuracy is analyzed in Section 8.5. The utility of two frequency receivers is discussed in Section 8.6. Carrier-smoothing of pseudorange observables is discussed in Section 8.7. Various methods for differential GPS processing are discussed in Section 8.8. Approaches to integer ambiguity resolution are discussed in Section 8.9.

8.1 GPS Overview

GPS is an all weather, worldwide, continuous coverage, satellite-based radio navigation system. GPS provides nearly uniform, worldwide accuracy. In addition, GPS receivers are available at very reasonable cost.

8.1.1 GPS System

GPS consists of three major segments: *space, control*, and *user*.

The *space segment* consists of the GPS satellites. The satellite vehicles (SVs) orbit Earth in six 12-hour (11 hr 58 minute) orbital planes with (nominally) four satellites in each plane. The orbits are nearly circular with inclination angles of 55 degrees and altitude above the Earth surface of approximately 20,200 km. The six orbit planes are equally spaced around the equator resulting in 60 degree separation. This constellation ensures that (barring obstructions) a user located anywhere on Earth has a direct line-of-sight to at least four satellites at any time. Since satellites are not in geosynchronous orbits, the geometric relation between the satellites is constantly changing with respect to a stationary receiver on Earth. This changing geometry results in changing, but predictable, position estimation accuracy for positions determined from a fixed set of satellites (see Section 8.5). The SVs emit coded radio signals which a GPS receiver will decode to determine important system parameters.

The *control segment* is responsible for monitoring the health and status of the space segment. The control segment consists of a system of tracking stations located around the world including six monitor stations, and a master control station. The ground monitoring stations measure signals from the SVs which are transmitted to the master control station. The master control station determines the orbital model and clock correction parameters for each satellite. These parameters (and other data) are transmitted to the satellites which then broadcast them to the user segment.

The *user segment* consists of antennas and receivers. Traditionally, receivers consist of three major stages. The *radio frequency (RF) front-end* performs three basic tasks on the input signal from the GPS antenna: amplification, filtering, and shifting of the GPS signal portion of the frequency spectrum to a lower frequency range. The *baseband* portion of the receiver has several channels that operate in parallel on the signal output by the RF front-end. Each channel tracks a signal from one satellite to determine certain basic channel variables and to determine the data bits carried by that signal. The basic channel variables determine the transit time and carrier phase. These are further described in Section 8.3. The data bits describe information needed to compute such items as SV position and velocity, SV clock error, and SV health. The *navigation* portion of the receiver uses the SV positions and the measured signal transit times from satellite to receiver to estimate position, velocity, and time for the user. The three stages described for a traditional receiver may be less clearly identifiable as such in advanced receivers. For example, an advanced receiver might use the navigation solution to aid in finding or tracking weaker signals from other satellites.

GPS provides two levels of service: a Standard Positioning Service (SPS) and a Precise Positioning Service (PPS). *SPS* is a positioning and timing service which is available to all GPS users on a continuous, worldwide basis with no direct charge. This level of service is (currently) provided by the L1 frequency through the course acquisition (C/A) code and a navigation data message. Full SPS operational capability was achieved in late 1993. Typical accuracy of GPS estimated position using SPS is discussed in Section 8.4.10. Precise positioning service *PPS* is a more accurate positioning, velocity, and timing service which is available only to users authorized by the U.S. government. Access to this service is controlled by a technique known as *anti-spoofing* (AS). Certain advanced GPS receivers are able to track the L2 signals without PPS authorization, with a mild performance penalty, thus allowing two frequency operation to civilian users.

GPS receivers operate passively (i.e., they do not transmit any signals); therefore, the GPS space segment can provide service to an unlimited number of users. The GPS system is a line-of-sight system. If the path between the receiver and a satellite is obstructed, then the satellite signal may not be received. For example, a typical GPS receiver will not function indoors, on underwater vehicles, or under significant foliage.

8.1.2 Original GPS Signal

Each GPS satellite continuously transmits using two carrier frequencies, L1 (1575.42 MHz) and L2 (1227.60 MHz). The carrier frequencies are modulated by data and spread spectrum signals to carry information to the user.

The L1 signal is modulated, in quadrature, by two Code Division Multiple Access (CDMA) signals: C/A and P(Y). The *coarse/acquisition* (C/A) code has a length of 1023 chips and a 1.023 MHz chip rate, resulting in a code period of one millisecond. Each chip has a value of ± 1. There is a different C/A PRN code for each satellite and each C/A PRN code is nearly orthogonal to all other C/A PRN codes. Although all satellites are broadcasting on the same two frequencies, a GPS receiver is able to lock-on to a specific satellite and discriminate between satellites by correlating an internally generated version of the known C/A code for that satellite with the received signal. Since the C/A codes for each satellite are unique and nearly orthogonal, the cross-satellite interference is small. The GPS space vehicles are often identified by their unique C/A PRN code numbers. The *precise (P) code* modulates both L1 and L2 carrier phases. The P-code is a very long (i.e., seven days) 10.23 MHz PRN code. In the *Anti-Spoofing* (AS) mode of operation, the P-code is encrypted into the *Y-code*. The encrypted Y-code requires a classified AS module for each receiver channel and is for use only by authorized users with cryptographic keys. Currently, the L2 signal is modulated only by the CDMA P(Y) signals. Therefore, it is intended to only be available to authorized users.

The L1 signal is also modulated, using binary phase shift keying, by the a 50 bit per second navigation message. The *navigation message* consists of data bits that a GPS receiver decodes to determine satellite orbit, clock correction, and other system parameters. Appendix C describes the means by which these parameters are used to calculate satellite position and velocity, clock corrections, and ionospheric corrections.

8.2 GPS Pseudorange

The electromagnetic signals broadcast by the satellites travels at the speed of light. If the transit time was measured, it would be converted to a distance by multiplying it by the speed of light c. Due to this scaling, each microsecond of timing error would result in a range error of about 300 meters; therefore, to achieve meter level range measurement accuracy would require clocks synchronized to a few nanoseconds. Such timing accuracy would make GPS receivers prohibitively expensive. Instead, typical GPS receivers use inexpensive free running clocks based on crystal oscillators and estimate the clock bias error during the position estimation process. Therefore, the measurement of the apparent transit time is biased by errors in the user and satellite clocks and is called a *pseudorange*.

If the clock errors were zero and the satellite positions were available, then each satellite to receiver range measurement would define a sphere of possible receiver locations. The intersection of two such spheres would define a circle of possible receiver locations. The intersection of three spheres would result in two possible receiver locations. With either a prior estimate of the receiver location or another range estimate, a single receiver position estimate would result. This geometrical example illustrates that (at least) three independent range measurements are necessary to determine the three position coordinates of the receiver antenna if the receiver clock error were zero. In realistic applications, the receiver clock error is non-zero. Therefore, in any set of simultaneous pseudorange measurements there are four unknowns and at least four pseudorange measurements will be required for a unique solution to exist.

Section 8.4 and Appendix C provide a more detailed description of the pseudorange measurement process.

8.2.1 GPS Pseudorange Notation

The geometric or "true" range between the receiver antenna located at $\mathbf{p}$ and the effective satellite antenna location $\mathbf{p}^i$ is

$$R(\mathbf{p}, \mathbf{p}^i) = \left\|\mathbf{p} - \mathbf{p}^i\right\|_2. \tag{8.1}$$

The superscript i notation refers the superscripted quantity to the i-th satellite.

The algorithm for computing SV position vector $\hat{\mathbf{p}}^i$ in ECEF coordinates uses data derived from the GPS navigation messages as described in Appendix C. The ranges $R^i = R(\mathbf{p}, \mathbf{p}^i)$ and $\hat{R}^i = R(\mathbf{p}, \hat{\mathbf{p}}^i)$ are distinct. The ephemeris error term $E^i = R^i - \hat{R}^i$ is derived in Section 8.4.5. In addition, there are time and reference frame issues that must be carefully addressed for accurate computation of $\hat{\mathbf{p}}^i$. These issues are discussed in Sections C.1 and C.3.

Ideally we want to measure R^i, but this is not possible due to various sources of error, one of which is receiver clock error. Denoting the receiver clock offset with respect to GPS system time as Δt_r, the true pseudorange is

$$\rho(\mathbf{x}, \mathbf{p}^i) = \left\|\mathbf{p} - \mathbf{p}^i\right\|_2 + c\Delta t_r \tag{8.2}$$

where $\mathbf{x} = \left[\mathbf{p}^\top, c\Delta t_r\right]^\top$. Therefore, the standard GPS positioning problem involves estimation of four unknown quantities: the three components of $\mathbf{p}$ and Δt_r.

The vector $\mathbf{x}$ is estimated based on pseudorange measurements which for the L1 pseudorange can each be modeled as

$$\tilde{\rho}^i_{r_1} = \rho(\mathbf{x}, \mathbf{p}^i) + c\delta t^i + \frac{f_2}{f_1} I^i_r + T^i_r + M^i_{\rho_1} + \nu^i_{\rho_1} \tag{8.3}$$

where c is the speed of light, f_1 is the L1 carrier frequency, and f_2 is the L2 carrier frequency. The subscript r on ρ refers to the receiver. Later in the chapter, when multiple receivers are considered, the subscript r will be used to count over the available receivers. The subsubscript 1 refers the measurement to the L1 signal. Using the same notational conventions, the L2 pseudorange can be modeled as

$$\tilde{\rho}^i_{r_2} = \rho(\mathbf{x}, \mathbf{p}^i) + c\delta t^i + \frac{f_1}{f_2} I^i_r + T^i_r + M^i_{\rho_2} + \nu^i_{\rho_2}. \tag{8.4}$$

The error models of eqns. (8.3–8.4) are derived in Section 8.4. The measurement error components are defined as follows:

δt^i – This symbol represents the residual satellite clock error after performing the corrections described in C.1. The satellite clock error is discussed in Section 8.4.2.

I_r^i – This symbol represents the error due to dispersive atmospheric effects described in Section 8.4.4.1.

T_r^i – This symbol represents the error due to non-dispersive atmospheric effects described in Section 8.4.4.2.

$M_{\rho_1}^i$, $M_{\rho_2}^i$ – These symbols represent the pseudorange multipath errors on the L1 and L2 pseudorange measurements. They are described in Section 8.4.7.

$\nu_{\rho_1}^i$, $\nu_{\rho_2}^i$ – These symbols represent random measurement noise on the L1 and L2 pseudorange measurements. They are discussed in Section 8.4.8.

The pseudorange measurement is described in greater depth in Sections 8.3 and Appendix C. The majority of the discussion of the next few sections will focus on the L1 signal and the notations $\tilde{\rho}_{r_1}^i$, $M_{\rho_1}^i$, and $\nu_{\rho_1}^i$ will be simplified to $\tilde{\rho}^i$,M_ρ^i, and ν_ρ^i.

Given a set of (at least) four pseudorange measurements, arbitrarily numbered one through four, define the vector of measurements as

$$\tilde{\boldsymbol{\rho}} = [\tilde{\rho}^1, \tilde{\rho}^2, \tilde{\rho}^3, \tilde{\rho}^4]^\top$$

and the vector of computed pseudoranges

$$\hat{\boldsymbol{\rho}}(\hat{\mathbf{x}}) = [\rho(\hat{\mathbf{x}}, \hat{\mathbf{p}}^1), \rho(\hat{\mathbf{x}}, \hat{\mathbf{p}}^2), \rho(\hat{\mathbf{x}}, \hat{\mathbf{p}}^3), \rho(\hat{\mathbf{x}}, \hat{\mathbf{p}}^4)]^\top,$$

the objective of the solution will then be to find the value of $\hat{\mathbf{x}}$ that minimizes the cost function

$$J(\hat{\mathbf{x}}) = \|\hat{\boldsymbol{\rho}}(\hat{\mathbf{x}}) - \tilde{\boldsymbol{\rho}}\|. \tag{8.5}$$

The solution to this problem is presented with a detailed example in Section 8.2.2.

The computed pseudoranges are functions of the computed satellite positions $\hat{\mathbf{p}}^i$ while the measured ranges are functions of the actual satellite positions. Therefore, the estimated value of $\mathbf{x}$ will be affected by the error $(\mathbf{p}_i - \hat{\mathbf{p}}^i)$. To make this dependence explicit in the model of the pseudorange, we manipulate eqn. (8.3) as shown below to obtain the L1 C/A pseudorange model of eqn. (8.6):

$$\begin{aligned}
\tilde{\rho}_r^i &= \rho(\mathbf{x}, \hat{\mathbf{p}}^i) + (\rho(\mathbf{x}, \mathbf{p}^i) - \rho(\mathbf{x}, \hat{\mathbf{p}}^i)) + c\delta t^i + \frac{f_2}{f_1} I_r^i + T_r^i + M_\rho^i + \nu_\rho^i \\
&= \rho(\mathbf{x}, \hat{\mathbf{p}}^i) + E^i + c\delta t^i + \frac{f_2}{f_1} I_r^i + T_r^i + M_\rho^i + \nu_\rho^i. \qquad (8.6)
\end{aligned}$$

The ephemeris error term $E^i = \left(\rho(\mathbf{x}, \mathbf{p}^i) - \rho(\mathbf{x}, \hat{\mathbf{p}}^i)\right)$ represents the error incurred due to using the computed satellite position in the solution process instead of the (unknown) actual satellite position. The derivation and characterization of this term is presented in Section 8.4.5.

The combined pseudorange measurement error will be denoted as

$$\chi^i = E^i + c\delta t^i + \frac{f_2}{f_1} I_r^i + T_r^i + M_\rho^i + \nu_\rho^i \tag{8.7}$$

where χ^i is called the User Equivalent Range Error (UERE) or User Range Error (URE). The vector of pseudorange measurement errors will be denoted by $\boldsymbol{\chi}$. Assuming that the errors for different satellites are independent,

$$cov(\boldsymbol{\chi}) = \sigma^2 \mathbf{I} \tag{8.8}$$

where the value of σ is discussed relative to Table 8.5. Portions of this total range error $\boldsymbol{\chi}$ are due to the space and control segments (i.e., $E^i + c\delta t^i$), to the path between the SV and receiver antenna (i.e., $I_r^i + T_r^i$), to the siting of the antenna (i.e., M_ρ^i), and to the receiver and antenna (ν_ρ^i). In later sections, it will often be convenient to simplify the notation by combining the multipath and measurement errors into a single error term

$$\eta^i = M_\rho^i + \nu_\rho^i. \tag{8.9}$$

The symbol η^i represents the non-common mode pseudorange error.

8.2.2 GPS Pseudorange Solution

Each pseudorange measurements represented in eqn. (8.6) is a nonlinear function of the receiver position $\mathbf{p}$. This section discusses the solution of a set of such measurement equations by an iterative method based on linearization of the measurement equations. When an accurate initial estimate of the receiver antenna position is available, possibly from a prior measurement epoch, then the solution may converge to the desired accuracy in a single iteration. If improved accuracy is desired, the procedure can be repeated iteratively with the result from one iteration serving as the linearization point for the next iteration. This method is straightforward and converges quickly.

A GPS independent review of this algorithm is presented in Section B.14. Example B.4 in that section has direct relevance to the discussion of GPS. The application of this algorithm to the solution of eqn. (8.6) for a set of (at least) four measurements is discussed below. The algorithm of eqn. (8.10) extends, without change, to the case where more than four measurements are available as long as the required matrix inverse exists.

At time t, assume that pseudorange measurements are available for satellites one through four and that initial (i.e., $k = 0$) estimates $\hat{\mathbf{p}}_0 =$

$[\hat{x}_0, \hat{y}_0, \hat{z}_0]^\top$ and $c\Delta\hat{t}_r(0)$ are available for the receiver position and clock bias. In the worst case, for a user near the surface of Earth, assume $\hat{\mathbf{p}}_0 = \mathbf{0}$ and $c\Delta\hat{t}_r(0) = 0$. We form the vector $\hat{\mathbf{x}}_k = \left[\hat{\mathbf{p}}_k^\top, c\Delta\hat{t}_r(k)\right]^\top$ where k is the iteration index at a fixed time t.

Our objective is to find the value of $\hat{\mathbf{x}}$ that minimizes the cost function of eqn. (8.5). The algorithm which is derived in eqn. (B.60) of Section B.14.2 is

$$\hat{\mathbf{x}}_{k+1} = \hat{\mathbf{x}}_k + \left(\mathbf{H}^\top\mathbf{H}\right)^{-1}\mathbf{H}^\top\left(\tilde{\boldsymbol{\rho}} - \hat{\boldsymbol{\rho}}(\hat{\mathbf{x}}_k)\right), \quad \text{for } k = 1, 2, 3 \ldots. \tag{8.10}$$

At the k-th iteration, the vector $\hat{\mathbf{x}}_k$ is the point of linearization for eqn. (8.2). The computed value of the pseudorange from eqn. (8.2) at $\hat{\mathbf{x}}$ (i.e., $\rho(\hat{\mathbf{x}}, \mathbf{p}^i)$) will be indicated as $\hat{\rho}^i(\hat{\mathbf{x}})$. Linearization of eqn. (8.2) for $\mathbf{x}$ near $\hat{\mathbf{x}}_k$ and $i = 1, \ldots, 4$ gives

$$\begin{bmatrix} \hat{\rho}^1(\mathbf{x}) \\ \hat{\rho}^2(\mathbf{x}) \\ \hat{\rho}^3(\mathbf{x}) \\ \hat{\rho}^4(\mathbf{x}) \end{bmatrix} = \begin{bmatrix} \hat{\rho}^1(\hat{\mathbf{x}}_k) \\ \hat{\rho}^2(\hat{\mathbf{x}}_k) \\ \hat{\rho}^3(\hat{\mathbf{x}}_k) \\ \hat{\rho}^4(\hat{\mathbf{x}}_k) \end{bmatrix} + \mathbf{H} \begin{bmatrix} (x - \hat{x}_k) \\ (y - \hat{y}_k) \\ (z - \hat{z}_k) \\ c(\Delta t_r - \Delta\hat{t}_r(k)) \end{bmatrix} \tag{8.11}$$

where

$$\mathbf{H} = \left.\begin{bmatrix} \frac{\delta\rho^1}{\delta\mathbf{x}} \\ \frac{\delta\rho^2}{\delta\mathbf{x}} \\ \frac{\delta\rho^3}{\delta\mathbf{x}} \\ \frac{\delta\rho^4}{\delta\mathbf{x}} \end{bmatrix}\right|_{\hat{\mathbf{x}}_k} = \begin{bmatrix} \frac{\hat{\mathbf{p}}_k - \hat{\mathbf{p}}^1}{\|\hat{\mathbf{p}}_k - \hat{\mathbf{p}}^1\|} & 1 \\ \frac{\hat{\mathbf{p}}_k - \hat{\mathbf{p}}^2}{\|\hat{\mathbf{p}}_k - \hat{\mathbf{p}}^2\|} & 1 \\ \frac{\hat{\mathbf{p}}_k - \hat{\mathbf{p}}^3}{\|\hat{\mathbf{p}}_k - \hat{\mathbf{p}}^3\|} & 1 \\ \frac{\hat{\mathbf{p}}_k - \hat{\mathbf{p}}^4}{\|\hat{\mathbf{p}}_k - \hat{\mathbf{p}}^4\|} & 1 \end{bmatrix}. \tag{8.12}$$

If each row of $\mathbf{H}$ is denoted as $\mathbf{h}^i$, then the first three components of each $\mathbf{h}^i$ define a unit vector pointing from the satellite to the point of linearization $\hat{\mathbf{x}}_k$. For each iteration of eqn. (8.10), the matrix $\mathbf{H}$ should be recalculated. This recalculation is most important when $\|\hat{\mathbf{p}}_k - \mathbf{p}\|$ is large. The iteration of eqn. (8.10) assumes that the rows of $\mathbf{H}$ are linearly independent so that the required matrix inverse exists.

The iteration of eqn. (8.10) is concluded when the change in the estimate $\hat{\mathbf{x}}_k$ given by

$$d\mathbf{x}_k = \left(\mathbf{H}^\top\mathbf{H}\right)^{-1}\mathbf{H}^\top\left(\tilde{\boldsymbol{\rho}} - \hat{\boldsymbol{\rho}}(\hat{\mathbf{x}}_k)\right) \tag{8.13}$$

in a given iteration has magnitude $\|d\mathbf{x}_k\|$ less than a prespecified threshold. The fact that $d\mathbf{x}_k$ is small at the conclusion of the iteration does not imply that

$$\delta\mathbf{x}_k = \mathbf{x} - \hat{\mathbf{x}}_k \tag{8.14}$$

is small. The estimate $\hat{\mathbf{x}}_k$ was selected to minimize the norm of the error between the computed pseudorange and measurements of the pseudorange. Therefore, the pseudorange measurement errors $\boldsymbol{\chi}$ defined in eqn. (8.7)

Satellite	$\hat{x}$, m	$\hat{y}$, m	$\hat{z}$, m
SV 2	+7766188.44	-21960535.34	+12522838.56
SV26	-25922679.66	-6629461.28	+31864.37
SV 4	-5743774.02	-25828319.92	+1692757.72
SV 7	-2786005.69	-15900725.80	+21302003.49

Table 8.1: ECEF satellite positions.

have a determinative role in the resulting characterization of the vector $\delta\mathbf{x}_k$. The relationship between χ, the matrix $\mathbf{H}$, and the characteristics of $\delta\mathbf{x}$ are discussed in Section 8.5.

Example 8.1 *At time t, a set of satellite positions are calculated using the equations in Appendix C. The resulting positions are shown in Table 8.1. At the same time t, the pseudoranges are measured to be*

$$\begin{aligned}\tilde{\rho}^2 &= 22228206.42m,\\ \tilde{\rho}^{26} &= 24096139.11m,\\ \tilde{\rho}^4 &= 21729070.63m,\\ \tilde{\rho}^7 &= 21259581.09m.\end{aligned}$$

What is the receiver location?

Table 8.2 shows the results of the iterative approach of eqn. (8.10) initialized at the ECEF origin $\mathbf{p} = [0, 0, 0]^\top$. *The position estimate converges to millimeter precision in five iterations.*

This convergence does not imply that the estimated position is accurate to millimeters. From prior measurements, the pre-surveyed ECEF antenna position is known to be

$$\mathbf{p} = [-2430829.17, \ -4702341.01, \ 3546604.39]^\top m.$$

At each iteration, the error between the estimated position and the known position is shown in Table 8.3. This table shows that while the cost function of eqn. (8.5) is minimized, the position error due to the measurement errors χ *remains. Note that the data for this example is from the era when selective availability was active. With selective availability now turned off, the magnitude of the error vector displayed in this example would be atypical.*

The measurement matrix $\mathbf{H}$ *for the last iteration is calculated from eqn. (8.12) to be*

$$\mathbf{H} = \begin{bmatrix} -0.4643 & 0.7858 & -0.4087 & 1.0000 \\ 0.9858 & 0.0809 & 0.1475 & 1.0000 \\ 0.1543 & 0.9842 & 0.0864 & 1.0000 \\ 0.0169 & 0.5333 & -0.8457 & 1.0000 \end{bmatrix}.$$

k	$\hat{x}$, m	$\hat{y}$, m	$\hat{z}$, m	$c\Delta\hat{t}_r$, m
0	0000000.000	0000000.000	0000000.000	000000.000
1	-2977571.476	-5635278.159	4304234.505	1625239.802
2	-2451728.534	-4730878.461	3573997.520	314070.732
3	-2430772.219	-4702375.802	3546603.872	264749.706
4	-2430745.096	-4702345.114	3546568.706	264691.129
5	-2430745.096	-4702345.114	3546568.706	264691.129

Table 8.2: Receiver position in ECEF coordinates as a function of iteration number.

k	δx, m	δy, m	δz, m
1	-546742.305	-932937.149	757630.116
2	-20899.364	-28537.451	27393.130
3	56.951	-34.792	-0.518
4	84.074	-4.104	-35.684
5	84.074	-4.104	-35.684

Table 8.3: Receiver position error in ECEF coordinates as a function of iteration number.

Iterative solution via eqn. (8.13) requires calculation of $\left(\mathbf{H}^\top\mathbf{H}\right)^{-1}\mathbf{H}^\top$*, which to four decimal accuracy is*

$$\left(\mathbf{H}^\top\mathbf{H}\right)^{-1}\mathbf{H}^\top = \begin{bmatrix} -1.9628 & -0.0862 & 1.0122 & 1.0368 \\ -1.7302 & -1.1479 & 2.0344 & 0.8437 \\ 1.1263 & 0.5680 & -0.0604 & -1.6339 \\ 1.9086 & 1.0941 & -1.1533 & -0.8493 \end{bmatrix}.$$

The first three rows of $\left(\mathbf{H}^\top\mathbf{H}\right)^{-1}\mathbf{H}^\top$ *transform the pseudoranges to the position estimates. Note that the elements of the first three rows sum to zero. Therefore, any bias that effects all the pseudoranges identically, such as the receiver clock error, will have not effect on the estimated position. The last row, which is the transformation of the pseudoranges to the estimate of the clock bias, sums to one. Therefore, any measurement error that effects all the pseudorange measurements identically will affect the estimate of the clock bias.* △

Example L1 C/A code position estimation results are shown in Figure 8.1. At each time step (1 second intervals), the position estimate is calculated independently using eqn. (8.10). The surveyed position is subtracted

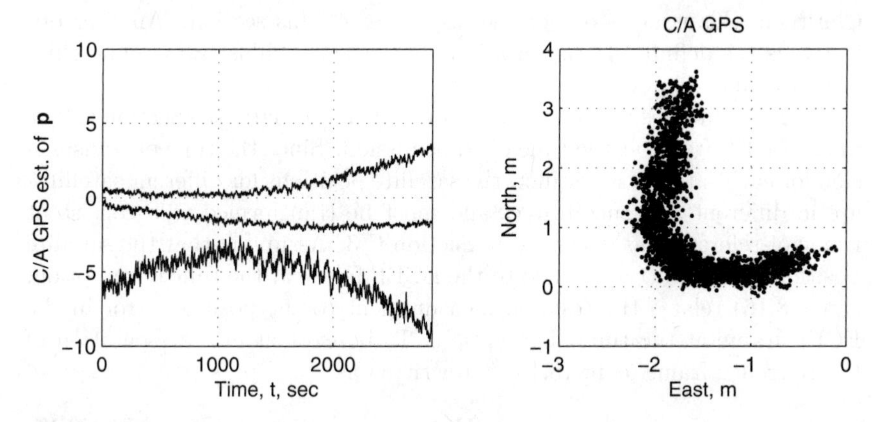

Figure 8.1: L1 C/A position estimate results. Left- Plot of the point position estimation error versus time. The top curve is north error. The middle curve is east error. The bottom curve is down error. Right- Scatter plot of the north and east estimation error.

from the estimated position so that the plots illustrate the estimation error. The same set of satellites (i.e., PRNS: 2, 4, 9, 12, 17, and 28) was used for the duration of the experiment. This figure illustrates the character of the positioning accuracy achievable by SPS GPS. First, note that the estimate of d is significantly noisier than the estimates of n and e. Also notice that the position estimation error has significant time correlation. This is due to the fact that various of the error sources (SV clock and ephemeris error, ionospheric error, and tropospheric error) are slowly time varying. This fact is used advantageously in differential GPS systems as discussed in Section 8.8. The sources and magnification of these errors are discussed in the Sections 8.4 and 8.5. Third, due to the fact that the error sources χ are time-correlated over the period of the experiment shown in Figure 8.1, the scatter plot does not have the characteristics ellipsoidal shape discussed in Section 4.9.1.2.

8.2.3 Satellite Azimuth and Elevation

By eqn. (8.11), the residual measurement

$$\delta\boldsymbol{\rho} = \tilde{\boldsymbol{\rho}} - \hat{\boldsymbol{\rho}}(\hat{\mathbf{x}}) \tag{8.15}$$

is related to the estimation error according to

$$\delta\boldsymbol{\rho} = \mathbf{H}\delta\mathbf{x} + \boldsymbol{\chi} \tag{8.16}$$

where $\delta\mathbf{x}$ is defined in eqn. (8.14) and $\boldsymbol{\chi}$ is defined in eqn. (8.7). Up to this point, the reference frame for $\mathbf{H}$ and $\delta\mathbf{x}$ has not been directly discussed.

Clarifying this issue is one of the objectives of this section. Another objective is the definition, computation, and graphical illustration of satellite elevation and azimuth.

The satellite position computed from ephemeris data, see Section C.2, is in the ECEF frame at the time of transmission. Since the time of transmission for each satellite is distinct, the satellite positions for different satellites are in different reference frames and must be transformed to a consistent frame-of-reference, as discussed in Section C.3. Assuming that the satellite positions are each transformed to the ECEF frame at the time of reception, eqn. (8.16) relates the residual measurement to the position error in the ECEF frame at the time of reception. To be explicit in representation of the reference frame, eqn. (8.16) is rewritten as

$$\delta\boldsymbol{\rho} = \mathbf{H}^e \delta\mathbf{x}^e + \boldsymbol{\chi}. \tag{8.17}$$

It is straightforward to transform this equation to any other frame-of-reference. For frame a,

$$\begin{aligned} \delta\boldsymbol{\rho} &= \left(\mathbf{H}^e \bar{\mathbf{R}}_a^e\right)\left(\bar{\mathbf{R}}_e^a \delta\mathbf{x}^e\right) + \boldsymbol{\chi} \\ &= \mathbf{H}^a \delta\mathbf{x}^a + \boldsymbol{\chi} \end{aligned} \tag{8.18}$$

where $\mathbf{H}^a = \mathbf{H}^e \bar{\mathbf{R}}_a^e$ and

$$\bar{\mathbf{R}}_e^a = \begin{bmatrix} \mathbf{R}_a^e & \mathbf{0}_{3\times 1} \\ \mathbf{0}_{1\times 3} & 1 \end{bmatrix}.$$

Note that the rotation matrix from the ECEF frame to the a-frame $\bar{\mathbf{R}}_e^a$ is orthonormal, as discussed in Section 2.4.

When the frame of interest is the tangent plane frame of the receiver, $\mathbf{H}^t = \mathbf{H}^e \bar{\mathbf{R}}_t^e$ has the useful interpretation:

$$\mathbf{H}^t = \begin{bmatrix} -\cos A^1 \cos E^1 & -\sin A^1 \cos E^1 & \sin E^1 & 1 \\ \vdots & \vdots & \vdots & \vdots \\ -\cos A^m \cos E^m & -\sin A^m \cos E^m & \sin E^m & 1 \end{bmatrix}. \tag{8.19}$$

This expression is useful for computation of the satellite azimuth A^i and elevation E^i:

$$A^i = \operatorname{atan2}(-h_2^i, -h_1^i) \tag{8.20}$$

$$E^i = \operatorname{asin}(h_3^i) \tag{8.21}$$

where $h^i = [h_1^i, h_2^i, h_3^i, 1]$ is the i-th row of $\mathbf{H}^t$. For users near the Earth surface, the horizon ensures that the vector from the satellite to the user always has a positive down component in the tangent frame; therefore, the third component h_3^i should always be positive.

Example 8.2 *The matrix* $\mathbf{H}^t$ *computed for Example 8.1 is*

$$\mathbf{H}^t = \begin{bmatrix} -0.0677 & -0.7733 & 0.6305 & 1.0000 \\ 0.4156 & 0.8385 & 0.3524 & 1.0000 \\ 0.6002 & -0.3149 & 0.7353 & 1.0000 \\ -0.4318 & -0.2299 & 0.8722 & 1.0000 \end{bmatrix}.$$

Using eqns. (8.20–8.21), the computed azimuth and elevation angles for the satellites are shown in Table 8.4. The satellite positions relative to the user can also be conveniently displayed as in Figure 8.2. This figure plots the satellite location as $[-h_2^i, -h_1^i]$ *which is a vector of length* $\cos E^i$*. Therefore the vector length indicates the satellite elevation with the origin being directly overhead (i.e.,* $E^i = 90°$*) and the circle of radius 1.0 being the horizon (i.e.,* $E^i = 0°$*).*

SV PRN	2	26	4	7
Azimuth, A, deg	85	-116	152	28
Elevation, E, deg	39	21	47	61

Table 8.4: Satellite azimuth and elevation angles for Example 8.1.

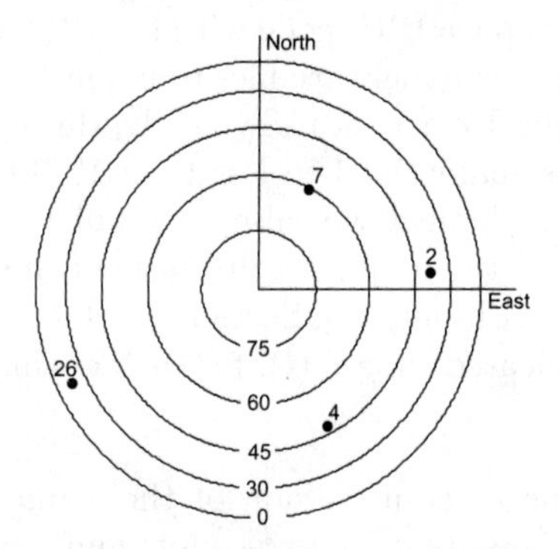

Figure 8.2: Graphical depiction of satellite positions relative to the GPS user. The center of the graph corresponds to elevation angle 90°. The elevation angle of each of the concentric circles is indicated at the bottom of the circle. The satellite locations are indicated by the black dots. Azimuth is measured clockwise from north.

△

8.3 GPS Receiver Overview

The purpose of this section is to provide an overview of GPS receiver operation. The objective is to provide a sufficient understanding to enable the discussion in the subsequent sections.

Modern receivers have several channels each capable of tracking a single CDMA signal. The receiver schedules measurements from all the tracking channels to occur simultaneously according to the receivers indicated time $\tau_r(t)$. Throughout this chapter the symbols $\tau_r(t)$ and $\tau^s(t)$ will be used to represent the receiver and satellite indicated times at GPS time t. When multiple satellites or receivers are being discussed, s and r can be considered as counters over the set of satellites or receivers.

The signal broadcast by GPS satellite s has the form

$$y(\tau^s) = C^s(\tau^s)D^s(\tau^s)\sin(2\pi f_c \tau^s)$$

where C^s is the PRN code assigned to the satellite, D^s is the bit sequence containing the 50 Hz data, and $\sin(2\pi f_c \tau^s)$ is the carrier signal. All of these signals are generated according to the satellite time $\tau^s(t)$ which is distinct for each satellite due to clock drift as discussed in Section 8.4.2.

When the PRN for an available satellite is assigned to a GPS receiver channel, the receiver channel performs the following functions: generates a replica of the PRN code sequence; searches for the code phase and Doppler frequency to align the replica PRN code with the PRN code of the incoming signal; and, then adjusts channel variables to maintain the alignment of the incoming and generated PRN codes. Typically, during tracking, the PRN code alignment can be maintained to less than 1% of one PRN code chip ($< 3m$). The receiver channel will also manipulate channel variables to acquire and track either the frequency or phase of the sinusoidal carrier of the signal at the channel input. See Section 8.4.9.

At the receiver indicated time $\tau_r(t)$, for each channel, the basic channel measurements are:

$\tau^s(t - \Delta t_r^s)$ – the time of transmission of the signal by the satellite, as marked on the signal (e.g., tow, z-count) and tracked by the phase of the code generator (i.e., integer and fractional chip count);

$\tau_r(t)$ – the receiver indicated time;

$\phi_{ca}(\tau_r(t))$ – the fractional portion of the phase carrier accumulator; and

$N_{ca}(\tau_r(t))$ – the whole cycle count of the phase carrier accumulator.

In the above expressions, t is the GPS system time and Δt_r^s is the signal transit time from the satellite to the receiver.

From these basic measurements, various other measurements can be computed. All measurements computed from the basic channel measurements will be referred to as observables.

The transit (pseudo) time, code pseudorange and phase observables are constructed as

$$\Delta\tilde{t}_r^s(\tau_r(t)) = \tau_r(t) - \tau^s(t - \Delta t_r^s), \quad seconds \tag{8.22}$$

$$\tilde{\rho}_r^s(\tau_r(t)) = c\,\Delta\tilde{t}_r^s(\tau_r(t)), \quad meters \tag{8.23}$$

$$\tilde{\phi}_r^s(\tau_r(t)) = N_{ca}(\tau_r(t)) + \phi_{ca}(\tau_r(t)), \quad cycles \tag{8.24}$$

where c is the speed of light in a vacuum. Note that the transit time Δt_r^s is not directly measured, but is computed from the basic measurements. For a two frequency GPS receiver each of the above observables can be computed for each of the carrier frequencies. Each carrier frequency for each satellite requires its own tracking channel (i.e., 12 channels to track both L1 and L2 signals for six SV's).

The pseudorange observable has been discussed extensively in Section 8.2. Its error model is presented in eqn. (8.6) and the use of the pseudorange observable to determine position is discussed in Section 8.2.2. Subsection 8.3.1 presents the error model for the carrier phase observable. Subsection 8.3.2 discusses another receiver output often called the Doppler or delta-pseudorange.

8.3.1 Carrier Phase Observables

The L1 and L2 carrier phase observables for the i-th satellite that is computed by eqn. (8.24) is converted to meters through multiplication by the wavelength and can be modeled as

$$\lambda_1\phi_{r_1}^i = \rho(\mathbf{x}, \hat{\mathbf{p}}^i) + E^i + c\delta t^i - \frac{f_2}{f_1}I_r^i + T_r^i + M_{\phi_1}^i + \nu_{\phi_1}^i + N_1^i\lambda_1 \tag{8.25}$$

$$\lambda_2\phi_{r_2}^i = \rho(\mathbf{x}, \hat{\mathbf{p}}^i) + E^i + c\delta t^i - \frac{f_1}{f_2}I_r^i + T_r^i + M_{\phi_2}^i + \nu_{\phi_2}^i + N_2^i\lambda_2 \tag{8.26}$$

where $\lambda_j = \frac{c}{f_j}$ is the wavelength of the carrier signal, f_j is the carrier frequency, and $j = 1,\ 2$. The remaining symbols are defined as follows:

$\phi_{r_j}^i$ – This is the receiver generated carrier phase observable. It is computed as in eqn. (8.24).

$M_{\phi_j}^i$ – This symbol represents the carrier signal multipath error described in Section 8.4.7.

$\nu_{\phi_j}^i$ – This symbol represents the random measurement noise described in Section 8.4.8.

N_j^i – This symbol represents the integer phase ambiguity which is defined in Section 8.4.9.

The symbols $\rho(\mathbf{x}, \hat{\mathbf{p}}^i)$, E^i, $c\delta t^i$, I_r^i, and T^i are defined following eqn. (8.3). The majority of the discussion to follow will focus on the L1 phase signal and will drop the subscript '1' to simplify the notation; however, the discussion directly extends to the L2 phase with appropriate changes to the ionospheric error term.

The common-mode errors

$$e_{cm}^i = E^i + c\delta t^i - \frac{f_2}{f_1} I_r^i + T_r^i \tag{8.27}$$

are essentially the same as those on the code observable, except that the ionospheric error I_r^i enters eqns. (8.6) and (8.25) with opposite signs. The interest in the carrier signal stems from the fact that the non-common mode errors M_ϕ^i and ν_ϕ^i are typically less that 1% of the magnitude of the respective errors for the code pseudorange. When the common-mode errors can be mitigated or removed, the carrier measurements allow position estimation at the centimeter level.

The integer phase ambiguity is a (usually large) unknown integer constant (barring cycle slips). To make use of carrier phase observable as a range estimate, the integer ambiguity must be determined (see Section 8.9). When the integer ambiguity N^i is known and the common mode errors are removed (see Section 8.8), then assuming that phase lock is maintained, eqn. (8.25) can be rewritten as

$$\lambda \left(\phi_r^i - N^i \right) = \rho^i + M_\phi^i + \nu_\phi^i \tag{8.28}$$

where the left-hand side is known. This equation has the same form as eqn. (8.6) with substantially fewer error terms. Since the structure of eqns. (8.6) and (8.28) are identical, given a vector of integer-resolved carrier phase measurements, the vector of equations with the form of eqn. (8.28) can be solved for the position and clock bias by the method of Section 8.2.2.

8.3.2 Delta Pseudorange Observable

The expression for the Doppler frequency:

$$f_r = f_T \left(1 - \frac{\dot{R}}{c} \right) \tag{8.29}$$

relates the frequency received by a user f_r to the rate of change of the range between the receiver and the transmitter. In this expression, f_T is the transmitted frequency and R is the geometric range between the user and transmitter. The Doppler shift is

$$f_r - f_T = -f_T \frac{\dot{R}}{c}. \tag{8.30}$$

In GPS receivers the delta pseudorange observable is a measurement of the quantity

$$\Delta\rho(\tau_r(t)) = \rho_r^s(\tau_r(t)) - \rho_r^s(\tau_r(t) - T) \tag{8.31}$$

where typically $T \leq 1.0$. This quantity is the change in range over the time interval $\tau_r \in [\tau_r(t) - T, \tau_r(t)]$:

$$\Delta\rho(\tau_r(t)) = \int_{\tau_r(t)-T}^{\tau_r(t)} \dot{\rho}_r^s(q)dq. \tag{8.32}$$

Therefore, if the delta pseudorange is divided by T it measures the average rate of change of the pseudorange over the indicate time interval. By the mean value theorem, there is a value of $q \in [\tau_r(t) - T, \tau_r(t)]$ such that $\dot{\rho}_r^s(q) = \frac{1}{T}\Delta\rho(\tau_r(t))$. By this reasoning the delta pseudorange observable is often modeled as the rate of change of the pseudorange at the midpoint of the interval $\tau_r \in [\tau_r(t) - T, \tau_r(t)]$ and referred to as the Doppler observable. The validity of this modeling assumption depends on the bandwidth of the vehicle B relative to T.

Eqn. (8.31) is the definition of the delta pseudorange observable, but the observable is not constructed by differences of the pseudorange observable. Instead, for a channel that has achieved phase lock for the carrier signal, the delta pseudorange observable is constructed from differences of the carrier phase observable over the period T. In this case, cycle slips or loss of phase lock during the interval, if not detected, will result in erroneous observables. For a channel that has only achieved frequency lock for the carrier signal, the delta pseudorange is constructed from the channel's replica Doppler frequency. In this case, loss of frequency lock during the interval will result in erroneous observables if not detected.

Assuming that the receiver has and maintains phase lock over the interval $\tau_r \in [\tau_r(t) - T, \tau_r(t)]$ that is of interest, then the Doppler observable can be computed as

$$\Delta\rho(\tau_r(t)) = \lambda\left(\phi_r^i(\tau_r(t)) - \phi_r^i(\tau_r(t) - T)\right). \tag{8.33}$$

The analysis to follow assumes that the product BT is sufficiently small that the vehicle velocity cannot change significantly over the period T. With this assumption it is reasonable to consider $D(\tau_r(t)) = \frac{\Delta\rho(\tau_r(t))}{\lambda T}$ as a measurement of the Doppler shift (in Hz) at the receiver indicated time. By eqn. (8.25) the Doppler measurement model is

$$\lambda T D_r^s(\tau_r) \quad = \quad (\rho_r^s(\tau_r) - \rho_r^s(\tau_r - T)) - c\Delta\dot{t}^i + \varepsilon(\tau_r)$$

where t arguments have been dropped to simplify notation, the temporal differences for E^i, I_r^i, and T_r^i have been dropped because they are small relative to the other terms, $\varepsilon(\tau_r)$ represents the measurement error due to multipath and receiver noise, and

$$\Delta\dot{t}^i = \Delta t^i(\tau_r) - \Delta t^i(\tau_r - T)$$

is the (uncorrected) satellite clock drift rate. The symbol $\Delta\dot{t}^i$ is used instead of $\delta\dot{t}^i$ as a reminder to the reader to correct the Doppler measurement for the satellite clock drift rate as described in eqn. (8.35).

Assuming that the line-of-sight vector from the satellite to the user $\vec{\mathbf{h}} = \frac{\mathbf{p}_r - \mathbf{p}^s}{\|\mathbf{p}_r - \mathbf{p}^s\|}$ is available from the position solution, a linearized model for the Doppler measurement is

$$\lambda D_r^s = \vec{\mathbf{h}}^\top(\mathbf{v_r} - \mathbf{v}^s) + c\Delta\dot{t}_r - c\Delta\dot{t}^i + \varepsilon. \tag{8.34}$$

The satellite velocity computation is described in Section C.4. The satellite clock drift rate $c\Delta\dot{t}^i$ is predicted by the broadcast model to be $a_{f1}T$, see Section C.1. The Doppler residual measurement is defined as

$$\lambda\delta D_r^s = \lambda D_r^s + \vec{\mathbf{h}}^\top\hat{\mathbf{v}}^s + ca_{f1}T, \tag{8.35}$$

which is the Doppler measurement corrected for satellite velocity and satellite clock drift rate. Substituting eqn. (8.34) into eqn. (8.35) yields the Doppler measurement error model

$$\begin{aligned}\lambda\delta D_r^s &= \vec{\mathbf{h}}^\top\mathbf{v}_r + c\Delta\dot{t}_r + \varepsilon \\ &= \mathbf{h}\dot{\mathbf{x}} + \varepsilon \end{aligned}\tag{8.36}$$

where the residual satellite clock error is small and has been dropped from the model.

8.4 GPS URE Characteristics

Eqns. (8.6) and (8.25) indicated that the GPS range and phase measurements can be represented as the sum of the geometric range from the user antenna position to the computed satellite position and various error terms. The cumulative effect of these errors is referred to as the User Equivalent Range Error (UERE) or User Range Error (URE). The objective of this section is to derive and discuss each of these error terms. The purpose is to clarify the source and characteristics of each error source so that each can be mitigated or modeled to the extent possible.

This section is not required reading the first time through the chapter. On the second reading the subsections are meant to be read in the order presented. They will start with the natural GPS measurements defined in eqns. (8.22–8.24) and work through the derivation of eqns. (8.6) and (8.25) one step at a time, but the presentation will need to touch on a variety of detailed topics that will seem to be a distraction to readers new to GPS.

From eqns. (8.22) and (8.23), the pseudorange observable is

$$\tilde{\rho}_r^s(\tau_r(t)) = c\left(\tau_r(t) - \tau^s(t - \Delta t_r^s)\right); \tag{8.37}$$

therefore, the discussion will begin with clock modeling.

8.4.1 Clocks

For centuries, clocks have been constructed using the cumulative phase angle from stable oscillators. Denote the nominal frequency of the stable oscillator as f_o and the actual frequency of the clock oscillation as

$$f_c(t) = f_o + \delta f(t). \tag{8.38}$$

The phase angle of the oscillator is the integral of the frequency

$$\dot{\phi}_c(t) = f_c(t); \tag{8.39}$$

therefore,

$$\begin{aligned}\phi_c(t) &= \phi_c(t_o) + \int_{t_o}^{t} f_c(q)dq \\ &= \phi_c(t_o) + f_o(t - t_o) + \int_{t_o}^{t} \delta f(q)dq.\end{aligned}$$

The clock time is computed as

$$\tau(t) = \frac{1}{f_o}\phi_c(t) \tag{8.40}$$

which is equivalent to

$$\begin{aligned}\tau(t) &= t + \tau(t_o) + \frac{1}{f_o}\int_{t_o}^{t} \delta f(q)dq \\ &= t + \Delta\tau(t)\end{aligned} \tag{8.41}$$

where $\tau(t_o) = \frac{\phi_c(t_o)}{f_o} - t_o$ and $\Delta\tau(t) = \tau(t_o) + \frac{1}{f_o}\int_{t_o}^{t} \delta f(q)dq$.

In the above analysis t represents true time, which for the purposes of this chapter is defined to be the GPS system time as maintained by the United States Naval Observatory (USNO). Eqn. (8.41) shows that the clock time can be modeled as true time plus a time varying error term that is determined by the initialization error $\tau(t_o)$ plus a time-varying (clock drift) error which is the integral of the frequency error $\delta f(q)$. Obviously, oscillators for which the frequency error is small and stable result in better clocks.

A pendulum clock nicely exemplifies the principle described above. The pendulum swings at a resonant frequency. A set of gears count the swings of the pendulum and display the cumulative phase of the oscillator using hands on the face of the clock. The user may be able to tune the resonant frequency by adjusting the length of the pendulum to decrease δf.

Modern, high accuracy clocks such as those used in GPS satellites and receivers use the same principle, but use very stable crystal oscillators with electronic counters to accumulate the oscillator phase.

Eqn. (8.41) applies to both the satellite and receiver clocks:

$$\tau^s(t) = t + \Delta\tau^s(t) \quad (8.42)$$

$$\tau_r(t) = t + \Delta\tau_r(t) \quad (8.43)$$

where $\Delta\tau^s(t)$ is the error in the clock of satellite s and $\Delta\tau_r(t)$ is the error in the clock of receiver r. Substituting eqn. (8.42–8.43) into eqn. (8.37) yields

$$\begin{aligned}\tilde{\rho}_r^s(\tau_r(t)) &= c\Big(t + \Delta\tau_r(t) - \big(t - \Delta t_r^s + \Delta\tau^s(t - \Delta t_r^s)\big)\Big) \\ &= c\big(\Delta t_r^s + \Delta\tau_r(t) - \Delta\tau^s(t - \Delta t_r^s)\big) \quad (8.44)\end{aligned}$$

which shows that the measured pseudorange $\tilde{\rho}_r^s(\tau_r(t))$ is proportional to the time of propagation Δt_r^s plus the receiver clock error $\Delta\tau_r(t)$ minus the satellite clock error $\Delta\tau^s(t - \Delta t_r^s)$ at the time of transmission. The time of propagation will be further analyzed in subsequent sections.

8.4.2 Satellite Clock Bias, $c\delta t^s$

The control segment monitors and fits a polynomial correction to $c\Delta t^s(t)$ for each satellite. Each satellite broadcasts the parameters for its clock correction model. The user reads these parameters and corrects for the predictable portion of the satellite clock error as discussed in Section C.1. The residual uncorrected satellite clock error is denote as δt^s. Its effect on the pseudorange measurement is scaled by the speed of light.

This residual satellite clock error will effect the pseudorange and carrier phase measurements as indicate in eqns. (8.6) and (8.25). The rate

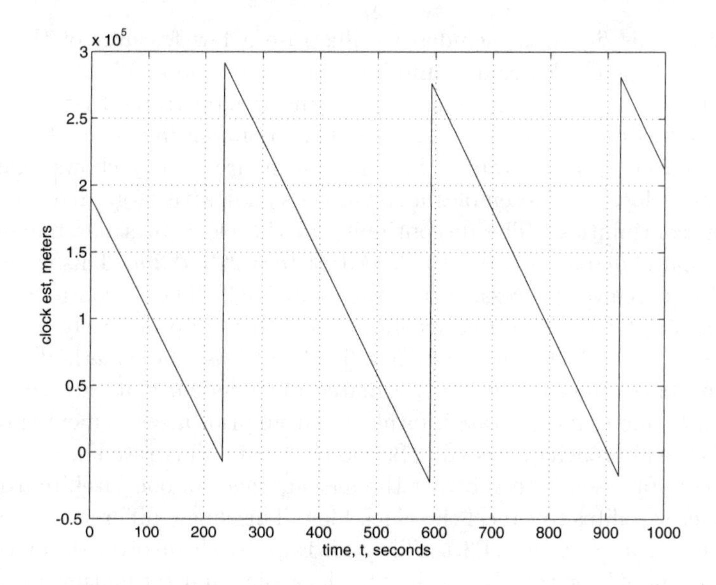

Figure 8.3: Estimated clock bias during L1 C/A code position estimation.

of change of the satellite clock error affects the Doppler measurement as described in eqn. (8.34).

The typical range error due to this term is shown in Table 8.5. Satellite clock error affects all users (i.e., those using C/A-code, P-code, and/or two frequency receivers) in the same fashion. The error is independent of the location of the user. Therefore, the satellite clock component of differential corrections is accurate for all users regardless of position.

8.4.3 Receiver Clock Error, $\Delta\tau_r$

Receiver clock bias is a time varying error that affects all simultaneous range measurements in the same fashion. Therefore, if at least four simultaneous satellite range measurements are available (one more than would be necessary to determine position based on true ranges), the clock bias and position can both be estimated. For this reason, clock bias error is not included in the position error budget discussed relative to Table 8.5.

Different GPS receivers handle the receiver clock error in different ways, but there are two basic approaches. In the one approach, the receiver clock error is allowed to accumulate. In the other approach, the receiver clock error is estimated and steered towards zero. A user working with raw observables should know which approach is being used to ensure that the clock bias is addressed appropriately.

Figure 8.3 shows a sequence of estimates of the receiver clock bias

$c\Delta\tau_r(t)$. This figure is included to illustrate a few features of the clock bias. First, the clock bias is ramp-like with stable slope (clock drift rate). This fact is in accord with eqn. (8.41) and demonstrates that while the oscillator frequency $f_c(t)$ does not equal the nominal frequency f_o the frequency offset $\delta f(t)$ is stable. Second, the figure clearly shows that the estimated clock bias resembles a sawtooth signal with large, fixed magnitude discontinuities. The discontinuity in the clock bias has magnitude approximately equal to $1.0 \times 10^{-3} s$ or $0.001c = 299,792m$. This is an artifact of the receiver processing, not the clock itself. The discontinuities are irrelevant as long as the user accommodates them appropriately.

The receiver clock bias can be handled by at least three methods. In the first approach, the clock bias is estimated independently at each time step without trying to incorporate information from prior measurement epoches. That is the approach discussed in Section 8.2.2 and illustrated in Figure 8.3. A second approach is to subtract the measurement of one satellite from all the other satellites to cancel the clock bias. This single difference approach is discussed in Section 8.4.3.1. The third approach involves developing a dynamic model for the change in the clock bias and estimating the clock model state via Kalman filtering. This third approach is presented Section 8.4.3.2.

8.4.3.1 Single Differences Across Satellites

This section discusses single differencing using the pseudorange observable. The same technique with appropriate modifications can be used for the Doppler and integer-resolved carrier phase measurements.

For simultaneous measurements from the same receiver, each pseudorange can be modeled as

$$\tilde{\rho}_r^i = R(\mathbf{p}, \hat{\mathbf{p}}^i) + c\Delta t_r + \chi^i \tag{8.45}$$

where $\sigma_\rho^2 = var(\chi^i)$ and time arguments have been dropped to simplify notation; however, it is critical to note that the receiver clock bias $c\Delta t_r(t)$ is the same for all satellites. Consider a set of measurements from m satellites. Subtracting the $\tilde{\rho}_r^m$ from each of the other measurements yields the new observable

$$\nabla\tilde{\rho}^{i,m} = \tilde{\rho}_r^i - \tilde{\rho}_r^m \tag{8.46}$$

that is modeled as

$$\nabla\tilde{\rho}^{i,m}(\mathbf{p}) = R(\mathbf{p}, \hat{\mathbf{p}}^i) - R(\mathbf{p}, \hat{\mathbf{p}}^m) + \chi^{i,m} \tag{8.47}$$

for $i = 1, \ldots, (m-1)$ where $\chi^{i,m} = (\chi^i - \chi^m)$. This shows that $\nabla\rho^{i,m}$ is unaffected by the receiver clock bias. However, the noise term $\chi^{i,m}$ is now larger and correlated between satellites with $var(\chi^{i,m}, \chi^{i,m}) = 2\sigma_\rho^2$ and $cov(\chi^{i,m}, \chi^{j,m}) = \sigma_\rho^2$ for $i, j = 1, 2, (m-1)$ and $i \neq j$.

Eqn. (8.47) contains three unknowns which are the components of $\mathbf{p}$. Therefore, at least three equations will be required to determine a solution. However, m must still be greater or equal to 4 because the m-th pseudorange is subtracted from the first $(m-1)$ pseudoranges.

For a set of $m \geq 4$ satellites, with the m-th measurement subtracted from each of the others, $(m-1)$ equations result. This set of $(m-1)$ equations can be solved as in Sections 8.2.2 and B.14.2, using

$$\hat{\mathbf{p}}_{k+1} = \hat{\mathbf{p}}_k + \left(\mathbf{H}^\top \mathbf{H}\right)^{-1} \mathbf{H}^\top \left(\nabla\tilde{\rho}^{i,m} - \left(R(\hat{\mathbf{p}}_k, \hat{\mathbf{p}}^i) - R(\hat{\mathbf{p}}_k, \hat{\mathbf{p}}^m)\right)\right)$$

where

$$\mathbf{H} = \left[\begin{array}{c} \frac{\delta}{\delta \mathbf{p}} \nabla\rho^{1,m} \\ \vdots \\ \frac{\delta}{\delta \mathbf{p}} \nabla\rho^{m-1,m} \end{array}\right]\Bigg|_{\hat{\mathbf{p}}_k} = \left[\begin{array}{c} \vec{\mathbf{h}}^1 - \vec{\mathbf{h}}^m \\ \vdots \\ \vec{\mathbf{h}}^{m-1} - \vec{\mathbf{h}}^m \end{array}\right] \tag{8.48}$$

where $\vec{\mathbf{h}}^i = \frac{\hat{\mathbf{p}}_k - \hat{\mathbf{p}}^i}{\|\mathbf{p}_k - \hat{\mathbf{p}}^i\|}$ is represented as a row vector. Alternatively, to properly account for the correlation in $\chi^{i,m}$, the algorithm described in Exercise B.17 should be used instead, with $\mathbf{R}$ defined as the (non-diagonal) measurement correlation matrix.

This section has discussed single differences across satellites for simultaneous measurements from a single receiver. Such satellite differencing, when used, is usually implemented in the form of double differencing. This technique is discussed in Section 8.8.3. Double differencing is one form of differential GPS, which is discussed more generally throughout Section 8.8.

8.4.3.2 Receiver Clock Bias Dynamic Model

Physically, as discussed relative to eqn. (8.41), the clock bias develops as the integral of the clock oscillator frequency error $\delta f(t)$. This physical process suggests a two state clock model described as

$$\left[\begin{array}{c} \dot{v}_1 \\ \dot{v}_2 \end{array}\right] = \left[\begin{array}{cc} 0 & 1 \\ 0 & 0 \end{array}\right] \left[\begin{array}{c} v_1 \\ v_2 \end{array}\right] + \left[\begin{array}{c} w_\phi \\ w_f \end{array}\right] \tag{8.49}$$

$$\Delta t_r = \left[\begin{array}{cc} 1 & 0 \end{array}\right] \mathbf{v} \tag{8.50}$$

where $\mathbf{v} = [v_1, v_2]^\top$ and w_ϕ and w_f are the process noise driving the clock phase and frequency error states, respectively. Assuming that the two noise processes are stationary, independent, and white with power spectral densities of S_ϕ and S_f, respectively, then for $\boldsymbol{\omega} = [w_\phi, w_f]^\top$

$$\mathbf{S}_{\boldsymbol{\omega}} = \mathbf{Q} = \left[\begin{array}{cc} S_\phi & 0 \\ 0 & S_f \end{array}\right]. \tag{8.51}$$

This two state model corresponds well with the data presented in Figure 8.3. The frequency error is modeled as a constant plus random walk (i.e., slowly changing constant). The phase error is modeled as the integral of the frequency error (i.e., ramp) plus a random walk. This is shown in block diagram form in Figure 8.4.

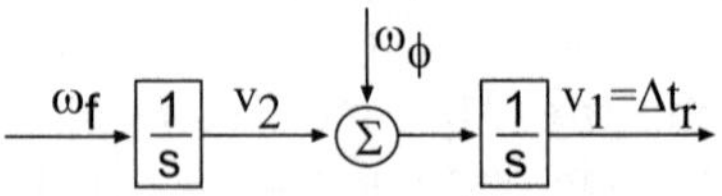

Figure 8.4: Continuous-time two state clock model.

For this clock model,

$$\mathbf{F} = \begin{bmatrix} 0 & 1 \\ 0 & 0 \end{bmatrix}.$$

For the discrete-time sampling period T_s, the state space model is

$$\begin{bmatrix} v_1(k+1) \\ v_2(k+1) \end{bmatrix} = \begin{bmatrix} 1 & T_s \\ 0 & 1 \end{bmatrix} \begin{bmatrix} v_1(k) \\ v_2(k) \end{bmatrix} + \boldsymbol{\omega}_d. \tag{8.52}$$

Because $\mathbf{F}^2$ is identically zero, $\mathbf{Qd}$ can be calculated exactly using the first three terms of eqn. (4.119):

$$cov(\boldsymbol{\omega}_d) = \mathbf{Qd} = \begin{bmatrix} S_\phi T_s + \frac{T_s^3}{3} S_f & \frac{T_s^2}{2} S_f \\ \frac{T_s^2}{2} S_f & S_f T_s \end{bmatrix}. \tag{8.53}$$

To complete the clock error model, the spectral densities of the driving noise processes must be specified. This can be accomplished by fitting the clock error variance

$$S_\phi T_s + \frac{T_s^3}{3} S_f \tag{8.54}$$

as specified in eqn. (8.53) to the Allan variance for the clock error. This is specified in [29, 130] to be

$$\frac{h_0}{2} T_s + 2h_{-1} T_s^2 + \frac{2}{3}\pi^2 h_{-2} T_s^3, \tag{8.55}$$

where h_0, h_{-1}, and h_{-2} are Allan variance parameters. Since the second order error model cannot fit the Allan variance exactly (in fact, an exact fit is not possible for any finite order, linear, state space model [29]), the parameters S_ϕ and S_f can be selected to optimize the fit in the vicinity of the known value of T_s. This method and the tradeoffs involved are thoroughly discussed in [29, 31, 130]. When S_ϕ and S_f are selected so that eqns. (8.54) and (8.55) match for two time intervals denoted T_1 and T_2 the least squares solution is

$$\begin{bmatrix} S_\phi \\ S_f \end{bmatrix} = \frac{\begin{bmatrix} T_2^3 & -T_1^3 \\ -3T_2 & 3T_1 \end{bmatrix}}{T_1 T_2^3 - T_2 T_1^3} \begin{bmatrix} T_1 & T_1^2 & T_1^3 \\ T_2 & T_2^2 & T_2^3 \end{bmatrix} \begin{bmatrix} \frac{h_0}{2} \\ 2h_{-1} \\ \frac{2}{3}\pi h_{-2} \end{bmatrix}. \tag{8.56}$$

Example 8.3 *For a temperature compensated crystal oscillator, the Allan variance parameters are* $h_0 = 2 \times 10^{-19} \frac{sec^2}{s}$, $h_{-1} = 7 \times 10^{-21} \frac{sec^2}{s^2}$, *and* $h_{-2} = 2 \times 10^{-20} \frac{sec^2}{s^3}$, *where the units sec and s have been used to distinguish between the two meanings of time in this example. Let* $T_1 = 1s$ *and* $T_2 = 10s$, *for an application where GPS measurements will be taken with* $T_s = 1s$. *Then, eqn. (8.56) results in* $S_\phi = 1.1 \times 10^{-19} \frac{sec^2}{s}$ *and* $S_f = 4.3 \times 10^{-20} \frac{sec^2}{s^3}$ *which can be converted to estimate clock bias in meters by multiplication by the square of the speed of light. The resulting value of the discrete-time process noise matrix, scaled to meters, is*

$$Q_d = \begin{bmatrix} 0.0114 & 0.0019 \\ 0.0019 & 0.0039 \end{bmatrix}. \tag{8.57}$$

△

8.4.4 Atmospheric Delay, $c\delta t_a^s$

The satellite signals travel distances of $20,000 - 26,000km$ to reach users near the surface of Earth. Typical scenarios are illustrated in Figure 8.5. A satellite $\mathbf{P}^1$ is near zenith with respect to the user. A second satellite $\mathbf{P}^2$ is near the horizon of the user. The atmosphere of Earth is the roughly 1000km wide layer of gases surrounding the planet. The figure is not to scale in the sense that the distance from the surface of Earth to the outer edge of the atmosphere (i.e., the dotted line) should be 5% of the length of the $\mathbf{v}_1$ vector. The figure illustrates several points that will be important in the discussion of this section. First, the length of the path of the signal through the atmosphere is clearly dependent on the satellite elevation angle. The minimum exposure to the atmosphere is for a satellite at zenith. This issue is addressed by *obliquity* or *slant factors* that are functions of the elevation angle. Second, the effect of the ionosphere is dependent on the path through the atmosphere which depends on both the satellite and user locations. The figure shows an asterisk at the midpoint of the portion of each path that is through the ionosphere. This point is called the *ionospheric pierce point.* The projection of this point onto the Earth surface can be far from the user location, especially for satellites at low elevation. When ionospheric models are used, the models are computed at an ionospheric pierce point.

Along the portion of the GPS signal path above the Earth atmosphere, the signal travels through a vacuum with speed $c = 2.99792458 \times 10^8 \frac{m}{s}$. Within the Earth atmosphere, the surrounding medium affects both the speed and path of signal propagation; however, the effect on the signal path is much less significant than the effect of the decreased speed of travel v. The ratio of c to v is the refractive index

$$\eta = \frac{c}{v}.$$

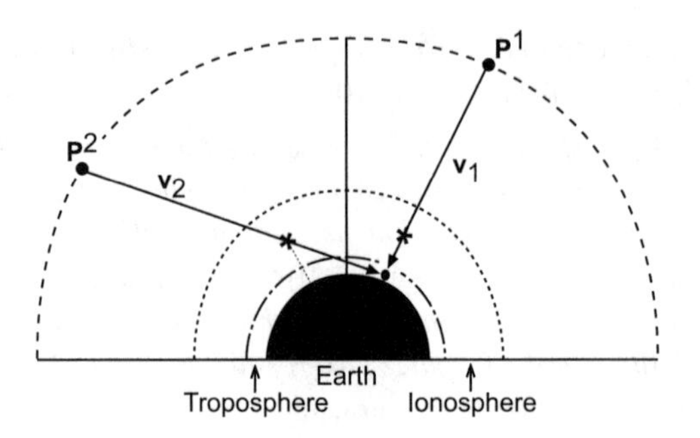

Figure 8.5: Simplified illustration of GPS signal propagation to Earth. The figure is not to scale.

The speed v and refractive index η are not constants; instead, they vary with the properties of the surrounding medium.

When the refractive index η is a function of the frequency f of the signal, then the medium is called dispersive. For the modeling of GPS atmospheric delays, it is sufficient to divide the atmosphere into two layers, the troposphere and the ionosphere, where the troposphere is non-dispersive and the ionosphere is dispersive. Each of these layers is discussed in one of the next two sections, after the derivation of eqns. (8.58-8.59).

Let the path of propagation from the satellite antenna to the user antenna be denoted by $\gamma(l)$ where l is the distance along the path. The time-of-travel from the satellite to the user is $\Delta t_r^s = \int_{\gamma(l)} \frac{1}{v(l)} dl$. Representing γ as

$$\gamma = \gamma_S + \gamma_I + \gamma_T$$

where γ_S is the portion of the path through space, γ_I is the portion of the path through the ionosphere, and γ_T is the portion of the path through the troposphere, the time-of-travel can be expressed as

$$\Delta t_r^s = \int_{\gamma_S(l)} \frac{1}{v(l)} dl + \int_{\gamma_I(l)} \frac{1}{v(l)} dl + \int_{\gamma_T(l)} \frac{1}{v(l)} dl.$$

The time-of-travel for a signal traveling the same path at the speed of light is

$$\Delta \hat{t}_r^s = \int_{\gamma_S(l)} \frac{1}{c} dl + \int_{\gamma_I(l)} \frac{1}{c} dl + \int_{\gamma_T(l)} \frac{1}{c} dl.$$

The delay due to the decreased speed through the atmosphere is

$$\begin{aligned} \delta t_a^s &= \Delta t_r^s - \Delta \hat{t}_r^s \\ &= \int_{\gamma_S} \left(\frac{1}{v(l)} - \frac{1}{c} \right) dl + \int_{\gamma_I} \left(\frac{1}{v(l)} - \frac{1}{c} \right) dl + \int_{\gamma_T} \left(\frac{1}{v(l)} - \frac{1}{c} \right) dl \\ &= \frac{1}{c} \int_{\gamma_I} (\eta(l) - 1)\, dl + \frac{1}{c} \int_{\gamma_T} (\eta(l) - 1)\, dl. \end{aligned} \tag{8.58}$$

The first term in eqn. (8.58) is the ionospheric delay. The second term in eqn. (8.58) is the tropospheric delay. The increased path length is

$$\delta\rho_a^s = c\delta t_a^s = \int_{\gamma_I} (\eta(l) - 1)\, dl + \int_{\gamma_T} (\eta(l) - 1)\, dl. \tag{8.59}$$

Eqns. (8.58–8.59) will be used in the following two subsections.

8.4.4.1 Ionospheric Delay, I_r^s

The ionosphere is the layer of the atmosphere with altitude between 50 and 1000 km that contains free electrons and positively charged molecules. The level of ionization is affected by solar activity, season, and time-of-day. Changes in the level of ionization affect the refractive index along the path through the ionosphere and therefore affect the travel time measured by the receiver.

For L-Band (e.g., GPS) signals, the ionosphere is a dispersive medium. For a modulated signal traveling through dispersive medium, the medium will affect the carrier and modulating signals differently. Let

$$y(t) = m(t)\sin(\omega_c t)$$

where $m(t)$ represents a modulating signal with bandwidth significantly smaller than the carrier frequency $\omega_c = 2\pi f_c$. The modulating signal $m(t)$ propagates at the *group velocity* as determined by the group refractive index $\eta_g(l)$ while the carrier signal $\sin(\omega_c t)$ propagates at the *phase velocity* determined by the phase refractive index $\eta_\phi(l)$. Detailed analyses are presented on p. 136 in [99] and p. 309 in [76]. To first order, the phase and group refractive indices can be modeled, respectively, as

$$\eta_\phi(l) = 1 - \frac{\kappa}{f_c^2} N_e(l) \tag{8.60}$$

$$\begin{aligned} \eta_g(l) &= \eta_\phi(l) + f_c \frac{d\eta_\phi}{df_c} \\ &= 1 + \frac{\kappa}{f_c^2} N_e(l) \end{aligned} \tag{8.61}$$

where $\kappa = 40.28$ and $N_e(l)$ is the density of free electrons at location $\gamma(l)$. Therefore, using the first term in eqn. (8.58), the delay experienced by the code is

$$\delta t_g = \frac{1}{c}\frac{\kappa}{f_c^2} \int_{\gamma_I} N_e(l) dl \tag{8.62}$$

and the delay experience by the carrier phase is

$$\delta t_\phi = -\frac{1}{c}\frac{\kappa}{f_c^2} \int_{\gamma_I} N_e(l) dl. \tag{8.63}$$

To first order, the group and phase experience the same delay with the same magnitude, but opposite in sign. The code is delayed while the phase is advanced. This phenomenon is referred to as *code carrier divergence.*

Defining the total electron count (TEC) along the path as

$$TEC = \int_{\gamma_I} N(l)dl$$

and defining $I_r^i = \frac{40.28}{f_1 f_2} TEC$, then at the L1 frequency the code and phase delays can be expressed simply as

$$\delta t_g = \frac{1}{c}\frac{f_2}{f_1} I_r^i \quad \text{and} \quad \delta t_\phi = -\frac{1}{c}\frac{f_2}{f_1} I_r^i. \tag{8.64}$$

At the L2 frequency the code and phase delays are similarly expressed as

$$\delta t_g = \frac{1}{c}\frac{f_1}{f_2} I_r^i \quad \text{and} \quad \delta t_\phi = -\frac{1}{c}\frac{f_1}{f_2} I_r^i. \tag{8.65}$$

Two frequency receivers are designed to take advantage of the frequency dependence to estimate I_r^i as is described in Section 8.6.

Single frequency receivers must rely on either differential operation or an ionospheric delay model. One such approach based on parameters broadcast by the satellite is described in Section C.5 of Appendix C. This Klobuchar model is expected to compensate for approximately 50% of the ionospheric delay [6]. Since TEC is path dependent, the ability of differential GPS techniques to compensate for ionospheric errors will depend on the user to reference station baseline vector.

8.4.4.2 Tropospheric Delay, T_r^s

The troposphere is the lower part of the atmosphere extending nominally to 50 km above the surface of the planet. The troposphere is composed essentially of electrically neutral particles and for L-Band signals it is non-dispersive. The troposphere experiences changes in temperature, pressure, and humidity associated with weather. Because these same variables affect the density of the air mass along the signal path and the index-of-refraction is a function of air mass density, tropospheric conditions affect the measured time-of-propagation with the measured value being larger than that the geometric range.

Tropospheric delays

$$T_r^s = \frac{1}{c}\int_{\gamma_T} (\eta(l) - 1)\, dl \tag{8.66}$$

can be quite considerable ($\sim 30m$) for satellites at low elevations. Tropospheric delay errors are consistent between the L1 and L2 signals and carrier and code signals. The tropospheric delay cannot be computed from the GPS observables; therefore, tropospheric effects are compensated via models. For most users, especially in navigation application, the input parameters for the models are average or typical values for the user location. Users with higher accuracy requirements can add meteorological instruments to sense the model input variables or may use differential GPS methods.

The refractive index is affected differently by the water vapor and by the dry components (e.g., nitrogen and oxygen) of the troposphere, so tropospheric models account for these wet and dry pressures separately. The wet component is difficult to predict due to local variations in the water vapor content of the troposphere and accounts for approximately 10% of the tropospheric delay. The dry component is relatively easier to predict and accounts for approximately 90% of the tropospheric delay.

Several models exist for the tropospheric wet and dry components [4, 104, 119, 122]. The models contain two parts. The first part of the model is the estimate of the zenith delay. The second part is a slant factor to account for the satellite elevation. For example, the Chao model is

$$\begin{aligned}
\delta\rho_{dry} &= 2.276 \times 10^{-5} P \\
F_{dry} &= \frac{1}{\sin(E) + \frac{0.00143}{\tan(E)+0.0445}} \\
\delta\rho_{wet} &= 4.70 \times 10^{2}\frac{e_o^{1.23}}{T^2} + 1.705 \times 10^{6}\alpha\frac{e_o^{1.46}}{T^3} \\
F_{wet} &= \frac{1}{\sin(E) + \frac{0.00035}{\tan(E)+0.017}} \\
\delta\hat{\rho}_T &= \delta\rho_{dry}F_{dry} + \delta\rho_{wet}F_{wet}
\end{aligned}$$

where $\delta\rho$ is the tropospheric delay expressed in meters, P is the atmospheric pressure in $\frac{N}{m^2}$, T is the temperature in °K, e_o is the partial pressure of water vapor in millibars, α is the temperature lapse rate in °K per meter, and E is the satellite elevation angle. The partial pressure e_o can be computed from T and the relative humidity. In most navigation applications, tropospheric delay is not compensated by equations such as this due to the expense involved in measuring the required formula input variables. Instead, simplified formulas have been determined which only depend on satellite elevation, receiver altitude, and satellite altitude.

The Magnavox and Collins algorithms are, respectively,

$$\begin{aligned}
\delta\hat{\rho}_{T_M} &= \frac{2.208}{\sin(E)}\left(e^{\frac{-h_r}{6900}} - e^{\frac{-h_s}{6900}}\right) \\
\delta\hat{\rho}_{T_C} &= \frac{2.4225}{0.026 + \sin(E)}e^{\frac{-h_r}{7492.8}}
\end{aligned}$$

where h_r and h_s are the receiver and satellite altitudes in meters and the rest of the variables are as previously defined. The Magnavox and Collins tropospheric correction models match each other and the Chao model (for standard assumptions) to within one meter for elevation angles greater than 15 degrees. The Magnavox model matches the Chao model more closely for elevation angles less than 5 degrees [119].

Each non-differential user should correct the measured range for tropospheric delay. The uncorrected portion of the tropospheric delay remains as the measurement error denoted as T_r^i.

Since tropospheric delay is dependent on local variables, receiver altitude, and the user-satellite line-of-sight, the ability of differential techniques to compensate for tropospheric effects will depend on the position of the user relative to the base station. The user of differential corrections must know whether the differential station is compensating for tropospheric delay in the broadcast corrections. If the broadcast corrections include tropospheric error and the user is at a different altitude than the reference station, the user can correct the broadcast corrections for tropospheric delay at the reference station and correct the measured range for tropospheric delay at the user location.

8.4.5 Ephemeris Errors, E^s

As illustrated in Figure 8.6, let the satellite position be $\mathbf{p}^s$ and the computed satellite position be $\hat{\mathbf{p}}^s$. The objective in this section is to determine the effective range error due to using $\hat{\mathbf{p}}^s$ instead of $\mathbf{p}^s$ in the range equations. Let $\mathbf{p}$ denote the receiver position. Figure 8.6 represents the plane determined by the three points $\mathbf{p}$, $\mathbf{p}^s$, and $\hat{\mathbf{p}}^s$.

The true range between the receiver and satellite is

$$R = \|\mathbf{p} - \mathbf{p}^s\| = \mathbf{h}^\top(\mathbf{p} - \mathbf{p}^s) \tag{8.67}$$

where $\mathbf{h}$ represents a unit vector in the direction $(\mathbf{p}-\mathbf{p}^s)$. The range between the receiver and computed satellite position is

$$\hat{R} = \|\mathbf{p} - \hat{\mathbf{p}}^s\| = \hat{\mathbf{h}}^\top(\mathbf{p} - \hat{\mathbf{p}}^s) \tag{8.68}$$

where $\hat{\mathbf{h}}$ represents a unit vector in the direction $(\mathbf{p} - \hat{\mathbf{p}}^s)$ and $\hat{\mathbf{h}}^\perp$ is a unit vector perpendicular to $\hat{\mathbf{h}}$ in the plane of the figure. Figure 8.6 illustrates that the computed range can be decomposed as follows

$$\begin{aligned}
\hat{R} &= \hat{\mathbf{h}}^\top(\mathbf{p} - \hat{\mathbf{p}}^s) \\
&= \hat{\mathbf{h}}^\top\left((\mathbf{p} - \mathbf{p}^s) + (\mathbf{p}^s - \hat{\mathbf{p}}^s)\right) \\
&= \hat{\mathbf{h}}^\top(\mathbf{p} - \mathbf{p}^s) + \hat{\mathbf{h}}^\top(\mathbf{p}^s - \hat{\mathbf{p}}^s).
\end{aligned} \tag{8.69}$$

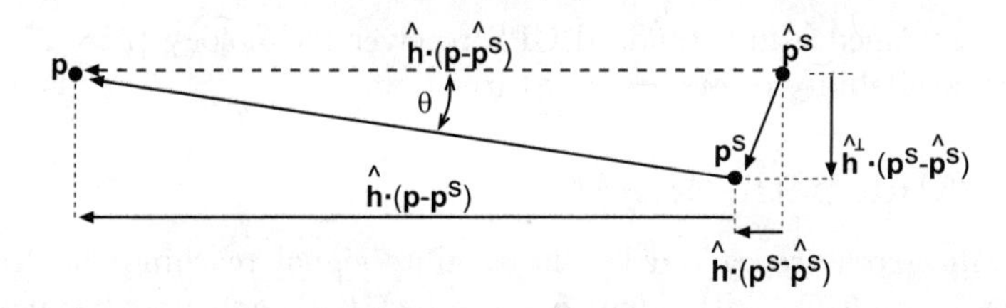

Figure 8.6: Geometry and notation for analysis of the effect of satellite position error on computed satellite range.

Using basic trigonometry, the term $\hat{\mathbf{h}}^\top(\mathbf{p}-\mathbf{p}^s)$ is

$$\begin{aligned} \hat{\mathbf{h}}^\top(\mathbf{p}-\mathbf{p}^s) &= R\cos(\theta) \\ &\approx R\left(1-\frac{\theta^2}{2}\right) \end{aligned} \tag{8.70}$$

where R was defined in eqn. (8.67) and $\theta = \operatorname{asin}\left(\frac{\hat{\mathbf{h}}^\perp\cdot(\mathbf{p}^s-\hat{\mathbf{p}}^s)}{R}\right)$.

The control segment monitors the satellite orbits and calculates the ephemeris parameters broadcast to the user by the satellites. The ephemeris model is a curve fit to the measured orbit, which allows the user receiver to compute $\hat{\mathbf{p}}^s$ such that $|\mathbf{p}^s-\hat{\mathbf{p}}^s|$ is significantly less than ten meters in magnitude; therefore, $|\theta| < \frac{10}{20\times10^6} = 5\times10^{-7}$ and the θ^2 term in (eqn. 8.70) can be safely ignored; therefore, by substituting eqn. (8.70) into eqn. (8.69) and rearranging, we arrive at

$$R - \hat{R} = E^i$$

where

$$E^i = \hat{\mathbf{h}}^\top(\hat{\mathbf{p}}^s-\mathbf{p}^s) \tag{8.71}$$

is typically less that $2m$.

The above error analysis related to Figure 8.6 was performed in the plane defined by the three points $\mathbf{p}$, $\mathbf{p}^s$, and $\hat{\mathbf{p}}^s$. The ephemeris error $(\mathbf{p}^s-\hat{\mathbf{p}}^s)$ is often discussed as having three components: *radial*, *tangential*, and *cross-track*. The radial component is $\left(\hat{\mathbf{h}}\cdot(\hat{\mathbf{p}}^s-\mathbf{p}^s)\right)\hat{\mathbf{h}}$. The vector $\left(\hat{\mathbf{h}}^\perp\cdot(\mathbf{p}-\mathbf{p}^s)\right)\hat{\mathbf{h}}^\perp$ is the vector sum of the tangential and cross-track components. Eqn. (8.71) shows that, to first order, $R-\hat{R}$ is affected only by the radial component.

8.4.6 Selective Availability, SA^s

Selective availability errors were artificially added to the satellite signals by the U.S. government to degrade the position and velocity accuracy that could be attained using standard GPS receiver technology [7, 8, 25, 109]. Selective availability errors are now turned off.

8.4.7 Multipath, M_ρ^s, M_ϕ^s

Multipath errors are caused by the satellite signal reaching the receiver antenna by multiple paths due to the direct path and possibly multiple signal reflections.

A basic understanding of the operation of a GPS receiver will aid the understanding of the corrupting effects of multipath. The receiver tracks the CDMA code by correlating an internally generated version of the satellite CDMA code with the received satellite signal. The internally generated code is shifted in time until maximum correlation occurs. The code is designed so that the correlation function has a symmetric triangular shape within one chip of its peak value. Multipath errors due to reflected signals shift the correlation peak and corrupt the theoretically symmetric receiver correlation envelope. Both of these changes to the correlation envelope result in erroneous pseudorange measurements.

Although reflected errors will always arrive after the direct path signal, the multipath range error can still be both positive or negative. Each reflected signal may interfere either constructively or destructively. Constructive interference adds positively to the correlation function possibly causing the correlation peak to shift to a later point, resulting in a positive range error. Destructive interference adds negatively to the correlation function possibly causing the correlation peak to shift to an earlier point, resulting in a negative range error.

Nominally, C/A multipath can result in errors of 0.1-3.0 meters depending on various design and antenna siting factors. Exceptional cases ($\sim$ 100m) have been reported [134]. L1 phase multipath error is expected to be less than 5 cm.

To decrease the effects of multipath, the user can take the following precautions:

- Typically the majority of reflective surfaces (e.g., earth, lakes, vehicle chassis) are below the antenna. In such cases, it is beneficial to select an antenna with low gain at small and negative elevations. The gain can further be decreased by the use of absorbent materials. One type of antenna designed to attenuate such signals is the choke ring.

- When possible, site the receiver antenna above the highest reflector. This is certainly possible in some stationary applications, but may not be feasible for mobile applications.

- Change the receiver settings to avoid using satellites at low elevations. Signals from low elevation satellites travel nearly parallel with the surface of the planet and therefore have a good chance of suffering from multiple reflections.

Due to the fact that the satellites orbit Earth twice per sidereal day (23 hr 56 min.), each satellite appears in the same location 4 minutes earlier each day. Therefore, the portion of multipath at a stationary receiver due to stationary objects will repeat with a period of 23 hr 56 min. (i.e., four minutes earlier each day).

For mobile applications, the antenna motion relative to reflecting surfaces may significantly decrease the time correlation of the multipath signal, thereby allowing it to be reduced via filtering.

Differential GPS reference stations should be carefully designed and sited to prevent base station multipath errors from being included in the 'corrections' broadcast to the users.

8.4.8 Receiver Noise, η^i_ρ, η^i_ϕ

The previous sections have considered issues that affect time of propagation (i.e., ionosphere and troposphere), range measurement error due to signal reflections (i.e., multipath), and the effects of using computed instead of actual satellite positions. In addition to those issues, various factors inside the antenna, cabling, and receiver affect each measurement. These issues include thermal fluctuations, extraneous RF signals and noise, cross-correlation between the CDMA codes, and signal quantization and sampling effects. These range errors due to these effects are cumulatively referred to as receiver noise. The magnitude of the ranging errors is different for each tracking channel and depends on the signal-to-noise ratio of the signal assigned to the channel.

The receiver noise is usually modeled as white and independent between both satellites and channels.

8.4.9 Carrier Tracking and Integer Ambiguity, N^i

The purpose of this section is to discuss the relation of the carrier phase observable to the psuedorange and explain the origin of the integer phase ambiguity. To achieve this goal will require a rather detailed discussion of GPS receivers including the basics of the Radio Frequency (RF) frontend and the phase lock loop in the baseband processor.

8.4.9.1 Receiver Overview

The purpose of the RF frontend is to amplify the antenna signal, to remove out-of-band noise, and to shift the frequency band of the desired signal to a lower frequency where it can be more easily processed. This lower frequency range is referred to as baseband. The purpose of the baseband

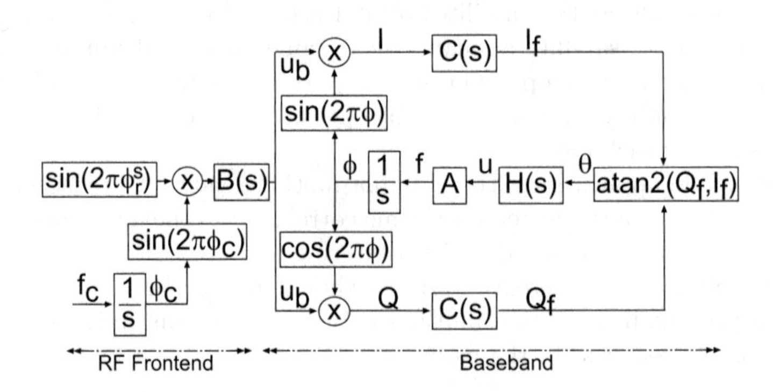

Figure 8.7: Block diagram for a basic phase lock loop.

is to track the PRN code, to track the carrier phase, and to strip the data bits. The following discussion focuses on the phase tracking objective. The discussion is simplified in that it assumes that the code and data bits have been completely removed.

A block diagram for a simplified RF frontend and phase lock loop (PLL) is shown in Figure 8.7. In the RF frontend, the antenna signal $\sin(\phi_r^s(t))$ is multiplied with a sine wave computed using the phase of the receiver clock $\phi_c(t)$. This product is

$$\sin(2\pi\phi_c)\sin(2\pi\phi_r^s) = \frac{1}{2}\Big(\cos\left(2\pi(\phi_c - \phi_r^s)\right) - \cos\left(2\pi(\phi_c + \phi_r^s)\right)\Big)$$

which is the input to the bandpass filter $B(s)$. This filter is designed to remove the second term which is oscillating at the frequency $f_c + f_r^s$. The amplified output of the bandpass filter is

$$u_b(t) = \cos\left(2\pi(\phi_c(t) - \phi_r^s(t))\right)$$

which has phase

$$\phi_b(t) = \phi_c(t) - \phi_r^s(t) \tag{8.72}$$

and is oscillating with frequency $f_b = f_c - f_r^s$, which may be time-varying. The signal $u_b(t)$ is the input to the baseband processor.

8.4.9.2 Baseband Processing

The portion of Figure 8.7 marked as the baseband is actually one channel of the baseband processor. Each channel would perform the same procedures as outlined below. Each channel includes a Numerically Controlled

Oscillator (NCO) which is represented in Figure 8.7 as the product $\frac{A}{s}$. The frequency f of the NCO is determined by the NCO input signal $u(t)$. The output of the NCO is the phase angle ϕ. The phase accumulator in the NCO tracks both the whole cycle count N_{ca} and the fractional portion of the phase ϕ_{ca} such that

$$\phi(\tau_r(t)) = N_{ca}(\tau_r(t)) + \phi_{ca}(\tau_r(t)). \tag{8.73}$$

Therefore, $\phi_r^s(\tau_r(t)) = \phi(\tau_r(t))$ can be computed at any time scheduled by the receiver as is required for eqn. (8.24). The relation of ϕ to the geometric distance between the user and satellite is derived below.

The baseband processor computes $\sin(\phi)$ and $\cos(\phi)$ which are multiplied with u_b to produce the signals

$$I = \sin(2\pi\phi_b)\sin(2\pi\phi) \quad \text{and} \quad Q = \sin(2\pi\phi_b)\cos(2\pi\phi).$$

As shown above for the RF frontend, the resulting I and Q signals are

$$\begin{aligned} I &= \frac{1}{2}\Big(\cos\big(2\pi(\phi_b - \phi)\big) - \cos\big(2\pi(\phi_b + \phi)\big)\Big) \\ Q &= \frac{1}{2}\Big(\sin\big(2\pi(\phi_b - \phi)\big) + \sin\big(2\pi(\phi_b + \phi)\big)\Big). \end{aligned}$$

The filter $C(s)$ is designed to remove the second term, which is oscillating at frequency $f_b + f$, from each of these equations. Therefore,

$$\begin{aligned} I_f(t) &= \frac{1}{2}\cos(2\pi\theta(t)) \\ Q_f(t) &= \frac{1}{2}\sin(2\pi\theta(t)) \end{aligned}$$

where $\theta(t) = \phi_b(t) - \phi(t)$. Given I_f and Q_f, θ can be computed by a variety of different methods. For this discussion, we choose the four quadrant arctangent function

$$\theta = \text{atan2}(Q_f, I_f).$$

The left portion of Figure 8.8 shows the vector $\mathbf{z} = [I_f, Q_f]$ and the angle θ in a two-dimensional plane. In general, $\mathbf{z}$ can point in any direction. The right portion of Figure 8.8 shows a simplified block diagram of the process described above wherein the detail for the process of computing the variable θ has been removed. The only item in Figure 8.7 that has not yet been discussed is the filter $H(s)$. The right portion of Figure 8.8 is a standard control loop. The filter $H(s)$ is designed to stabilize this loop while achieving desirable transient and noise properties. The stability of the loop means that the variable $\theta(t)$ is forced towards zero. When this is achieved, then the direction of the vector $\mathbf{z} = [I_f, Q_f]$ is essentially along

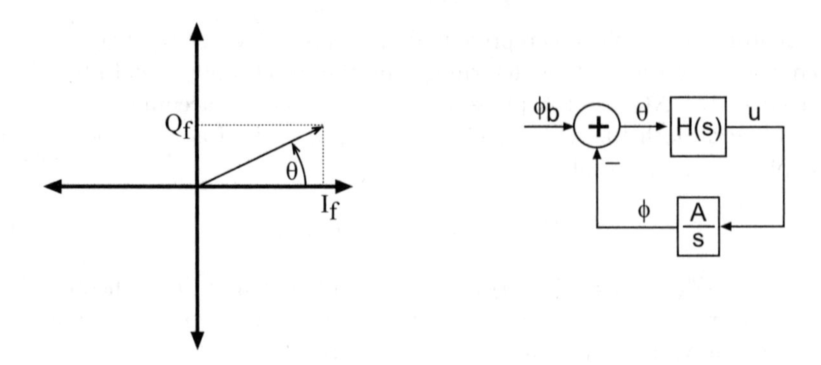

Figure 8.8: Simplified phase lock loop variables and block diagram. Left – Depiction of the vector $[I_f, Q_f]$ and the angle θ. Right – Block diagram of simplified block diagram for a phase lock loop.

the I axis with the Q_f component near zero. Under these conditions $\phi(t)$ is accurately tracking $\phi_b(t)$.

Figure 8.9 shows the phase tracking variables I, Q, and $\theta = \text{atan2}(Q, I)$ versus time for a phase lock loop. Initially, the loop is not phase-locked. Slightly after $t = 1$, the variable θ converges to and stays within the threshold value of 0.1 cycles until approximately $t = 9$. For the time interval, $t \in [1.5, 9]$ the system has phase lock. Slightly after $t = 9$, some event occurs that causes θ to exceed the lock threshold. The system is seen to be converging back towards phase lock at $t = 10$, but a cycle slip may have occurred.

8.4.9.3 Integer Ambiguity

Although the right portion of Figure 8.8 indicates that $\theta = \phi_b - \phi$, this is not strictly true because the four quadrant arctangent function always returns a value $-0.5 < \theta \leq 0.5$. The arctangent function cannot distinguish between θ and $\theta + N$ for integer values of N. Because the arctangent function cannot tell one cycle from another, the relation is actually

$$\theta(t) = \phi_b(t) - \phi(t) + N(t) \tag{8.74}$$

where angles are expressed in cycles and $N(t)$ is the unique integer that results in $\theta(t) \in (-0.5, 0.5]$. The job of the control system is to stabilize the control loop of the PLL, which maintains $\theta(t)$ near zero. For reasonable high signal-to-noise ratios, GPS receivers typically maintain phase lock and track the incoming signal with an accuracy of better than 1% of a cycle. Under such circumstances, $N(t)$ is constant and the linearized model $\theta = \phi_b - \phi$ is sufficient for the control system design and analysis; however, the integer N and the full model of eqn. (8.74) play critical roles in the use of the phase

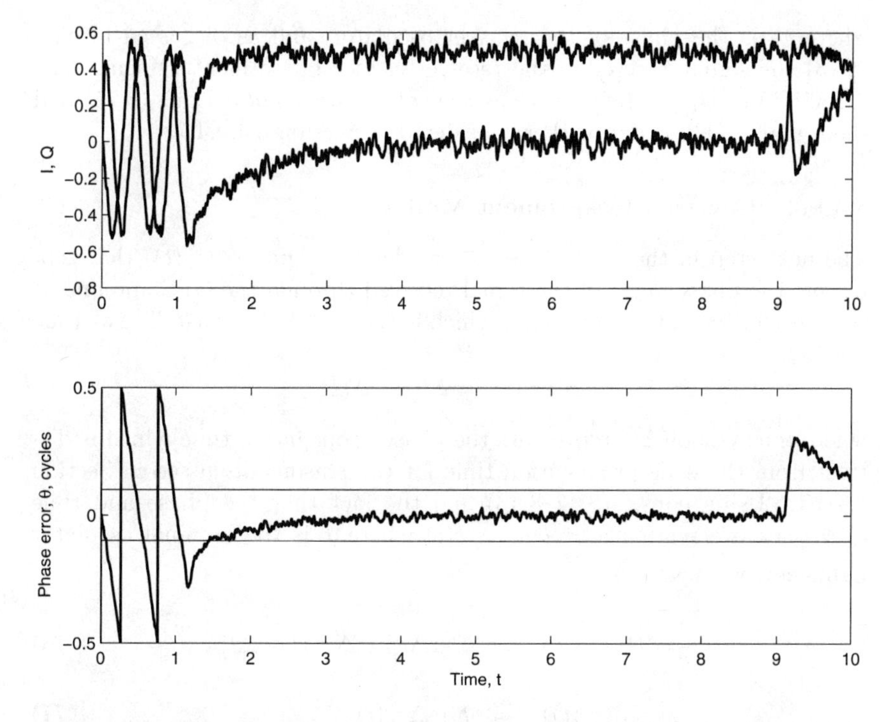

Figure 8.9: Phase tracking variables. Top – Plot of filtered I (magnitude near 0.5) and Q (magnitude near 0.0) versus time. Bottom – Plot of $\theta = atan2(Q, I)$ versus time.

angle ϕ as a ranging signal. Referring back to Figure 8.9, for $t \in (0, 1)$, the integer $N(t)$ changed twice. For $t \in (9, 10)$, the integer may have changed. For $t \in [1.5, 9.0]$ the integer $N(t)$ is constant. The ability to reliably detect loss of lock and cycle slips based on (I_f, Q_f) or θ depends in part on the signal to noise ratio. The actual value of $N(t)$ is not known and is irrelevant to the PLL.

For general discussion, let t_l be the time at which lock is achieved and t_u be the time at which lock is lost. At the instant the lock is achieved,

$$\phi(\tau_r(t_l)) = \phi_b(\tau_r(t_l)) - \theta(\tau_r(t_l)) + N(\tau_r(t_l))$$

where the phase angles are now expressed in cycles. During the time interval $t \in [t_l, t_u]$, the tracking error θ is small and $N(t)$ is constant; therefore, substituting eqn. (8.72) into eqn. (8.74) yields the measurement model

$$\begin{aligned} \phi(\tau_r(t)) &= \phi_b(\tau_r(t)) - \theta(\tau_r(t)) + N(\tau_r(t_l)) \\ &= \phi_c(\tau_r(t)) - \phi_r^s(\tau_r(t)) - \theta(\tau_r(t)) + N(\tau_r(t_l)) \end{aligned} \quad (8.75)$$

where ϕ_c is the phase of the receiver oscillator, and as discussed in eqn. (8.73) the signal $\phi(\tau_r(t))$ is the receiver carrier phase measurement. The variable $N(\tau_r(t_l))$ is referred to as the *integer ambiguity*. Eqn. (8.75) and the signal $\phi_r^s(\tau_r(t))$ are further considered in Section 8.4.9.4.

8.4.9.4 Carrier Measurement Model

The next step in the analysis is to consider the signal $\phi_r^s(\tau_r(t))$ that represents the phase angle of the signal received through the GPS antenna at the receiver indicated time $\tau_r(t)$, which is the phase of the satellite at time $t - \Delta\bar{t}_r^s$:

$$\phi_r^s(\tau_r(t)) = \phi^s(t - \Delta\bar{t}_r^s)$$

where the symbol $\Delta\bar{t}_r^s$ represents the phase propagation time which is distinct from the code propagation time for the reasons discussed in Section 8.4.4.1. Using eqns. (8.42–8.43) and the fact that the phase and time of a clock are related as $\tau(t) = \frac{1}{f}\phi(t)$ where f is the nominal oscillator frequency, we have that

$$\begin{aligned} \frac{1}{f}\phi^s(t - \Delta\bar{t}_r^s) &= t - \Delta\bar{t}_r^s + \Delta\tau^s(t - \Delta\bar{t}_r^s) && (8.76) \\ \frac{1}{f}\phi_c(\tau_r(t)) &= t + \Delta\tau_r(t). && (8.77) \end{aligned}$$

Proceeding from eqn. (8.75), we have

$$\begin{aligned} \frac{1}{f}\phi(\tau_r(t)) &= t + \Delta\tau_r - (t - \Delta\bar{t}_r^s + \Delta\tau^s(t - \Delta\bar{t}_r^s)) - \frac{\theta}{f} + \frac{1}{f}N(\tau_r(t_l)) \\ &= \Delta\bar{t}_r^s + \Delta\tau_r - \Delta\tau^s(t - \Delta\bar{t}_r^s) - \frac{\theta}{f} + \frac{1}{f}N(\tau_r(t_l)). && (8.78) \end{aligned}$$

Combining eqns. (8.58), (8.64), and eqn. (8.66) we have that

$$c\Delta\bar{t}_r^s = \|\mathbf{p} - \mathbf{p}^s\| + c\delta t_\phi^s + T_r^s. \qquad (8.79)$$

Substituting eqn. (8.79) into eqn. (8.78), accounting for multipath measurement errors, accounting for ephemeris errors E^s as in eqn. (8.71), and recognizing that ϕ in eqn. (8.73) is identical to ϕ_r^s in eqn. (8.24), we obtain the measurement model of eqn. (8.25) which is written here explicitly for both L1 and L2:

$$\begin{aligned} \lambda_1\tilde{\phi}_{r_1}^s &= \rho(\mathbf{x}, \hat{\mathbf{p}}^s) + E^s - c\delta\tau^s - \frac{f_2}{f_1}I_r^s + T_r^s + M_{\phi_1}^s + \nu_{\phi_1}^s + \lambda_1 N_{r_1}^s \\ \lambda_2\tilde{\phi}_{r_2}^s &= \rho(\mathbf{x}, \hat{\mathbf{p}}^s) + E^s - c\delta\tau^s - \frac{f_1}{f_2}I_r^s + T_r^s + M_{\phi_2}^s + \nu_{\phi_2}^s + \lambda_2 N_{r_2}^s \end{aligned}$$

Common Mode Errors	L1 C/A, σ, m	L1/L2, σ, m
Ionosphere	7-10	–
Troposphere	1	1
SV Clock	2	2
SV Ephemeris	2	2
Non-common Mode Errors		
Receiver Noise	0.1 - 0.7	0.1 - 0.7
Multipath	0.1 - 3.0	0.1 - 3.0
URE	8-11	3-4

Table 8.5: User pseudorange error standard deviation.

where the numeric subscripts 1 and 2 indicate the frequencies L1 and L2 respectively and we have used the fact that for $i = 1, 2$ $\lambda_i = \frac{c}{f_i}$. In subsequent sections, to simplify notation, the symbols β_1^s and β_2^s will be used to represent the sum of the noise and multipath terms.

8.4.10 Summary

Table 8.5 lists standard deviations for the various error components that corrupt the GPS pseudorange observables. Common-mode error refers to those error sources that would be common to every receiver operating within a limited ($\approx 15km$) geographic region. Noncommon-mode errors refer to those errors that would be distinct to receivers operating even in close proximity. For the phase observables, the magnitude of the common-mode errors would be unchanged and the magnitude of the noncommon-mode errors would be decreased by about two orders of magnitude.

The values stated in Table 8.5 should be considered as reasonable estimates, not known values. The user's actual performance will depend on various factors including receiver technology, antenna technology, and the user environment. With the reasonable assumption that the error components are independent, the User Range Error (URE), also called the User Equivalent Range Error (UERE) is the root of the sum of the square (RSS) of the individual error components.

Based on the standard deviation estimates in Table 8.5, standard GPS has a URE of 8-11 m. The dominant error source is the ionospheric error. The right column shows that for a two frequency receiver able to eliminate the ionospheric error, multipath and SV clock and ephemeris errors become the dominant error sources. The translation of the range error into an estimate of the position error covariance is the topic of Section 8.5. The relationships between range error standard deviation and other performance metrics (e.g., CEP, R95, 2drms, etc.) are discussed in Section 4.9.1.

In Section 8.8, differential GPS is discussed. Differential operation achieves significant accuracy improvements by canceling the effect of the common mode errors. The resulting URE would be a few 0.1-3m for differential pseudorange or a few centimeters for differential phase.

8.5 Geometric Dilution of Precision

This section derives dilution of precision (DOP) factors that related the URE to the expected position estimation accuracy. In the discussion of this section, it is assumed that $\mathbf{H}$ includes a rotation matrix so that the vector $\mathbf{p}$, which contains the first three states of $\mathbf{x}$, represents horizontal position and altitude (i.e., north, east, and down).

After convergence of the GPS update of eqn. (8.13), we expect that $d\mathbf{x} = \mathbf{0}$ which implies that

$$\left(\mathbf{H}^\top \mathbf{H}\right)^{-1} \mathbf{H}^\top \left(\tilde{\rho} - \hat{\rho}(\hat{\mathbf{x}})\right) = \mathbf{0}.$$

Also, from eqns. (8.6) and (8.11), we have that

$$\tilde{\rho} - \hat{\rho}(\hat{\mathbf{x}}) = \mathbf{H}\delta\mathbf{x} + \chi$$

where χ is defined in eqn. (8.7). Combining these equations yields

$$\delta\mathbf{x} = -\left(\mathbf{H}^\top \mathbf{H}\right)^{-1} \mathbf{H}^\top \chi \tag{8.80}$$

where the number of measurements m is greater than three and the inverse matrix exists.

Assuming that the error vector χ is zero mean with covariance $\sigma^2\mathbf{I}$, we can determine the expected value and covariance of the position error. The expected value of $\delta\mathbf{x}$ is

$$E\langle\delta\mathbf{x}\rangle = E\left\langle \left(\mathbf{H}^\top \mathbf{H}\right)^{-1} \mathbf{H}^\top \chi \right\rangle = \mathbf{0};$$

therefore, the estimate of the receiver antenna position and receiver clock bias is expected to be unbiased. This statement should be interpreted with caution. The ensemble average has an expected value of zero. In fact, if a temporal sequence of positions was averaged long enough, the position error should converge to zero. However, the temporal average has significant time correlation due to the range errors (E^i, T^i_r, I^i_r, M^i_ρ, and $c\delta t^i$) all being time correlated. Therefore short-term averages can contain significantly biased errors.

Define $\mathbf{P} = cov(\delta\hat{\mathbf{x}})$. From eqn. (8.80) the covariance matrix $\mathbf{P}$ can be computed as shown in eqn. (8.81):

$$\begin{aligned}
\mathbf{P} &= E\langle\delta\mathbf{x}\delta\mathbf{x}^\top\rangle \\
&= E\left\langle\left(\mathbf{H}^\top\mathbf{H}\right)^{-1}\mathbf{H}^\top\chi\chi^\top\mathbf{H}\left(\mathbf{H}^\top\mathbf{H}\right)^{-1}\right\rangle \\
&= \left(\mathbf{H}^\top\mathbf{H}\right)^{-1}\mathbf{H}^\top\left(\sigma^2\mathbf{I}\right)\mathbf{H}\left(\mathbf{H}^\top\mathbf{H}\right)^{-1} \\
&= \left(\mathbf{H}^T\mathbf{H}\right)^{-1}\sigma^2. \qquad (8.81)
\end{aligned}$$

This $\mathbf{P}$ matrix could be used with the methods of Section 4.9.1 to define probability ellipses. Define the matrix $\mathbf{G} = \left(\mathbf{H}^T\mathbf{H}\right)^{-1}$. Based on eqn. (8.81), $\mathbf{P} = \mathbf{G}\sigma^2$ which shows that the variance of the $\mathbf{x}$ estimation error is determined by the product of the URE variance σ^2 and the matrix $\mathbf{G}$ which is determined by the geometry of the satellites relative to the user. Due to the structure of the covariance matrix, as discussed in eqn. (4.24), the variance of the position estimates and various useful error metrics can be computed, for example:

$$\begin{aligned}
var(\hat{n}) &= G_{11}\sigma^2, & var(\hat{e}) &= G_{22}\sigma^2, \\
var(\hat{d}) &= G_{33}\sigma^2, & var(c\Delta\hat{t}_r) &= G_{44}\sigma^2,
\end{aligned}$$

and $E\langle\delta\mathbf{p}^\top\delta\mathbf{p}\rangle = (G_{11} + G_{22} + G_{33})\,\sigma^2$. The matrix $\mathbf{G}$ defines various *dilution of precision* (DOP) factors that are convenient for specifying the amplification of the URE in the estimation of specific portions of $\mathbf{x}$. Typically used DOP factors are

$$\begin{aligned}
VDOP &= \sqrt{\mathbf{G}_{33}} \\
HDOP &= \sqrt{\mathbf{G}_{11} + \mathbf{G}_{22}} \\
PDOP &= \sqrt{\mathbf{G}_{11} + \mathbf{G}_{22} + \mathbf{G}_{33}} \\
GDOP &= \sqrt{\mathbf{G}_{11} + \mathbf{G}_{22} + \mathbf{G}_{33} + \mathbf{G}_{44}} \\
&= \sqrt{trace\left(\mathbf{H}^T\mathbf{H}\right)^{-1}} \qquad (8.82)
\end{aligned}$$

where $VDOP$ quantifies the magnification of URE in estimating altitude, $HDOP$ quantifies the magnification of URE in estimating the horizontal position $[n, e]^\top$, etc.

Most GPS receivers calculate and can be configured to output the various DOP factors and URE. This allows the user (or real-time software) to monitor the expected instantaneous estimation accuracy. GPS planning software is also available which allows users to analyze the DOP factors as a function of time and location to determine either satisfactory or optimal times for performing GPS related activities (see Chapter 7 in [76]).

The matrix $\mathbf{H}$ is a function of the number of satellites in view and the geometry of the line-of-sight vectors $\vec{\mathbf{h}}^i$ from the satellites to the user. The

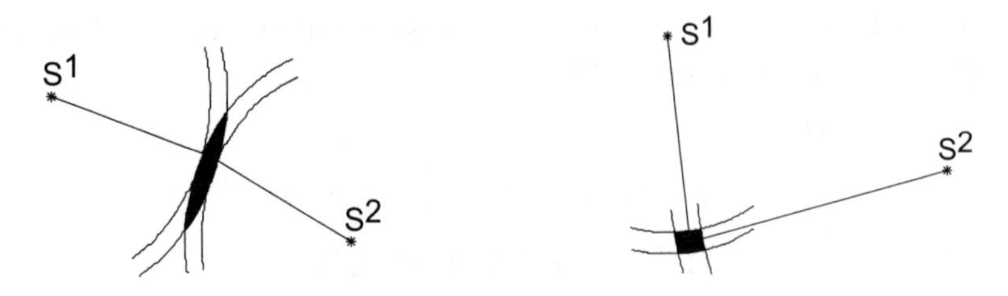

Figure 8.10: Two dimensional illustration of the relationship between satellite-to-user geometry and the resulting position accuracy.

effect of satellite-to-user geometry on position estimation accuracy is illustrated in Figure 8.10 using two dimensions. This figure shows two possible user-satellite configurations. In each configuration, a receiver measures the range to two satellites at the indicated positions. Due to measurement errors, the range measurement is not known exactly, but the receiver position is expected to lie between the two indicated concentric circles. In the left half of the figure the receiver vectors $\vec{\mathbf{h}}^1$ and $\vec{\mathbf{h}}^2$ are nearly collinear; the intersection of the two concentric circles for the two satellites results in a long thin region of possible positions with the largest uncertainty direction orthogonal to the satellite vectors. In this case, the DOP factor would be large. In the right half of the figure the vectors $\vec{\mathbf{h}}^1$ and $\vec{\mathbf{h}}^2$ are nearly orthogonal; the intersection of the concentric circles for each satellite results in a much more equally proportioned uncertainty region.

The measurement matrix $\mathbf{H}$ is not constant in time because of the GPS satellite positions change as they orbit Earth. Large DOP values result when the rows of $\mathbf{H}$ are nearly linearly dependent. This could be measured by the condition of the matrix $\mathbf{H}^\top\mathbf{H}$, but this approach is not often used or necessary. There may be times when for a given set of four satellites the $\mathbf{H}$ matrix approaches singularity and GDOP approaches infinity. This condition is called a *GDOP chimney*, see Figure 8.11. This was a difficulty in the early days of GPS, when satellites had fewer tracking channels or a limited number of satellites were in view. Fortunately, there are usually more than four satellites in view and modern receivers track more than the minimum number of satellites; hence, either all or an optimal set of the tracked satellites can be selected to minimize a desired DOP URE amplification factor.

Example 8.4 *Consider the following case for which six satellites are in view of the receiver antenna. The rows of the* $\mathbf{H}$ *matrix for each satellite*

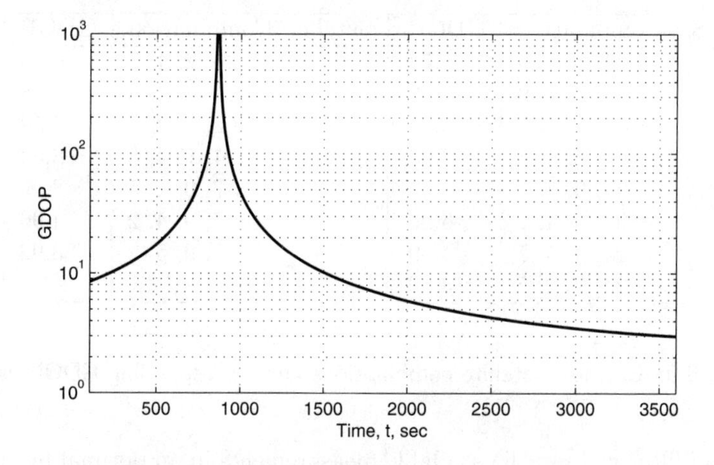

Figure 8.11: GDOP chimney for a set of four satellites.

are:

$SV2$	-0.557466	$+0.829830$	-0.024781	$+1.000000$
$SV24$	$+0.469913$	$+0.860463$	$+0.196942$	$+1.000000$
$SV4$	$+0.086117$	$+0.936539$	-0.339823	$+1.000000$
$SV5$	$+0.661510$	-0.318625	-0.678884	$+1.000000$
$SV7$	-0.337536	$+0.461389$	-0.820482	$+1.000000$
$SV9$	$+0.762094$	$+0.267539$	-0.589606	$+1.000000$

For this set of six satellites, there are fifteen different combinations of four satellites, as shown in Table 8.6. The GDOP figures for the various sets of four satellites range from 3.58 *to* 16.03. *Instead of selecting the best four satellites, an all-in-view approach using all six satellites would result in a GDOP of 2.96.* △

A comparison of the above discussion with the discussion of observability in Section 3.6.1 shows that the various DOP factors are measures the degree of observability of a selected set of position states are at a given time instant (and location) for a given set of satellites. If observability is temporarily lost (e.g., due to satellite occlusion), some of the DOP factors become infinite and a full position solution is not possible by stand-alone GPS. In an aided navigation system, the integration of the high rate sensors through the vehicle kinematic equations will maintain an estimate of the vehicle state which includes the position. In addition, the navigation system will maintain an estimate of the state error covariance matrix denoted as $\mathbf{P}$. During the time period when PDOP is infinite, some linear combination of the states will be unobservable from the available GPS measurements. The

SV Combination	GDOP	Satellite Combination	GDOP
24, 2, 4, 5	3.72	24, 2, 4, 7	6.25
24, 2, 4, 9	4.86	24, 2, 7, 5	4.48
24, 2, 9, 5	5.09	2, 7, 4, 5	3.58
2, 9, 4, 5	16.03	24, 7, 4, 5	8.63
24, 9, 4, 5	5.45	7, 9, 4, 5	8.59
24, 9, 4, 7	6.96	7, 9, 4, 2	4.96
24, 9, 5, 7	4.89	7, 9, 5, 2	3.92
24, 9, 2, 7	13.10		

Table 8.6: Example satellite combinations and corresponding GDOP values.

Kalman filter will use the available measurements in an optimal fashion to maintain the estimate accuracy to the extent possible. The unobservable portions of the state vector will be indicated by growth in the corresponding directions of the **P** matrix. This could be analyzed via a SVD or eigen-decomposition of **P**.

This section has worked entirely with covariance of position error. Other performance metrics (e.g., CEP, R95, 2drms, etc.) are discussed in Section 4.9.1.

8.6 Two Frequency Receivers

A major reason for the design of multi-frequency receivers is to allow estimation and compensation of the ionospheric delay.

Simultaneous L1 and L2 pseudorange observables from the same satellite and receiver can be modeled as

$$\tilde{\rho}_1 = \rho + E_{cm} + \frac{f_2}{f_1} I_a + \eta_1 \tag{8.83}$$

$$\tilde{\rho}_2 = \rho + E_{cm} + \frac{f_1}{f_2} I_a + \eta_2 \tag{8.84}$$

where E_{cm} is the lumped common mode errors other that the ionospheric delay $I_a = \frac{40.3}{f_1 f_2} TEC$, η_i represents the sum of the receiver noise and multipath errors, and f_1 and f_2 are the carrier frequencies. In the analysis to follow, we will assume that η_1 and η_2 are independent, Gaussian random variables with variance σ_ρ^2.

Eqns. (8.83–8.84) can be manipulated to provide an ionospheric free pseudorange observable as

$$\bar{\rho} = \frac{f_1^2 \tilde{\rho}_1 - f_2^2 \tilde{\rho}_2}{(f_1^2 - f_2^2)} \tag{8.85}$$

which has the measurement model

$$\bar{\rho} = \rho + E_{cm} + \frac{f_1^2}{f_1^2 - f_2^2}\eta_1 - \frac{f_2^2}{f_1^2 - f_2^2}\eta_2. \tag{8.86}$$

Eqn. (8.85) is not usually used directly, because as shown in eqn. (8.86), the noise variance (i.e., effect of η_1 and η_2 on $\bar{\rho}$) is approximately $9\sigma_\rho^2$. The following two paragraphs discuss alternative approaches to the construction of ionospheric free pseudorange observables.

An estimate of the ionospheric delay can be computed as

$$\hat{I}_{a_\rho} = \frac{f_1 f_2}{f_1^2 - f_2^2}\left(\tilde{\rho}_2 - \tilde{\rho}_1\right). \tag{8.87}$$

Direct substitution of eqns. (8.83–8.84) into eqn. (8.87) shows that

$$\hat{I}_{a_\rho} = I_a + \frac{f_1 f_2}{f_1^2 - f_2^2}\left(\eta_2 - \eta_1\right) \tag{8.88}$$

$$= I_a + 1.984\left(\eta_2 - \eta_1\right) \tag{8.89}$$

which shows that $\hat{I}_{a_\rho}$ is unbiased. The variance of $\hat{I}_{a_\rho}$ at each epoch is approximately $8\sigma_\rho^2$. Because of the magnification of the receiver noise, the ionospheric estimate of eqn. (8.87) is not used directly to compensate the pseudorange measurement. Instead, because I_a changes slowly with a correlation time of a few hours, while η_1 and η_2 have much shorter correlation times, $\hat{I}_{a_\rho}$ could for example be low pass filtered by a filter with a time constant of several minutes to greatly decrease the effect of η_1 and η_2 while maintaining the time variation of $\hat{I}_{a_\rho}$. If we denote the filtered version of $\hat{I}_{a_\rho}$ as $\bar{I}_{a_\rho}$ then an ionosphere free pseudorange can be computed as

$$\bar{\rho} = \tilde{\rho}_1 - \frac{f_2}{f_1}\bar{I}_{a_\rho} \tag{8.90}$$

which has the error model

$$\bar{\rho} = \rho + E_{cm} + \eta_1$$

where the ionospheric error has been (essentially) removed and the measurement noise has not been amplified. The remaining common-mode errors could be removed via differential processing. Another approach is discussed subsequently.

Simultaneous L1 and L2 phase observables from the same satellite and receiver can be modeled as

$$\lambda_1\tilde{\phi}_1 = \rho + E_{cm} - \frac{f_2}{f_1}I_a + \beta_1 + \lambda_1 N_1 \tag{8.91}$$

$$\lambda_2\tilde{\phi}_2 = \rho + E_{cm} - \frac{f_1}{f_2}I_a + \beta_2 + \lambda_2 N_2 \tag{8.92}$$

where $\lambda_1 = \frac{c}{f_1}$ and $\lambda_2 = \frac{c}{f_2}$ are the wavelength of the L1 and L2 carrier signals, respectively, and β_i is the combined effect of multipath and receiver noise. The subsequent analysis assumes that β_1 and β_2 are independent Gaussian processes with variance σ_ϕ^2. Using the phase observables, the ionospheric delay can be estimated as

$$\hat{I}_{a_\phi} = \frac{f_1 f_2}{f_1^2 - f_2^2}\left(\lambda_1\tilde{\phi}_1 - \lambda_2\tilde{\phi}_2\right). \tag{8.93}$$

Direct substitution of eqns. (8.91) and (8.92) into eqn. (8.93) shows that

$$\begin{aligned}\hat{I}_{a_\phi} &= I_a + \frac{f_1 f_2}{f_1^2 - f_2^2}\left(\beta_1 - \beta_2 + N_1\lambda_1 - N_2\lambda_2\right) && (8.94)\\ &= I_a + 1.984\left(\beta_2 - \beta_1\right) + \frac{f_1 f_2}{f_1^2 - f_2^2}\left(N_1\lambda_1 - N_2\lambda_2\right). && (8.95)\end{aligned}$$

Therefore, the variance of $\hat{I}_{a_\rho}$ at each epoch is approximately $8\sigma_\phi^2$; however, the estimate contains the significant constant bias

$$B = \frac{f_1 f_2}{f_1^2 - f_2^2}\left(N_1\lambda_1 - N_2\lambda_2\right).$$

Eqns. (8.89) and (8.95) have the form

$$\begin{aligned}\hat{I}_{a_\rho} &= I_a + n_1 && (8.96)\\ \hat{I}_{a_\phi} &= I_a + n_2 + B. && (8.97)\end{aligned}$$

Eqns. (8.96) and (8.97) are in the exact form considered in Example 5.4. Therefore, the solution is

$$\hat{I}_a(k) = \hat{I}_{a_\phi}(k) - \hat{B}(k) \tag{8.98}$$

where

$$\hat{B}_k = \hat{B}_{k-1} + \frac{1}{k}\left(\left(\hat{I}_{a_\phi} - \hat{I}_{a_\rho}\right) - \hat{B}_{k-1}\right)$$

starting at $k = 1$ with $B_0 = 0$. The estimation error variance for $\hat{B}$ is $P_B = 8(1+\mu^2)\frac{\sigma_\rho^2}{k}$ and for $\hat{I}_a(k)$ is $P_{I_a} = 8\left(\frac{1-\mu^2}{k} + \mu^2\right)\sigma_\rho^2$, where $\sigma_\phi = \mu\sigma_\rho$ and $\mu = 0.01$ is the ratio of the phase noise to the pseudorange noise.

The desire to estimate the ionospheric delay was a primary motivation for the GPS system to incorporate signals on two frequencies. The specification of the two frequencies involved a tradeoff related to the frequency spacing. If the frequency separation was too small, then measurement errors would be significantly magnified as shown in eqn. (8.88) and (8.94). However, if the frequency separation were too large, then separate antennas would be required to receive the two signals.

8.6.1 Wide and Narrow Lane Observables

This section develops the equations for the *narrow-lane* and *wide-lane* observables. These variables are synthesized as linear combinations of the L1 and L2 measurements. The interest in the wide-lane signal is that its wavelength is large enough that the wide-lane variable is often used to facilitate the problem of integer ambiguity resolution, see Section 8.9.

The phase measurements of eqns. (8.91–8.92) can be modeled as

$$\tilde{\phi}_1 - N_1 = \frac{f_1}{c}(\rho + E_{cm} + \beta_1) - \frac{f_2}{c} I_a \tag{8.99}$$

$$\tilde{\phi}_2 - N_2 = \frac{f_2}{c}(\rho + E_{cm} + \beta_2) - \frac{f_1}{c} I_a. \tag{8.100}$$

Forming the sum and difference of eqns. (8.99) and (8.100) results in

$$\begin{aligned}\left(\tilde{\phi}_1 + \tilde{\phi}_2\right) &= \left(\frac{f_1}{c} + \frac{f_2}{c}\right)\rho - \left(\frac{f_2}{c} + \frac{f_1}{c}\right) I_a + (N_2 + N_1) \\ &\quad + \left(\frac{f_1}{c} + \frac{f_2}{c}\right) E_{cm} + \frac{f_1}{c}\beta_1 + \frac{f_2}{c}\beta_2\end{aligned} \tag{8.101}$$

$$\begin{aligned}\left(\tilde{\phi}_1 - \tilde{\phi}_2\right) &= \left(\frac{f_1}{c} - \frac{f_2}{c}\right)\rho - \left(\frac{f_2}{c} - \frac{f_1}{c}\right) I_a + (N_1 - N_2) \\ &\quad + \left(\frac{f_1}{c} - \frac{f_2}{c}\right) E_{cm} + \frac{f_1}{c}\beta_1 - \frac{f_2}{c}\beta_2.\end{aligned} \tag{8.102}$$

By defining the wide and narrow lane wavelengths as

$$\lambda_w = \frac{c}{f1 - f2} \tag{8.103}$$

$$\lambda_n = \frac{c}{f1 + f2}, \tag{8.104}$$

eqns. (8.101) and (8.102) can be written in meters as

$$\begin{aligned}(\tilde{\phi}_1 + \tilde{\phi}_2)\lambda_n &= \rho + E_{cm} - I_a + (N_2 + N_1)\lambda_n \\ &\quad + \frac{\lambda_n}{\lambda_1}\beta_1 + \frac{\lambda_n}{\lambda_2}\beta_2\end{aligned} \tag{8.105}$$

$$\begin{aligned}(\tilde{\phi}_1 - \tilde{\phi}_2)\lambda_w &= \rho + E_{cm} + I_a + (N_1 - N_2)\lambda_w \\ &\quad + \frac{\lambda_w}{\lambda_1}\beta_1 - \frac{\lambda_w}{\lambda_2}\beta_2\end{aligned} \tag{8.106}$$

where $\frac{\lambda_w}{\lambda_1} = \frac{154}{34} \approx 4.5$ and $\frac{\lambda_w}{\lambda_2} = \frac{120}{34} \approx 3.5$; therefore, the standard deviation of the widelane noise (i.e., $\frac{\lambda_w}{\lambda_1}\beta_1 - \frac{\lambda_w}{\lambda_2}\beta_2$) is approximately 5.7 times the standard deviation of the L1 or L2 phase noise (i.e. β_1 or β_2). The fact that the wide-lane phase has a wavelength of approximately $86cm$ can simplify integer ambiguity resolution in differential GPS applications.

The pseudorange estimates can be processed similarly yielding

$$\left(\frac{\tilde{\rho}_1}{\lambda_1}-\frac{\tilde{\rho}_2}{\lambda_2}\right)\lambda_w = \rho + E_{cm} - I_a + \frac{\lambda_w}{\lambda_1}\eta_1 - \frac{\lambda_w}{\lambda_2}\eta_2 \quad (8.107)$$

$$\left(\frac{\tilde{\rho}_1}{\lambda_1}+\frac{\tilde{\rho}_2}{\lambda_2}\right)\lambda_n = \rho + E_{cm} + I_a + \frac{\lambda_n}{\lambda_1}\eta_1 + \frac{\lambda_n}{\lambda_2}\eta_2. \quad (8.108)$$

Both the code and phase narrow-lane observables have noise reduced by $\sqrt{2}$.

8.7 Carrier-Smoothed Code

The code observables provide a noisy but complete measurement of the pseudorange while the phase observables provide a relatively noise-free but biased measurement of the pseudorange. Carrier-smoothing is one approach that has been to achieve an unbiased and smooth pseudorange estimate. In this section, we present the approach for non-differential wide-lane phase and narrow-lane code observables. This approach can also be used directly with L1 code and phase for a single frequency receiver, but the designer should be careful due to the issue of code-carrier divergence.

Eqns. (8.106) and (8.108) have the form

$$\left(\frac{\tilde{\rho}_1}{\lambda_1}+\frac{\tilde{\rho}_2}{\lambda_2}\right)\lambda_n = r + n_1 \quad (8.109)$$

$$(\tilde{\phi}_1-\tilde{\phi}_2)\lambda_w = r + C + n_2 \quad (8.110)$$

where $r = \rho + E_{cm} + I_a$, $n_1 = \frac{\lambda_n}{\lambda_1}\eta_1 + \frac{\lambda_n}{\lambda_2}\eta_2$, $n_2 = \frac{\lambda_w}{\lambda_1}\beta_1 - \frac{\lambda_w}{\lambda_2}\beta_2$ and $C = (N_1 - N_2)\lambda_w$. Note the C is a constant and that $\frac{C}{\lambda_w}$ is an integer. Eqns. (8.109–8.110) are (again) of the exact form considered in Example 5.4; therefore, C can be estimated as

$$\hat{C}_k = \hat{C}_{k-1} + \frac{1}{k}\left(\left(\lambda_w\tilde{\phi}_w(k) - \tilde{\rho}_n(k)\right) - \hat{C}_{k-1}\right) \quad (8.111)$$

where $\tilde{\phi}_w = \tilde{\phi}_1 - \tilde{\phi}_2$ and $\tilde{\rho}_n = \left(\frac{\tilde{\rho}_1}{\lambda_1}+\frac{\tilde{\rho}_2}{\lambda_2}\right)\lambda_n$. The variance for $\hat{C}_k$ is $P_C = (1+\mu^2)\frac{\sigma_\rho^2}{k}$ where $\mu = \frac{\sigma_\phi}{\sigma_\rho} = 0.01$. The carrier-smoothed pseudorange estimate is

$$\hat{r}_k = \lambda_w\tilde{\phi}_w(k) - \hat{C}_k \quad (8.112)$$

where $P_r = (1-\mu^2)\frac{\sigma_\rho^2}{k} + \mu^2\sigma_\rho^2$ so the code noise and multipath effects are significantly reduced over time; however, by its definition following eqn.

(8.110), $\hat{r}_k$ is still (fully) affected by I_a and E_{cm}. The ionospheric term can be estimated as in eqn. (8.98) and compensated. This leaves the tropospheric, ephemeris, and SV clock errors as the dominant elements of the range error.

The carrier-smoothing approach is also applicable to differentially corrected pseudorange and phase measurements, in which case the residual tropospheric, ephemeris, and SV clock errors would be small.

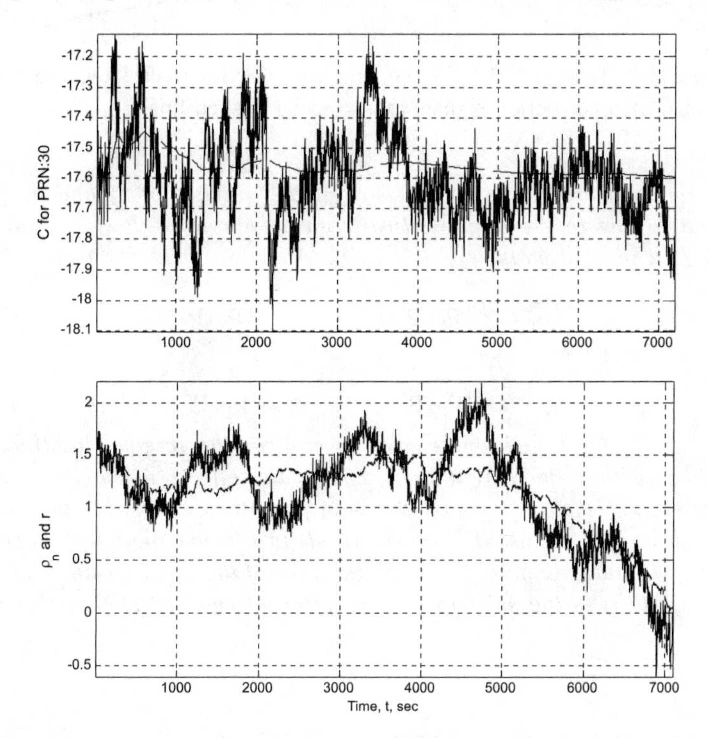

Figure 8.12: Carrier smoothed code results for Example 8.5. Top – Results for PRN 30. The solid line is $\left(\lambda_w \tilde{\phi}_w - \tilde{\rho}_n\right)$. The dashed line is $\hat{C}$. Bottom – The solid line is $\tilde{\rho}_n$. The dashed line is $\hat{r}$.

Example 8.5 *Figure 8.12 displays results of the carrier smoothed code operation. The top graph displays* $\left(\lambda_w \tilde{\phi}_w - \tilde{\rho}_n\right)$ *(solid) and* $\hat{C}$ *(dashed). The estimate of the constant* $\hat{C}$ *is significantly smoother than the instantaneous value of* $\left(\lambda_w \tilde{\phi}_w - \tilde{\rho}_n\right)$.

The variable r is the pseudorange plus the ionosphere and common mode errors. In particular, for the i-th SV, r contains the range $R(\mathbf{p}, \mathbf{p}^i)$ *and the receiver clock error which are large and change rapidly relative to the other*

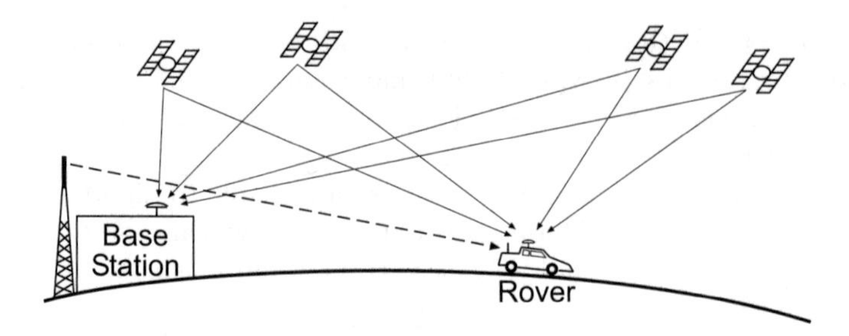

Figure 8.13: Differential GPS scenario with GPS signals indicated by solid lines and DGPS correction signals indicated by dashed lines.

terms in r. Therefore, a comparison of the graphes of r^i and ρ^i_n is not informative due to the large vertical scale. Therefore, the lower graph in Figure 8.12 instead plots

$$\left(\tilde{\rho}^i_n - R(\mathbf{p}_o, \hat{\mathbf{p}}^i)\right) - \left(\tilde{\rho}^2_n - R(\mathbf{p}_o, \hat{\mathbf{p}}^2)\right) \tag{8.113}$$

and

$$\left(r^i - R(\mathbf{p}_o, \hat{\mathbf{p}}^i)\right) - \left(r^2 - R(\mathbf{p}_o, \hat{\mathbf{p}}^2)\right) \tag{8.114}$$

where $\mathbf{p}_o$ is a fixed position selected to be near the receiver location. The differences within the parentheses remove the effects of satellite motion. The difference between the two terms in parentheses removes the effect of the receiver clock. These differences result in a convenient scale factor for the plot, without changing the characteristics of the comparison. The graph of eqn. (8.113) is the solid line. The graph of eqn. (8.114) is the dashed line. △

8.8 Differential GPS

The error sources previously described limit the accuracy attainable using GPS. Several of the sources of error (ionospheric delay, tropospheric delay, satellite ephemeris and clock error), referred to as the common mode errors, are spatially and temporally correlated. Therefore, if the errors could be estimated by one base receiver and promptly broadcast to other roving receivers, then the GPS positioning accuracy at each of the roving receivers could be substantially improved. This is the basic principle of how Differential GPS (DGPS) works. A typical DGPS setup is illustrated in Figure 8.13.

The following subsections will discuss a few DGPS methods. In each case, DGPS will involve a GPS receiver/antenna at a location $\mathbf{p}_o$, a receiver/antenna at an unknown possibly changing position $\mathbf{p}_r$, a satellite at the calculated position $\hat{\mathbf{p}}^i$, and a communication medium from the first receiver to the second receiver. In the discussion, the former receiver will be referred to as the *base*. The latter receiver will be referred to as the *rover*. Although the discussion is phrased in terms of one base and one rover, the number of roving receivers is not limited.

8.8.1 Relative DGPS

This section has three parts. Section 8.8.1.1 considers relative DGPS based on the pseudorange observables. Section 8.8.1.2 considers relative DGPS based on the phase observables. Section 8.8.1.3 analyzes the effect of base-rover separation on the computation of the range difference via projection of the base-rover offset vector onto the satellite line-of-sight vector.

8.8.1.1 Pseudorange

The correction equations of this subsection are derived for a single satellite, but the process is identical for all satellites of interest. The GPS pseudorange observables, measured simultaneously at the base and rover are, respectively,

$$\tilde{\rho}_o^i = R(\mathbf{p}_o, \hat{\mathbf{p}}^i) + c\Delta t_{r_o} + \eta_o^i + c\delta t^i + c\delta t_{a_o}^i + E_o^i \tag{8.115}$$

$$\tilde{\rho}_r^i = R(\mathbf{p}_r, \hat{\mathbf{p}}^i) + c\Delta t_{r_r} + \eta_r^i + c\delta t^i + c\delta t_{a_r}^i + E_r^i \tag{8.116}$$

where $\delta t_{a_r}^i$ represents the sum of the ionospheric and tropospheric errors, the other error terms are all defined as in eqn. (8.6), and the subscripts o and r denote the error terms corresponding to the base and rover, respectively.

If the base pseudorange observables were communicated to the rover, then the single-differential measurements

$$\Delta\tilde{\rho}_{ro}^i = \tilde{\rho}_r^i - \tilde{\rho}_o^i \tag{8.117}$$

could be formed. From eqns. (8.115–8.116), this differential measurement is modeled as

$$\begin{aligned} \Delta\tilde{\rho}_{ro}^i &= (R(\mathbf{p}_r, \hat{\mathbf{p}}^i) + c\Delta t_{r_r} + \eta_r^i + c\delta t^i + c\delta t_{a_r}^i + E_o^i) \\ &\quad -(R(\mathbf{p}_o, \hat{\mathbf{p}}^i) + c\Delta t_{r_o} + \eta_o^i + c\delta t^i + c\delta t_{a_o}^i + E_r^i) \\ &= R(\mathbf{p}_r, \hat{\mathbf{p}}^i) - R(\mathbf{p}_o, \hat{\mathbf{p}}^i) + (c\Delta t_{r_r} - c\Delta t_{r_o}) + \eta \qquad (8.118) \\ &\approx \hat{\mathbf{h}}_r^i \Delta\mathbf{x}_{ro} + \eta \qquad (8.119) \end{aligned}$$

where $\Delta\mathbf{x}_{ro} = \left[(\mathbf{p}_r - \mathbf{p}_o)^\top, (c\Delta t_{r_r} - c\Delta t_{r_o})\right]^\top$, $\hat{\mathbf{h}}_r^i = \left[\frac{\mathbf{p}_r - \hat{\mathbf{p}}^i}{\|\mathbf{p}_r - \hat{\mathbf{p}}^i\|}, 1\right]$, and

$$\eta = (c\delta t_{a_r}^i - c\delta t_{a_o}^i) + (E_r^i - E_o^i) + (\eta_r - \eta_o).$$

The transition from eqn. (8.118) to eqn. (8.119) is often invoked, but it is not exact. The analysis of the incurred error is presented in Section 8.8.1.3. The satellite clock errors are identical for all receivers making simultaneous measurements; therefore, they cancel in eqns. (8.118–8.119). The correlation of the atmospheric error term between the base and receiver locations depends on the various factors, the most important of which is the base-to-rover distance $d = \|\mathbf{p}_r - \mathbf{p}_o\|$. The satellite position error terms decorrelate with distance due to the change of the satellite position error that projects onto the line-of-sight vector. The effect of satellite position error on differential measurements is bounded (see p. 3-50 in [108]) by $\frac{d\,\delta r}{r}$ where r is the range to the satellite and δr is the magnitude of the satellite position error. This bound is about $2cm$ for $d = 40km$, and $\delta r = 10m$.

Although, for a vector of measurements, eqn. (8.119) would appear to have the form of a least squares estimation problem, strictly speaking it is not a LS problem due to the dependence of $\mathbf{h}^i$ on $\mathbf{p}_r$. However, eqn. (8.118) is easily solved by a few iterations of the algorithm described in Section 8.2.2. After convergence, the result $\Delta\mathbf{x}_{ro}$ is the relative position and relative receiver clock bias between the rover and the base receivers. In navigation applications, the relative clock error is usually inconsequential.

For some applications, for example maneuvering relative to a landing strip or formation flight, relative position may be sufficient. In such applications, the base station position $\mathbf{p_o}$ is not required. In other applications where the Earth relative position is required and the base position is known, it is straightforward to compute the Earth relative rover position as the sum of the base position and the estimated relative position vector. Note that errors in the estimate of the base position vector directly affect the estimated rover position vector.

Since the non-common mode noise η has range errors between 0.1 and $3.0m$, depending on receiver design and multipath mitigation techniques, DGPS position accuracy is much better than SPS GPS accuracy. Cancellation of the common mode noise sources in eqn. (8.119) assumes that the rover is sufficiently near (within 10-50 mi.) the base station and that the corrections are available at the rover in a timely fashion.

8.8.1.2 Carrier Phase

In non-differential operations, the common mode errors corrupting the phase observables limit the utility of the phase observables. In differential mode, the common mode errors are essentially removed. Assuming that the integer ambiguities can be identified, the carrier observables then enable position estimation with accuracy approximately 100 times better than is possible with pseudorange alone.

For the carrier phase observables, the relative DGPS process is nearly identical to that described for the pseudorange in Section 8.8.1.1. Let the phase observables be modeled as

$$\lambda\tilde{\phi}_o^i = R(\mathbf{p}_o, \hat{\mathbf{p}}^i) + c\Delta t_{r_o} + \beta_o^i + \lambda N_o^i + c\delta t^i + c\bar{\delta t}_{a_o}^i + E_o^i \quad (8.120)$$

$$\lambda\tilde{\phi}_r^i = R(\mathbf{p}_r, \hat{\mathbf{p}}^i) + c\Delta t_{r_r} + \beta_r^i + \lambda N_r^i + c\delta t^i + c\bar{\delta t}_{a_r}^i + E_r^i \quad (8.121)$$

where the symbols $c\bar{\delta t}_{a_o}^i$ and $c\bar{\delta t}_{a_r}^i$ are used to distinguish the phase atmospheric delays from the similarly denoted pseudorange atmospheric delays in Section 8.8.1.1. The code and phase atmospheric delays are not the same as discussed in Section 8.4.4.1.

If the base observables were communicated to the rover, then by methods similar to those in Section 8.8.1.1, the differential measurements

$$\Delta\tilde{\phi}_{ro}^i = \tilde{\phi}_r^i - \tilde{\phi}_o^i \quad (8.122)$$

could be formed and modeled as

$$\lambda\Delta\tilde{\phi}_{ro}^i = R(\mathbf{p}_r, \hat{\mathbf{p}}^i) - R(\mathbf{p}_o, \hat{\mathbf{p}}^i) + (c\Delta t_{r_r} - c\Delta t_{r_o}) + \beta + \lambda N_{ro}^i \quad (8.123)$$

$$\approx \hat{\mathbf{h}}^i \Delta\mathbf{x}_{ro} + \beta + \lambda N_{ro}^i \quad (8.124)$$

where $N_{ro}^i = N_r - N_o$ is still an integer and

$$\beta = (c\delta t_{a_r}^i - c\delta t_{a_o}^i) + (E_r^i - E_o^i) + (\beta_r - \beta_o).$$

The common mode errors cancel in the same sense as was discussed in Section 8.8.1.1. Often the residual ionospheric error is modeled as an additional term in eqns. (8.123–8.124) when its effect is expected to be significant, as can be the case in integer ambiguity resolution over between receivers separated by large distances.

The η term in eqns. (8.118–8.119) is expected to have magnitude on the meter scale, while the β term of eqns. (8.123–8.124) is expected to have magnitude on the centimeter scale. However, the challenge to using eqns. (8.123–8.124) as a range estimate for precise point positioning is that the integer N_{ro} is unknown. Integer ambiguity resolution is discussed in Section 8.9. Carrier-smoothing between eqns. (8.118) and (8.123) is also an option during the time interval that the integer search is in progress.

8.8.1.3 Rover-Base Separation

The transition from eqn. (8.118) to eqn. (8.119) involves the approximation

$$R(\mathbf{p}_r, \hat{\mathbf{p}}^i) - R(\mathbf{p}_o, \hat{\mathbf{p}}^i) = \hat{\mathbf{h}}^i \Delta\mathbf{x}_{ro}. \quad (8.125)$$

This section first motivates the approximation from a physical perspective and then shows the result mathematically.

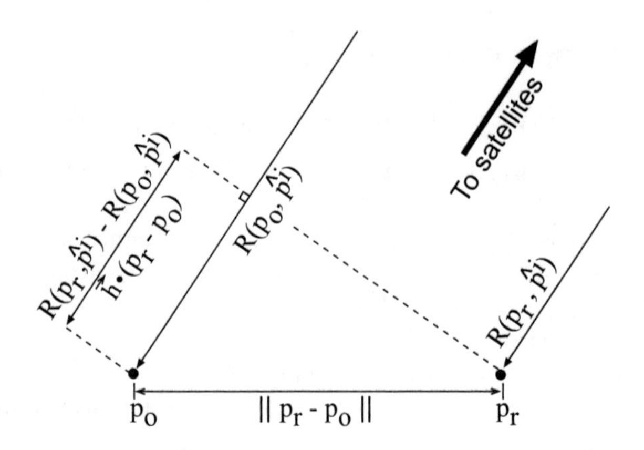

Figure 8.14: Relative GPS range.

For the physical description, consider the geometry depicted in Figure 8.14. The figure shows signals arriving at the base and rover locations. The geometry of the figure assumes that (1) the satellite is sufficiently far away for the wave fronts to appear as planes that are indicated by dashed lines; and, (2) the line-of-sight vectors for the rover and base are identical. From the figure geometry it is clear that subject to the two assumptions, eqn. (8.125) is an accurate approximation. The analysis below shows that the accuracy of the approximations is approximately $\frac{d^2}{2R_o}$ which has magnitude $2.5cm$ for $d = 1km$ and $2.5m$ at $d = 10km$.

The following mathematical derivation will allow quantitative analysis. The derivation will start from the following facts:

$$\begin{aligned} R(\mathbf{p}_r, \hat{\mathbf{p}}^i) &= \vec{\mathbf{h}}_r \left(\mathbf{p}_r - \hat{\mathbf{p}}^i\right) \\ R(\mathbf{p}_o, \hat{\mathbf{p}}^i) &= \vec{\mathbf{h}}_o \left(\mathbf{p}_o - \hat{\mathbf{p}}^i\right) \end{aligned}$$

where $\vec{\mathbf{h}}_r = \frac{(\mathbf{p}_r - \hat{\mathbf{p}}^i)}{\|\mathbf{p}_r - \hat{\mathbf{p}}^i\|}$ and $\vec{\mathbf{h}}_o = \frac{(\mathbf{p}_o - \hat{\mathbf{p}}^i)}{\|(\mathbf{p}_o - \hat{\mathbf{p}}^i)\|}$ are represented as row vectors so that the superscript $\top$ for transpose can be dropped. To decrease the notation slightly, we will use the following symbols: $R_o = R(\mathbf{p}_o, \hat{\mathbf{p}}^i)$ and $R_r = R(\mathbf{p}_r, \hat{\mathbf{p}}^i)$. The analysis is as follows:

$$\begin{aligned} R_r - R_o &= \vec{\mathbf{h}}_r \left(\mathbf{p}_r - \hat{\mathbf{p}}^i\right) - \vec{\mathbf{h}}_o \left(\mathbf{p}_o - \hat{\mathbf{p}}^i\right) + \vec{\mathbf{h}}_r \left(\mathbf{p}_o - \mathbf{p}_o\right) \\ &= \vec{\mathbf{h}}_r \left(\mathbf{p}_r - \mathbf{p}_o\right) + \vec{\mathbf{h}}_r \left(\mathbf{p}_o - \hat{\mathbf{p}}^i\right) - \vec{\mathbf{h}}_o \left(\mathbf{p}_o - \hat{\mathbf{p}}^i\right). \end{aligned} \quad (8.126)$$

Figure 8.15 shows the notation and geometry that will be useful for the analysis of the last two terms in eqn. (8.126). From the figure it is clear that

$$\vec{\mathbf{h}}_r \left(\mathbf{p}_o - \hat{\mathbf{p}}^i\right) = \vec{\mathbf{h}}_o \left(\mathbf{p}_o - \hat{\mathbf{p}}^i\right) \cos(\theta). \quad (8.127)$$

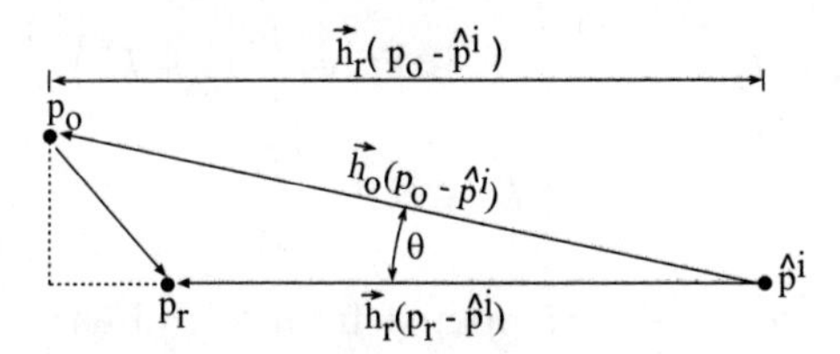

Figure 8.15: Notation and geometry for analysis of rover to base separation.

Therefore,

$$\begin{aligned} R_r - R_o &= \vec{\mathbf{h}}_r(\mathbf{p}_r - \mathbf{p}_o) + (\cos(\theta) - 1)\,\vec{\mathbf{h}}_o(\mathbf{p}_o - \vec{\mathbf{p}}^i) \\ &\approx \vec{\mathbf{h}}_r(\mathbf{p}_r - \mathbf{p}_o) + \frac{\theta^2}{2} R_o. \end{aligned} \quad (8.128)$$

Using the approximation $\theta \approx \sin(\theta)$ (accurate to third order in θ) we have

$$|\theta| \leq \frac{\|\mathbf{p}_r - \mathbf{p}_o\|}{\vec{\mathbf{h}}_o(\mathbf{p}_o - \hat{\mathbf{p}}^i)} = \frac{d}{R_o}.$$

Using this information, eqn. (8.128) reduces to

$$R_r - R_o \approx \vec{\mathbf{h}}_r(\mathbf{p}_r - \mathbf{p}_o) + \frac{d^2}{2R_o}. \quad (8.129)$$

In the transition from eqn. (8.118) to eqn. (8.119) this error term is incorporated into the measurement error. The error can be substantial for $d > 1km$.

8.8.2 Differential GPS

The common mode noise sources are continuous and slowly time-varying (see Figure 8.16). Due to this significant short-term correlation, it is often more efficient to send corrections instead of the full measurements that were discussed in Sections 8.8.1.1 and 8.8.1.2.

8.8.2.1 Differential Corrections

For the L1 and L2, code and phase observables, the common mode, additive, ideal error corrections are

$$\Delta\rho_1^i(t) = -\left(E^i + c\delta t^i + \frac{f_2}{f_1} I_r^i + T_r^i\right) \quad (8.130)$$

$$\Delta\rho_2^i(t) = -\left(E^i + c\delta t^i + \frac{f_1}{f_2} I_r^i + T_r^i\right) \quad (8.131)$$

$$\Delta\phi_1^i(t) = -\left(E^i + c\delta t^i - \frac{f_2}{f_1} I_r^i + T_r^i\right)\frac{1}{\lambda_1} \quad (8.132)$$

$$\Delta\phi_2^i(t) = -\left(E^i + c\delta t^i - \frac{f_1}{f_2} I_r^i + T_r^i\right)\frac{1}{\lambda_2}. \quad (8.133)$$

These corrections are easily computed from the observables at the base station as

$$\hat{\Delta}\rho_1^i(t) = \hat{R}_o^i(t) - \tilde{\rho}_{o_1}^i(t) \quad (8.134)$$

$$\hat{\Delta}\rho_2^i(t) = \hat{R}_o^i(t) - \tilde{\rho}_{o_2}^i(t) \quad (8.135)$$

$$\hat{\Delta}\phi_1^i(t) = \hat{R}_o^i(t)\frac{1}{\lambda_1} - \tilde{\phi}_{o_1}^i(t) \quad (8.136)$$

$$\hat{\Delta}\phi_2^i(t) = \hat{R}_o^i(t)\frac{1}{\lambda_2} - \tilde{\phi}_{o_2}^i(t). \quad (8.137)$$

where $\hat{R}_o^i(t)$ is computed in eqn. (8.1) using the known base location $\mathbf{p}_o$ and the computed satellite location $\hat{\mathbf{p}}^i$.

It is useful to consider the error models appropriate for the computed corrections. For the computed pseudorange correction $\hat{\Delta}\rho_1^i$, the measurement model is

$$\hat{\Delta}\rho_1^i(t) = \Delta\rho_1^i(t) + c\Delta t_o(t) + \eta_{\rho_o}^i(t). \quad (8.138)$$

First, it is important to note that multipath errors and receiver noise directly affect the computed corrections. Therefore, a high quality base receiver-antenna pair and care in antenna siting are critical to the accuracy that will ultimately be attained by the rover. Second, the sum $(\Delta\rho_1^i(t) + \eta_{\rho_o}^i(t))$ is expected to be on the order of 10 meters. The base station clock bias term $c\Delta t_o$ is unneeded by the rover and can be as large as $3 \times 10^5 m$. This term significantly increases both the magnitude and rate of change of the computed corrections. For these reasons, it is common practice to subtract an estimate of the base clock bias from each of the computed corrections. The main issues are that the rover clock bias estimate should be a smooth function of time and that the exact same clock bias estimate should be subtracted from each of the simultaneous differential corrections of eqns. (8.134–8.137).

For the computed phase correction, $\hat{\Delta}\phi_1(t)$, the measurement model is

$$\hat{\Delta}\phi_1^i(t) = \Delta\phi_1^i(t) + \frac{1}{\lambda_1}\left(c\Delta t_o(t) + \nu_{\phi_o}(t)\right) + N_{o_1}^i.$$

The base receiver clock bias was discussed in the previous paragraph. For phase corrections, the integer ambiguity is also important because it is large and its specific value is not important to the rover computations. Therefore, let t_o^i represent the first time instant after the i-th satellite rises for which $\hat{\Delta}\phi_1^i(t)$ will be broadcast. At time t_o^i the base can select an arbitrary integer

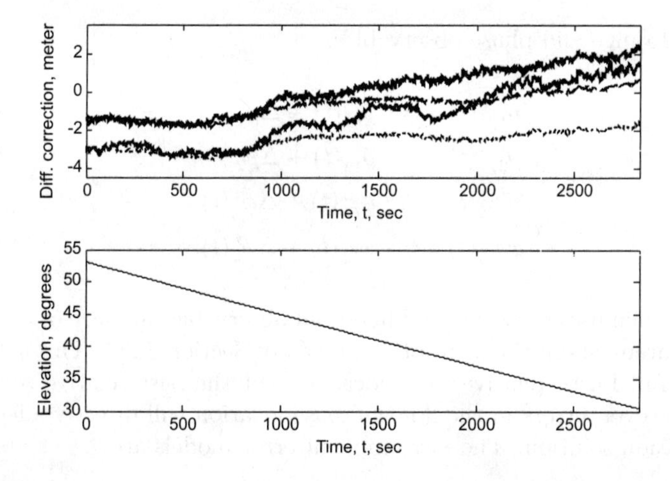

Figure 8.16: Differential corrections from Example 8.6 as computed by eqns. (8.134–8.137).

number of cycles to add to the correction. This integer must then be kept constant and added to all subsequently broadcast corrections for satellite i until the satellite is no longer available to the base receiver. Two common approaches are to select the integer so that the initial phase correction is either near zero or near the corresponding code correction. The actual value of the integer is not important, as long as it is kept constant.

If the base receiver loses lock to a satellite, it must have a means to communicate that fact to the rovers. So that each rover can recompute the integer ambiguities for that satellite.

Example 8.6 *Figure 8.16 shows example differential GPS corrections as computed by eqns. (8.134–8.137) after removing an estimate of the base receiver clock bias and selecting the initial base integer ambiguities such that the code and phase measurements are approximately the same at* $t = 0$. *At* $t = 0$, *the L1 corrections start near -1.5 m and the L2 corrections start near -3.0 m. The phase corrections are drawn with a narrower line than the pseudorange corrections. The time correlated nature of the corrections is clearly shown in the figure. In addition, over the duration of this experiment, as the satellite descends, the divergence between the code and carrier corrections is clearly exhibited.* △

8.8.2.2 Rover DGPS Computations

At each rover, the differential corrections for each of the observables is added to the corresponding rover observables to compute single-differenced

L1 and L2 range and phase observables:

$$\rho^i_{1_{ro}}(t) = \tilde{\rho}_{r_1}(t) + \hat{\Delta}\rho^i_1(t) \tag{8.139}$$
$$\rho^i_{2_{ro}}(t) = \tilde{\rho}_{r_2}(t) + \hat{\Delta}\rho^i_2(t) \tag{8.140}$$
$$\phi^i_{1_{ro}}(t) = \tilde{\phi}_{r_1}(t) + \hat{\Delta}\phi^i_1(t) \tag{8.141}$$
$$\phi^i_{2_{ro}}(t) = \tilde{\phi}_{r_2}(t) + \hat{\Delta}\phi^i_2(t). \tag{8.142}$$

The rover then uses these single-differenced observables in place of the original measurements in the solution approach of Section 8.2.2. This solution gives the full Earth relative rover location, not the base relative rover location; however, errors in the 'known' base location will directly affect the rover position solution. The measurement error models are

$$\rho^i_{1_{ro}}(t) = R(\mathbf{p}_r, \hat{\mathbf{p}}^i) + \frac{f_2}{f_1}\delta I^i_r + c\Delta t_{ro} + \eta^i_{ro} \tag{8.143}$$
$$\rho^i_{2_{ro}}(t) = R(\mathbf{p}_r, \hat{\mathbf{p}}^i) + \frac{f_1}{f_2}\delta I^i_r + c\Delta t_{ro} + \eta^i_{ro} \tag{8.144}$$
$$\lambda_1\phi^i_{1_{ro}}(t) = R(\mathbf{p}_r, \hat{\mathbf{p}}^i) - \frac{f_2}{f_1}\delta I^i_r + c\Delta t_{ro} + \beta^i_{ro} + N^i_1\lambda_1 \tag{8.145}$$
$$\lambda_2\phi^i_{2_{ro}}(t) = R(\mathbf{p}_r, \hat{\mathbf{p}}^i) - \frac{f_1}{f_2}\delta I^i_r + c\Delta t_{ro} + \beta^i_{ro} + N^i_2\lambda_2 \tag{8.146}$$

where N^i_1 and N^i_2 are unknown integers and δI^i_r is the residual ionospheric error between the rover and base.

To successfully implement a range space DGPS system, the base station must, at a minimum, broadcast the following set of information for each satellite: satellite id, range and phase corrections, ephemeris set identifier, and a reference time. A roving receiver then selects the most appropriate set of satellites for its circumstances. The ephemeris set identifier is crucial because the satellite clock and position errors are distinct for each ephemeris set. When a new ephemeris set becomes available, for practical reasons, it is also standard practice to send corrections for both the new and old ephemeris set for some short period of time to allow rovers to download the new ephemeris set.

Example 8.7 *This example demonstrates the accuracy difference between GPS and range space DGPS processing. In this example, the rover and base receivers are connected to the same antenna; therefore, multipath errors will cancel. In addition, the origin of the local tangent plane coordinate system is the antenna location; hence, the correct position estimate is* $(0, 0, 0)$.

The measurement matrix $\mathbf{H}$ *for this data epoch is calculated to be*

$$\hat{\mathbf{H}} = \begin{bmatrix} .4513 & -.3543 & -.8190 & 1.0000 \\ -.5018 & .5502 & -.6674 & 1.0000 \\ -.6827 & -.6594 & -.3147 & 1.0000 \\ -.3505 & -.4867 & -.8001 & 1.0000 \end{bmatrix}.$$

The pseudoranges corrected for satellite clock bias and the expected range to the origin of the tangent plane reference frame are

$$\delta\boldsymbol{\rho}_r = \begin{bmatrix} 3316.75 \\ 3268.47 \\ 3330.98 \\ 3346.88 \end{bmatrix}$$

where $\delta\boldsymbol{\rho} = \tilde{\rho} - \hat{\rho}|_{\mathbf{p}_r=\mathbf{0}}$. *Using the equation*

$$\mathbf{x}_r = \left(\mathbf{H}^\top\mathbf{H}\right)^{-1}\mathbf{H}^\top\delta\boldsymbol{\rho}_r,$$

the GPS estimates of tangent plane position and clock bias are

$$\begin{bmatrix} (x - x_o) \\ (y - y_o) \\ (z - z_o) \\ c\Delta t_r \end{bmatrix} = \begin{bmatrix} -27.84 \\ -69.86 \\ -76.67 \\ 3241.77 \end{bmatrix}.$$

Note these positioning results are from the era when selective availability was active.

The L1 pseudorange differential corrections received at the rover from the base station are

$$\hat{\Delta\boldsymbol{\rho}}_1 = \begin{bmatrix} 12.52 \\ 60.86 \\ 0.93 \\ -17.24 \end{bmatrix}.$$

Using the equation

$$\Delta\mathbf{x}_r = \left(\mathbf{H}^\top\mathbf{H}\right)^{-1}\mathbf{H}(\delta\boldsymbol{\rho}_r + \Delta\boldsymbol{\rho}_1),$$

to recalculate the local tangent plane position and clock bias error for the rover receiver yields

$$\begin{bmatrix} (x - x_o) \\ (y - y_o) \\ (z - z_o) \\ c\Delta t_r \end{bmatrix} = \begin{bmatrix} -0.37 \\ -0.41 \\ 0.44 \\ 3229.67 \end{bmatrix}.$$

△

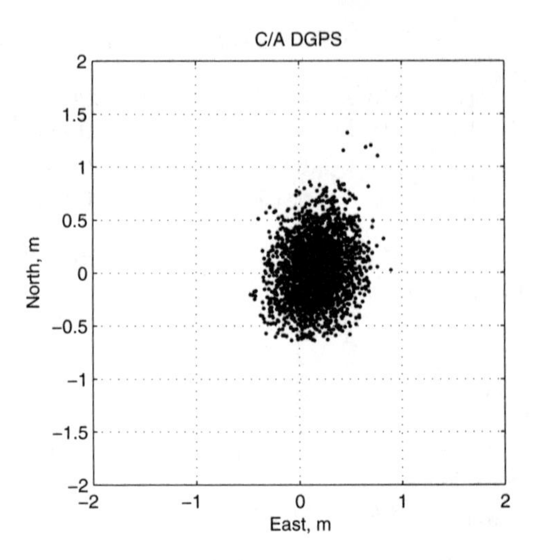

Figure 8.17: Single difference DGPS L1 C/A position estimate scatter plot.

Example 8.8 *Figure 8.1 presented GPS position estimation results. For the same set of data, a scatter plot of the differential L1 pseudorange position estimation results is shown in Figure 8.17. The north, east, and down position estimation error mean vector is* $[0.02, 0.16, -0.08]m$ *the error standard deviation vector is* $[0.3, 0.2, 0.7]m$.

A scatter plot of the differential L1 (integer-resolved) phase position estimation results is shown in Figure 8.18. The north, east, and down position estimation error mean vector is $[0.008, -0.001, -0.014]m$ *the error standard deviation vector is* $[0.004, 0.002, 0.009]$. *As expected the phase estimation standard deviation is approximately 100 times improved relative to the pseudorange estimation standard deviation.*

△

8.8.2.3 Differential Correction Sources

The implementation of a DGPS system requires at least two receivers, and a means to communicate the corrections from the base to the rover in a timely fashion [18, 51, 142]. As an alternative to implementing a private DGPS system, corrections are already available from a variety of public and private sources. Examples include: Continuously Operating Reference Stations (CORS), LandStar, OmniStar, StarFire, U.S. Coast Guard [109], and Wide Area Augmentation System (WAAS). In addition, in network-based DGPS the provider accumulates GPS data from various ground stations. The provider either computes corrections specifically for a user location

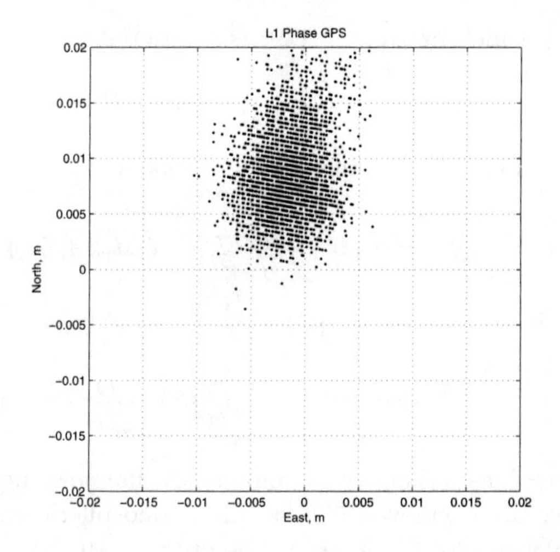

Figure 8.18: Single difference DGPS L1 phase position estimate scatter plot.

or computes data in a form such that the user can easily compute corrections for their location. This information is broadcast to the user over any standard network communication medium (e.g., cell phone).

Since uniformity of DGPS service across providers will simplify distribution and acceptance, the Radio Technical Commission for Marine Services has defined and maintains standards for the implementation and message format of DGPS services. This RTCM protocol is described in [7]. In addition, there are various proprietary protocols.

8.8.3 Double Differences

Section 8.8.2 has discussed the DGPS single-difference observables. Such single differences are free from common-mode errors, but are corrupted by the clock errors from both receivers. In addition, various small receiver biases remain. Finally, affects remain from the relative rotation between the rover and satellites due to the circular polarization of the signal.

All of these artifacts can be eliminated through a process referred to as *double differencing.* The double difference method is the same for the various observables. For brevity, in this section we only consider the L1 pseudorange and phase double differences. The derivations of this section will work from the equations in Section 8.8.2. The double-difference approach can similarly be defined on the basis of the single-differenced observables defined in Section 8.8.1.

Consider the difference between the single-differenced pseudoranges for satellites i and j, which by eqn. (8.139) is computed as

$$\nabla\Delta\rho^{i,j} = \rho^i_{1_{ro}}(t) - \rho^j_{1_{ro}}(t)$$

which by eqn. (8.143) has the measurement model

$$\begin{aligned}\nabla\Delta\rho^{i,j} &= (R(\mathbf{p}_r, \hat{\mathbf{p}}^i) + \frac{f_2}{f_1}\delta I^i_{ro} + c\Delta t_{ro} + \eta^i_{ro}) \\ &\quad -(R(\mathbf{p}_r, \hat{\mathbf{p}}^j) + \frac{f_2}{f_1}\delta I^j_{ro} + c\Delta t_{ro} + \eta^j_{ro}) \\ &= (R(\mathbf{p}_r, \hat{\mathbf{p}}^i) - R(\mathbf{p}_r, \hat{\mathbf{p}}^j) + \frac{f_2}{f_1}\delta I^{ij}_{ro} + \eta^{ij}_{ro}. \end{aligned} \quad (8.147)$$

Note that receiver biases that are common to all channels, including receiver clock error, have been removed. The residual ionospheric error denoted by δI^{ij}_{ro} is typically small for baselines on the order of 10km.

Eqn. (8.147) can be solved using the approach described in Section 8.2.2. The main difference is that the **H** matrix will now contain rows defined by $(\vec{\mathbf{h}}^i - \vec{\mathbf{h}}^j)$ where $\vec{\mathbf{h}}^i \in R^3$ is a vector pointing from satellite i to the receiver. In contrast, to the GPS solution, **H** does not include a column for the receiver clock bias, because the receiver bias terms are canceled in the double-difference operation.

At least three double differences are required to estimate position. This will still require at least four satellites. While the noise on each satellite range measurement is independent, if any satellite is used in more than one double difference, the double-differencing operation results in the double-differenced ranges having correlated noise processes. This measurement noise correlation must be accounted for if optimal filtering is expected.

The base station for a double-difference implementation broadcasts the same information as required for a single-difference base station. The rover processing is the same except for the additional step of forming the double differences from the single differences and forming the double-difference **H** matrix. The main advantages of double-differencing techniques is that clock modeling is not required and various small receiver error terms are removed. The drawbacks are the potential processing of correlated measurements, and the possible increase in range noise.

The double-differenced phase measurements formed from eqn. (8.141) are

$$\nabla\Delta\phi^{i,j} = \phi^i_{1_{ro}}(t) - \phi^j_{1_{ro}}(t)$$

which by eqn. (8.145) has the error model

$$\lambda(\nabla\Delta\phi^{i,j}) = (R(\mathbf{p}_r, \hat{\mathbf{p}}^i) - R(\mathbf{p}_r, \hat{\mathbf{p}}^j) - \frac{f_2}{f_1}\delta I^{ij}_{ro} + \beta^{ij}_{ro} + \lambda N^{ij}_{ro}. \quad (8.148)$$

where $N_{ro}^{ij} = (N_r^i - N_r^j) - (N_o^i - N_o^j)$ which is still an unknown integer. If the integer ambiguities have been identified, then Eqn. (8.148) can be solved for $\mathbf{p}_r$ in the same manner as was discussed for eqn. (8.147).

Due to the fact that satellites rise and set, the common SV will change with time. However, changing the common SV does not necessitate researching for the integer ambiguities. To see why this is true, assume that the double-difference approach is implemented with the m-th satellite being subtracted from the other $m-1$ satellites and that N_{ro}^{im} has been found for $i = 1, \ldots, m-1$. If at some later time, it is desired to instead subtract satellite p from all satellites instead of satellite m. Assuming that N_{ro}^{im} and N_{ro}^{pm} are available, then

$$\begin{aligned} N_{ro}^{ip} &= (N_r^i - N_r^p) - (N_o^i - N_o^p) \\ &= (N_r^i - N_r^m + N_r^m - N_r^p) - (N_o^i - N_o^m + N_o^m - N_o^p) \\ &= (N_r^i - N_r^m) - (N_o^i - N_o^m) + (N_r^m - N_r^p) - (N_o^m - N_o^p) \\ &= N_{ro}^{im} + N_{ro}^{mp} \\ &= N_{ro}^{im} - N_{ro}^{pm}. \end{aligned}$$

This shows that when the common satellite is changed from m to p, as long as the integer N_{ro}^{pm} is known, then the double-difference integer relative to the new common satellite N_{ro}^{ip} can be easily computed from the double-difference integers for the old common satellite N_{ro}^{im}.

8.9 Integer Ambiguity Resolution

This section considers the integer ambiguity resolution problem. The problem has the general format described as follows. Two sets of vector measurements are available with the form

$$\tilde{\boldsymbol{\rho}} = \boldsymbol{\rho} + \boldsymbol{\eta} \tag{8.149}$$

$$\tilde{\boldsymbol{\phi}}\lambda = \boldsymbol{\rho} + \boldsymbol{\beta} + \mathbf{N}\lambda \tag{8.150}$$

where $\mathbf{N}$ is known to be an integer vector, the wavelength λ is accurately known, and $\boldsymbol{\eta}$ and $\boldsymbol{\beta}$ are mutually uncorrelated measurement noise processes with $\sigma_\eta \approx 100\sigma_\beta$. The question to be resolved is how to determine $\mathbf{N}$ so that $(\tilde{\boldsymbol{\phi}} - \mathbf{N})\lambda$ can be used as an unbiased measurement of $\boldsymbol{\rho}$ instead of $\tilde{\boldsymbol{\rho}}$.

Throughout the discussion, we will only consider differential GPS data, with the assumption that common-mode errors are small. Failure to satisfy this assumption will affect performance. In the case of single differences, the covariance matrix for $\boldsymbol{\beta}$ denoted as $\boldsymbol{\Sigma}_{\boldsymbol{\beta}} = \sigma_\beta^2\mathbf{I}$ is diagonal, $\boldsymbol{\Sigma}_{\boldsymbol{\eta}} = \sigma_\eta^2\mathbf{I} \approx 100\boldsymbol{\Sigma}_{\boldsymbol{\beta}}$, and m satellites will yield a set of m measurements. The majority of the discussion will focus on double-differenced measurements so that the receiver and satellite biases are eliminated and the equations simplify to

$$\tilde{\rho} = \mathbf{R} + \boldsymbol{\eta} \quad (8.151)$$

$$\tilde{\phi}\lambda = \mathbf{R} + \boldsymbol{\beta} + \mathbf{N}\lambda \quad (8.152)$$

where $\mathbf{R}$ is the vector of user-to-satellite geometric ranges. In the case of double differences, the covariance matrix for $\boldsymbol{\beta}$ denoted as $\boldsymbol{\Sigma}_{\boldsymbol{\beta}}$ is not diagonal; instead,

$$[\boldsymbol{\Sigma}_{\boldsymbol{\beta}}]_{ij} = \left\{ \begin{array}{ll} 2\sigma_{\beta}^2 & \text{for } i = j, \\ \sigma_{\beta}^2 & \text{for } i \neq j \end{array} \right\} \quad (8.153)$$

and a set of $(m + 1)$ satellites will be required to generate a set of m measurements.

The integer ambiguity resolution problem can be applied in various scenarios: L1, L2, L1 and L2, and widelane. The number of possible scenarios increases further with the advent of a third frequency in the modernized signal structure. Therefore, in this section, we will discuss general characteristics of the problem and means for its solution.

The solution process will be structured as follows.

1. Initialization – The main objective of this step is to provide an initial estimate $\hat{\mathbf{p}}_0$ of the position that is sufficiently accurate so that the h.o.t's of the approximation

$$\delta\boldsymbol{\phi}\lambda = \mathbf{H}\delta\mathbf{p} + \boldsymbol{\beta} + \mathbf{N}\lambda \quad (8.154)$$

are negligible, where $\delta\boldsymbol{\phi} = \tilde{\boldsymbol{\phi}} - \frac{1}{\lambda}\hat{\mathbf{R}}(\mathbf{p}_0)$ and $\mathbf{H}$ is the Jacobian of eqn. (B.59) defined at $\hat{\mathbf{p}}_0$. The initial estimate of $\hat{\mathbf{p}}_0$ may, for example, come from the solution of eqn. (8.151).

2. Candidate Generation – The main objective of this step is to generate candidate integer vectors. For the i–th SV, a real-valued estimate of the integer is

$$\bar{N}^i = \tilde{\phi}_i - \frac{\tilde{\rho}^i}{\lambda}$$

which has a standard deviation of approximately $\sigma_N = \frac{\sigma_\rho}{\lambda}$. A candidate integer vector can be formed by the operation $N^i = \lfloor \bar{N}^i \rceil$ where the notation $\lfloor x \rceil$ indicates rounding x to the nearest integer. Given a set of m satellites, there would be approximately $\lfloor 2\sigma_N + 1 \rceil^m$ candidate integers within one sigma of $\bar{\mathbf{N}}$. For $m = 8$, with $\sigma_\rho = 1$, for the L1 signal, with $\lambda \approx 0.2$ meters, this is approximately 11^8 candidates. For the widelane signal, with $\lambda \approx 0.9$ meters, there would be approximately 3^8 candidates. The benefit of decreasing the number of candidates is obvious and will be discussed in Section 8.9.1.

3. Candidate Evaluation – The main objective for this step is to define a cost function useful for ranking the candidate integer vectors. For each candidate integer vector $\hat{\mathbf{N}}$, the solution to eqn. (8.154) provides an estimate of the position error vector:

$$\delta\hat{\mathbf{p}} = \lambda \left(\mathbf{H}^\top \mathbf{H}\right)^{-1} \mathbf{H}^\top \left(\delta\boldsymbol{\phi} - \hat{\mathbf{N}}\right). \tag{8.155}$$

For this $\delta\hat{\mathbf{p}}$, the predicted phase measurement is

$$\delta\hat{\boldsymbol{\phi}} = \mathbf{R}\left(\delta\boldsymbol{\phi} - \hat{\mathbf{N}}\right) + \hat{\mathbf{N}} \tag{8.156}$$

where $\mathbf{R} = \mathbf{H}\left(\mathbf{H}^\top\mathbf{H}\right)^{-1}\mathbf{H}^\top \in \mathbb{R}^{m\times m}$ and the residual phase measurement $\mathbf{r} = \delta\boldsymbol{\phi} - \delta\hat{\boldsymbol{\phi}}$ computed for the candidate integer vector $\hat{\mathbf{N}}$ is

$$\mathbf{r}_{\hat{N}} = \mathbf{Q}\left(\delta\boldsymbol{\phi} - \hat{\mathbf{N}}\right) \tag{8.157}$$

where $\mathbf{Q} = \mathbf{I} - \mathbf{R} \in \mathbb{R}^{m\times m}$. Both $\mathbf{Q}$ and $\mathbf{R}$ are projection matrices and are idempotent. For a vector $\mathbf{v}$, the vector $\mathbf{z} = \mathbf{R}\mathbf{v}$ will be in the range space of $\mathbf{H}$ and the vector $\mathbf{z}^\perp = \mathbf{Q}\mathbf{v}$ will be in the subspace that is orthogonal to the range space of $\mathbf{H}$. For the correct integer vector $\|\mathbf{r}_{\hat{N}}\|^2$ should be small; therefore, we have an obvious cost function

$$\begin{aligned} J(\hat{\mathbf{N}}) &= \|\mathbf{r}_{\hat{N}}\|^2 = \left(\delta\boldsymbol{\phi} - \hat{\mathbf{N}}\right)^\top \mathbf{Q}^\top\mathbf{Q}\left(\delta\boldsymbol{\phi} - \hat{\mathbf{N}}\right) \\ &= \left(\delta\boldsymbol{\phi} - \hat{\mathbf{N}}\right)^\top \mathbf{Q}\left(\delta\boldsymbol{\phi} - \hat{\mathbf{N}}\right) \end{aligned} \tag{8.158}$$

where the minimum is attained for $\mathbf{Q}\delta\boldsymbol{\phi} = \mathbf{Q}\hat{\mathbf{N}}$. Solutions to this equation are considered in Section 8.9.1.

4. Candidate Validation – The main objective of this step is to define criteria to test the validity of the candidate integer vector both at the time of the search and later to detect cycle slips. The main test is that $\|\mathbf{r}_{\hat{N}}\|$ is small (i.e., the same order of magnitude as σ_β). In addition, ideally $\|\mathbf{r}_{\hat{N}}\|$ for the best candidate should be significantly smaller than the norm of the residual for any other candidate.

   When the search is using the widelane phase, then additional validation criteria can be defined. Given a candidate $\hat{\mathbf{N}}_w$ and the equations

$$\mathbf{N}_w = \mathbf{N}_1 - \mathbf{N}_2 \tag{8.159}$$

$$\lambda_1\delta\boldsymbol{\phi}_1 = \mathbf{H}\delta\mathbf{p} + \mathbf{N}_1\lambda_1 + \boldsymbol{\beta}_1 \tag{8.160}$$

$$\lambda_2\delta\boldsymbol{\phi}_2 = \mathbf{H}\delta\mathbf{p} + \mathbf{N}_2\lambda_2 + \boldsymbol{\beta}_2, \tag{8.161}$$

integer vectors $\mathbf{N}_1$ and $\mathbf{N}_2$ can be computed as

$$\hat{\mathbf{N}}_1 = \left\lfloor \frac{1}{\lambda_2 - \lambda_1} \left(\lambda_2(\delta\boldsymbol{\phi}_2 + \hat{\mathbf{N}}_w) - \lambda_1 \delta\boldsymbol{\phi}_1 \right) \right\rceil \quad (8.162)$$

$$\hat{\mathbf{N}}_2 = \hat{\mathbf{N}}_1 - \hat{\mathbf{N}}_w. \quad (8.163)$$

Given $\hat{\mathbf{N}}_w$, $\hat{\mathbf{N}}_2$, $\hat{\mathbf{N}}_1$, in addition to eqn. (8.157) computed for the widelane phase, the L1 and L2 phase residuals

$$\mathbf{r}_{\hat{N}_1} = \mathbf{Q}\left(\delta\boldsymbol{\phi} - \hat{\mathbf{N}}_1\right)$$
$$\mathbf{r}_{\hat{N}_2} = \mathbf{Q}\left(\delta\boldsymbol{\phi} - \hat{\mathbf{N}}_2\right)$$

can be computed. For the correct $\hat{\mathbf{N}}_2$ and $\hat{\mathbf{N}}_1$ these residuals should also be small. Note that the noise component of

$$\bar{\mathbf{N}}_1 = \frac{1}{\lambda_2 - \lambda_1} \left(\lambda_2(\delta\boldsymbol{\phi}_2 + \hat{\mathbf{N}}_w) - \lambda_1 \delta\boldsymbol{\phi}_1 \right)$$

is $(\lambda_2\boldsymbol{\beta}_2 - \lambda_1\boldsymbol{\beta}_1)/(\lambda_2 - \lambda_1)$ which is about 5.7 times the standard deviation of the of the L1 phase measurement error. If the $\delta\boldsymbol{\phi}$ error, from multipath or residual common-mode errors, reaches approximately 0.1 cycles then the error in $\bar{\mathbf{N}}_1$ could exceed a half cycle; therefore, the use of the computed $\hat{\mathbf{N}}_1$ and $\hat{\mathbf{N}}_2$ to judge the validity of $\hat{\mathbf{N}}_w$ is not without risk.

8.9.1 Decreasing the Search Space

Although eqn. (8.154) appears to contain $(m + 3)$ free variables (i.e., $\delta\mathbf{p}$ and $\mathbf{N}$), a very useful insight for reduction of the search space is the fact that there are only three degrees of freedom [64]. To see that this is true, note that if $\delta\mathbf{p}$ were known, then $\mathbf{N}$ could be directly computed. Similarly, if three elements of $\mathbf{N}$ were known, then $\delta\mathbf{p}$ and the remainder of $\mathbf{N}$ could be computed. Therefore, instead of testing on the order of $\lfloor 2\sigma_N + 1 \rceil^m$ candidate integers, we instead only need to test on the order of $\lfloor 2\sigma_N + 1 \rceil^3$ candidate integers. A important remaining question is how to appropriately generate these candidate integers. The method presented below relates to the Local Minima Search algorithm presented in [112].

Consider the $\mathbf{LU}$ decomposition of the matrix $\mathbf{Q}$ defined in eqn. (8.157). As discussed following the definition of $\mathbf{Q}$ in eqn. (8.157), the matrix $\mathbf{H}$ has m rows and 3 columns and is rank 3; therefore, its range space has dimension 3. The subspace orthogonal to the range space of $\mathbf{H}$ has dimension $(m-3)$; therefore, $\mathbf{Q}$ has rank equal to $(m - 3)$. In the decomposition $\mathbf{Q} = \mathbf{LU}$ with $\mathbf{L} \in \mathbb{R}^{m\times m}$ lower triangular and $\mathbf{U} \in \mathbb{R}^{m\times m}$ upper triangular, due to

$$
\begin{array}{l}
\mathbf{A} = \mathbf{C}^{-1}\mathbf{D} \\
\text{for } i = -d : d \\
\quad \text{for } j = -d : d \\
\quad\quad \text{for } k = -d : d \\
\quad\quad\quad \mathbf{N}_D = [i, j, k]^\top \\
\quad\quad\quad \bar{\mathbf{N}}_C = \delta\boldsymbol{\phi}_C + \mathbf{A}\,(\delta\boldsymbol{\phi}_D - \mathbf{N}_D) \\
\quad\quad\quad \vdots
\end{array}
$$

Figure 8.19: Triple 'for' loop to compute $\bar{\mathbf{N}}_C$.

the fact that $\mathbf{Q}$ has rank equal to $(m-3)$, there exists an ordering of the satellites such that $\mathbf{L}$ is invertible and the matrix $\mathbf{U}$ can be represented as

$$\mathbf{U} = \begin{bmatrix} \mathbf{C} & \mathbf{D} \\ \mathbf{0} & \mathbf{0} \end{bmatrix}$$

where $\mathbf{D} \in \mathbb{R}^{(m-3)\times 3}$ and $\mathbf{C} \in \mathbb{R}^{(m-3)\times(m-3)}$ is upper triangular and invertible (see Section B.10). Decomposing the vector $\hat{\mathbf{N}}$ into two subvectors $\hat{\mathbf{N}} = [\mathbf{N}_C^\top, \mathbf{N}_D^\top]^\top$ with $\mathbf{N}_C \in \mathbb{R}^{(m-3)}$ and $\mathbf{N}_D \in \mathbb{R}^3$, the equation $\mathbf{Q}\delta\boldsymbol{\phi} = \mathbf{QN}$ from eqn. (8.158) can be equivalently written as

$$
\begin{aligned}
\mathbf{LUN} &= \mathbf{Q}\delta\boldsymbol{\phi} \\
\begin{bmatrix} \mathbf{C} & \mathbf{D} \\ \mathbf{0} & \mathbf{0} \end{bmatrix}\begin{bmatrix} \mathbf{N}_C \\ \mathbf{N}_D \end{bmatrix} &= \mathbf{L}^{-1}\mathbf{Q}\delta\boldsymbol{\phi} \\
\begin{bmatrix} \mathbf{C} \\ \mathbf{0} \end{bmatrix}\mathbf{N}_C &= \mathbf{U}\delta\boldsymbol{\phi} - \begin{bmatrix} \mathbf{D} \\ \mathbf{0} \end{bmatrix}\mathbf{N}_D \\
\mathbf{CN}_C &= \begin{bmatrix} \mathbf{C} & \mathbf{D} \end{bmatrix}\delta\boldsymbol{\phi} - \mathbf{DN}_D \\
\bar{\mathbf{N}}_C &= \delta\boldsymbol{\phi}_C + \mathbf{C}^{-1}\mathbf{D}\,(\delta\boldsymbol{\phi}_D - \mathbf{N}_D) \qquad (8.164)
\end{aligned}
$$

where $\bar{\mathbf{N}}_C$ denotes the real-valued estimate of $\mathbf{N}_C$ and $\delta\boldsymbol{\phi} = [\delta\boldsymbol{\phi}_C^\top, \delta\boldsymbol{\phi}_D^\top]^\top$ has been broken into two subvectors with $\delta\boldsymbol{\phi}_C \in \mathbb{R}^{(m-3)}$ and $\delta\boldsymbol{\phi}_D \in \mathbb{R}^3$.

As illustrated in the pseudocode in Figure 8.19, integer candidate vectors $\mathbf{N}_D$ can be generated using a three 'for' loop. For each candidate vector $\mathbf{N}_D$, the entire right hand side of eqn. (8.164) is known. The matrix product $\mathbf{C}^{-1}\mathbf{D}$ is computed once and reused for all candidate vectors $\mathbf{N}_D$. Therefore, the real-valued solutions $\bar{\mathbf{N}}_C$ can be efficiently computed for each of the $(2d+1)^3$ candidate $\mathbf{N}_D$ vectors.

8.9.2 Selection of Optimal Integers

The straightforward approach to produce an integer estimate from $\bar{\mathbf{N}}_C$ is to select

$$\hat{\mathbf{N}}_C = \lfloor \bar{\mathbf{N}}_C \rceil; \qquad (8.165)$$

however, as discussed in [123, 124], this may not be the correct solution in situations where the elements of $\bar{\mathbf{N}}_C$ are strongly cross-correlated. The LAMBDA method was developed to address this issue [123, 124]. In this section, we use ideas from the derivation of the LAMBDA method to motivate the issues and solution ideas, but do not present the LAMBDA method or a detailed solution to this issue.

The covariance matrix $\mathbf{\Sigma} = cov\left(\bar{\mathbf{N}}_C\right)$ computed from eqn. (8.164) is

$$\mathbf{\Sigma} = \mathbf{\Sigma}_{CC} + \mathbf{C}^{-1}\mathbf{D}\mathbf{\Sigma}_{CD} + \mathbf{\Sigma}_{DC}\mathbf{D}^\top\mathbf{C}^{-\top} + \mathbf{C}^{-1}\mathbf{D}\mathbf{\Sigma}_{DD}\mathbf{D}^\top\mathbf{C}^{-\top}$$

where $\mathbf{\Sigma}_{CC} = cov\,(\delta\boldsymbol{\phi}_C)$, $\mathbf{\Sigma}_{CD} = \mathbf{\Sigma}_{DC}^\top = cov\,(\delta\boldsymbol{\phi}_C, \delta\boldsymbol{\phi}_D)$, and $\mathbf{\Sigma}_{DD} = cov\,(\delta\boldsymbol{\phi}_D)$ which are each easily computable as submatrices of $\mathbf{\Sigma}_{\boldsymbol{\beta}}$ as defined in eqn. (8.153). The cost function

$$J(\mathbf{N}_C) = \left(\mathbf{N}_C - \bar{\mathbf{N}}_C\right)^\top \mathbf{\Sigma}^{-1}\left(\mathbf{N}_C - \bar{\mathbf{N}}_C\right) \tag{8.166}$$

is a measure of the distance between the real-valued $\bar{\mathbf{N}}_C$ and the integer vector $\mathbf{N}_C$. Due to the fact that $\mathbf{\Sigma}$ is not diagonal, the ellipsoidal level curves of $J(\mathbf{N}_C)$ (for real valued $\mathbf{N}_C$) could be elongated ellipsoids whose axes are not aligned with the coordinate axes. In such cases, the integer vector that minimizes eqn. (8.166) may be distinct from the result of the rounding operation that results from eqn. (8.165); in fact, the integer vector $\mathbf{N}_C$ that minimizes eqn. (8.166) may be several integer perturbations away from $\lfloor\bar{\mathbf{N}}_C\rceil$.

To find the integer vector that minimizes eqn. (8.166), it is useful to define a nonsingular matrix transformation $\mathbf{Z}$ with the property that both $\mathbf{Z}$ and $\mathbf{Z}^{-1}$ are matrices whose $(m-r)\times(m-r)$ elements are each integers. With these properties, $\mathbf{Z}$ and $\mathbf{Z}^{-1}$ map integer vectors to integer vectors. Let $\mathbf{M}_C = \mathbf{Z}\mathbf{N}_\mathbf{C}$. In this situation, the cost function written in terms of $\mathbf{M}_C$ is

$$J(\mathbf{M}_C) = \left(\mathbf{M}_C - \bar{\mathbf{M}}_C\right)^\top (\mathbf{Z}\mathbf{\Sigma}\mathbf{Z}^\top)^{-1}\left(\mathbf{M}_C - \bar{\mathbf{M}}_C\right) \tag{8.167}$$

where $\bar{\mathbf{M}}_C = \mathbf{Z}\bar{\mathbf{N}}_C$ and $\mathbf{M}_C$ is an integer vector to be selected. The solution would proceed in three steps:

1. Find a matrix $\mathbf{Z}$ such that $\mathbf{Z}\mathbf{\Sigma}\mathbf{Z}^\top$ is nearly diagonal.

2. Compute the real-valued vector $\bar{\mathbf{M}}_C = \mathbf{Z}\bar{\mathbf{N}}_C$.

3. Compute the integer vector $\hat{\mathbf{N}}_C = \mathbf{Z}^{-1}\lfloor\bar{\mathbf{M}}_\mathbf{C}\rceil$.

The intricate step in the procedure is the selection of $\mathbf{Z}$ in Step 1. The LAMBDA method itself and methods to construct $\mathbf{Z}$ are discussed in [123, 124].

8.9.3 Modernized GPS Signal

The GPS signal is currently in a modernization phase that is expected to be complete in approximately 2015, with incremental benefits accruing in the intervening years. There are several aspects of the modernization, which promise enhanced resistance to radio frequency interference and improved performance in navigation applications.

Existing two-frequency civilian receivers use proprietary techniques such as semi-codeless tracking to obtain the L2 measurements. The cost of using such tricks are lower levels of accuracy, longer signal acquisition times, and fragile tracking. The modernized GPS signal will contain a civilian signal on the L2 carrier and will introduce a new civilian signal on a carrier referred to as L5 (or Lc) broadcast with a 1176.45 MHz carrier frequency, which is in a protected aviation band. In addition to the addition of a new carrier frequency and two new civilian signals, the modernization program has resulted in higher broadcast power and selective availability being turned off.

A summary of the benefits to the navigation user is as follows:

- Increased availability. The modernized signal structure will enhance availability in a few different ways.
  - The increased signal power will enhance the ability to track GPS signals and will increase the ability to rapidly acquire GPS satellites.
  - The increased number of civilian signals decreases the likelihood that all GPS signals will be lost due to radio frequency interference. This likelihood is further decreased by the L5 carrier being selected within a protected aviation band.
  - The availability of civilian signals on multiple frequencies eliminates the current need for manufacturer proprietary L2 signal tracking techniques. Correspondingly, this eliminates the associated performance penalties and enhances the ability to track L2 signals at lower elevations.
- Increased integrity. Integrity is enhanced for at least two reasons. First, the new L5 carrier is in a protected aviation band. Therefore, the likelihood of accidental radio frequency interference is low. Second, given three related signals they can be directly compared to detect faulty signals.
- Increased accuracy.
  - The availability of civilian signals on the L2 and L5 carriers will allow calibration of the ionospheric errors using methods

similar to those discussed in Section 8.6. For ionospheric delay calibration, the L1/L5 pairing is preferred as it has the largest frequency separation.

- For differential carrier phase positioning, the difference between the L2 and L5 phase measurements is of interest. By analysis similar to that in Section 8.6.1, it is straightforward to show that this difference has a wavelength of $\lambda_x = 5.86m$. This wavelength is large enough to allow resolution of its integer N_x directly from the corresponding pseudorange combination

$$N_x = (\tilde{\phi}_2 - \tilde{\phi}_5) - \frac{f_2\tilde{\rho}_2 + f_5\tilde{\rho}_5}{\lambda_x(f_2 + f_5)}.$$

The disadvantage of the resulting pseudorange observable

$$\hat{\phi}_x = (\tilde{\phi}_2 - \tilde{\phi}_5 - N_x)\lambda_x$$

is that it amplifies the measurement noise of $\tilde{\phi}_2$ and $\tilde{\phi}_5$ by factors of 24 and 23, respectively. However, depending on receiver to reference station separation and receiver quality, it may be possible to use $\hat{\phi}_x$ to directly acquire the widelane and subsequently the L1 integer ambiguities using range space methods.

8.10 GPS Summary

The GPS system is a worldwide asset that is provided by the U.S. government and available to the user at the cost of a receiver. The uniform worldwide accuracy and high reliability of the system are motivations for the use of GPS in numerous applications.

The range accuracies for various GPS operating scenarios are presented in Table 8.7. Position standard deviations can be calculated from the range accuracies in the table based on assumed PDOP values as described in Section 8.5. The standard deviations in the table are based on the error standard deviations presented in Table 8.5, which are from [8]. The typical multipath column uses the maximum values for multipath and receiver noise errors from Table 8.5. The minimum multipath column uses the minimum values for multipath and receiver noise errors from Table 8.5. This column would correspond to a good receiver, with special precautions taken to reduce multipath effects. For the carrier phase row, the receiver noise is assumed to have a standard deviation of a millimeter and the phase multipath used values of 0.1 and 0.01 m.

The presentation of this chapter has not included Kalman filtering, but has presented and compared various GPS processing techniques in a unified format. With the understanding of GPS techniques that have been

GPS Mode	Typ. Range STD. (m.)	Min. Range STD. (m.)
C/A Code, SA off	9.0	8.0
C/A Code DGPS	3.0	0.1
L1 Phase DGPS	0.10	0.01

Table 8.7: Range accuracies for alternative GPS operating scenarios.

developed in this chapter, Chapters 9 and 11 will present applications that could use GPS as an aiding sensor.

8.11 References and Further Reading

The main references for this chapter were [31, 76, 99, 108]. The main references for the description of the GPS system were [22, 75, 76, 99]. This chapter has not describe GPS history, satellite message formats, or receiver pseudorange tracking and decoding algorithms. Several recent books discuss these topics in detail (e.g., [1, 54, 68, 76, 99, 108, 128]). The history of the GPS system as well as detailed chapters on many GPS related topics and applications can be found in [108]. In depth discussions of receiver tracking are contained in [54, 76, 99, 128]. GPS receiver operation and message decoding are discussed in [76, 99].

Various methods have been suggested to utilize carrier phase and Doppler information (e.g., [21, 35, 55, 63, 64, 70, 81, 89]). Modeling of the Doppler or delta pseudorange observables is discussed in greater detail in [32, 57, 56, 76]. The discussion of two frequency receivers follows the presentation of [63, 64]. The main sources for the discussion of integer ambiguity resolution were [62, 64, 81, 99, 112, 123, 124].

This chapter has also limited its discussion of GPS error modeling to the extent necessary to understand the nature of the error and the extent that the error can be reduced. The main references for the the error modeling sections were [68, 73, 76, 99]. Dynamic error models for clocks are discussed in greater depth in [29, 31, 130]. For an in-depth discussion of atmospheric error, interested readers can refer to [68, 76, 99]. For an in-depth discussion of multipath and its effect on receiver observables, readers can refer to [76].

GPS modernization and its effects are discussed in significantly more detail than presented here in [65, 99].

Chapter 9

GPS Aided Encoder-Based Dead-Reckoning

Encoder-based navigation is a form of odometry or dead-reckoning for land vehicles. The basic physical setup is illustrated in Figure 9.1. On an axle of length L, two wheels are equipped with encoders that measure discrete changes in wheel rotation. The two wheels are separately actuated so that each can rotate independently on the axle. The goal is to use the encoder measurements with GPS aiding to maintain an estimate of the tangent plane position $\mathbf{p}^t = [n, e, h]^\top$ of the center of the axle.

The general three-dimensional navigation problem will be discussed briefly in Section 9.7. In the main body of this chapter, we consider a vehicle maneuvering on a planar surface. We assume that the plane is horizontal with respect to the Earth surface at a point $\mathbf{p}_0$ that will serve as the origin of the tangent plane navigation frame. The height of the vehicle is known to be h_r; therefore, $h = h_r$ and the navigation position estimation problem is two dimensional. The horizontal position $\mathbf{p} = [n, e]^\top$ and the tangent frame velocity $\mathbf{v}^t = [\dot{n}, \dot{e}]^\top$ vectors have only two components, and the pitch and roll angles are identically zero. The only attitude variable is the vehicle yaw angle ψ. The kinematic state dimension is three: n, e, and ψ. The kinematic inputs are the (change of) encoder pulse counts $\Delta e_L(k)$ and $\Delta e_R(k)$ for encoders attached to the left and right axles, respectively, during a time interval dT. The vector of calibration coefficients contains the wheel radii $[R_R, R_L]^\top$. Thus the state vector is $\mathbf{x} = [n, e, \psi, R_R, R_L]^\top$.

Due to the horizontal planar surface assumption the altitude h is known. This approach easily extends to non-horizontal planar surfaces and can be extended to smooth non-planar surfaces with roll and pitch being input

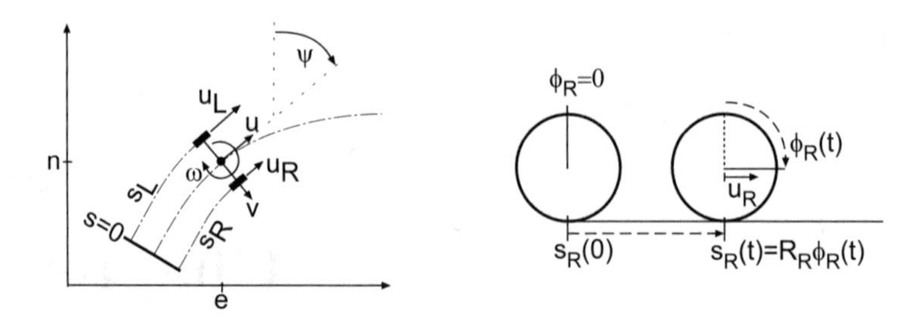

Figure 9.1: Variable definitions for encoder-based navigation. In the left image, the center of the axle has linear velocity $\mathbf{v} = [u, v]^\top$ and angular rate ω about the vehicle vertical axis. The vehicle yaw angle relative to north is ψ. The velocity and arc length of the left wheel are denoted by u_L and s_L, respectively. The velocity and arc length of the right wheel are denoted by u_R and s_R, respectively. The image on the right shows the relationships between the angular rotation ϕ_R of the right wheel and the variables u_R and s_R.

variables determined by additional sensors. For accuracy of the estimate, an important issue is that the wheels do not slip.

The outline for this chapter is as follows. Section 9.1 presents a brief discussion of encoders with the encoder model that will be used in this chapter. The vehicle kinematic model is presented in Section 9.2. Section 9.3 presents both continuous-time and discrete-time navigation mechanization equations. The continuous-time equations are useful for derivation of the error state dynamic equations. The discrete-time equations are suitable for navigation implementation. The error state dynamic model is derived in Section 9.4. Two alternative GPS measurement prediction and residual measurement model equations are presented and compared in Section 9.5. The system observability is considered in Section 9.6. Finally, Section 9.7 considers an extension of the method presented in this chapter to remove the planar surface assumption.

9.1 Encoder Model

Encoders are often implemented as a thin disk attached to the axle with narrow slits cut radially through the disk. A light source and light detector are placed on opposite sides of the disk. As the axle and disk rotate, the light detector senses light pulses. Using multiple light detectors and arranging the slits at different distances from the disk center, it is also possible to sense the direction of rotation. Various types of encoders exist, but the common performance factors are that the encoder output is an integer proportional to the angular rotation of the axle. In the case where

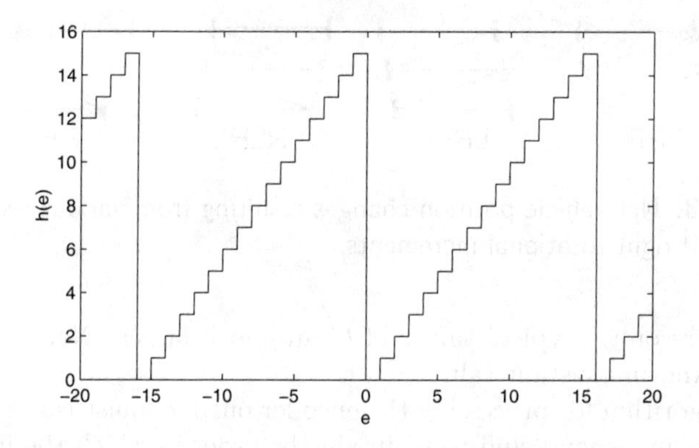

Figure 9.2: Example absolute encoder output with scale factor $C = 16$ counts per revolution.

incremental encoders are used, the design would need to include either interrupt driven or hardware accumulators of the increments to utilize the methods described herein. An important issue is to ensure that no encoder increments are missed.

Although the encoder outputs are discrete in nature, the wheel angular rotations are continuous. Because the continuous-time and continuous state representation is convenient for development of the error models, this section will model continuous valued encoder signals e_L and e_R. The encoder sensor output signals will be quantized versions of the continuous signal $\tilde{e}_L$ and $\tilde{e}_R$. The continuous signal will be used for the development of the error model equations while only the quantized sensor output can be used in the implemented navigation equations.

Let C denote the encoder scale factor in units of counts per revolution. The variables (ϕ_L, ϕ_R) denote the radian angular rotation about the axle of the left and right wheels, respectively. Then, the shaft angle and the (continuous) encoder signal are related by

$$\left.\begin{aligned} e_L(t) &= \tfrac{C}{2\pi}\phi_L(t) \\ e_R(t) &= \tfrac{C}{2\pi}\phi_R(t) \end{aligned}\right\} \tag{9.1}$$

where the signals e_L and e_R are real-valued. The encoder sensor outputs are quantized to integer values and reflected back to always lie in the range $[0, C-1]$:

$$\left.\begin{aligned} \tilde{e}_L &= h(e_L) \\ \tilde{e}_R &= h(e_R). \end{aligned}\right\} \tag{9.2}$$

A typical form of the nonlinearity h is shown in Figure 9.2 for a scale factor of $C = 16$ counts per revolution. This value of C is selected for clarity

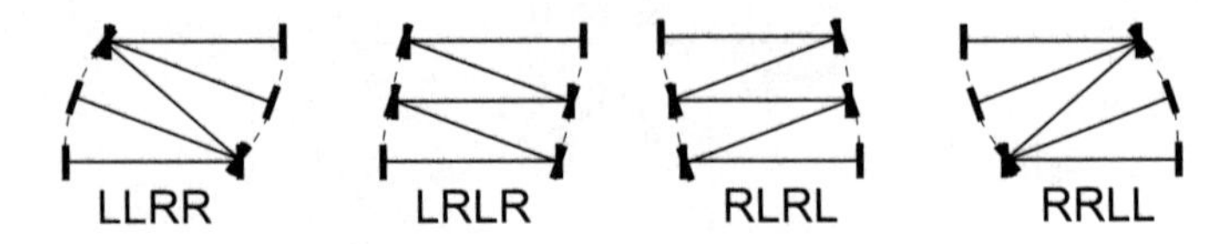

Figure 9.3: Net vehicle position changes resulting from various sequences of two left and right rotational increments.

of the figure only. Typical values of C are much larger. Note that C is a perfectly known constant value.

The algorithm for processing the encoder outputs must run fast enough to not miss increments and to maintain the order in which the increments occur. Figure 9.3 illustrates four possible scenarios involving two left increments indicated by an 'L' and four right increments indicated by an 'R'. The only differences between the scenarios are the ordering of occurrence of the incremental wheel rotations. Note that, as shown in Figure 9.3, a LRLR sequence of increments is distinct from a RRLL sequence. While both result in the vehicle having the same yaw angle, the first moves the vehicle slightly to the right, while the second moves the vehicle slightly to the left. The algorithm must also be designed to correctly accommodate the overflow that occurs at integer multiples of C counts.

9.2 Kinematic Model

Assuming that there is no wheel slip and that the lateral vehicle velocity is zero, the vehicle kinematics are described by

$$\left.\begin{aligned} \dot{\mathbf{p}} &= \begin{bmatrix} \dot{n} \\ \dot{e} \end{bmatrix} = \begin{bmatrix} \cos(\psi) \\ \sin(\psi) \end{bmatrix} u \\ \dot{\psi} &= \omega \end{aligned}\right\} \tag{9.3}$$

where ω is the body frame yaw rate and $\mathbf{v}^b = [u, v]^\top$ is the body frame velocity vector. The body and tangent frame velocity vectors are related by $\mathbf{v}^t = \mathbf{R}_b^t \mathbf{v}^b$, where

$$\mathbf{R}_b^t = \begin{bmatrix} \cos(\psi) & -\sin(\psi) \\ \sin(\psi) & \cos(\psi) \end{bmatrix}.$$

By the zero lateral velocity assumption, the lateral velocity $v = 0$.

Using the kinematic relationship $\mathbf{v}_w = \mathbf{v}^b + \boldsymbol{\omega}_{tb}^b \times \mathbf{R}$ where the superscript w denotes wheel and the subscript b denotes body, the body frame linear velocity of the center of each wheel is

$$\left.\begin{aligned} u_L &= u + \tfrac{L}{2}\omega \\ u_R &= u - \tfrac{L}{2}\omega. \end{aligned}\right\} \tag{9.4}$$

These linear velocities are illustrated in Figure 9.1. The function in eqn. (9.4) from (u, ω) to (u_L, u_R) can be inverted so that the speed and angular rate can be computed when the wheel velocities are known:

$$\left.\begin{aligned} u &= \tfrac{1}{2}\left(u_L + u_R\right) \\ \omega &= \tfrac{1}{L}\left(u_L - u_R\right). \end{aligned}\right\} \tag{9.5}$$

Let (s_L, s_R) denote the arc length traveled by the left and right wheels, respectively. Due to the no slip assumption and assuming that $s_L(0) = s_R(0) = \phi_L(0) = \phi_R(0) = 0$, we have the conditions

$$\left.\begin{aligned} s_L &= R_L \phi_L & \phi_L &= \frac{1}{R_L} s_L \\ s_R &= R_R \phi_R & \phi_R &= \frac{1}{R_R} s_R \end{aligned}\right\} \tag{9.6}$$

where the angular rotation of the left and right axles (ϕ_L, ϕ_R) are defined in eqn. (9.1). By the definition of arc length, we have

$$\left.\begin{aligned} \dot{s}_L &= u_L \\ \dot{s}_R &= u_R. \end{aligned}\right\} \tag{9.7}$$

Therefore, by substituting eqn. (9.7) into the derivative of eqn. (9.6) and solving for (u_L, u_R) we obtain

$$\left.\begin{aligned} u_L &= R_L \dot{\phi}_L \\ u_R &= R_R \dot{\phi}_R. \end{aligned}\right\} \tag{9.8}$$

Finally, combining eqns. (9.1), (9.3), (9.5) and (9.8) we obtain

$$\left.\begin{aligned} \dot{n} &= \tfrac{1}{2}\tfrac{2\pi}{C}\left(R_L \dot{e}_L + R_R \dot{e}_R\right)\cos(\psi) + \omega_n \\ \dot{e} &= \tfrac{1}{2}\tfrac{2\pi}{C}\left(R_L \dot{e}_L + R_R \dot{e}_R\right)\sin(\psi) + \omega_e \\ \dot{\psi} &= \tfrac{1}{L}\tfrac{2\pi}{C}\left(R_L \dot{e}_L - R_R \dot{e}_R\right) + \omega_\psi. \end{aligned}\right\} \tag{9.9}$$

The signals ω_n, ω_e, and ω_ψ represent errors due to, for example, violation of the no-slip assumption. These errors are not necessarily stationary and mutually uncorrelated random signals; however, unless some form of slip detection algorithm is incorporated, ω_n, ω_e, and ω_ψ would typically be modeled as independent random processes.

The length of the wheel radii R_L and R_R vary relative to their nominal values. The variation depends on the type of tire as well as various unpredictable factors. The wheel radii are modeled as

$$R_L = R_0 + \delta R_L \tag{9.10}$$

$$R_R = R_0 + \delta R_R \tag{9.11}$$

where R_0 is the known nominal value of the wheel radii. The error in the wheel radii will be modeled as scalar Gauss-Markov processes

$$\delta\dot{R}_L = -\lambda_R \delta R_L + \omega_L \tag{9.12}$$

$$\delta\dot{R}_R = -\lambda_R \delta R_R + \omega_R \tag{9.13}$$

with $\lambda_R > 0$, $E\langle \delta R_L(0) \rangle = E\langle \delta R_R(0) \rangle = 0$, and

$$var(\delta R_R(0)) = var(\delta R_L(0)) = \bar{P}_R > 0.$$

The correlation time $\frac{1}{\lambda_R}$ is selected large (i.e., a correlation time of 10 hours), because the wheel radii are very slowly time varying. The signals ω_L and ω_R are each white, Gaussian noise processes with PSD of $\sigma_R^2 = 2\lambda_R \bar{P}_R$. This value of σ_R^2 is selected so that the steady state solution of eqn. (4.100) when applied to eqns. (9.12-9.13) is $\bar{P}_R$.

9.3 Encoder Navigation Equations

This section presents the navigation mechanization equations in both continuous and discrete-time. The continuous-time equations will be used to develop the navigation error equations. The discrete-time equations will be used for implementation of the dead-reckoning navigation system.

9.3.1 Continuous-Time: Theory

Based on augmented state vector

$$\hat{\mathbf{x}} = [\hat{n}, \hat{e}, \hat{\psi}, \hat{R}_l, \hat{R}_r]^\top, \tag{9.14}$$

the continuous-time navigation system mechanization equations are

$$\left.\begin{aligned} \dot{\hat{n}} &= \frac{1}{2}\frac{2\pi}{C}\left(\hat{R}_L \dot{e}_L + \hat{R}_R \dot{e}_R\right)\cos(\hat{\psi}) \\ \dot{\hat{e}} &= \frac{1}{2}\frac{2\pi}{C}\left(\hat{R}_L \dot{e}_L + \hat{R}_R \dot{e}_R\right)\sin(\hat{\psi}) \\ \dot{\hat{\psi}} &= \frac{1}{L}\frac{2\pi}{C}\left(\hat{R}_L \dot{e}_L - \hat{R}_R \dot{e}_R\right) \end{aligned}\right\} \tag{9.15}$$

with

$$\left.\begin{aligned} \hat{R}_R &= R_0 + \delta\hat{R}_R \\ \hat{R}_L &= R_0 + \delta\hat{R}_L \end{aligned}\right\} \tag{9.16}$$

where $\delta\hat{R}_L(0)$ and $\delta\hat{R}_R(0)$ will be estimated by the Kalman filter. Eqns. (9.15) and (9.16) will be used in Section 9.4 to derive the state space error model.

9.3.2 Discrete-Time: Implementation

This section considers the integration of eqns. (9.15) over a time increment ΔT. Depending on the type of encoder used, the increment ΔT can be realized by different means. For a pair of incremental encoders, the update would ideally be performed for each encoder increment. In this case, the time increment would not be constant, but would be determined by the vehicle motion. Alternatively, the cumulative encoder increments could be read at a fixed rate. The main criteria in the selection of that sample rate is that the accumulated encoder increments must each be small and none can be missed. The encoder increments will be denoted as

$$\begin{aligned}\Delta e_L(k) &= e_L(t_k) - e_L(t_{k-1})\\ \Delta e_R(k) &= e_R(t_k) - e_R(t_{k-1})\end{aligned}$$

over the time interval $t \in (t_{k-1}, t_k]$ for $\Delta T_k = t_k - t_{k-1}$.

From the third line of eqn. (9.15), given $\hat{\psi}_{k-1}$, the updated yaw is

$$\begin{aligned}\hat{\psi}_k &= \hat{\psi}_{k-1} + \frac{1}{L}\frac{2\pi}{C}\left(\hat{R}_L\frac{\Delta e_L(k)}{\Delta T_k} - \hat{R}_R\frac{\Delta e_R(k)}{\Delta T_k}\right)\Delta T_k\\ &= \hat{\psi}_{k-1} + \frac{1}{L}\frac{2\pi}{C}\left(\hat{R}_L\Delta e_L(k) - \hat{R}_R\Delta e_R(k)\right). \end{aligned} \tag{9.17}$$

Similarly, from the first two lines of eqn. (9.15), given the last estimate of the position $\hat{\mathbf{p}}_{k-1}$, the updated vehicle position is

$$\begin{aligned}\hat{\mathbf{p}}_k &= \hat{\mathbf{p}}_{k-1} + \begin{bmatrix}\cos(\hat{\psi}_k)\\ \sin(\hat{\psi}_k)\end{bmatrix}\frac{\pi}{C}\left[\hat{R}_l\frac{\Delta e_L(k)}{\Delta T_k} + \hat{R}_r\frac{\Delta e_R(k)}{\Delta T_k}\right]\Delta T_k\\ &= \hat{\mathbf{p}}_{k-1} + \begin{bmatrix}\cos(\hat{\psi}_k)\\ \sin(\hat{\psi}_k)\end{bmatrix}\frac{\pi}{C}\left[\hat{R}_l\Delta e_L(k) + \hat{R}_r\Delta e_R(k)\right]. \end{aligned} \tag{9.18}$$

Due to the integer nature of $\Delta e_L(k)$ and $\Delta e_R(k)$, if the assumptions were correct (i.e., planar surface, no slip, and high update rate), then the results of eqns. (9.17) and (9.18) would be very accurate. Also, eqn. (9.18) could be implemented with either $\hat{\psi}_{k-1}$ or $\frac{1}{2}(\hat{\psi}_{k-1} + \hat{\psi}_k)$ used in place of $\hat{\psi}_k$ or the position update could be performed using both $\hat{\psi}_{k-1}$ and $\hat{\psi}_k$ in a predictor-corrector implementation.

9.4 Error State Dynamic Model

The error state vector is $\delta\mathbf{x} = [\delta n, \delta e, \delta\psi, \delta R_L, \delta R_R]^\top$, where each error term is defined as $\delta x = x - \hat{x}$. In this section, we will also use the following variables to decrease the complexity of the resulting equations:

$$\hat{u}_L = \hat{R}_L\dot{\hat{\phi}}_L, \quad \hat{u}_R = \hat{R}_R\dot{\hat{\phi}}_R, \quad \text{and } \hat{u} = (\hat{u}_L + \hat{u}_R)/2.$$

Each of these quantities is computable as an average over any time increment based on the encoder readings.

Forming the difference of the Taylor series expansion of eqn. (9.9) and eqn. (9.15) results in the error state equations

$$\delta \dot{n} = \frac{1}{2}\cos(\hat{\psi})\left(\frac{\hat{u}_L}{\hat{R}_L}\delta R_L + \frac{\hat{u}_R}{\hat{R}_R}\delta R_R\right) - \hat{u}\sin(\hat{\psi})\delta\psi + \omega_n \quad (9.19)$$

$$\delta \dot{e} = \frac{1}{2}\sin(\hat{\psi})\left(\frac{\hat{u}_L}{\hat{R}_L}\delta R_L + \frac{\hat{u}_R}{\hat{R}_R}\delta R_R\right) + \hat{u}\cos(\hat{\psi})\delta\psi + \omega_e \quad (9.20)$$

$$\delta \dot{\psi} = \frac{1}{L}\left(\frac{\hat{u}_L}{\hat{R}_L}\delta R_L - \frac{\hat{u}_R}{\hat{R}_R}\delta R_R\right) + \omega_\psi. \quad (9.21)$$

The error models for R_R and R_L are given by eqns. (9.12–9.13).

The error state equations can be put into be in the standard form of eqn. (4.57) with

$$\mathbf{F}(t) = \begin{bmatrix} 0 & 0 & -\hat{u}\sin(\hat{\psi}) & \frac{\hat{u}_L}{2\hat{R}_L}\cos(\hat{\psi}) & \frac{\hat{u}_R}{2\hat{R}_R}\cos(\hat{\psi}) \\ 0 & 0 & \hat{u}\cos(\hat{\psi}) & \frac{\hat{u}_L}{2\hat{R}_L}\sin(\hat{\psi}) & \frac{\hat{u}_R}{2\hat{R}_R}\sin(\hat{\psi}) \\ 0 & 0 & 0 & \frac{1}{L}\frac{\hat{u}_L}{\hat{R}_L} & -\frac{1}{L}\frac{\hat{u}_R}{\hat{R}_R} \\ 0 & 0 & 0 & -\lambda_R & 0 \\ 0 & 0 & 0 & 0 & -\lambda_R \end{bmatrix} \quad (9.22)$$

and $\mathbf{\Gamma} = \mathbf{I}_5$ where $\boldsymbol{\omega} = [\omega_n, \omega_e, \omega_\psi, \omega_L, \omega_R]$.

Let σ_1^2, σ_2^2, and σ_3^2 denote the PSD's of ω_n, ω_e, and ω_ψ, respectively. These process driving noise terms are included in the model due to the fact that the assumptions will not be perfectly satisfied. For example, the wheels may slip or the surface may not be perfectly planar. The wheel slip is a function of wheel speed or acceleration. Similarly, the effect of a non-planar surface on the computed position is a function of the speed. Therefore, the quantities σ_1^2, σ_2^2, and σ_3^2 and hence the $\mathbf{Q}$ matrix are sometimes defined to be increasing functions of the speed u and/or the acceleration (i.e. as indicated by the change u).

9.5 GPS Aiding

Using the navigation state vector defined in eqn. (9.14), this section presents the GPS measurement prediction and the GPS measurement residual equations that would be used by the Kalman filter.

Using the estimate of the vehicle position $\hat{\mathbf{p}}$ and the ECEF position of the i-th satellite $\hat{p}^i$, which is computed from the ephemeris data as discussed in Appendix C, the computed range to the i-th satellite vehicle (SV) is

$$\hat{R}(\hat{\mathbf{x}}, \hat{\mathbf{p}}^i) = \left\|\hat{\mathbf{p}}^e - \hat{\mathbf{p}}^i\right\|_2. \quad (9.23)$$

In this computation,

$$\hat{\mathbf{p}}^e = \mathbf{p}_0^e + \mathbf{R}_t^e \hat{\mathbf{p}}^t \tag{9.24}$$

where $\hat{\mathbf{p}}^e$ is the estimated ECEF position of the system, $\mathbf{p}_0^e$ is the ECEF position of the origin of the t-frame, and $\hat{\mathbf{p}}^t = [\hat{n}, \hat{e}, h_r]^\top$ is the estimated tangent plane position of the system. The matrix $\mathbf{R}_t^e$ is the constant rotation matrix from the t-frame to the ECEF frame.

The following discussion will assume that there is information available about the common mode errors related to the i-th satellite. It will be denoted as $\hat{\chi}^i$ for pseudorange (see eqn. (8.7)) and as $\hat{\Upsilon}^i$ for carrier phase (see eqn. (8.27)). This information could be obtained, for example, from differential corrections. If there is no information available about the common mode errors, then $\hat{\chi}^i = 0$, $\hat{\Upsilon}^i = 0$, and methods such as the Schmidt-Kalman filter or state augmentation could be used to accommodate the GPS time correlated errors.

The pseudorange is a function of both the range $R(\mathbf{x}, \mathbf{p})$ and the receiver clock bias $c\Delta t_r$ (see Section 8.4.3). The designer can choose between (at least) two approaches to handling the receiver clock error. These choices are discussed in the following two subsections.

9.5.1 Receiver Clock Modeling

In this approach, the receiver clock error is augmented to the navigation state and estimated. The augmented model is

$$\begin{aligned}
\mathbf{x} &= [n, e, \psi, R_L, R_R, c\Delta t_r, c\Delta \dot{t}_r]^\top \\
\delta\mathbf{x} &= [\delta n, \delta e, \delta\psi, \delta R_L, \delta R_R, c\delta t_r, c\delta \dot{t}_r]^\top.
\end{aligned}$$

According to eqn. (8.49), the clock estimate is propagated through time as

$$c\Delta\hat{t}_r(k) = c\Delta\hat{t}_r(k-1) + c\Delta\dot{\hat{t}}_r(k-1)\Delta T_k. \tag{9.25}$$

The pseudorange and carrier phase measurements are defined in eqns. (8.6) and (8.25). Based on those equations, the pseudorange and carrier phase measurements can be predicted from the navigation state as

$$\hat{\rho}^i = \hat{R}(\hat{\mathbf{x}}, \hat{\mathbf{p}}^i) + c\Delta\hat{t}_r + \hat{\chi}^i \tag{9.26}$$

$$\lambda\hat{\phi}^i = \hat{R}(\hat{\mathbf{x}}, \hat{\mathbf{p}}^i) + c\Delta\hat{t}_r + \hat{\Upsilon}^i + \hat{N}^i\lambda \tag{9.27}$$

where the measurements are assumed to be simultaneous and the time arguments have been dropped to simplify the notation. Note that prediction of the carrier phase measurements assumes that estimates of the carrier phase integer ambiguities $\hat{N}^i$ are available. The discussion below assumes that the integer estimates are correct and available. If the integer ambiguities are incorrect, the phase measurements would be biased by an integer number of wavelengths. If the integer ambiguities are not available, then the phase measurements are not used. Alternatively, the phase measurements could be used either to smooth the pseudoranges or through temporal differencing to estimate the velocity.

In this approach, the measurement residuals for the i-th satellite are computed as

$$\delta\rho^i(t) = \tilde{\rho}^i(t) - \hat{\rho}^i(t) \tag{9.28}$$

$$\delta\phi^i(t) = \tilde{\phi}^i(t) - \hat{\phi}^i(t). \tag{9.29}$$

The error models for the measurement residuals are

$$\delta\rho^i(t) = \mathbf{h}_g^i \delta\mathbf{x} + \eta^i \tag{9.30}$$

$$\delta\phi^i(t) = \mathbf{h}_g^i \delta\mathbf{x} + \beta^i \tag{9.31}$$

where $\eta^i \sim N(0, \sigma_\rho^2)$ and $\beta^i \sim N(0, \sigma_\phi^2)$. In the above expression, the errors χ^i and Υ^i have been absorbed into η^i and β^i, respectively. The measurement matrix for the i-th satellite is

$$\mathbf{h}_g^i = [\vec{h}_1^t, \vec{h}_2^t, 0, 0, 0, 1, 0] \tag{9.32}$$

where $\vec{\mathbf{h}}^t = \vec{\mathbf{h}}^e \bar{\mathbf{R}}_t^e$, $\vec{h}_1^t$ and $\vec{h}_2^t$ denote the first two components of $\vec{\mathbf{h}}^t$, and $\vec{\mathbf{h}}^e = \frac{(\hat{\mathbf{p}}^e - \hat{\mathbf{p}}^i)}{\|\hat{\mathbf{p}}^e - \hat{\mathbf{p}}^i\|_2}$. The definition of $\vec{\mathbf{h}}^t$ is related to the discussion in eqn. (8.18) with the a-frame selected as the tangent frame.

The augmented continuous-time error state model has

$$\mathbf{F}(t) = \begin{bmatrix} 0 & 0 & -\hat{u}\sin(\hat{\psi}) & \frac{\hat{u}_L}{2\hat{R}_L}\cos(\hat{\psi}) & \frac{\hat{u}_R}{2\hat{R}_R}\cos(\hat{\psi}) & 0 & 0 \\ 0 & 0 & \hat{u}\cos(\hat{\psi}) & \frac{\hat{u}_L}{2\hat{R}_L}\sin(\hat{\psi}) & \frac{\hat{u}_R}{2\hat{R}_R}\sin(\hat{\psi}) & 0 & 0 \\ 0 & 0 & 0 & \frac{1}{L}\frac{\hat{u}_L}{\hat{R}_L} & -\frac{1}{L}\frac{\hat{u}_R}{\hat{R}_R} & 0 & 0 \\ 0 & 0 & 0 & -\lambda_R & 0 & 0 & 0 \\ 0 & 0 & 0 & 0 & -\lambda_R & 0 & 0 \\ 0 & 0 & 0 & 0 & 0 & 0 & 1 \\ 0 & 0 & 0 & 0 & 0 & 0 & 0 \end{bmatrix}, \tag{9.33}$$

$\mathbf{\Gamma} = \mathbf{I}_7$, and $\boldsymbol{\omega} = [\omega_n, \omega_e, \omega_\psi, \omega_L, \omega_R, \omega_\phi, \omega_f]$. The PSD of ω_ϕ and ω_f are S_ϕ and S_f, respectively, as discussed in Section 8.4.3.2.

9.5.2 Measurement Differencing

In this approach, the measurement from one satellite is subtracted from the measurements of all other satellites to eliminate the receiver clock bias.

Therefore, the state and error state remain

$$\begin{aligned}\mathbf{x} &= [n, e, \psi, R_L, R_R]^\top \\ \delta\mathbf{x} &= [\delta n, \delta e, \delta\psi, \delta R_L, \delta R_R]^\top\end{aligned}$$

and $\mathbf{F}(t)$, $\mathbf{\Gamma}$, and $\boldsymbol{\omega}$ remain as defined in eqn. (9.22).

Assuming that m satellites are available and that measurements from the first satellite are subtracted from all the others, the differenced pseudorange and carrier phase measurements can be predicted from the navigation state as

$$\nabla\hat{\rho}^i = \left(\hat{R}(\hat{\mathbf{x}}, \hat{\mathbf{p}}^i) + \hat{\chi}^i\right) - \left(\hat{R}(\hat{\mathbf{x}}, \hat{\mathbf{p}}^1) + \hat{\chi}^1\right) \tag{9.34}$$

$$\lambda\nabla\hat{\phi}^i = \left(\hat{R}(\hat{\mathbf{x}}, \hat{\mathbf{p}}^i) + \hat{\Upsilon}^i\right) - \left(\hat{R}(\hat{\mathbf{x}}, \hat{\mathbf{p}}^1) + \hat{\Upsilon}^1\right) + \left(\hat{N}^i - \hat{N}^1\right)\lambda \tag{9.35}$$

for $i = 2, \ldots, m$. The same assumptions and comments apply as were discussed following eqns. (9.26-9.27).

In this approach, the measurement residuals for the i-th satellite would be computed as

$$\delta\rho^i(t) = \left(\tilde{\rho}^i(t) - \tilde{\rho}^1(t)\right) - \nabla\hat{\rho}^i(t) \tag{9.36}$$

$$\delta\phi^i(t) = \left(\tilde{\phi}^i(t) - \tilde{\phi}^1(t)\right) - \nabla\hat{\phi}^i(t). \tag{9.37}$$

For $i = 2, \ldots, m$, the error models for the measurement residuals are

$$\delta\rho^i(t) = \bar{\mathbf{h}}_g^i \delta\mathbf{x} + \left(\eta^i - \eta^1\right) \tag{9.38}$$

$$\delta\phi^i(t) = \bar{\mathbf{h}}_g^i \delta\mathbf{x} + \left(\beta^i - \beta^1\right) \tag{9.39}$$

where $\bar{\mathbf{h}}_g^i$ is equal to the first five elements of $\left(\mathbf{h}_g^i - \mathbf{h}_g^1\right)$. The quantities η^i, β^i, and $\mathbf{h}_g^i$ are defined for $i = 1, \ldots, m$ in eqns. (9.30–9.32).

9.5.3 Comparison

The previous two sections presented two approaches to handling the GPS receiver clock error. This section briefly compares the two approaches.

In the approach of Section 9.5.1, the clock error and its rate of change are augmented to the state vector. This results in a larger state vector and requires time propagation of the clock error. The augmented state vector allows the estimated clock error to be removed from each satellite's pseudorange and phase measurement. The additive noise on the measurement residuals for different satellites is uncorrelated. The residuals can be used by the Kalman filter as scalar measurements as described in Section 5.6.1.

In the approach of Section 9.5.2 simultaneous measurements from different satellites are subtracted, which eliminates the common clock error from the difference. The difference affects the measurement matrix, increases the amount of noise per measurement, and causes the additive noise on the measurement residuals for different satellites to be correlated. Consider[1] the variance of the measurement noise for the residual pseudorange measurements of eqn. (9.38):

$$var(\eta^i - \eta^1) = E\langle(\eta^i - \eta^1)(\eta^i - \eta^1)\rangle = 2\sigma_\rho^2. \tag{9.40}$$

Consider the covariance between the measurement noise for the residual pseudorange measurements for satellites 2 and 3:

$$cov((\eta^2 - \eta^1), (\eta^3 - \eta^1)) = E\langle(\eta^2 - \eta^1)(\eta^3 - \eta^1)\rangle = \sigma_\rho^2. \tag{9.41}$$

Therefore, when the measurements of SV 1 are subtracted from the measurements of all other SV's, the Kalman filter measurement covariance is not diagonal, but has the structure

$$\mathbf{R} = \begin{bmatrix} 2 & 1 & 1 & \dots & 1 \\ 1 & 2 & 1 & \dots & 1 \\ \vdots & \vdots & \vdots & \ddots & \vdots \\ 1 & 1 & 1 & \dots & 2 \end{bmatrix} \sigma_\rho^2 \in \mathbb{R}^{(m-1)\times(m-1)} \tag{9.42}$$

where m is the number of available satellites. Due to correlated measurement noise, the residuals could either be processed as a vector or as described in Section 5.6.2.

9.6 Performance Analysis

The material in the previous sections allow the GPS aided encoder navigation system to be implemented. This section presents a basic performance analysis to determine if, or when, the error state is observable and to allow an analysis of expected performance.

Given that measurements from at least three non-coplanar GPS satellites are available, then the position error will be observable. The specific satellite configuration will affect the quality of the position estimate, but not the observability.

9.6.1 Observability

To avoid specific assumptions related to the GPS satellite configuration, for the discussion of observability, we will assume that we can measure position

[1] The following analysis assumes that the noise covariance is the same for all satellites. In practice, this is not true due to issues such as satellite elevation. This fact influences the choice of the satellite that is used in the difference.

directly. The residual measurement is

$$\delta \mathbf{y} = \mathbf{H}\delta \mathbf{x} + \eta \tag{9.43}$$

with

$$\mathbf{H} = \begin{bmatrix} 1 & 0 & 0 & 0 & 0 \\ 0 & 1 & 0 & 0 & 0 \end{bmatrix}.$$

The error state and $\mathbf{F}$ matrix are defined as in Section 9.4. With these assumptions, the observability matrix is

$$\mathcal{O} = \begin{bmatrix} \mathbf{H} \\ \mathbf{HF} \\ \mathbf{HF}^2 \end{bmatrix} = \begin{bmatrix} 1 & 0 & 0 & 0 & 0 \\ 0 & 1 & 0 & 0 & 0 \\ 0 & 0 & -\hat{u}\sin(\hat{\psi}) & \frac{\hat{u}_L}{2\hat{R}_L}\cos(\hat{\psi}) & \frac{\hat{u}_R}{2\hat{R}_R}\cos(\hat{\psi}) \\ 0 & 0 & \hat{u}\cos(\hat{\psi}) & \frac{\hat{u}_L}{2\hat{R}_L}\sin(\hat{\psi}) & \frac{\hat{u}_R}{2\hat{R}_R}\sin(\hat{\psi}) \\ 0 & 0 & 0 & \mathcal{O}_{54} & \mathcal{O}_{55} \\ 0 & 0 & 0 & \mathcal{O}_{64} & \mathcal{O}_{65} \end{bmatrix}$$

where

$$\begin{aligned}
\mathcal{O}_{54} &= \frac{\hat{u}_L}{\hat{R}_L}\left(-\frac{\hat{u}}{L}\sin(\hat{\psi}) - \frac{\lambda_R}{2}\cos(\hat{\psi})\right) \\
\mathcal{O}_{55} &= \frac{\hat{u}_R}{\hat{R}_R}\left(\frac{\hat{u}}{L}\sin(\hat{\psi}) - \frac{\lambda_R}{2}\cos(\hat{\psi})\right) \\
\mathcal{O}_{64} &= \frac{\hat{u}_L}{\hat{R}_L}\left(\frac{\hat{u}}{L}\cos(\hat{\psi}) - \frac{\lambda_R}{2}\sin(\hat{\psi})\right) \\
\mathcal{O}_{65} &= \frac{\hat{u}_R}{\hat{R}_R}\left(-\frac{\hat{u}}{L}\cos(\hat{\psi}) - \frac{\lambda_R}{2}\sin(\hat{\psi})\right).
\end{aligned}$$

Several specific cases can be considered.

For $u = \omega = 0$: In this case $u_L = u_R = 0$. Therefore, the third through fifth columns of $\mathcal{O}$ are zero. In this case, the unobservable subspace is three dimensional and includes the yaw angle and wheel radii errors.

For $u = 0$: In this case the third column of $\mathcal{O}$ is zero which shows that at least the yaw angle is not observable. If $\omega \neq 0$, then $u_L \neq 0$, $u_R \neq 0$, and ψ is changing with time. Under these conditions, the last two columns of $\mathcal{O}$ are linearly independent of each other and of the first two columns, which shows that the rank of $\mathcal{O}$ is four. Therefore, the wheel radii and position are observable.

For $u_L = 0$: In this case the fourth column of $\mathcal{O}$ is zero which shows that at least the R_L is not observable.

For $u_R = 0$: In this case the fifth column of $\mathcal{O}$ is zero which shows that at least the R_R is not observable.

Over time intervals where u_L and u_R are time varying, the state vector may not be observable at a given time instant; nonetheless, the state vector can be observable over the time interval. This is easily demonstrated via covariance analysis.

9.6.2 Covariance Analysis

The main purpose of this section is to present an covariance analysis demonstrating that the state can be observable over an interval of time even though the observability matrix is not full rank at any time instant during that interval. The covariance analysis results are presented in Figure 9.4.

During this simulation, GPS aiding measurements are assumed to occur at a one Hertz rate with $t_k = kT$ and $T = 1$s. The vehicle forward speed is $u(t) = 1\frac{m}{s}$ for all t. The angular rate ω is periodic with a period of $10s$. For $t \in [0, 10]$, the angular rate is defined as

$$\omega(t) = \begin{cases} 0\ \frac{rad}{s} & \text{for } t \in [0, 8)s \\ \frac{\pi}{2}\frac{rad}{s} & \text{for } t \in [8, 10)s. \end{cases}$$

The resulting trajectory is oval-shaped with $8m$ long straight edges in the north-south directions connected by semicircles at each end. At the measurement time instants during this simulation, the rank of the observability matrix is 4; therefore, the error state is not observable at any instant of time.

The parameter definitions for this covariance analysis were $L = 1.0m$, $R_0 = 0.1m$, $\bar{P}_R = (1.0 \times 10^{-4}m)^2$, and $\lambda_R = 1/36000sec^{-1}$. The covariance of the measurement noise η in eqn. (9.43) was $\mathbf{R} = (0.3)^2\mathbf{I}$ meters. The initial error state covariance matrix was $\mathbf{P} = diag([1, 1, (.1)^2, 0.1\bar{P}_R, 0.1\bar{P}_R])$. The PSD matrix for the vector $\boldsymbol{\omega}(t)$ was defined as

$$\mathbf{Q}(t) = diag([(0.001u(t))^2, (0.001u(t))^2, (0.0001u(t))^2, 2\bar{P}_R\lambda_R, 2\bar{P}_R\lambda_R]).$$

For the covariance analysis, the $\mathbf{\Phi}$ and $\mathbf{Qd}$ matrices were computed (see Section 4.7) and the error covariance matrix $\mathbf{P}$ was propagated at $100Hz$.

The $10s$ period of the trajectory is most evident in the plot of the standard deviation σ_n of the estimate of n in the top plot. During the first $8s$ of each $10s$ period, the value of σ_n decreases. The standard deviation σ_e of the estimate of e attains a steady state oscillation. Note that for $t \in [0, 8]s$ or $t \in [10, 18]s$, the growth of σ_e is much more significant than the growth of σ_n. During these time intervals, the vehicle is traveling either north or south and the yaw error $\delta\psi$ has a much larger effect on the lateral variable e than the longitudinal variable n. When the vehicle is rotating

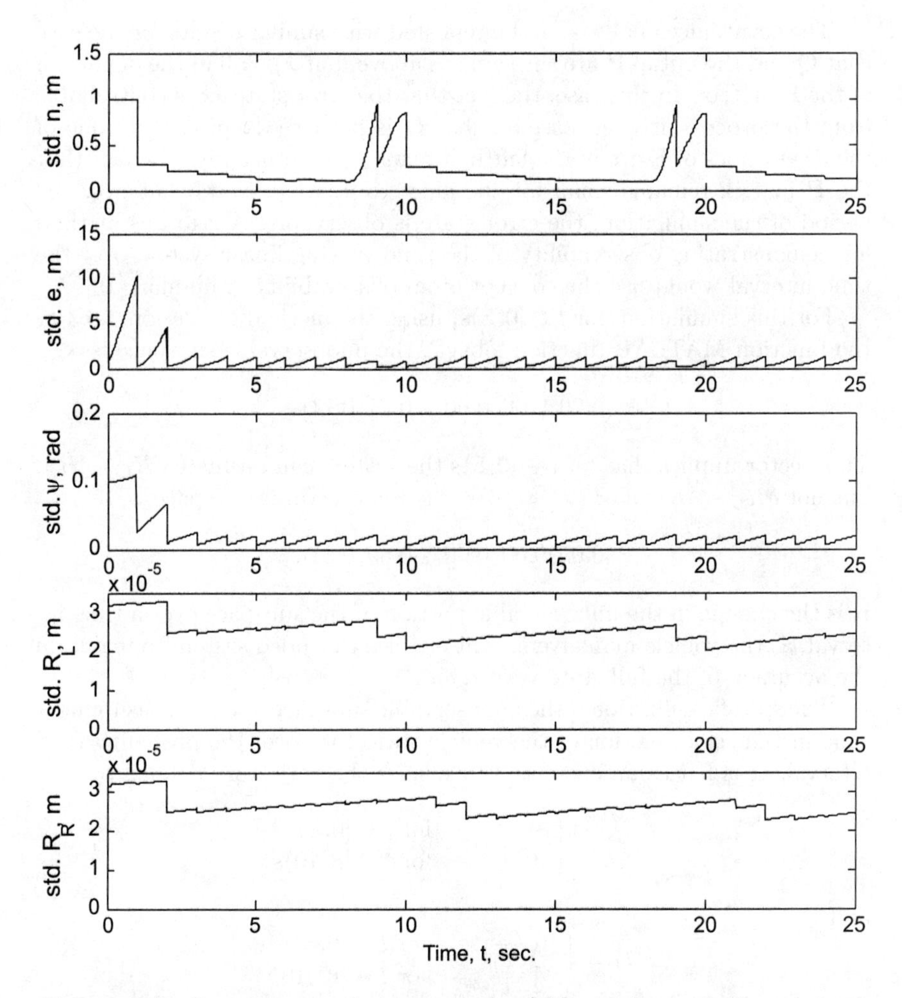

Figure 9.4: Covariance analysis results for GPS aided encoder-based dead-reckoning. The operating scenario is described in Section 9.6.2.

(i.e., $t \in [8, 10]s$ or $t \in [18, 20]s$), the effect of the yaw error on the north position error is clearly indicated. The maximum rate of growth of the north position error is at $t = 9s$ when the angle $\psi = \frac{\pi}{2} rad$. At the same time, the magnitude of the rate of growth of the east position error is near zero.

The covariance analysis can be repeated with similar results for the case that $\mathbf{Q}$ and the initial $\mathbf{P}$ are the same as above, but $\lambda_R = 0$ in the definition of the $\mathbf{F}$ matrix. In this case, the fact that the error state $\delta\mathbf{x}$ is controllable from the process driving noise $\boldsymbol{\omega}$, that $\mathbf{Q}$ is positive definite, and none of the eigenvalues of $\mathbf{F}$ are in the left half plane — together with the fact that the $\mathbf{P}$ matrix remains small for the aided system — shows that over the period of the simulation, the error state is observable. A rigorous method for demonstrating observability of the time varying linear system over the time interval would use the concept of an observability grammian [19].

For this simulation, for $t \in [0, 8)s$, using the method of Section 3.6.3 or the function MATLAB function 'obsvf', the unobservable subspace is

$$[0.00, 0.00, 0.00, -0.71, 0.71].$$

This vector implies that for $t \in [0, 8)s$ the system can estimate $\delta R_L + \delta R_R$, but not $\delta R_L - \delta R_R$. For $t \in [8, 10)s$, the unobservable subspace is

$$[0.00, 0.00, 0.00, -0.59, 0.81].$$

It is the change in the unobservable portion of the subspace over a time interval, as the vehicle maneuvers, that enables the aided system to maintain the accuracy of the full state vector.

The specific definition if the unobservable subspace at each measurement time instant is a function of the vehicle trajectory over the preceding time interval. It is left as an exercise, to consider the trajectory defined by

$$u(t) = \begin{cases} 1 \frac{m}{s} & \text{for } t \in [0, 8)s \\ 0 \frac{m}{s} & \text{for } t \in [8, 10)s \end{cases}$$

and

$$\omega(t) = \begin{cases} 0 \frac{rad}{s} & \text{for } t \in [0, 8)s \\ \frac{\pi}{2} \frac{rad}{s} & \text{for } t \in [8, 10)s. \end{cases}$$

These conditions result in the vehicle driving $8m$ north, performing a π radian rotation without changing position, driving $8m$ south, and performing another π radian rotation without changing position. The vehicle is now back at its original location and orientation. The process repeats every 10 seconds. In this case, for $t \in [8, 10)s$ the unobservable space is two dimensional with basis vectors defined by

$$[0.00, 0.00, 1.00, 0.00, 0.00]$$

and

$$[0.00, 0.00, 0.00, 0.71, 0.71].$$

The predicted navigation accuracy for the same parameter settings is essentially the same as that indicated in Figure 9.4.

9.7 General 3-d Problem

The main body of this chapter has presented a GPS aided encoder-based dead-reckoning approach built on the assumption of a planar operating surface. The approach is easily generalized to nonplanar surfaces when roll angle ϕ and the pitch angle θ can be measured. The main changes to the approach would occur in the kinematic model derived in Section 9.3.

The kinematic model of eqn. (9.3) would be replaced by

$$\dot{\mathbf{p}} = \begin{bmatrix} \cos(\psi)\cos(\theta) \\ \sin(\psi)\cos(\theta) \\ -\sin(\theta) \end{bmatrix} u \qquad (9.44)$$

$$\dot{\psi} = \frac{\cos(\phi)}{\cos(\theta)}\omega. \qquad (9.45)$$

Eqn. (9.44) was derived using $\mathbf{v}^t = \mathbf{R}_b^t \mathbf{v}^b$ with the assumption that $\mathbf{v}^b = [u, 0, 0]^\top$. Eqn. (9.45) is derived using eqn. (2.74) and the assumption that the product $\frac{\sin(\phi)}{\cos(\theta)} q$ is negligibly small where q represents the angular rate about the body frame v-axis. In addition to accounting for the $\cos(\theta)$ terms in eqns. (9.9) and (9.18), an additional equation was required to propagate the vertical position component h.

Note that attitude sensors provide the roll and pitch relative to the geodetic frame. When the region of operation is small, the difference between the tangent frame and the geodetic frame would be inconsequential for this approach. When the area of operation is large enough for the difference to be important, the designer can either compute the tangent frame relative attitude from the geodetic frame relative attitude or modify the approach of this chapter to work in the geodetic frame.

Chapter 10

AHRS

An attitude and heading reference system (AHRS) is a combination of instruments capable of maintaining an accurate estimate of the vehicle roll ϕ, pitch θ, and yaw ψ as the vehicle maneuvers. Various AHRS design approaches are reviewed in Chapter 9 of [78]. This chapter considers a single approach wherein gyro outputs are integrated through the attitude kinematic equations and aided with accelerometer (i.e., gravity direction in body frame) and magnetometer (i.e., Earth magnetic field in body frame) measurements.

The vehicle attitude describes the relative orientation of the axes of the body (b-frame) and navigation (n-frame) frames-of-reference. This book describes three equivalent attitude representations. Each attitude representation could be used to implement the method of this chapter: integration of eqn. (2.54) yields the direction cosine matrix from which the Euler angles could be computed by eqns. (2.45–2.47); integration of eqn. (2.74) yields the Euler angles directly; or, integration of eqn. (D.28) yields the quaternion representation of the rotational transformation from which the Euler angles could be computed using eqns. (D.16–D.18). All three approaches are theoretically equivalent. In this chapter we choose to use the quaternion approach due to its computational efficiency and lack of singularities. Various useful quaternion related results are presented in Appendix D.

A block diagram illustration of the approach described in this chapter is shown in Figure 10.1. A triad of gyros will measure the angular rate vector $\boldsymbol{\omega}_{ib}^{b}$ of the b-frame relative to the i-frame represented in the b-frame. Integration of the gyro measurements through the attitude kinematics yields the rotation matrix $\mathbf{R}_{b}^{n}$ from which the Euler angles can be computed. Body frame accelerometer and magnetometer measurements are transformed to the navigation frame to predict the known Earth gravitational acceleration and magnetic field vectors. A Kalman filter uses the residual measurements to estimate the attitude error and sensor calibration factors. Due to the

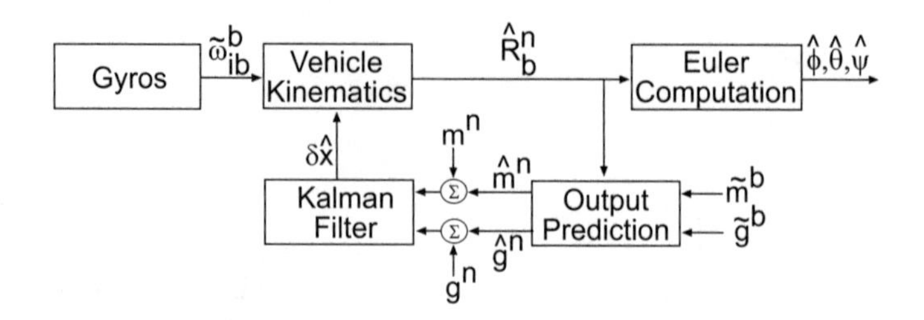

Figure 10.1: Block diagram for attitude and heading reference system.

fact that the body frame accelerometer measures specific force (i.e., kinematic acceleration minus gravitational acceleration) the approach will be designed to detect when the vehicle is accelerating and to de-weight or ignore the accelerometer measurements during such time periods (see Section 10.5.4 and eqn. (10.70)).

10.1 Kinematic Model

The angular rate of the n-frame with respect to the b-frame represented in the b-frame can be decomposed as

$$\boldsymbol{\omega}^b_{bn} = \boldsymbol{\omega}^b_{in} - \boldsymbol{\omega}^b_{ib}$$

where $\boldsymbol{\omega}^b_{ib} = [p, q, r]^\top$ is the vector of body rates relative to the inertial frame. Denote the components of $\boldsymbol{\omega}^b_{bn} = [\omega_1, \omega_2, \omega_3]^\top$.

By eqn. (D.28), the kinematic equation for the quaternion $\mathbf{b}$ representing the rotation from n-frame to b-frame is

$$\dot{\mathbf{b}} = \frac{1}{2}\begin{bmatrix} -b_2 & -b_3 & -b_4 \\ b_1 & b_4 & -b_3 \\ -b_4 & b_1 & b_2 \\ b_3 & -b_2 & -b_1 \end{bmatrix}\begin{bmatrix} \omega_1 \\ \omega_2 \\ \omega_3 \end{bmatrix}. \tag{10.1}$$

Initial conditions for this set of differential equations are defined in Section 10.3.

Given the quaternion $\mathbf{b}$ at time t, according to eqn. (D.13), the rotational transformation from the n-frame to the b-frame at time t is

$$\mathbf{R}^b_n = \begin{bmatrix} b_1^2+b_2^2-b_3^2-b_4^2 & 2(b_2b_3-b_1b_4) & 2(b_1b_3+b_2b_4) \\ 2(b_2b_3+b_1b_4) & b_1^2-b_2^2+b_3^2-b_4^2 & 2(b_3b_4-b_1b_2) \\ 2(b_2b_4-b_1b_3) & 2(b_1b_2+b_3b_4) & b_1^2-b_2^2-b_3^2+b_4^2 \end{bmatrix} \tag{10.2}$$

and from eqns. (D.16–D.18) the Euler angles at the same time instant are

$$\phi = atan2\left(2(b_3b_4-b_1b_2), 1-2(b_2^2+b_3^2)\right) \tag{10.3}$$

$$\theta = asin\left(-2(b_2b_4 + b_1b_3)\right) \tag{10.4}$$

$$\psi = atan2\left(2(b_2b_3 - b_1b_4), 1 - 2(b_3^2 + b_4^2)\right). \tag{10.5}$$

The b-frame specific force vector is

$$\mathbf{f}^b = \mathbf{a}_{ib}^b - \mathbf{g}^b$$

where $\mathbf{a}_{ib}^b$ is the acceleration of the b-frame relative to the inertial frame represented in the b-frame and $\mathbf{g}$ is the gravity vector in the b-frame. If the IMU position was available, then $\mathbf{g}^n$ could be computed by eqn. (2.8). When the IMU position is unknown, as is assumed in this chapter, then $\mathbf{g}^n = [0, 0, g_e]^\top$ where g_e is the Earth gravity constant. Using the definition of $\mathbf{R}_n^b$ from eqn. (2.43), the relationship between the gravity vector in the b and the n-frames is

$$\mathbf{g}^b = \mathbf{R}_n^b \mathbf{g}^n = \begin{bmatrix} -\sin(\theta) \\ \sin(\phi)\cos(\theta) \\ \cos(\phi)\cos(\theta) \end{bmatrix} g_e. \tag{10.6}$$

The b-frame Earth magnetic field vector is

$$\mathbf{m}^b = \mathbf{R}_n^b \mathbf{m}^n \tag{10.7}$$

where $\mathbf{m}^n$ is the Earth magnetic field represented in the navigation frame.

10.2 Sensor Models

Let $\mathbf{u}$ denote the gyro measurement. We assume that the gyro measurement model is

$$\mathbf{u} = \boldsymbol{\omega}_{ib}^b + \mathbf{x}_g + \boldsymbol{\nu}_g \tag{10.8}$$

where $\boldsymbol{\nu}_g$ is Gaussian white noise with PSD $\sigma_{\nu_g}^2$ and $\mathbf{x}_g$ is an additive error. For the purpose of discussion in this chapter, we assume that the gyro error is a bias modeled as a first-order Gauss-Markov process:

$$\dot{\mathbf{x}}_g = \mathbf{F}_g \mathbf{x}_g + \boldsymbol{\omega}_g \tag{10.9}$$

where $\mathbf{F}_g = -\lambda_g \mathbf{I}$ and $\boldsymbol{\omega}_g$ is a Gaussian white noise process with PSD $\sigma_{\omega_g}^2$.

Denote the accelerometer measurement as $\mathbf{y}_a$. The accelerometer measurement model is

$$\mathbf{y}_a = \mathbf{a}_{ib}^b - \mathbf{g}^b + \mathbf{x}_a + \boldsymbol{\nu}_a' \tag{10.10}$$

where $\boldsymbol{\nu}_a'$ is Gaussian white noise with PSD $\sigma_{\nu_a'}^2$ and $\mathbf{x}_a$ is an additive error. For the purpose of discussion in this chapter, we assume that the accelerometer error is a bias modeled as a first-order Gauss-Markov process:

$$\dot{\mathbf{x}}_a = \mathbf{F}_a \mathbf{x}_a + \boldsymbol{\omega}_a \tag{10.11}$$

where $\mathbf{F}_a = -\lambda_a \mathbf{I}$ and $\boldsymbol{\omega}_a$ is a Gaussian white noise process with PSD $\sigma^2_{\omega_a}$.

Denote the magnetometer measurement as $\mathbf{y}_m$. The magnetometer measurement model is

$$\mathbf{y}_m = \mathbf{m}^b_e + \mathbf{m}^b_b + \boldsymbol{\nu}_m \tag{10.12}$$

where $\boldsymbol{\nu}_m$ is Gaussian white noise with PSD $\sigma^2_{\nu_m}$, $\mathbf{m}^b_e$ is the Earth magnetic field vector represented in the b-frame, and $\mathbf{m}^b_b$ is the magnetic field vector generated by the vehicle represented in body frame. We will assume that the magnetometer is either isolated from or compensated for the vehicle magnetic field, so that the $\mathbf{m}^b_b$ term can be assumed to be zero. We will also assume that the compass has been compensated for the local magnetic field deviation from true north so that $\mathbf{m}^n = [m_e, 0, 0]^\top$ where m_e is the magnitude of Earth magnetic field. Finally, we assume that the compass is mounted such that its axes align with the IMU axes. With these assumptions, the magnetometer measurement is a function of the vehicle yaw angle relative to true north as defined in eqn. (10.7).

10.3 Initialization

This section discusses initialization of the AHRS state vector. Initialization of the error state covariance matrix is discussed in Section 10.5.5.

10.3.1 State Initialization: Approach 1

During some initial period of time $t \in [0, T]$ when the vehicle is known to be non-accelerating, perhaps detected by program initialization or $\|\mathbf{y}_a\| \approx g_e$, the accelerometer outputs can be averaged to yield an estimate of the gravity vector in body frame. Based on eqn. (10.10) and assuming temporarily that $\mathbf{a}^b_{ib} = \mathbf{0}$ and $\mathbf{x}_a = \mathbf{0}$, the estimate of the body frame gravity vector is

$$\bar{\mathbf{g}}^b = \frac{1}{T}\int_0^T -\mathbf{y}_a(\tau)d\tau. \tag{10.13}$$

This estimate of the gravity vector is affected by accelerometer noise, accelerometer biases, and by vehicle acceleration. Sensor noise and vehicle vibrations are expected to average out during the period of integration. Sensor bias will not. For a system that in nominally level during the initialization process, each milli-g of sensor bias contributes about 1 mrad of error to the initial attitude estimate.[1]

[1]This is straightforward to show. For example, by eqn. (10.14), if $\mathbf{x}_{a_2}$ represents the bias on the v-accelerometer (i.e., the second component of $\mathbf{y}_a$), then $\frac{d\phi}{d\mathbf{x}_{a_2}} = \frac{d}{d\bar{g}_2} atan\left(\frac{\bar{g}_2}{\bar{g}_3}\right)\frac{d\bar{g}_2}{d\mathbf{x}_{a_2}}$ determines the effect of the bias $\mathbf{x}_{a_2}$ on ϕ. If $\mathbf{x}_{a_2}$ is constant over

Denote the components of $\bar{\mathbf{g}}^b$ as $[\bar{g}_1, \bar{g}_2, \bar{g}_3]^\top$. Using eqn. (10.6), the initial roll and pitch values can be computed as

$$\hat{\phi}(T) = atan2(\bar{g}_2, \bar{g}_3) \tag{10.14}$$

$$\hat{\theta}(T) = atan2\left(-\bar{g}_1, \sqrt{\bar{g}_2^2 + \bar{g}_3^2}\right). \tag{10.15}$$

Decomposing the b-frame to n-frame rotation into its three plane rotations as defined in Section 2.5.3, for the Earth magnetic field in the b-frame and the n-frame, we have

$$\begin{aligned} \mathbf{m}^b &= [\phi]_1[\theta]_2[\psi]_3\mathbf{m}^n \\ [\theta]_2^\top[\phi]_1^\top\mathbf{m}^b &= [\psi]_3\mathbf{m}^n \\ \begin{bmatrix} \cos(\theta) & \sin(\theta)\sin(\phi) & \sin(\theta)\cos(\phi) \\ 0 & \cos(\phi) & -\sin(\phi) \\ -\sin(\theta) & \cos(\theta)\sin(\phi) & \cos(\theta)\cos(\phi) \end{bmatrix}\mathbf{m}^b &= \begin{bmatrix} \cos(\psi) \\ -\sin(\psi) \\ 0 \end{bmatrix} m_e \end{aligned}$$

where we have used the facts that $\mathbf{m}^n = [m_e, 0, 0]^\top$, $[\phi]_1^{-1} = [\phi]_1^\top$, and $[\theta]_2^{-1} = [\theta]_2^\top$. Given the initial conditions for ϕ and θ are available from eqns. (10.14–10.15), the left-hand side of the above equation can be computed based on the body frame magnetometer reading. We define the vector $\bar{\mathbf{m}}^w = [\bar{m}_1^w, \bar{m}_2^w, \bar{m}_3^w]$ as

$$\bar{\mathbf{m}}^w = \begin{bmatrix} \cos(\theta) & \sin(\theta)\sin(\phi) & \sin(\theta)\cos(\phi) \\ 0 & \cos(\phi) & -\sin(\phi) \\ -\sin(\theta) & \cos(\theta)\sin(\phi) & \cos(\theta)\cos(\phi) \end{bmatrix}\bar{\mathbf{m}}^b, \tag{10.16}$$

where

$$\bar{\mathbf{m}}^b = \frac{1}{T}\int_0^T \mathbf{y}_m(\tau)d\tau. \tag{10.17}$$

The w-frame is an intermediate frame of reference defined by the projection of the vehicle u-axis onto the Earth tangent plane. From the above analysis, the rotation matrix $\mathbf{R}_n^w = [\psi]_3$; the rotation matrix $\mathbf{R}_b^w = [\theta]_2^\top[\phi]_1^\top$; and, $\mathbf{R}_b^w\mathbf{R}_n^b = \mathbf{R}_n^w$. The initial yaw angle can be compute from the first two components of $\bar{\mathbf{m}}^w$ as

$$\psi(T) = atan2(-\bar{m}_2^w, \bar{m}_1^w). \tag{10.18}$$

Given the initial values of the Euler angles from eqns. (10.14), (10.15), and (10.18), the initial $\mathbf{R}_n^b$ is then computed using eqn. (2.43). The initial value of the quaternion $\hat{\mathbf{b}}(0)$ is computed from eqn. (D.15).

the period of integration, then $\frac{d\bar{g}_2}{d\mathbf{x}_{a_2}} = 1$. Also, since the system is nominally level, $\bar{g}_2 \ll \bar{g}_3 = g_e$ and $\frac{d}{d\bar{g}_2}atan\left(\frac{\bar{g}_2}{\bar{g}_3}\right) \approx \frac{1}{g_e}$. Therefore, $\frac{d\phi}{d\mathbf{x}_{a_2}} \approx \frac{1}{g_e}$.

During this same initialization period while the vehicle is assumed to be non-accelerating, the gyro outputs can be averaged to provide initial estimates of the gyro biases. Ideally the earth rate $\boldsymbol{\omega}_{ie}^b$ would be subtracted from the gyro measurements; however, the gyros used in AHRS applications may not be sufficiently accurate to resolve the Earth rate.

10.3.2 State Initialization: Approach 2

Given that the Earth gravity $\mathbf{g}$ and magnetic field $\mathbf{m}$ vectors are not have the same direction, an alternative approach to attitude initialization can be developed by defining the vector $\mathbf{r} = \mathbf{g} \times \mathbf{m}$ which is orthogonal to both $\mathbf{g}$ and $\mathbf{m}$. In the navigation frame, we form the matrix

$$\mathbf{A}^n = [\mathbf{m}^n, \mathbf{g}^n \times \mathbf{m}^n, \mathbf{g}^n] = \begin{bmatrix} m_e & 0 & 0 \\ 0 & m_e g_e & 0 \\ 0 & 0 & g_e \end{bmatrix}. \tag{10.19}$$

which is known. Because $\mathbf{m}^b = \mathbf{R}_n^b \mathbf{m}^n$, $\mathbf{r}^b = \mathbf{R}_n^b \mathbf{r}^n$, and $\mathbf{g}^b = \mathbf{R}_n^b \mathbf{g}^n$, we have that

$$\mathbf{A}^b = \mathbf{R}_n^b \mathbf{A}^n \tag{10.20}$$

where

$$\mathbf{A}^b = \left[\mathbf{m}^b, \mathbf{g}^b \times \mathbf{m}^b, \mathbf{g}^b\right]. \tag{10.21}$$

Given the vectors $\bar{\mathbf{g}}^b$ and $\bar{\mathbf{m}}^b$ defined in eqns. (10.13) and (10.17), respectively, we can compute $\bar{\mathbf{A}}^b = [\bar{\mathbf{m}}^b, \bar{\mathbf{g}}^b \times \bar{\mathbf{m}}^b, \bar{\mathbf{g}}^b]$ from the accelerometer and magnetometer measurements. From eqn. (10.20) with $\bar{\mathbf{A}}^b$ and $\mathbf{A}^n$ known, the initial value of the rotation matrix can be computed as

$$\hat{\mathbf{R}}_n^b = \bar{\mathbf{A}}^b (\mathbf{A}^n)^{-1} = \left[\frac{1}{m_e}\, \bar{\mathbf{m}}^b, \frac{1}{m_e g_e}\, \bar{\mathbf{g}}^b \times \bar{\mathbf{m}}^b, \frac{1}{g_e}\, \bar{\mathbf{g}}^b\right]. \tag{10.22}$$

With $\hat{\mathbf{R}}_n^b$ as defined in eqn. (10.22), the initial quaternion and Euler angles can be computed by eqn. (D.15) and eqns. (2.45–2.47), respectively.

10.4 AHRS Mechanization Equations

The state vector is $\hat{\mathbf{x}} = [\hat{\mathbf{b}}^\top, \hat{\mathbf{x}}_g^\top, \hat{\mathbf{x}}_a^\top]^\top$. Given the initial estimate of $\hat{\mathbf{b}}(0)$ as discussed in Section 10.3, the estimate of $\mathbf{b}(t)$ is maintained by integration of the equation

$$\dot{\hat{\mathbf{b}}} = \frac{1}{2} \begin{bmatrix} -\hat{b}_2 & -\hat{b}_3 & -\hat{b}_4 \\ \hat{b}_1 & \hat{b}_4 & -\hat{b}_3 \\ -\hat{b}_4 & \hat{b}_1 & \hat{b}_2 \\ \hat{b}_3 & -\hat{b}_2 & -\hat{b}_1 \end{bmatrix} \hat{\omega}_{bn}^b \tag{10.23}$$

where

$$\hat{\boldsymbol{\omega}}_{bn}^{b} = \hat{\boldsymbol{\omega}}_{in}^{b} - \hat{\boldsymbol{\omega}}_{ib}^{b}. \tag{10.24}$$

If an external estimate of the velocity was available, then $\hat{\omega}_{in}^{b}$ could be computed using eqns. (2.57) and (2.39) for use in eqn. (10.24). In this chapter, we assume that such external input is not available. Without additional information we define $\hat{\omega}_{in}^{b} = \mathbf{0}$. Based on eqn. (10.8) and the gyro output $\mathbf{u}$, the body angular rate is computed as

$$\hat{\boldsymbol{\omega}}_{ib}^{b} = \mathbf{u} - \hat{\mathbf{x}}_g. \tag{10.25}$$

Given $\hat{\mathbf{b}}$ at time t, the Euler angles (ϕ, θ, ψ) can be computed for the same time using eqns. (10.3–10.5) and the rotational transformation $\hat{\mathbf{R}}_n^b$ can be computed as in eqn. (10.2).

Based on $\hat{\mathbf{R}}_n^b$, the measured accelerometer $\mathbf{y}_a$ and magnetometer $\mathbf{y}_m$ vectors can be used to predict the navigation frame vectors according to

$$\hat{\mathbf{g}}^n = \hat{\mathbf{R}}_b^n (\hat{\mathbf{x}}_a - \mathbf{y}_a) \tag{10.26}$$

$$\hat{\mathbf{m}}^n = \hat{\mathbf{R}}_b^n \mathbf{y}_m \tag{10.27}$$

where the true values are known to be $\mathbf{g}^n = [0, 0, g_e]^\top$ and $\mathbf{m}^n = [m_e, 0, 0]^\top$. The residual between the true and computed values will be used in a Kalman filter to estimate the error state.

At some periodic rate, the quaternion should be normalized to have magnitude one, as $\hat{\mathbf{b}} \doteq \hat{\mathbf{b}}/\|\hat{\mathbf{b}}\|$, where the symbol $\doteq$ should be interpreted in the algorithmic sense of "the value is replaced by."

10.5 Error Models

This section derives both the measurement error models and the dynamic model for the state error. The error state vector is defined as

$$\delta\mathbf{x} = [\boldsymbol{\rho}^\top, \delta\mathbf{x}_g^\top, \delta\mathbf{x}_a^\top]^\top$$

where the vector $\boldsymbol{\rho}$ is defined below.

Due to misalignment, measurement, computation, and initialization errors, the body-to-navigation frame transformation will be in error. The vector $\boldsymbol{\rho} = [\epsilon_N, \epsilon_E, \epsilon_D]^\top$ contain the small-angle rotations defined with respect to the navigation frame to rotate the navigation frame to align with the computed navigation frame. Of the components of $\boldsymbol{\rho}$, ϵ_N and ϵ_E are referred to as tilt errors and ϵ_D is referred to as the yaw error.

Three attitude mechanization approaches (i.e., direction cosine, Euler angle, and quaternion) have previously been discussed. The vector $\boldsymbol{\rho}$ is not a new attitude mechanization approach. It is a vector convenient for modeling and correcting the attitude error. Because the three attitude mechanization approaches are all theoretically equivalent, the $\boldsymbol{\rho}$ attitude error representation can be used with any of the three mechanization approaches.

For error analysis, the transformation from the actual to the computed navigation frame can be represented via a multiplicative small angle transformation $(\mathbf{I} - \mathbf{P})$ where the skew-symmetric matrix $\mathbf{P}$ has the representation

$$\mathbf{P} = [\boldsymbol{\rho}\times] = \begin{bmatrix} 0 & -\epsilon_D & \epsilon_E \\ \epsilon_D & 0 & -\epsilon_N \\ -\epsilon_E & \epsilon_N & 0 \end{bmatrix}. \tag{10.28}$$

With the above definitions, we have that

$$\hat{\mathbf{R}}_b^n = (\mathbf{I} - \mathbf{P})\mathbf{R}_b^n. \tag{10.29}$$

The following relations are derived directly from eqn. (10.29) and will be useful in the subsequent analysis:

$$\hat{\mathbf{R}}_n^b = \mathbf{R}_n^b(\mathbf{I} + \mathbf{P}) \tag{10.30}$$

$$\mathbf{R}_n^b = \hat{\mathbf{R}}_n^b(\mathbf{I} - \mathbf{P}) \tag{10.31}$$

$$\mathbf{R}_b^n = (\mathbf{I} + \mathbf{P})\hat{\mathbf{R}}_b^n. \tag{10.32}$$

Because $\mathbf{R}_b^n$ and $\hat{\mathbf{R}}_b^n$ are both orthogonal, $\mathbf{P}^\top = -\mathbf{P}$, and $(\mathbf{I}-\mathbf{P})^{-1} = (\mathbf{I}+\mathbf{P})$ to first order, eqns. (10.30–10.32) are accurate to first order.

10.5.1 Measurement Error Model

For both the magnetometer and the accelerometer, a vector measurement is obtained in the body frame for which a value is known in the navigation frame. Section 10.5.1.1 will consider that measurement scenario in general, then Sections 10.5.1.2 and 10.5.1.3 will specialize the result to the accelerometer and magnetometer measurements.

10.5.1.1 General Analysis

Let a vector $\mathbf{r}$ have the known value $\mathbf{y} = \mathbf{r}^n$ in navigation frame. A measurement of $\mathbf{y}$ is available in body frame that is modeled as

$$\tilde{\mathbf{y}}^b = \mathbf{r}^b + \mathbf{x}_y + \boldsymbol{\nu}_y \tag{10.33}$$

where $\mathbf{x}_y$ represented sensor errors that will be calibrated and $\boldsymbol{\nu}_y$ represents other sensor errors that will be modeled as white noise. The computed value of the measurement in navigation frame is

$$\hat{\mathbf{y}}^n = \hat{\mathbf{R}}_b^n \left(\tilde{\mathbf{y}}^b - \hat{\mathbf{x}}_y\right) \tag{10.34}$$

where $\hat{\mathbf{x}}_y$ is an estimate of $\mathbf{x}_y$ if one is available. The error model for $\mathbf{y}^n$ is

$$\begin{aligned} \delta \mathbf{y}^n &= \mathbf{y}^n - \hat{\mathbf{y}}^n & (10.35) \\ &= \mathbf{y}^n - (\mathbf{I} - \mathbf{P})\mathbf{R}_b^n \left(\mathbf{r}^b + \mathbf{x}_y + \boldsymbol{\nu}_y - \hat{\mathbf{x}}_y\right) \\ &= \mathbf{P}\mathbf{r}^n - \mathbf{R}_b^n \left(\delta \mathbf{x}_y + \boldsymbol{\nu}_y\right) + \mathbf{P}\mathbf{R}_b^n \left(\delta \mathbf{x}_y + \boldsymbol{\nu}_y\right) \\ &= -\left[\mathbf{r}^n \times\right] \boldsymbol{\rho} - \mathbf{R}_b^n \left(\delta \mathbf{x}_y + \boldsymbol{\nu}_y\right) & (10.36) \end{aligned}$$

where we have used the fact that $\mathbf{P}\mathbf{r}^n = -\left[\mathbf{r}^n \times\right] \boldsymbol{\rho}$. As defined in eqn. (B.15), the matrix $[\mathbf{r}^n \times]$ is the skew symmetric matrix form of the vector $\mathbf{r}^n$. The term $\mathbf{P}\mathbf{R}_b^n \left(\delta \mathbf{x}_y + \boldsymbol{\nu}_y\right)$ is dropped because it is second order in the error quantities.

10.5.1.2 Accelerometer Analysis

By eqn. (10.10), the accelerometer measurement is

$$\mathbf{y}_a = \mathbf{a}_{ib}^b - \mathbf{g}^b + \mathbf{x}_a + \boldsymbol{\nu}_a'. \tag{10.37}$$

For the AHRS application, we are interested in using the gravity vector to estimate attitude. We are not interested in the acceleration vector $\mathbf{a}_{ib}$. Therefore, the body frame acceleration $\mathbf{a}_{ib}^b$ is an error source, which can change rapidly. Without additional information, $\mathbf{a}_{ib}^b$ cannot be estimated; therefore, we treat it as a measurement error by defining $\boldsymbol{\nu}_a = \boldsymbol{\nu}_a' + \mathbf{a}_{ib}^b$ so that

$$\mathbf{y}_a = -\mathbf{g}^b + \mathbf{x}_a + \boldsymbol{\nu}_a. \tag{10.38}$$

From eqn. (10.38), the body frame estimate of the gravity vector is

$$\hat{\mathbf{g}}^b = \hat{\mathbf{x}}_a - \mathbf{y}_a. \tag{10.39}$$

The measurement "noise" vector $\boldsymbol{\nu}_a$ is not white, nor is it stationary. The measurement noise vector $\boldsymbol{\nu}_a'$ is white and stationary. Therefore, the strategy that is used in the proposed Kalman filter is to select time instants t_{a_i} when $\mathbf{a}_{ib}(t_{a_i}) \approx \mathbf{0}$ and to use accelerometer aiding only at those time instants. By this approach $\boldsymbol{\nu}_a(t_{a_i}) \approx \boldsymbol{\nu}_a'(t_{a_i})$ and the measurement noise at the instants of the accelerometer aiding can be modeled as white and stationary. The method for defining the times t_{a_i} is defined in Section 10.5.4 and eqn. (10.70).

Because the accelerometer measurement of eqn. (10.38) includes $-\mathbf{g}^b$, the analysis of the residual measurement of eqn. (10.36) changes slightly, but the result is similar. The residual gravity measurement in navigation frame defined as

$$\delta \mathbf{g}^n = \mathbf{g}^n - \hat{\mathbf{g}}^n$$

where $\hat{\mathbf{g}}^n = \hat{\mathbf{R}}_b^n \hat{\mathbf{g}}^b$. The residual has the error model

$$\delta \mathbf{g}^n = \mathbf{H}_a \delta \mathbf{x} + \mathbf{R}_b^n \boldsymbol{\nu}_a$$

where $\mathbf{H}_a$ is derived as follows:

$$\begin{aligned}
\delta \mathbf{g}^n &= \mathbf{g}^n - \hat{\mathbf{R}}_b^n \hat{\mathbf{g}}^b \qquad (10.40)\\
&= \mathbf{g}^n - (\mathbf{I}-\mathbf{P})\mathbf{R}_b^n\left(\hat{\mathbf{x}}_a - \mathbf{y}_a\right)\\
&= \mathbf{g}^n + (\mathbf{P}-\mathbf{I})\mathbf{R}_b^n\left(\mathbf{g}^b - \delta\mathbf{x}_a - \boldsymbol{\nu}_a\right)\\
&= -[\mathbf{g}^n\times]\boldsymbol{\rho} + \mathbf{R}_b^n\left(\delta\mathbf{x}_a + \boldsymbol{\nu}_a\right)\\
&= \begin{bmatrix} 0 & g_e & 0 \\ -g_e & 0 & 0 \\ 0 & 0 & 0 \end{bmatrix} \boldsymbol{\rho} + \mathbf{R}_b^n\left(\delta\mathbf{x}_a + \boldsymbol{\nu}_a\right); \qquad (10.41)
\end{aligned}$$

therefore, the measurement matrix is

$$\mathbf{H}_a = \left[\ [-\mathbf{g}^n\times] \quad \mathbf{0} \quad \mathbf{R}_b^n\ \right]. \qquad (10.42)$$

This model indicates that this measurement does not provide useful information concerning the yaw error ϵ_D; but may allow estimation of the tilt errors ϵ_N and ϵ_E.

10.5.1.3 Magnetometer Analysis

For the residual magnetometer measurement defined as

$$\delta\mathbf{m}^n = \mathbf{m}^n - \hat{\mathbf{m}}^n$$

where $\hat{\mathbf{m}}^n = \hat{\mathbf{R}}_b^n \mathbf{y}_m$. For the magnetometer measurement, eqn. (10.36) specializes to

$$\delta\mathbf{m}^n = \begin{bmatrix} 0 & 0 & 0 \\ 0 & 0 & m_e \\ 0 & -m_e & 0 \end{bmatrix} \boldsymbol{\rho} - \mathbf{R}_b^n \boldsymbol{\nu}_m. \qquad (10.43)$$

Therefore, the magnetometer residual is modeled as

$$\delta\mathbf{m}^n = \mathbf{H}_m \delta\mathbf{x} - \mathbf{R}_b^n \boldsymbol{\nu}_m$$

where the measurement matrix is

$$\mathbf{H}_m = \left[\ -[\mathbf{m}^n\times] \quad \mathbf{0} \quad \mathbf{0}\ \right]. \qquad (10.44)$$

This model indicates that magnetometer measurement might allow calibration of ϵ_E and ϵ_D. However, the accuracy of estimation of ϵ_E relies on the accuracy of the vertical component of $\mathbf{m}^n$ which is location dependent. When position information is not available, the magnetometer measurement matrix will be reduced to

$$\mathbf{H}_m = \left[\ [0, 0, m_e] \quad \mathbf{0} \quad \mathbf{0}\ \right] \qquad (10.45)$$

which will allow calibration of the yaw ϵ_D.

10.5.2 Attitude Error Dynamics

Section 10.5.1 derived the error state measurement models. To complete the analysis we require a model for the dynamics of the attitude error vector ρ. The general equation is derived in Section 10.5.2.1 and will be used in Chapter 11. The general equation is simplified for the AHRS application in Section 10.5.2.2.

The analysis that follows uses the navigation frame yaw and tilt error vector $\boldsymbol{\rho}$. It could also be derived using a quaternion error vector as defined in Exercise D.4.

10.5.2.1 General Analysis

From Section 2.6, the differential equation for the direction cosine matrix is

$$\dot{\mathbf{R}}_b^n = \mathbf{R}_b^n \boldsymbol{\Omega}_{nb}^b. \tag{10.46}$$

Based on the kinematics of eqn. (10.46), the estimate $\hat{\mathbf{R}}_b^n$ could be computed as

$$\dot{\hat{\mathbf{R}}}_b^n = \hat{\mathbf{R}}_b^n(\hat{\boldsymbol{\Omega}}_{ib}^b - \hat{\boldsymbol{\Omega}}_{in}^b) \tag{10.47}$$

where $\hat{\boldsymbol{\Omega}}_{ib}^b = [\hat{\boldsymbol{\omega}}_{ib}^b \times]$ with $\hat{\boldsymbol{\omega}}_{ib}^b = \mathbf{u} - \hat{\mathbf{x}}_g$ calculated based on the gyro outputs, $\hat{\boldsymbol{\Omega}}_{in}^b = [\hat{\boldsymbol{\omega}}_{in}^b \times]$, $\hat{\boldsymbol{\omega}}_{in}^b = \mathbf{R}_n^b \hat{\boldsymbol{\omega}}_{in}^n$, and $\hat{\boldsymbol{\omega}}_{in}^n$ is calculated based on quantities in the navigation system as described in eqn. (2.57).

To determine a dynamic model for $\boldsymbol{\rho}$ (or $\mathbf{P}$), set the derivative of eqn. (10.32) equal to the left side of eqn. (10.46):

$$\dot{\mathbf{P}}\hat{\mathbf{R}}_b^n + (\mathbf{I}+\mathbf{P})\dot{\hat{\mathbf{R}}}_b^n = (\mathbf{I}+\mathbf{P})\hat{\mathbf{R}}_b^n(\boldsymbol{\Omega}_{ib}^b - \boldsymbol{\Omega}_{in}^b). \tag{10.48}$$

Using the definitions

$$\begin{aligned} \delta\boldsymbol{\Omega}_{ib}^b &= \boldsymbol{\Omega}_{ib}^b - \hat{\boldsymbol{\Omega}}_{ib}^b \\ \delta\boldsymbol{\Omega}_{in}^b &= \boldsymbol{\Omega}_{in}^b - \hat{\boldsymbol{\Omega}}_{in}^b, \end{aligned}$$

eqn. (10.48) simplifies as follows,

$$\begin{aligned} \dot{\mathbf{P}}\hat{\mathbf{R}}_b^n + (\mathbf{I}+\mathbf{P})\dot{\hat{\mathbf{R}}}_b^n &= (\mathbf{I}+\mathbf{P})\hat{\mathbf{R}}_b^n(\hat{\boldsymbol{\Omega}}_{ib}^b - \hat{\boldsymbol{\Omega}}_{in}^b + \delta\boldsymbol{\Omega}_{ib}^b - \delta\boldsymbol{\Omega}_{in}^b) \\ &= (\mathbf{I}+\mathbf{P})\dot{\hat{\mathbf{R}}}_b^n + (\mathbf{I}+\mathbf{P})\hat{\mathbf{R}}_b^n(\delta\boldsymbol{\Omega}_{ib}^b - \delta\boldsymbol{\Omega}_{in}^b) \\ \dot{\mathbf{P}} &= \hat{\mathbf{R}}_b^n(\delta\boldsymbol{\Omega}_{ib}^b - \delta\boldsymbol{\Omega}_{in}^b)\hat{\mathbf{R}}_n^b. \end{aligned} \tag{10.49}$$

The term $\mathbf{P}\hat{\mathbf{R}}_b^n(\delta\boldsymbol{\Omega}_{ib}^b - \delta\boldsymbol{\Omega}_{in}^b)\hat{\mathbf{R}}_n^b$ has been dropped as it is second order in the error quantities. In eqn. (10.49), $\dot{\mathbf{P}} = [\dot{\boldsymbol{\rho}}\times]$; therefore, the equation contains only three independent terms. Using eqns. (2.23) and (B.15), eqn. (10.49) can be written in the vector form,

$$\dot{\boldsymbol{\rho}} = \hat{\mathbf{R}}_b^n(\delta\boldsymbol{\omega}_{ib}^b - \delta\boldsymbol{\omega}_{in}^b) \tag{10.50}$$

where $\delta\boldsymbol{\omega}_{ib}^b$ is the error in the gyro measurement of the body-frame inertial-relative angular rate. From eqn. (2.57), the vector $\boldsymbol{\omega}_{in}^n$ is defined as

$$\boldsymbol{\omega}_{in}^n = \begin{bmatrix} (\dot{\lambda} + \omega_{ie})\cos(\phi) \\ -\dot{\phi} \\ -(\dot{\lambda} + \omega_{ie})\sin(\phi) \end{bmatrix} \tag{10.51}$$

where $\dot{\lambda} = \frac{v_e}{\cos(\phi)(R_N+h)}$ and $\dot{\phi} = \frac{v_n}{R_M+h}$. Therefore, the vector $\delta\boldsymbol{\omega}_{in}^b$ depends on the error in position and velocity, which are not available to AHRS applications.

Further specialization of eqn. (10.50) is considered in Section 11.4.2 in relation to INS applications. The following section considers the special case relevant to AHRS applications.

10.5.2.2 Special Case: AHRS

For an AHRS application in which position and velocity information are not available, $\hat{\boldsymbol{\omega}}_{in}^n = \mathbf{0}$. In this case, eqn. (10.46) is mechanized as

$$\dot{\hat{\mathbf{R}}}_b^n = \hat{\mathbf{R}}_b^n \hat{\boldsymbol{\Omega}}_{ib}^b. \tag{10.52}$$

Following a method similar to that of Section 10.5.2.1, to derive the dynamic model for $\mathbf{P}$, set the derivative of eqn. (10.32) equal to the left side of eqn. (10.46):

$$\dot{\mathbf{P}}\hat{\mathbf{R}}_b^n + (\mathbf{I}+\mathbf{P})\dot{\hat{\mathbf{R}}}_b^n = (\mathbf{I}+\mathbf{P})\hat{\mathbf{R}}_b^n(\boldsymbol{\Omega}_{ib}^b - \boldsymbol{\Omega}_{in}^b) \tag{10.53}$$

which simplifies as follows,

$$\begin{aligned} \dot{\mathbf{P}}\hat{\mathbf{R}}_b^n + (\mathbf{I}+\mathbf{P})\dot{\hat{\mathbf{R}}}_b^n &= (\mathbf{I}+\mathbf{P})\hat{\mathbf{R}}_b^n(\hat{\boldsymbol{\Omega}}_{ib}^b - \boldsymbol{\Omega}_{in}^b + \delta\boldsymbol{\Omega}_{ib}^b) \\ &= (\mathbf{I}+\mathbf{P})\dot{\hat{\mathbf{R}}}_b^n + (\mathbf{I}+\mathbf{P})\hat{\mathbf{R}}_b^n\delta\boldsymbol{\Omega}_{ib}^b - \hat{\mathbf{R}}_b^n\boldsymbol{\Omega}_{in}^b \\ \dot{\mathbf{P}} &= \hat{\mathbf{R}}_b^n\delta\boldsymbol{\Omega}_{ib}^b\hat{\mathbf{R}}_n^b - \hat{\mathbf{R}}_b^n\boldsymbol{\Omega}_{in}^b\hat{\mathbf{R}}_n^b \end{aligned} \tag{10.54}$$

to first order. In vector form (using eqns. (2.23) and (B.15)), eqn. (10.54) is

$$\dot{\boldsymbol{\rho}} = \hat{\mathbf{R}}_b^n\left(\delta\boldsymbol{\omega}_{ib}^b - \boldsymbol{\omega}_{in}^n\right) \tag{10.55}$$

where $\delta\boldsymbol{\omega}_{ib}^b$ is the error in the gyro measurement of the body-frame inertial-relative angular rate vector. By its definition as $\delta\boldsymbol{\omega}_{ib}^b = \boldsymbol{\omega}_{ib}^b - \hat{\boldsymbol{\omega}}_{ib}^b$, the definition of $\hat{\omega}^b_{ib}$ in eqn. (10.25), and the definition of the gyro measurement in eqn. (10.8), the gyro measurement error is modeled as

$$\delta\boldsymbol{\omega}^b_{ib} = -\delta\mathbf{x}_g - \boldsymbol{\nu}_g \tag{10.56}$$

with

$$\delta\dot{\mathbf{x}}_g = \mathbf{F}_g\delta\mathbf{x}_g + \boldsymbol{\omega}_g.$$

Substituting eqn. (10.56) into eqn. (10.55) yields

$$\dot{\boldsymbol{\rho}} = -\hat{\mathbf{R}}^n_b\delta\mathbf{x}_g - \hat{\mathbf{R}}^n_b\boldsymbol{\nu}_g - \hat{\mathbf{R}}^n_b\boldsymbol{\omega}^n_{in}. \tag{10.57}$$

The vector $\boldsymbol{\omega}^n_{in}$, derived on p. 52 and reprinted in eqn. (10.51), is the rotation rate of the navigation frame relative to the inertial frame represented in navigation frame. Eqn. (10.57) shows that the lack of position and velocity information affects the AHRS accuracy by the accumulation of attitude error at rates related to $\dot{\lambda}$ and $\dot{\phi}$. In the description that follows, $\boldsymbol{\omega}^n_{in}$ is treated as a white noise term driving the error state model. In fact, this term is not white or even stationary. Improved handling of this term could lead to improved performance. The accumulated error is intended to be estimated and removed by the accelerometer and magnetometer aiding sensors.

10.5.3 AHRS State Space Error Model

Based on the above analysis, with the error state vector defined as $\delta\mathbf{x} = [\boldsymbol{\rho}^\top, \delta\mathbf{x}_g^\top, \delta\mathbf{x}_a^\top]^\top$, we combine eqns. (10.9), (10.11), and (10.57) to obtain the dynamic model for the state vector as

$$\begin{bmatrix} \dot{\boldsymbol{\rho}} \\ \delta\dot{\mathbf{x}}_g \\ \delta\dot{\mathbf{x}}_a \end{bmatrix} = \begin{bmatrix} \mathbf{0} & -\hat{\mathbf{R}}^n_b & \mathbf{0} \\ \mathbf{0} & \mathbf{F}_g & \mathbf{0} \\ \mathbf{0} & \mathbf{0} & \mathbf{F}_a \end{bmatrix} \begin{bmatrix} \boldsymbol{\rho} \\ \delta\mathbf{x}_g \\ \delta\mathbf{x}_a \end{bmatrix} + \begin{bmatrix} -\hat{\mathbf{R}}^n_b & \mathbf{0} & -\hat{\mathbf{R}}^n_b & \mathbf{0} \\ \mathbf{0} & \mathbf{I} & \mathbf{0} & \mathbf{0} \\ \mathbf{0} & \mathbf{0} & \mathbf{0} & \mathbf{I} \end{bmatrix} \begin{bmatrix} \boldsymbol{\omega}^n_{in} \\ \boldsymbol{\omega}_g \\ \boldsymbol{\nu}_g \\ \boldsymbol{\omega}_a \end{bmatrix}.$$

We combine eqns. (10.41) and (10.43) to obtain the error state measurement models as

$$\delta\mathbf{g}^n = \mathbf{H}_a\delta\mathbf{x} + \mathbf{R}^n_b\boldsymbol{\nu}_a \tag{10.58}$$

$$\delta\mathbf{m}^n = \mathbf{H}_m\delta\mathbf{x} - \mathbf{R}^n_b\boldsymbol{\nu}_m \tag{10.59}$$

with $\mathbf{H}_a$ and $\mathbf{H}_m$ defined by eqns. (10.42) and (10.45), respectively.

10.5.4 Measurement Noise Covariance

The Kalman filter measurement update computations require the variance of the discrete-time aiding measurements. This is straightforward for the magnetometer measurements:

$$R_m = var(\boldsymbol{\nu}_m) = \sigma^2_{\nu_m}.$$

The accelerometer aiding is more interesting. In the subsequent discussion, let R_a denote the accelerometer measurement variance that will be used in the computation of the Kalman filter gains used for accelerometer aiding. Note that R_a is not necessarily the same as $var(\boldsymbol{\nu}_a)$, which is not known.

The challenge in using the accelerometer measurements for AHRS aiding is to limit the possibilities for the body frame acceleration $\mathbf{a}^b_{ib}$ to cause inaccuracy in the attitude estimate. Consider the signal

$$\mu(t) = \|\mathbf{y}_a(t)\| - g_e. \tag{10.60}$$

The signal $\mu(t)$ differs from zero due to noise $\boldsymbol{\nu}'_a$, biases $\mathbf{x}_a$, and acceleration $\mathbf{a}_{ib}$. Typical values of $\boldsymbol{\nu}'_a$ and $\mathbf{x}_a$ can be characterized based on specifications from the accelerometer manufacturer. Therefore, it is possible to use $\mu(t)$ as an indicator of the magnitude of $\mathbf{a}_{ib}$. The indicator $\mu(t)$ could be used in at least two ways.

1. The indicator $\mu(t)$ could be used to influence R_a. An example might be
$$R_a = \sigma^2_{\nu'_a} + R_\mu. \tag{10.61}$$
If R_μ is an increasing function of μ, then the Kalman gain for the accelerometer corrections would decrease as μ (i.e., the acceleration) increases.

2. In an asynchronous measurement aiding approach, the indicator $\mu(t)$ could be used, together with additional conditions, to select time instants t_{a_i} when the acceleration should be small. Such time instants are expected to be appropriate for accelerometer aiding.

The second approach is somewhat more consistent with the stochastic character of the Kalman filter design.

The application presented in Section 10.8 uses the asynchronous aiding approach and computes R_a as in eqn. (10.61). The condition used to select the accelerometer aiding times is defined in eqn. (10.70). Many alternative conditions could be defined. There is of course a tradeoff. If the conditions are made too stringent, then accelerometer aiding will rarely if ever occur and the tilt errors may grow with time. If the conditions are made too loose, then the vehicle acceleration may corrupt the attitude estimate through the accelerometer aiding.

10.5.5 Initial Error Covariance Matrix

Given the initial estimates of Section 10.3, we can approximate the initial covariance matrix of the error state $\delta\mathbf{x} = [\boldsymbol{\rho}^\top, \delta\mathbf{x}_g^\top, \delta\mathbf{x}_a^\top]^\top$. The initialization procedure assumed that the vehicle was stationary, non-accelerating, and non-rotating, over some initial period of time $t \in [t_1, t_2]$ with $T = t_2 - t_1$.

With the definition of the error covariance matrix there is a conflict between the symbols used to indicate $[\boldsymbol{\rho}\times]$ and error covariance. In the balance of this chapter, $\mathbf{P}$ with no subscript or superscript denotes $[\boldsymbol{\rho}\times]$ while any use of $\mathbf{P}$ to denote an error covariance will contain a subscript of a superscript. See for example eqns. (10.63) or (10.66).

The initial values of the gyro biases are estimated by averaging the gyro outputs

$$\hat{\mathbf{x}}_g^b(t_2) = \frac{1}{T}\int_0^T \mathbf{u}(\tau)d\tau. \tag{10.62}$$

With $\boldsymbol{\omega}_{ib}^b = \mathbf{0}$ and T small relative to $\frac{1}{\lambda_g}$ so that $\mathbf{x}_g$ can be considered constant over the initialization period, then

$$\begin{aligned} E\langle\hat{\mathbf{x}}_g^b(t_2)\rangle &= \mathbf{x}_g^b(t_2) \\ var\left(\hat{\mathbf{x}}_g^b(t_2)\right) &= \frac{\sigma_{\nu_g}^2}{T}\mathbf{I} \end{aligned} \tag{10.63}$$

where $\mathbf{I}$ is the identity in $\mathbb{R}^{3\times3}$. Eqn. (10.63) is derived in the same manner as was eqn. (4.85). The shorthand notation $\mathbf{P}_{gg} = \frac{\sigma_{\nu_g}^2}{T}\mathbf{I}$ will be used subsequently.

With $\mathbf{a}_{ib}^b = \mathbf{0}$ and T small relative to $\frac{1}{\lambda_a}$ so that $\mathbf{x}_a$ can be considered constant over the initialization period, then with $\bar{\mathbf{g}}^b(t_2)$ defined as $\frac{1}{T}\int_0^T -\mathbf{y}_a(\tau)d\tau$ we have that

$$\bar{\mathbf{g}}^b(t_2) = \mathbf{g}^b - \mathbf{x}_a - \frac{1}{T}\int_0^T \boldsymbol{\nu}_a'(\tau)d\tau;$$

which has a mean value of $\left(\mathbf{g}^b - \mathbf{x}_a\right)$ and $\mathbf{P}_f = cov(\bar{\mathbf{g}}^b(t_2)) = \frac{\sigma_{\nu_a'}^2}{T}\mathbf{I}$. Given the available information, there is no means to distinguish the correct values of $\mathbf{g}^b$ and $\mathbf{x}_a$. Therefore, the initial value assignments

$$\hat{\mathbf{g}}^b(t_2) = \bar{\mathbf{g}}^b(t_2) \tag{10.64}$$

$$\hat{\mathbf{x}}_a(t_2) = \mathbf{0} \tag{10.65}$$

are a design choice which has the correct mean value (i.e., $\hat{\mathbf{g}}^b(t_2) - \hat{\mathbf{x}}_a(t_2) = \bar{\mathbf{g}}^b(t_2)$) and corresponds to the expectation that the dominant initialization error will be due to attitude not accelerometer bias. This choice was discussed in Example 4.9 on page 118.

Assuming that the attitude is initialized by the method of Section 10.3.1, then based on eqns. (10.14), (10.15), and (10.18) we can define a function

$$\mathbf{h}(\mathbf{g},\mathbf{m}) = \begin{bmatrix} atan2\,(g_2, g_3) \\ atan2\left(-g_1, \sqrt{g_2^2+g_3^2}\right) \\ atan2(-m_2^w, m_1^w) \end{bmatrix}$$

such that the initial values for the Euler angles are computed as

$$\hat{\mathbf{E}} = [\hat{\phi}, \hat{\theta}, \hat{\psi}]^\top = \mathbf{h}(\bar{\mathbf{g}}, \bar{\mathbf{m}}).$$

Using analysis similar to that described in Exercise 4.11 the covariance of the initial condition error vector $[\delta\mathbf{E}^\top, \delta\mathbf{x}_a^\top]^\top$ is

$$\mathbf{P}_E = \begin{bmatrix} \mathbf{H}_g(\mathbf{P}_a+\mathbf{P}_f)\mathbf{H}_g^\top + \mathbf{H}_m\mathbf{P}_m\mathbf{H}_m^\top & -\mathbf{H}_g\mathbf{P}_a \\ -\mathbf{P}_a\mathbf{H}_g^\top & \mathbf{P}_a \end{bmatrix}$$

where $\mathbf{P}_a$ denotes the prior covariance of the accelerometer bias $\mathbf{x}_a$, $\mathbf{H}_g = \frac{\partial \mathbf{h}}{\partial \bar{\mathbf{g}}}$, $\mathbf{H}_m = \frac{\partial \mathbf{h}}{\partial \bar{\mathbf{m}}}$, and $\mathbf{P}_m$ is the covariance of the averaged magnetometer noise over the initialization period.

Also as described in Exercise 4.11, the covariance of the initial condition error vector $[\delta\boldsymbol{\rho}^\top, \delta\mathbf{x}_a^\top]^\top$ is

$$\mathbf{P}_E = \begin{bmatrix} \boldsymbol{\Omega}_T\left(\mathbf{H}_g(\mathbf{P}_a+\mathbf{P}_f)\mathbf{H}_g^\top + \mathbf{H}_m\mathbf{P}_m\mathbf{H}_m^\top\right)\boldsymbol{\Omega}_T^\top & -\boldsymbol{\Omega}_T\mathbf{H}_g\mathbf{P}_a \\ -\mathbf{P}_a\mathbf{H}_g^\top\boldsymbol{\Omega}_T^\top & \mathbf{P}_a \end{bmatrix}$$

where $\boldsymbol{\Omega}_T$ is defined in eqn. (2.80). If we define

$$\begin{aligned} \mathbf{P}_{\rho\rho} &= \boldsymbol{\Omega}_T\left(\mathbf{H}_g(\mathbf{P}_a+\mathbf{P}_f)\mathbf{H}_g^\top + \mathbf{H}_m\mathbf{P}_m\mathbf{H}_m^\top\right)\boldsymbol{\Omega}_T^\top \\ \mathbf{P}_{\rho a} &= -\boldsymbol{\Omega}_T\mathbf{H}_g\mathbf{P}_a = \mathbf{P}_{a\rho}^\top, \end{aligned}$$

then the initial covariance matrix for the error state $\delta\mathbf{x}$ is

$$\mathbf{P}^+(t_2) = \mathbf{P}^-(t_2) = \begin{bmatrix} \mathbf{P}_{\rho\rho} & \mathbf{0} & \mathbf{P}_{\rho a} \\ \mathbf{0} & \mathbf{P}_{gg} & \mathbf{0} \\ \mathbf{P}_{a\rho} & \mathbf{0} & \mathbf{P}_{aa} \end{bmatrix}. \tag{10.66}$$

10.6 AHRS Approach Summary

Using either approach of Section 10.3, the initial values for the quaternion $\hat{\mathbf{b}}(t_2)$, the gyro errors $\hat{\mathbf{x}}_g(t_2)$, and the accelerometer errors $\hat{\mathbf{x}}_a(t_2)$ can be defined from the sensor data. Given these initial conditions, the mechanization equations of Section 10.4 would be integrated through time to provide the reference trajectory $\hat{\mathbf{b}}(t)$ for $t > t_2$. In addition, at any time t that they are required, the quaternion can be used to compute values for either the Euler angles $(\phi(t), \theta(t), \psi(t))$ or the rotation matrices $\mathbf{R}_n^b(t)$ and $\mathbf{R}_b^n(t)$. The aiding with magnetometer measurements can be accomplished at a designer specified periodic rate. The aiding via the accelerometer measurements will be performed asynchronously at selected time instants where the acceleration vector is deemed to be small as discussed in Section 10.5.4.

If the symbol t_k for $k = 1, 2, 3, \ldots$ is used to indicate times at which Kalman filter corrections occur, then the error covariance matrix is propagated between these time instants according to

$$\mathbf{P}_k^- = \mathbf{\Phi}(t_k, t_{k-1})\mathbf{P}_{k-1}^+ \mathbf{\Phi}^\top(t_k, t_{k-1}) + \mathbf{Qd}_{k-1}.$$

Because the accelerometer measurements are asynchronous, the matrices $\mathbf{\Phi}$ and $\mathbf{Qd}_k$ must be computed online. For $t \in [t_{k-1}, t_k]$ the state transition matrix $\mathbf{\Phi}(t, t_{k-1})$ is accumulated as described in eqn. (7.20) and the discrete-time process noise matrix $\mathbf{Qd}_k$ is accumulated as described in eqn. (7.21). The integration of the state transition matrix and the discrete-time process noise matrix are initiated at t_{k-1} according to: $\mathbf{\Phi}(t_{k-1}, t_{k-1}) = \mathbf{I}$ and $\mathbf{Q}(t_{k-1}, t_{k-1}) = \mathbf{0}$.

At time t_k, the Kalman gain $\mathbf{K}_k$, the state correction $\delta\hat{\mathbf{x}}_k^+$, and the error covariance of matrix $\mathbf{P}_k^+$, are each computed using the standard Kalman filter equations. With $\delta\hat{\mathbf{x}}_k^+$ available, it is necessary to initialize the state vector for the next period of integration. The accelerometer and gyro bias states are straightforward:

$$\begin{aligned} \hat{\mathbf{x}}_a(t_k)^+ &= \hat{\mathbf{x}}_a(t_k)^- + \delta\hat{\mathbf{x}}_a(t_k)^+ \\ \hat{\mathbf{x}}_g(t_k)^+ &= \hat{\mathbf{x}}_g(t_k)^- + \delta\hat{\mathbf{x}}_g(t_k)^+. \end{aligned}$$

The first three components of $\delta\mathbf{x}_k^+$ define $\boldsymbol{\rho}^+(t_k)$, according to eqn. (10.31) the corrected rotation matrix is

$$\left(\hat{\mathbf{R}}_n^b(t_k)\right)^+ = \left(\hat{\mathbf{R}}_n^b(t_k)\right)^- (\mathbf{I} - \hat{\mathbf{P}}) \tag{10.67}$$

where $\hat{\mathbf{P}} = [\hat{\boldsymbol{\rho}}^+(t_k)\times]$. The quaternion $\hat{\mathbf{b}}^+(t_k)$ can be computed from $\left(\hat{\mathbf{R}}_n^b(t_k)\right)^+$. The variables $\hat{\mathbf{b}}^+(t_k)$, $\hat{\mathbf{x}}_a(t_k)^+$, and $\hat{\mathbf{x}}_g(t_k)^+$ serve as the initial condition for the mechanization equations over the next period of integration. Given that the initial condition $\hat{\mathbf{x}}^+(t_k)$ has been corrected for $\delta\hat{\mathbf{x}}_k^+$, the new best estimate for $\delta\mathbf{x}_k^+$ is now the zero vector (see Section 5.10.5.3).

10.7 Observability and Performance Analysis

Given the linearized error model of Section 10.5.3, by observability analysis it is straightforward to show that even over intervals containing both magnetometer and accelerometer measurements, the full error state is not observable unless the system is rotating. The few interesting questions relate to levels of observability, performance analysis, and tradeoff studies are:

- For a stationary system, what is the expected level of performance?
- For a system moving with lateral velocity, how does the lack of knowledge concerning $\boldsymbol{\omega}_{in}$ affect the system performance? How does this limit the applicability of this AHRS approach?
- How much performance would be lost by eliminating the accelerometer biases from the model?
- How much performance could be gained by estimating scale factor errors?

These questions are left as exercises for the reader.

10.8 Pitch and Roll Application

This section contains application results from laboratory bench testing. During the experiment, magnetometer measurements were not available. Therefore, in this experiment, only pitch θ and roll ϕ estimates are of interest. The IMU sampling rate is approximately 172 Hz. The IMU is manipulated by hand through a sequence of maneuvers. For the first and last few seconds, the IMU is stationary in the same position with the same orientation. The raw IMU data is shown in Figures 10.2 and 10.3. The non-smooth nature of the data should not be interpreted as noise. The standard deviation of the gyro and accelerometer measurements are

$$4.0 \times 10^{-3} \frac{rad}{s} \quad \text{and} \quad 2.2 \times 10^{-2} \frac{m}{s^2}.$$

This level of noise is too small to be resolved at the scale used in these figures. Instead, the lack of smoothness is caused by the gyro handling. The gyros are specified for input rates of $\pm 90 \frac{deg}{s}$; therefore, the roll gyro measuring p saturates for $t \in [148.27, 148.37]$. This saturation causes error in the compute attitude, which the system is able to estimate and remove.

The curves in the top graph in Figure 10.4 show the roll and pitch angles as computed by the AHRS approach presented in this chapter. It is natural to consider whether eqns. (10.14–10.15) could be used throughout instead of the presented approach. For comparison, the curves in the bottom graph of Figure 10.4 show the roll and pitch angles computed using eqns. (10.14–10.15). Figure 10.5 magnifies a portion of the time axis to allow a more detailed comparison. Notice the following:

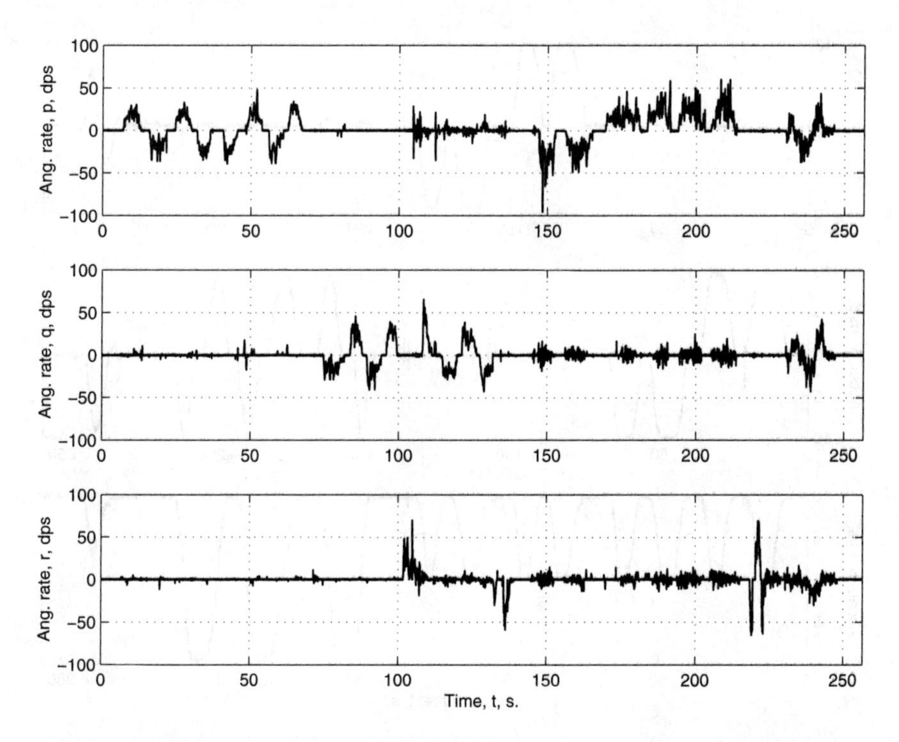

Figure 10.2: Angular rates while the IMU is being manipulated by hand as described in Section 10.8. The abbreviation 'dps' indicates degrees per second.

- The estimates attained by integrating the angular rates are significantly less noisy than those computed from

$$\hat{\phi}(t) = atan2\,(g_2, g_3) \tag{10.68}$$

$$\hat{\theta}(t) = atan2\left(-g_1, \sqrt{g_2^2 + g_3^2}\right) \tag{10.69}$$

  where $[g_1, g_2, g_3]^\top = \hat{\mathbf{x}}_a(t) - \mathbf{y}_a(t)$.

- The presented approach works well even near 90 degrees pitch, where eqns. (10.68–10.69) become nearly singular.

- To the extent that the designer is able to limit the Kalman filter measurement updates to times where the acceleration is nearly zero, the estimates obtained by integrating the angular rates are not affected by vehicle acceleration.

Figure 10.6 displays information pertaining to the initialization of the AHRS system. Throughout the time period of operation, the system maintains a flag which indicates *true* (i.e., 1) when the vehicle is considered to

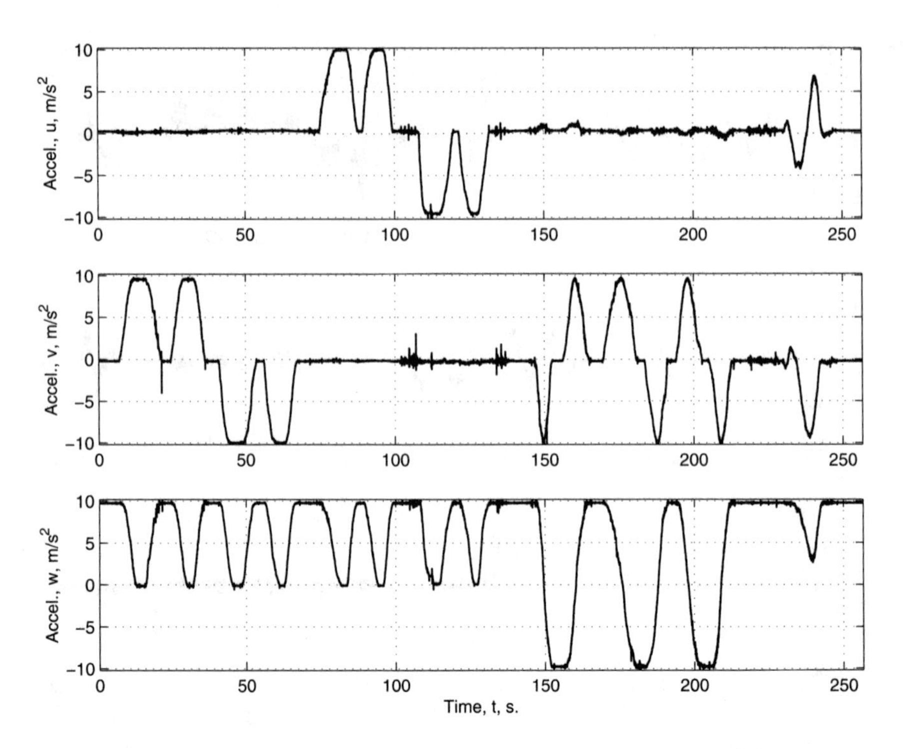

Figure 10.3: Gravity estimate $\mathbf{g}^b = -\mathbf{y}_a$.

be non-accelerating and *false* (i.e., 0) when the vehicle is considered to be accelerating. At time t, based on the IMU data, this flag is defined to be true when the logical condition

$$\left(\mu(t) < \beta_\mu\right) \text{ and } \left(\|\mathbf{u}(t)\| < \beta_g\right) \text{ and } \left(\mu_f(t) < \beta_f\right) \tag{10.70}$$

is true. The signal $\mu(t)$ is defined in eqn. (10.60). The signal $\mu_f(t)$ is a bandpass filtered version of $\mu(t)$. The signal $\mathbf{u}$ is defined in eqn. (10.8). For the results shown $\beta_\mu = 0.1\frac{m}{s^2}$, $\beta_g = 1\frac{deg}{s}$, and $\beta_f = 0.02\frac{m}{s^2}$. This flag is plotted in the top graph of Figure 10.6. For $t \in [0, 2.47]s$, the system is considered to be stationary and the system is in the initialization mode described in Section 10.3.1; however, due to the lack of magnetometer readings, the yaw angle is arbitrarily initialized to zero. At time t=2.47, motion is detected for the first time and the system switches to the mechanization defined in Section 10.4. For $t > 2.47$, the mechanization of Section 10.4 and the Kalman filter are in operation. Note that for the purpose of producing the figure, the residuals (and their standard deviation $\sqrt{\mathbf{HPH}^\top + \sigma^2_{\nu'_a}\mathbf{I}}$) are computed at the IMU rate of 172 Hz; however, Kalman filter measurement updates are only implemented at selected time instants as discussed

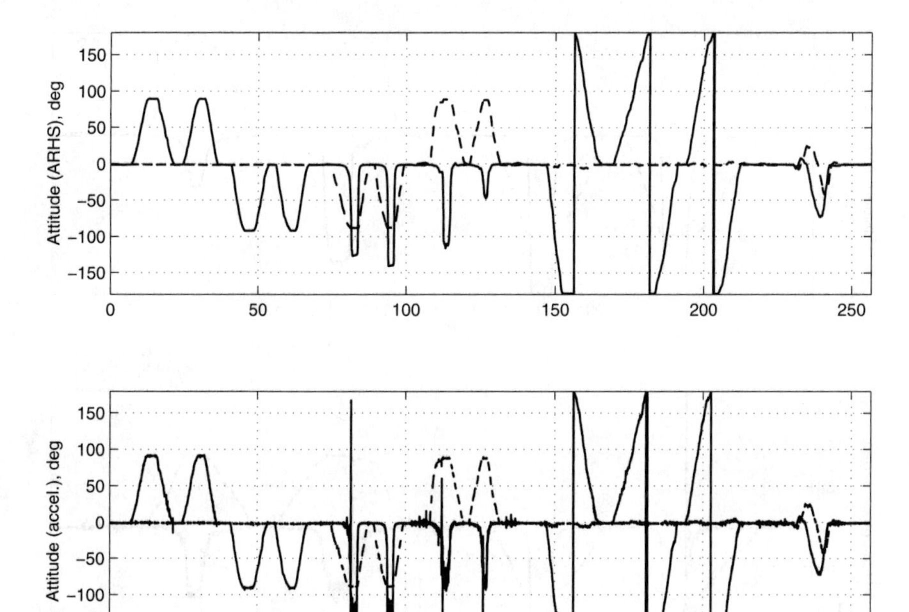

Figure 10.4: Attitude for $t \in [0, 256]s$. Roll ϕ is plotted as a solid curve. Pitch θ is plotted as a dashed curve. Top – Roll and pitch as estimated by the presented AHRS approach. Bottom – Roll and pitch as computed by eqns. (10.68–10.69).

in Section 10.5.4. For this application, the time instants for Kalman filter updates are selected when the non-accelerating flag is 1 and when at least $0.1s$ has elapsed since the last measurement update. For $t \in [2.47, 7.5]s$ there are forty-three measurement updates. For each measurement update, R_a is selected according to eqn. (10.61) with

$$R_\mu(t) = (10\mu_f(t))^2 . \tag{10.71}$$

This formula is designed to decrease the Kalman gain as the acceleration magnitude increases. For $t \in [6.98, 7.5]$, the flag indicates that the system is accelerating, which inhibits Kalman filter corrections during that time frame. Also in that time frame, the effect of the motion is clearly evident at least in the graph of the second residual. By inhibiting the Kalman filter while accelerating, the acceleration "noise" on the gravity estimate does not affect the AHRS attitude estimate.

Because we do not know the true roll and pitch angles, to assess the estimation accuracy, we compare the estimated angles at three times when

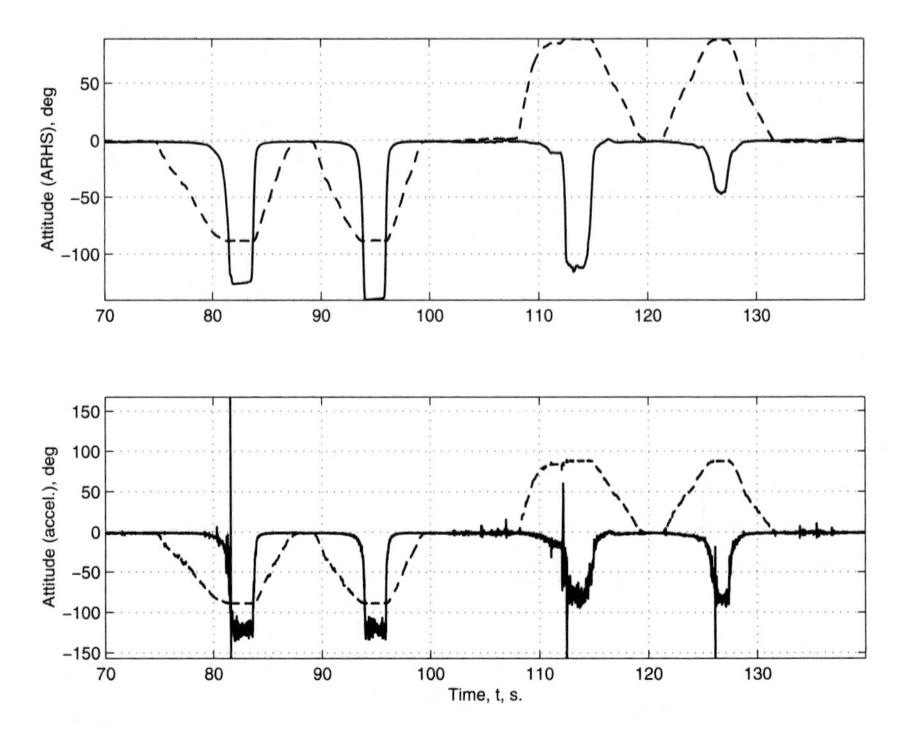

Figure 10.5: Attitude for $t \in [70, 140]s$. Roll ϕ is plotted as a solid curve. Pitch θ is plotted as a dashed curve. Top – Roll and pitch as estimated by the presented AHRS approach. Bottom – Roll and pitch as computed by eqns. (10.68–10.69).

the IMU is placed in the same location with the same orientation as it was initialized. Figures 10.7 and 10.8 plot the attitude angles for two such time ranges. For Figure 10.7 the IMU is not being handled for $t \in [140, 144]s$. For Figure 10.8 the IMU is not being handled for $t \in [248, 255]s$. The mean and standard deviation of the roll and pitch in each of these time intervals is displayed in Table 10.1. Based on those results, it is reasonable to conclude that the system is accurate to better than 0.5 degree during this experiment, after initialization. The only evidence to the contrary is the near one degree roll correction at $t = 138.34$ seconds.

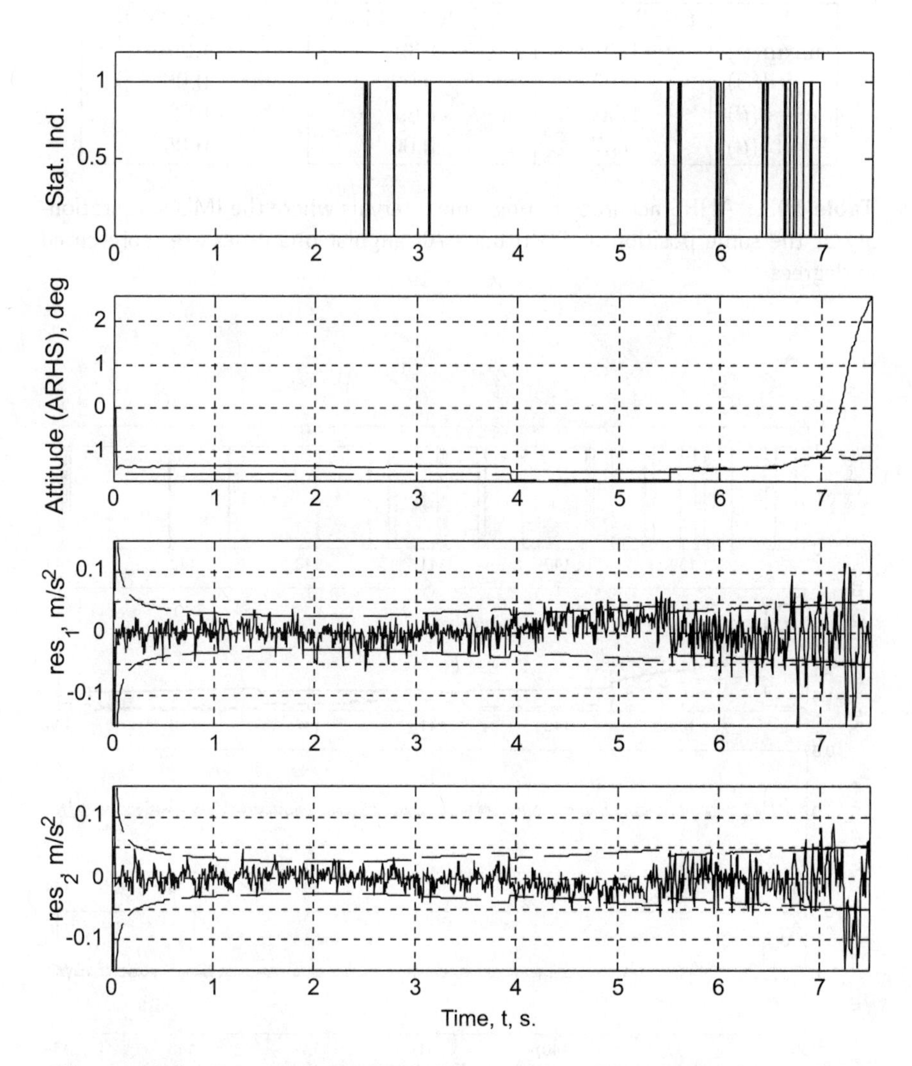

Figure 10.6: Data during initialization. Top – Flag designed to indicate whether or not the vehicle is accelerating. Second – Roll ϕ (solid) and pitch θ (dashed). Third and Fourth – Residuals computed from the specific force oriented along the u and v axes (solid), and the theoretical standard deviation of that residual $\pm\left(\sqrt{\mathbf{HPH}^\top + \sigma^2_{\nu'_a}\mathbf{I}}\right)$ (dashed).

	$t \in [2.5, 6.0]s$	$t \in [140.0, 144.0]s$	$t \in [248.0, 255.0]s$
mean(ϕ)	-1.39	-1.39	-1.29
std(ϕ)	0.05	0.02	0.06
mean(θ)	-1.56	-1.69	-1.62
std(θ)	0.07	0.06	0.05

Table 10.1: AHRS accuracy during time intervals where the IMU was stationary at the same position and attitude. All angular quantities are represented in degrees.

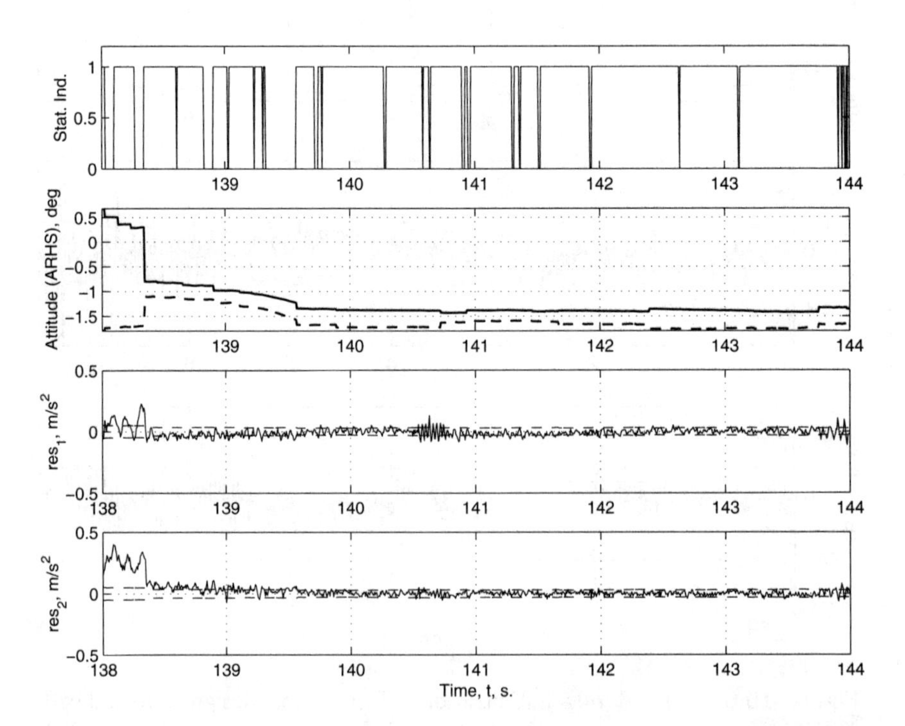

Figure 10.7: Data during the experiment. Top – Flag designed to indicate whether or not the vehicle is accelerating. Second – Roll ϕ (solid) and pitch θ (dashed). Third and Fourth – Residuals computed from the specific force oriented along the u and v axes (solid), and the theoretical standard deviation of that residual $\pm \left(\sqrt{\mathbf{HPH}^\top + \sigma^2_{\nu'_a} \mathbf{I}} \right)$ (dashed).

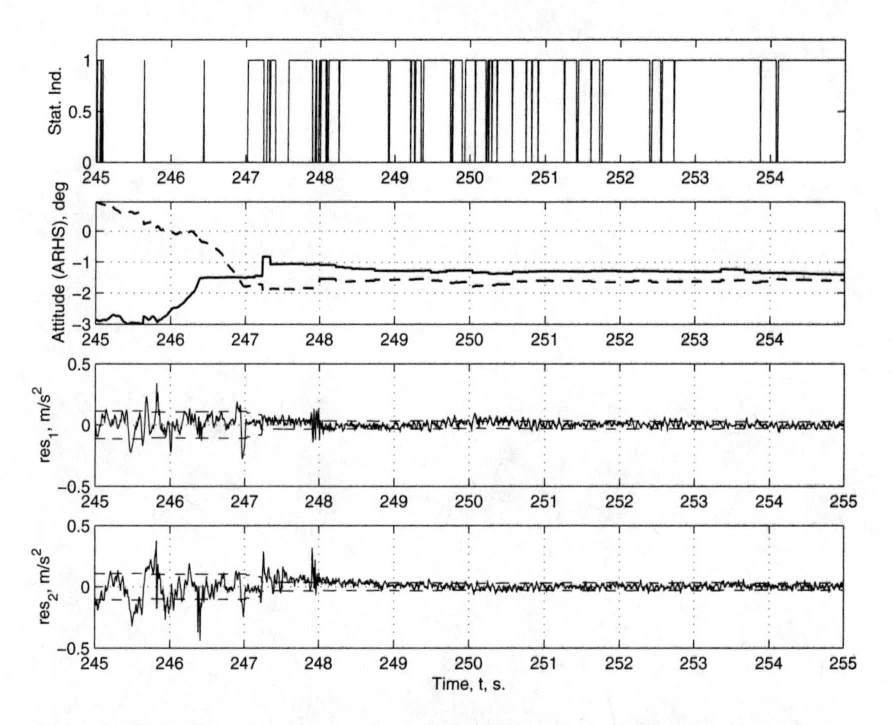

Figure 10.8: Data at termination. Top – Flag designed to indicate whether or not the vehicle is accelerating. Second – Roll ϕ (solid) and pitch θ (dashed). Third and Fourth – Residuals computed from the specific force oriented along the u and v axes (solid), and the theoretical standard deviation of that residual $\pm\left(\sqrt{\mathbf{HPH}^\top + \sigma^2_{\nu'_a}\mathbf{I}}\right)$ (dashed).

10.9 References and Further Reading

The topic of attitude estimation is widely studied, e.g. [14, 40, 41, 42, 45, 46, 49, 82, 85, 91, 120, 132, 135, 136, 145]. The main references for this chapter were [78, 143, 144]. Because the approach of this chapter uses the Earth's gravitational and magnetic fields, successful operation is limited to the vicinity of the Earth. Similar approaches using star sightings can be developed for space vehicles.

References and Further Reading

Chapter 11

Aided Inertial Navigation

The topic of this chapter is aided inertial navigation. Inertial navigation itself is a large topic with a significant body of related literature. The presentation herein begins with a discussion of gravity and definition of the concept of specific force. Both of these concepts and their interrelation are critical to inertial navigation. Next, the chapter discusses the kinematics of inertial navigation in various reference frames. The input variables to the kinematic models are accelerations and angular rates. The acceleration vector is computed from a measured specific force vector by compensating for gravity and instrument errors. The dynamic evolution of the INS error state is a topic of detailed study. The chapter includes a detailed derivation of a state-space model, analytic analysis of simplified models, simulation analysis of the full linear state-space model, and derivation and discussion of instrument error models. Finally, the chapter concludes with sections that discuss initialization, INS aiding, and error state observability.

11.1 Gravitation and Specific Force

The purpose of this section is to distinguish between inertial and non-inertial forces and to define the concept of a specific force.

11.1.1 Gravitation

Newton's law of gravitation states that the force of gravitational attraction of mass m_1 on m_2 is defined by

$$\mathbf{F}_{12} = -\frac{Gm_1m_2}{\|\mathbf{p}_{12}\|^3}\mathbf{p}_{12} \tag{11.1}$$

where G is the universal gravitational constant, and $\mathbf{p}_{12} = \mathbf{p}_2 - \mathbf{p}_1$ is the position of center of mass $\mathbf{p}_2$ with respect to the center of mass $\mathbf{p}_1$. For the purposes of this book we consider inertial and gravitational masses to be identical. The negative sign indicates that the direction of the force is opposite that of $\mathbf{p}_{12}$ (i.e., m_2 is attracted towards m_1).

In particular, if $m_1 = M_e$ represents the mass of Earth and the subscript on m_2 is dropped, then the gravitational attraction of Earth on $m_2 = m$ is

$$\mathbf{F}_{e2} = -\frac{GM_e m}{\|\mathbf{p}_{e2}\|^3}\mathbf{p}_{e2} \tag{11.2}$$

and the gravitational attraction of m on Earth is

$$\mathbf{F}_{2e} = -\frac{GM_e m}{\|\mathbf{p}_{2e}\|^3}\mathbf{p}_{2e}. \tag{11.3}$$

Using Newton's second law, eqns. (11.2) and (11.3) yield the following two differential equations, respectively

$$\begin{aligned}\ddot{\mathbf{p}}_e^i &= -\frac{Gm}{\|\mathbf{p}_{2e}\|^3}\mathbf{p}_{2e}^i \\ \ddot{\mathbf{p}}_2^i &= -\frac{GM_e}{\|\mathbf{p}_{e2}\|^3}\mathbf{p}_{e2}^i;\end{aligned}$$

therefore, because $\mathbf{p}_{e2} = \mathbf{p}_2 - \mathbf{p}_e = -\mathbf{p}_{2e}$ we have

$$\begin{aligned}\ddot{\mathbf{p}}_{e2}^i &= -\frac{G(M_e + m)}{\|\mathbf{p}_{e2}\|^3}\mathbf{p}_{e2}^i \\ &\approx -\frac{GM_e}{\|\mathbf{p}_{e2}\|^3}\mathbf{p}_{e2}^i = \mathbf{G}^i\end{aligned}$$

where $\mathbf{G}$ is the gravitational acceleration defined as

$$\mathbf{G} = -\frac{GM_e}{\|\mathbf{p}_{e2}\|^3}\mathbf{p}_{e2}.$$

This relatively simple model of gravitation is derived for a central force field. It would be approximately valid for vehicles in space. For vehicles near the surface of Earth more detailed gravitational models are required, see Section 2.3.2.2.

11.1.2 Specific Force

Newtonian physics applies to inertial reference frames. An inertial reference frame is non-accelerating, non-rotating, and has no gravitational field. According to Newton's second law, in an inertial reference frame, the acceleration $\ddot{\mathbf{p}}^i$ of a mass m is proportional to the inertial (i.e., physically applied) forces $\mathbf{F}_I$

$$\mathbf{F}_I = m\ddot{\mathbf{p}}^i. \tag{11.4}$$

The quantity $\mathbf{f} = \frac{\mathbf{F}_I}{m}$ has units of acceleration and is referred to as the *specific force.* The specific force is the inertial force per unit mass required to produce the acceleration $\ddot{\mathbf{p}}^i$.

Examples of inertial forces include spring forces, friction, lift, thrust. An example that is important to the understanding of accelerometer operation is the support force applied by a mechanical structure to the case of an accelerometer. The force of gravity is not an inertial force.

When inertial forces are applied to a mass in the presence of the Earth gravitational field, the dynamic model for the position $\mathbf{p}$ of the mass m becomes

$$\begin{aligned} m\ddot{\mathbf{p}}^i &= \mathbf{F}_I - \frac{GM_e m}{\|\mathbf{p}\|^3}\mathbf{p}^i \\ \ddot{\mathbf{p}}^i &= \mathbf{f}^i + \mathbf{G}^i. \end{aligned} \tag{11.5}$$

Eqn. (11.5) is referred to as the fundamental equation of inertial navigation in the inertial reference frame. As will be discussed in Section 11.1.3, accelerometers measure the specific force vector

$$\mathbf{f}^i = \ddot{\mathbf{p}}^i - \mathbf{G}^i. \tag{11.6}$$

The accelerometer output is represented in the accelerometer frame of reference.

The accelerometer measurement of specific force, after transformation from accelerometer frame to inertial frame, can be used as an input to eqn. (11.5). Integration of eqn. (11.5) with $\mathbf{f}^i$ as an input would compute the velocity $\dot{\mathbf{p}}^i$ and the position $\mathbf{p}^i$. Sections 11.2 and 12.1.2 will derive equations similar to eqn. (11.5) that are applicable in alternative reference frames that are more convenient for navigation applications.

11.1.3 Accelerometers

The objective of this section is to describe the basic operation of an ideal accelerometer. Readers interested in more detailed discussion of real accelerometer designs should consult references such as [37, 73, 126].

Consider an accelerometer constructed via a spring-mass-damper system as depicted in Figure 11.1. The position vector of the mass m is $\mathbf{p}$. The position of the accelerometer case is $\mathbf{p}_c$. The case relative position of the mass is $\delta\mathbf{p}_c = \mathbf{p} - \mathbf{p}_c$. For this discussion, we assume that $\delta\mathbf{p}_c$ can be perfectly measured. The equilibrium position of mass m is $\delta\mathbf{p}_c = \mathbf{0}$.

The following discussion distinguishes between inertial (i.e., physically applied) and kinematic forces (e.g., gravity). In Figure 11.1, $\mathbf{F}_I$ represents a force physically applied to the accelerometer case.

In an inertial reference frames (i.e., the frame is not accelerating, not rotating, and has no gravitational field) by Newton's laws, the dynamic

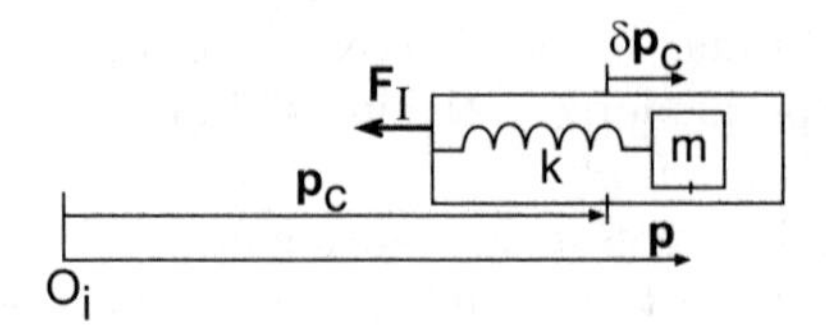

Figure 11.1: Basic accelerometer.

equation for the inertial acceleration of the mass m is

$$\ddot{\mathbf{p}} = -\frac{k}{m}\delta\mathbf{p}_c - \frac{b}{m}\delta\dot{\mathbf{p}}_c \tag{11.7}$$

where k is the spring constant and b is the viscous damping constant. Defining the accelerometer output as $\mathbf{f} = -\frac{k}{m}\delta\mathbf{p_c}$, then eqn. (11.7) can be rewritten as

$$\alpha\dot{\mathbf{f}} = -\mathbf{f} + \ddot{\mathbf{p}} \tag{11.8}$$

where the parameter $\alpha = \frac{b}{k}$ is the time constant of the sensor. The bandwidth of the accelerometer is determined by the parameter α. When the bandwidth of the acceleration signal $\ddot{\mathbf{p}}(t)$ is less that the sensor bandwidth, then $\alpha\dot{\mathbf{f}}(t)$ is small and the sensor maintains the condition

$$\mathbf{f}(t) = \ddot{\mathbf{p}}(t). \tag{11.9}$$

Note that $\mathbf{f}$ is a specific force with units of acceleration.

In the presence of a gravitational field, the accelerometer dynamic equation is

$$\ddot{\mathbf{p}} = -\frac{k}{m}\delta\mathbf{p}_c - \frac{b}{m}\delta\dot{\mathbf{p}}_c + \mathbf{G} \tag{11.10}$$

where $\mathbf{G}$ represents the position dependent gravitational acceleration. It can be shown by manipulations similar to those shown above that

$$\alpha\dot{\mathbf{f}} = -\mathbf{f} + \ddot{\mathbf{p}} - \mathbf{G}(\mathbf{p}). \tag{11.11}$$

When the acceleration signal $\ddot{\mathbf{p}}(t)$ is within the sensor bandwidth, the *specific force* output $\mathbf{f}$ is

$$\mathbf{f} = \ddot{\mathbf{p}} - \mathbf{G}(\mathbf{p}). \tag{11.12}$$

Eqn. (11.12) represents the accelerometer output equation (neglecting bandwidth effects). The equation does not make any assumptions about the accelerometer trajectory, but does recognize that the gravitational acceleration is location dependent.

The above discussion states that an accelerometer measures specific force or the relative acceleration between the case and the mass m. It does not detect accelerations that affect the case and mass m identically. It is useful to consider a few special cases.

- A (non-rotating) accelerometer with no applied forces is in free-fall with $\ddot{\mathbf{p}} = \mathbf{G}$; therefore, the accelerometer output is $\mathbf{f} = \mathbf{0}$.

- An accelerometer in stable orbit around Earth, is also in free-fall. It is constantly accelerating towards Earth with acceleration $\ddot{\mathbf{p}} = \mathbf{G}$; therefore, the accelerometer output is again $\mathbf{f} = \mathbf{0}$.

- Consider an accelerometer at rest on the Earth surface. In this case, the accelerometer is subject to the Earth's gravitational field and is caused to rotate about the Earth at the Earth rate ω_{ie}. Defining the origin of the inertial frame to be coincident with the Earth center of mass, we have that $\ddot{\mathbf{p}} = \mathbf{\Omega}_{ie}\mathbf{\Omega}_{ie}\mathbf{p}$ where $\mathbf{\Omega}_{ie} = [\boldsymbol{\omega}_{ie}\times]$; therefore, the accelerometer output is

$$\mathbf{f} = \mathbf{\Omega}_{ie}\mathbf{\Omega}_{ie}\mathbf{p} - \mathbf{G}, \tag{11.13}$$

  which is the inertial force applied by the supporting structure to the case to maintain the case in a stationary Earth relative position.

When using an accelerometer, the user must compensate the specific force output for the effects of gravity. To prepare for the subsequent analysis, define the local gravity vector as

$$\mathbf{g} = \mathbf{G}(\mathbf{p}) - \mathbf{\Omega}_{ie}\mathbf{\Omega}_{ie}\mathbf{p}. \tag{11.14}$$

The local gravity vector indicates the direction that a weight would hang at a location indicated by $\mathbf{p}$. Figure 11.2 shows the relation of these three accelerations. The vector magnitudes are not drawn to scale. In reality $\|\mathbf{\Omega}_{ie}\mathbf{\Omega}_{ie}\mathbf{p}\| \approx \frac{1}{300}\|\mathbf{G}\|$. In this figure, ϕ and ϕ_c denotes geodetic and geocentric latitude, respectively.

The specific force vector can be represented in any desired frame of reference. When this is done, the analyst should be careful with terminology and interpretation. For example, the specific force vector in the ECEF

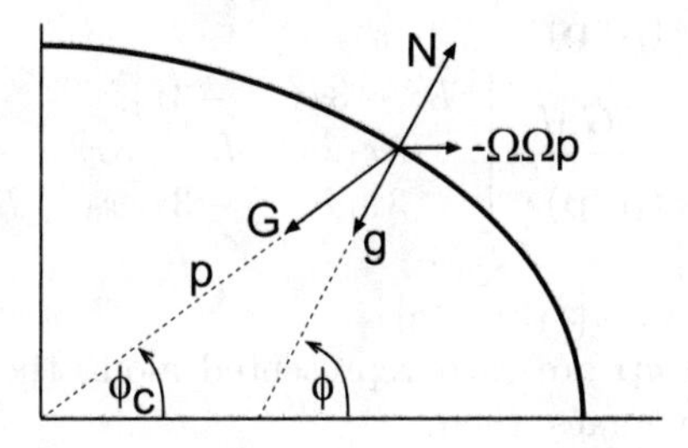

Figure 11.2: Effective gravity vector.

frame is

$$\begin{aligned}\mathbf{f}^e &= \mathbf{R}_i^e\left(\ddot{\mathbf{p}}^i - \mathbf{g}^i\right) \\ &= \mathbf{R}_i^e\ddot{\mathbf{p}}^i - \mathbf{g}^e\end{aligned}$$

which is the inertial acceleration minus gravity both represented in ECEF frame. If we let $\mathbf{p}^e$ represent the position vector in the ECEF frame, it is important to note that $\mathbf{R}_i^e\ddot{\mathbf{p}}^i \neq \frac{d^2}{dt^2}\mathbf{p}^e$, see Section 11.2.2.

11.1.4 Gravity Error

The purpose of this section is to derive an expression for the effect of position error on the computed navigation frame gravity vector. The second term in eqn. (11.14) is straightforward; therefore, this section focuses on the term $\mathbf{G}(\mathbf{p}) = -\frac{GM}{\|\mathbf{p}\|^3}\mathbf{p}$.

The analysis of this section will be used later in this chapter; however, on the first reading of the chapter, the majority of this section may be skipped. In that case, it is recommended that the reader proceed directly to the final result of eqn. (11.22) and the discussion that follows it.

Denote the actual and computed vehicle locations relative to the Earth center of mass by $\mathbf{p}$ and $\hat{\mathbf{p}}$, respectively. The error between these locations is $\delta\mathbf{p} = \mathbf{p} - \hat{\mathbf{p}}$. Using the first two terms in the Taylor's series expansion, we obtain the linear error model

$$\delta\mathbf{g} = \mathbf{G}(\mathbf{p}) - \mathbf{G}(\hat{\mathbf{p}}) = \left.\frac{\partial\mathbf{G}}{\partial\mathbf{p}}\right|_{\hat{\mathbf{p}}}\delta\mathbf{p}. \tag{11.15}$$

where

$$\begin{aligned}\left.\frac{\partial\mathbf{G}}{\partial\mathbf{p}}\right|_{\hat{\mathbf{p}}} &= -GM\left[\frac{1}{(\hat{\mathbf{p}}^\top\hat{\mathbf{p}})^{\frac{3}{2}}}\mathbf{I} - \frac{3}{(\hat{\mathbf{p}}^\top\hat{\mathbf{p}})^{\frac{5}{2}}}\hat{\mathbf{p}}\hat{\mathbf{p}}^\top\right] & (11.16)\\ &= \frac{-GM}{(\hat{\mathbf{p}}^\top\hat{\mathbf{p}})^{\frac{5}{2}}}\left[(\hat{\mathbf{p}}^\top\hat{\mathbf{p}})\,\mathbf{I} - 3\hat{\mathbf{p}}\hat{\mathbf{p}}^\top\right] & (11.17)\\ &= \frac{-GM}{(\hat{\mathbf{p}}^\top\hat{\mathbf{p}})^{\frac{5}{2}}}\begin{bmatrix} R^2 - 3\hat{x}_1^2 & -3\hat{x}_1\hat{x}_2 & -3\hat{x}_1\hat{x}_3 \\ -3\hat{x}_1\hat{x}_2 & R^2 - 3\hat{x}_2^2 & -3\hat{x}_2\hat{x}_3 \\ -3\hat{x}_1\hat{x}_3 & -3\hat{x}_2\hat{x}_3 & R^2 - 3\hat{x}_3^2\end{bmatrix} & (11.18)\end{aligned}$$

where $R = \|\hat{\mathbf{p}}\|$ and $\hat{\mathbf{p}} = [\hat{x}_1, \hat{x}_2, \hat{x}_3]^\top$.

If $\mathbf{G}$, $\mathbf{p}$, $\hat{\mathbf{p}}$, and $\delta\mathbf{p}$ are each represented along the geographic frame north, east, and down axes, then

$$\hat{\mathbf{p}}^n = [0, 0, -R]^\top \tag{11.19}$$

and eqn. (11.18) reduces to

$$\left.\frac{\partial \mathbf{G}^n}{\partial \mathbf{p}^n}\right|_{\hat{\mathbf{p}}^n} = \frac{-GM}{R^3}\begin{bmatrix} 1 & 0 & 0 \\ 0 & 1 & 0 \\ 0 & 0 & -2 \end{bmatrix}. \tag{11.20}$$

Continuing to neglect the $\mathbf{\Omega}_{ie}\mathbf{\Omega}_{ie}\mathbf{p}$ term from eqn. (11.14), we have

$$\delta \mathbf{g}^n = \frac{-g}{R}\begin{bmatrix} 1 & 0 & 0 \\ 0 & 1 & 0 \\ 0 & 0 & -2 \end{bmatrix}\delta \mathbf{p}^n \tag{11.21}$$

where $g = \frac{GM}{R^2}$ and $\delta \mathbf{g}^n = [\delta g_n, \delta g_e, \delta g_d]^\top$.

The effect of the gravity error on the velocity error dynamics is considered in detail in Section 11.4.3. In that section, the vertical component of the position error is height above the ellipse δh, not error in the down component δd of the geodetic vector. The fact that $\delta h = -\delta d$ inserts one extra negative sign in the error analysis of that section

$$\delta \mathbf{g}^n = \frac{-g}{R}\begin{bmatrix} 1 & 0 & 0 \\ 0 & 1 & 0 \\ 0 & 0 & 2 \end{bmatrix}\begin{bmatrix} \delta n \\ \delta e \\ \delta h \end{bmatrix}. \tag{11.22}$$

The following two paragraphs briefly discuss simple models for the horizontal and vertical error dynamics. The purpose is to introduce the important effects that gravity has on the inertial system navigation systems.

In the navigation frame, the lateral position error differential equations can be represented as

$$\begin{aligned} \delta \ddot{n} &= \delta g_n + \alpha_n \\ \delta \ddot{e} &= \delta g_e + \alpha_e \end{aligned}$$

where all error sources other than those due to gravity are combined into the terms α_n and α_e. The full error model that defines α_n and α_e will be discussed in Section 11.4.1. From eqn. (11.22), $\delta g_n = \frac{-g}{R}\delta n$ and $\delta g_e = \frac{-g}{R}\delta e$; therefore, the dependence of the lateral position error on the gravity error has the form

$$\begin{aligned} \delta \ddot{n} + \frac{g}{R}\delta n &= \alpha_n \\ \delta \ddot{e} + \frac{g}{R}\delta e &= \alpha_e \end{aligned}$$

which are forced harmonic oscillators with natural frequency equal to the *Schuler frequency* $\omega_s = \sqrt{\frac{g}{R}}$. If the gravity error terms were absent, then the lateral position errors would be the pure integral of the errors α_n and

α_e. We see that the gravity error terms in the lateral dynamics have a stabilizing effect. Because they are small and stable, they are often dropped in the error analysis.

For the vertical error dynamics, the situation is different. The vertical position error differential equation can be written as

$$\delta\ddot{d} = \delta g_d + \alpha_d$$

where α_d represent all other error sources except for those due to gravity and from eqn. (11.21) we have $\delta g_d = 2\frac{g}{R}\delta d$. Therefore, the vertical position error model is

$$\delta\ddot{d} - 2\frac{g}{R}\delta d = \alpha_d.$$

This system has eigenvalues at $\pm\sqrt{2\frac{g}{R}}$ where the positive eigenvalue shows that the vertical error dynamics due to gravity error are unstable. Even though this eigenvalue is small, because the effect of the gravity error on the system is destabilizing, the effect of gravity error on the vertical dynamics cannot be neglected in the overall error analysis in Section 11.4.3.

11.2 INS Kinematic Equations

Let $\mathbf{p}$ represent the vector from the i-frame origin to the point $\mathbf{P}$. The a-frame is another arbitrary reference frame. The relative rate of rotation of frame a with respect to frame i represented in frame a is $\boldsymbol{\omega}^a_{ia}$. The notation $\boldsymbol{\Omega}^a_{ia} = [\boldsymbol{\omega}^a_{ia}\times]$ is defined in eqn. (B.15). It should be reviewed now as it will be used frequently throughout the chapter.

From eqn. (2.61) in Section 2.6.2, the second derivative of $\mathbf{p}$ in the i and a frames are related by

$$\ddot{\mathbf{p}}^i = \mathbf{R}^i_a\left[2\boldsymbol{\Omega}^a_{ia}\mathbf{v}^a + \boldsymbol{\Omega}^a_{ia}\boldsymbol{\Omega}^a_{ia}\mathbf{r}^a + \dot{\boldsymbol{\Omega}}^a_{ia}\mathbf{r}^a + \frac{d^2\mathbf{r}^a}{dt^2}\right] \tag{11.23}$$

where $\mathbf{r}$ represents the vector from the a-frame origin to the point $\mathbf{P}$. The $\ddot{\boldsymbol{\rho}}^i$ term from eqn. (2.61) has been dropped in eqn. (11.23). In eqn. (2.61) $\boldsymbol{\rho}$ represented the vector from the i-frame origin to the a-frame origin. In the instances in which this equation is used in this chapter, the origins of the a and i frames will be coincident, so that $\boldsymbol{\rho} = \mathbf{0}$. The symbol $\boldsymbol{\rho}$ will be reserved for another use later in the chapter.

From eqn. (11.5) we have that $\ddot{\mathbf{p}}^i = \mathbf{f}^i + \mathbf{G}^i$. Solving eqn. (11.23) for the second derivative of $\mathbf{r}^a$ yields

$$\frac{d^2\mathbf{r}^a}{dt^2} = \mathbf{f}^a + \mathbf{G}^a - 2\boldsymbol{\Omega}^a_{ia}\mathbf{v}^a - (\boldsymbol{\Omega}^a_{ia}\boldsymbol{\Omega}^a_{ia} + \dot{\boldsymbol{\Omega}}^a_{ia})\mathbf{r}^a \tag{11.24}$$

$$= \mathbf{R}^a_b\mathbf{f}^b + \mathbf{G}^a - 2\boldsymbol{\Omega}^a_{ia}\mathbf{v}^a - (\boldsymbol{\Omega}^a_{ia}\boldsymbol{\Omega}^a_{ia} + \dot{\boldsymbol{\Omega}}^a_{ia})\mathbf{r}^a. \tag{11.25}$$

These two equation allow a brief comparison between mechanized and strapdown inertial navigation systems.

In a *mechanized INS*, the accelerometers and gyros are attached to a platform that is designed to maintain its alignment with the a-frame as the vehicle maneuvers. Assuming that the platform is initially aligned with the a-frame, a main idea of the design is to apply torques to the platform so that the outputs of the gyros remain at zero (i.e., platform to a-frame alignment is maintained). This approach requires that the gyro biases be accurately calibrated. With the mechanized approach, the accelerometers measure the specific force in the a-frame. Given initial conditions $\mathbf{v}^a(0)$ and $\mathbf{p}^a(0)$ and alignment of the platform frame with the a-frame, eqn. (11.24) can be integrated to compute $\mathbf{v}^a(t)$ and $\mathbf{p}^a(t)$. The vehicle attitude with respect to the a-frame could be determined by measuring the angles between the platform and the vehicle.

In a *strapdown INS* the accelerometers and gyros are mounted on a platform that is rigidly attached to the vehicle. In this approach, the gyros experience the full rotational rate of the vehicle as it maneuvers; hence gyro scale factor accuracy (as well as bias accuracy) is important. In the strapdown implementation, the accelerometers measure the body frame (b-frame) specific force $\mathbf{f}^b$ which can be transformed into the a-frame using $\mathbf{f}^a = \mathbf{R}_b^a \mathbf{f}^b$. The strapdown INS approach requires initial conditions $\mathbf{v}^a(0)$ and $\mathbf{p}^a(0)$ and $\mathbf{R}_b^a(0)$. The gyro outputs (and navigation state) are used to compute $\boldsymbol{\omega}_{ab}^b(t)$ so that the INS can maintain $\mathbf{R}_b^a(t)$ computationally by integrating

$$\dot{\mathbf{R}}_b^a = \mathbf{R}_b^a \mathbf{\Omega}_{ab}^b. \tag{11.26}$$

The strapdown INS integrates eqn. (11.25) to compute $\mathbf{v}^a(t)$ and $\mathbf{p}^a(t)$. The vehicle attitude is represented by $\mathbf{R}_b^a$.

This chapter will focus on strapdown inertial navigation. Throughout the presentation, the direction cosine representation of attitude will be assumed. Other representations exist (e.g., Euler angles, quaternions). Readers are encouraged to consider the quaternion representation (see Appendix D). It may have a steeper learning curve, but provides a computationally efficient and singularity-free attitude representation approach. The error modeling approach presented in this chapter is applicable to any of the attitude representations.

Equations for specific choices of the a-frame will be considered in the following subsections. For each frame, a differential equation will be derived for each of the position, velocity, and attitude.

11.2.1 Inertial Frame

If the a-frame is selected to be the inertial frame, eqn. (11.24) reduces to

$$\frac{d^2\mathbf{r}^i}{dt^2} = \mathbf{f}^i + \mathbf{G}^i(\mathbf{r}) \tag{11.27}$$

because $\omega_{ii} = \mathbf{0}$. Eqn. (11.27) is the same as eqn. (11.5). With knowledge of $\mathbf{r}^i(0)$, $\mathbf{v}^i(0)$, and $\mathbf{G}^i(\mathbf{r})$, the specific force vector $\mathbf{f}^i$ could be integrated to determine $\mathbf{r}^i(t)$ and $\mathbf{v}^i(t)$.

For strapdown inertial systems, the specific force vector would be measured in body frame and transformed into the i-frame. This requires that the INS maintain the rotational transformation $\mathbf{R}_b^i$ based on integration of the gyro measurements. Then eqn. (11.27) can be computed as

$$\frac{d^2\mathbf{r}^i}{dt^2} = \mathbf{R}_b^i\mathbf{f}^b + \mathbf{G}^i(\mathbf{r}).$$

Choosing the direction cosine attitude representation, the inertial frame kinematic equations are

$$\dot{\mathbf{r}}^i = \mathbf{v}^i \tag{11.28}$$

$$\dot{\mathbf{v}}^i = \mathbf{R}_b^i\mathbf{f}^b + \mathbf{G}^i(\mathbf{r}) \tag{11.29}$$

$$\dot{\mathbf{R}}_b^i = \mathbf{R}_b^i\mathbf{\Omega}_{ib}^b. \tag{11.30}$$

The inputs are $\mathbf{f}^b$ and $\boldsymbol{\omega}_{ib}^b$.

11.2.2 ECEF Frame

Let the a-frame be selected to represent the ECEF frame of reference and let the i-frame and e-frame origins be coincident. In this case, eqn. (11.25) reduces to

$$\frac{d^2\mathbf{r}^e}{dt^2} = \mathbf{R}_b^e\mathbf{f}^b + \mathbf{G}^e(\mathbf{r}) - 2\mathbf{\Omega}_{ie}^e\mathbf{v}^e - \mathbf{\Omega}_{ie}^e\mathbf{\Omega}_{ie}^e\mathbf{r}^e$$

where we have used the fact that $\dot{\omega}_{ie}^e$ can be considered to be zero for the purpose of navigation applications. Again, choosing the direction cosine attitude representation, the e-frame kinematic equations are

$$\dot{\mathbf{r}}^e = \mathbf{v}^e \tag{11.31}$$

$$\dot{\mathbf{v}}^e = \mathbf{R}_b^e\mathbf{f}^b + \mathbf{g}^e - 2\mathbf{\Omega}_{ie}^e\mathbf{v}^e \tag{11.32}$$

$$\dot{\mathbf{R}}_b^e = \mathbf{R}_b^e(\mathbf{\Omega}_{ib}^b - \mathbf{\Omega}_{ie}^b) \tag{11.33}$$

where $\boldsymbol{\omega}_{ie}^b = \mathbf{R}_e^b\boldsymbol{\omega}_{ie}^e$ and the local gravity vector is $\mathbf{g}^e = \mathbf{G}^e(\mathbf{r}) - \mathbf{\Omega}_{ie}^e\mathbf{\Omega}_{ie}^e\mathbf{r}^e$. Again, the inputs are $\mathbf{f}^b$ and $\boldsymbol{\omega}_{ib}^b$. Initial conditions would be required for $\mathbf{r}^e(0)$, $\mathbf{v}^e(0)$, and $\mathbf{R}_b^e(0)$.

The vector $\mathbf{v}^e$ is the Earth relative velocity vector represented in the ECEF frame. This vector can be expressed in various frames-of-reference. To avoid the confusion of double superscripts, let the symbol $\mathbf{v}_e$ represent the Earth relative velocity. In the ECEF frame, we have that $\mathbf{v}_e^e = \mathbf{v}^e$. In the tangent and geographic[1] frames, we have that

$$\mathbf{v}_e^t = \mathbf{R}_e^t \mathbf{v}^e \tag{11.34}$$
$$\mathbf{v}_e^n = \mathbf{R}_e^n \mathbf{v}^e. \tag{11.35}$$

These equations will be used in Sections 11.2.3 and Section 11.2.4.

11.2.3 Tangent Frame

The tangent frame has its origin at a fixed location on the Earth. The tangent frame position is the integral of the Earth relative velocity represented in tangent frame:

$$\dot{\mathbf{p}}^t = \mathbf{v}_e^t. \tag{11.36}$$

The rate of change of $\mathbf{v}_e^t$ is computed by the law of Coriolis (see p. 54) applied to eqn. (11.34):

$$\dot{\mathbf{v}}_e^t = \mathbf{R}_e^t \left(\mathbf{\Omega}_{te}^e \mathbf{v}_e + \dot{\mathbf{v}}^e \right). \tag{11.37}$$

Using eqn. (11.32) to eliminate $\dot{\mathbf{v}}^e$ from eqn. (11.37), using the fact that $\boldsymbol{\omega}_{te} = \mathbf{0}$, and using eqn. (2.23), we obtain

$$\dot{\mathbf{v}}_e^t = \mathbf{R}_b^t \mathbf{f}^b + \mathbf{g}^t - 2\mathbf{\Omega}_{ie}^t \mathbf{v}^t.$$

Therefore, with the direction cosine attitude representation, the tangent frame kinematic equations are

$$\dot{\mathbf{p}}^t = \mathbf{v}_e^t \tag{11.38}$$
$$\dot{\mathbf{v}}_e^t = \mathbf{R}_b^t \mathbf{f}^b + \mathbf{g}^t - 2\mathbf{\Omega}_{ie}^t \mathbf{v}^t \tag{11.39}$$
$$\dot{\mathbf{R}}_b^t = \mathbf{R}_b^t (\mathbf{\Omega}_{ib}^b - \mathbf{\Omega}_{ie}^b) \tag{11.40}$$

where $\boldsymbol{\omega}_{ie} = \boldsymbol{\omega}_{it}$ because $\boldsymbol{\omega}_{et}^b = \mathbf{0}$. The inputs are again $\mathbf{f}^b$ and $\mathbf{\Omega}_{ib}^b$. Initial conditions are required for $\mathbf{p}^t(0)$, $\mathbf{v}_e^t(0)$, and $\mathbf{R}_b^t(0)$.

11.2.4 Geographic Frame

The origin of the geographic frame is the projection of the vehicle frame onto the Earth reference ellipsoid. Therefore, the horizontal components of the vehicle position vector in geographic frame are always zero. The Earth

---

[1] The geographic frame is denoted by a superscript n for navigation. This allows the superscript g to be reserved for use later in the chapter to represent the gyro frame.

relative vehicle position is described by the latitude ϕ and longitude λ of the geographic frame origin and the height h of the vehicle relative to the Earth ellipsoid. By eqn. (2.39),

$$\begin{bmatrix} \dot{\phi} \\ \dot{\lambda} \\ \dot{h} \end{bmatrix} = \begin{bmatrix} \frac{v_n}{R_M+h} \\ \frac{v_e}{\cos(\phi)(R_N+h)} \\ -v_d \end{bmatrix} \tag{11.41}$$

where $\mathbf{v}_e^n = [v_n, v_e, v_d]^\top$. The rate of change of $\mathbf{v}_e^n$ is computed by the law of Coriolis applied to eqn. (11.35):

$$\dot{\mathbf{v}}_e^n = \mathbf{R}_e^n \left(\mathbf{\Omega}_{ne}^e \mathbf{v}^e + \dot{\mathbf{v}}^e \right). \tag{11.42}$$

Using eqn. (11.32) to eliminate $\dot{\mathbf{v}}^e$ from eqn. (11.42) and using eqn. (2.23), we obtain

$$\dot{\mathbf{v}}_e^n = \mathbf{R}_b^n \mathbf{f}^b + \mathbf{g}^n - \left(\mathbf{\Omega}_{en}^n + 2\mathbf{\Omega}_{ie}^n \right) \mathbf{v}^n \tag{11.43}$$

where, as shown in Example 2.6 on page 52,

$$\boldsymbol{\omega}_{en}^n = \begin{bmatrix} \dot{\lambda}\cos(\phi) \\ -\dot{\phi} \\ -\dot{\lambda}\sin(\phi) \end{bmatrix} \text{ and } \boldsymbol{\omega}_{ie}^n = \begin{bmatrix} \omega_{ie}\cos(\phi) \\ 0 \\ -\omega_{ie}\sin(\phi) \end{bmatrix}.$$

The rate of change of the direction cosine matrix is

$$\dot{\mathbf{R}}_b^n = \mathbf{R}_b^n \mathbf{\Omega}_{nb}^b = \mathbf{R}_b^n \left(\mathbf{\Omega}_{ib}^b - \mathbf{\Omega}_{in}^b \right) \tag{11.44}$$

where $\mathbf{\Omega}_{in}^b = [\boldsymbol{\omega}_{in}^b \times]$, $\boldsymbol{\omega}_{in}^b = \mathbf{R}_b^n \boldsymbol{\omega}_{in}^n$ and $\boldsymbol{\omega}_{in}^n = \boldsymbol{\omega}_{ie}^n + \boldsymbol{\omega}_{en}^n$, which yields

$$\boldsymbol{\omega}_{in}^n = \begin{bmatrix} (\dot{\lambda} + \omega_{ie})\cos(\phi) \\ -\dot{\phi} \\ -(\dot{\lambda} + \omega_{ie})\sin(\phi) \end{bmatrix}. \tag{11.45}$$

The kinematic equations for the geographic frame are eqns. (11.41), (11.43), and (11.44). The system of differential equations has inputs $\mathbf{f}^b$ and $\boldsymbol{\omega}_{ib}^b$. Initial conditions are required for $\mathbf{p}(0) = [\phi(0), \lambda(0), h(0)]^\top$, $\mathbf{v}_e^n(0)$, and $\mathbf{R}_b^n(0)$.

11.3 INS Mechanization Equations

Section 11.2 derived the kinematic equations applicable in various reference frames for strapdown inertial navigation. This section provides slightly more detail for the mechanization of the geographic frame implementation.

In the following presentation, we assume that the accelerometer and gyro measurements are modeled as

$$\tilde{\mathbf{f}}^b = \mathbf{f}^b + \Delta\mathbf{f}^b \tag{11.46}$$

$$\tilde{\boldsymbol{\omega}}_{ib}^b = \boldsymbol{\omega}_{ib}^b + \Delta\boldsymbol{\omega}_{ib}^b \tag{11.47}$$

where $\Delta\mathbf{f}^b$ and $\Delta\boldsymbol{\omega}_{ib}^b$ represent specific force and inertial relative angular rate measurements errors. These measurement errors include both random noise terms and instrumentation calibration factors. The instrument calibration factors that can be estimated and compensated. Given the measurements $\tilde{\mathbf{f}}^b$ and $\tilde{\boldsymbol{\omega}}_{ib}^b$ and estimates of the calibration factors $\Delta\hat{\mathbf{f}}^b$ and $\Delta\hat{\boldsymbol{\omega}}_{ib}^b$, the specific force and angular rate vectors for use in the navigation equations are computed as

$$\hat{\mathbf{f}}^b = \tilde{\mathbf{f}}^b - \Delta\hat{\mathbf{f}}^b \tag{11.48}$$

$$\hat{\boldsymbol{\omega}}_{ib}^b = \tilde{\boldsymbol{\omega}}_{ib}^b - \Delta\hat{\boldsymbol{\omega}}_{ib}^b. \tag{11.49}$$

Possible definitions for $\Delta\mathbf{f}^b$ and $\Delta\boldsymbol{\omega}_{ib}^b$ are discussed in Section 11.6.

For the geographic frame, $\mathbf{p} = [\phi, \lambda, h]^\top$ and the velocity vector is $\mathbf{v}^n = \mathbf{v}_e^n = [v_n, v_e, v_d]^\top$. The geodetic position of the geographic frame origin and the vehicle ellipsoidal height that are represented in $\mathbf{p}$ are computed as

$$\begin{bmatrix} \dot{\hat{\phi}} \\ \dot{\hat{\lambda}} \\ \dot{\hat{h}} \end{bmatrix} = \begin{bmatrix} \frac{1}{R_M+\hat{h}} & 0 & 0 \\ 0 & \frac{1}{\cos(\hat{\phi})(R_N+\hat{h})} & 0 \\ 0 & 0 & -1 \end{bmatrix} \begin{bmatrix} \hat{v}_n \\ \hat{v}_e \\ \hat{v}_d \end{bmatrix} \tag{11.50}$$

and the Earth relative velocity in geographic frame is computed as

$$\begin{bmatrix} \dot{\hat{v}}_n \\ \dot{\hat{v}}_e \\ \dot{\hat{v}}_d \end{bmatrix} = \begin{bmatrix} \hat{f}_n \\ \hat{f}_e \\ \hat{f}_d \end{bmatrix} + \mathbf{g}^n - \begin{bmatrix} 0 & -\omega_d & \omega_e \\ \omega_d & 0 & -\omega_n \\ -\omega_e & \omega_n & 0 \end{bmatrix} \begin{bmatrix} \hat{v}_n \\ \hat{v}_e \\ \hat{v}_d \end{bmatrix} \tag{11.51}$$

where

$$\begin{bmatrix} \omega_n \\ \omega_e \\ \omega_d \end{bmatrix} = (\boldsymbol{\omega}_{en}^n + 2\boldsymbol{\omega}_{ie}^n) = \begin{bmatrix} \left(\dot{\hat{\lambda}} + 2\omega_{ie}\right)\cos(\hat{\phi}) \\ -\dot{\hat{\phi}} \\ -\left(\dot{\hat{\lambda}} + 2\omega_{ie}\right)\sin(\hat{\phi}) \end{bmatrix}, \tag{11.52}$$

$$\begin{bmatrix} \hat{f}_n \\ \hat{f}_e \\ \hat{f}_d \end{bmatrix} = \hat{\mathbf{f}}^n = \hat{\mathbf{R}}_b^n \hat{\mathbf{f}}^b \tag{11.53}$$

and $\hat{\mathbf{f}}^b$ is defined in eqn. (11.48).

The direction cosine matrix is computed by integration of

$$\dot{\hat{\mathbf{R}}}_b^n = \hat{\mathbf{R}}_b^n \left(\hat{\mathbf{\Omega}}_{ib}^b - \hat{\mathbf{\Omega}}_{in}^b\right) \tag{11.54}$$

where $\hat{\mathbf{\Omega}} = [\hat{\boldsymbol{\omega}}\times]$, $\hat{\boldsymbol{\omega}}_{ib}^b$ is defined in eqn. (11.49), $\hat{\boldsymbol{\omega}}_{in}^b = \hat{\mathbf{R}}_n^b \hat{\boldsymbol{\omega}}_{in}^n$, and

$$\hat{\boldsymbol{\omega}}_{in}^n = \begin{bmatrix} \omega_N \\ \omega_E \\ \omega_D \end{bmatrix} = \begin{bmatrix} \left(\dot{\hat{\lambda}} + \omega_{ie}\right)\cos(\hat{\phi}) \\ -\dot{\hat{\phi}} \\ -\left(\dot{\hat{\lambda}} + \omega_{ie}\right)\sin(\hat{\phi}) \end{bmatrix}. \tag{11.55}$$

11.4 INS Error State Dynamic Equations

Navigation error arises due to initial condition error and the accumulation of instrumentation errors through the integration process. The purpose of this section is to derive a state space model for the navigation error vector. This section focuses on model development for the geographic frame mechanization approach as defined in Section 11.3.

The model developed in this section will have the form

$$\begin{bmatrix} \delta\dot{\mathbf{p}} \\ \delta\dot{\mathbf{v}} \\ \dot{\boldsymbol{\rho}} \end{bmatrix} = \begin{bmatrix} \mathbf{F}_{pp} & \mathbf{F}_{pv} & \mathbf{F}_{p\rho} \\ \mathbf{F}_{vp} & \mathbf{F}_{vv} & \mathbf{F}_{v\rho} \\ \mathbf{F}_{\rho p} & \mathbf{F}_{\rho v} & \mathbf{F}_{\rho\rho} \end{bmatrix} \begin{bmatrix} \delta\mathbf{p} \\ \delta\mathbf{v} \\ \boldsymbol{\rho} \end{bmatrix} + \begin{bmatrix} \mathbf{0} & \mathbf{0} \\ -\hat{\mathbf{R}}_b^n & \mathbf{0} \\ \mathbf{0} & \hat{\mathbf{R}}_b^n \end{bmatrix} \begin{bmatrix} \delta\mathbf{f}^b \\ \delta\boldsymbol{\omega}_{ib}^b \end{bmatrix}. \tag{11.56}$$

The error vector is defined as $\delta\mathbf{x} = [\delta\mathbf{p}, \delta\mathbf{v}, \boldsymbol{\rho}]^\top$, where

$$\delta\mathbf{p} = [\delta\phi,\ \delta\lambda,\ \delta h]^\top \tag{11.57}$$
$$\delta\mathbf{v} = [\delta v_n, \delta v_e, \delta v_d]^\top \tag{11.58}$$
$$\boldsymbol{\rho} = [\epsilon_N,\ \epsilon_E,\ \epsilon_D]^\top. \tag{11.59}$$

The role of the attitude error vector $\boldsymbol{\rho}$ is discussed in Section 10.5. The remaining error quantities are defined to be the actual value minus the calculated quantity:

$$\delta\mathbf{p} = \mathbf{p} - \hat{\mathbf{p}} \tag{11.60}$$
$$\delta\mathbf{v} = \mathbf{v}_e^n - \hat{\mathbf{v}}_e^n \tag{11.61}$$
$$\delta\mathbf{f}^b = \Delta\mathbf{f}^b - \Delta\hat{\mathbf{f}}^b \tag{11.62}$$
$$\delta\boldsymbol{\omega}_{ib}^b = \Delta\boldsymbol{\omega}_{ib}^b - \Delta\hat{\boldsymbol{\omega}}_{ib}^b. \tag{11.63}$$

The subcomponents of the $\mathbf{F}$ matrix are each matrices in $\mathbb{R}^{3\times 3}$. These submatrices will be defined in the following subsections.

11.4.1 Position Error Linearization

The dynamic equations for the geodetic positions and position estimates are given by eqns. (11.41) and (11.50), respectively. These equations can be written as

$$\dot{\mathbf{p}} = \mathbf{f}_p(\mathbf{p}, \mathbf{v}) \tag{11.64}$$

$$\dot{\hat{\mathbf{p}}} = \mathbf{f}_p(\hat{\mathbf{p}}, \hat{\mathbf{v}}) \tag{11.65}$$

where $\mathbf{f}_p$ is the right-hand side of eqn. (11.41). By linearization of the Taylor series expansion of eqn. (11.64) around the solution of eqn. (11.65), we obtain

$$\dot{\mathbf{p}} = \mathbf{f}_p(\hat{\mathbf{p}}, \hat{\mathbf{v}}) + \mathbf{F}_{pp}\delta\mathbf{p} + \mathbf{F}_{pv}\delta\mathbf{v} + \mathbf{F}_{p\rho}\boldsymbol{\rho} \tag{11.66}$$

where

$$\mathbf{F}_{pp} = \begin{bmatrix} 0 & 0 & \frac{-\hat{v}_n}{(R_M+\hat{h})^2} \\ \frac{\hat{v}_e \sin(\hat{\phi})}{\left((R_N+\hat{h})\cos(\hat{\phi})^2\right)} & 0 & \frac{-\hat{v}_e}{\left((R_N+\hat{h})^2\cos(\hat{\phi})\right)} \\ 0 & 0 & 0 \end{bmatrix}, \tag{11.67}$$

$$\mathbf{F}_{pv} = \begin{bmatrix} \frac{1}{(R_M+\hat{h})} & 0 & 0 \\ 0 & \frac{1}{((R_N+\hat{h})\cos(\hat{\phi}))} & 0 \\ 0 & 0 & -1 \end{bmatrix}, \text{ and} \tag{11.68}$$

$$\mathbf{F}_{p\rho} = \mathbf{0} \tag{11.69}$$

where $\frac{dR_M}{d\phi}$ and $\frac{dR_N}{d\phi}$ terms have been neglected. Because $\delta\dot{\mathbf{p}} = \dot{\mathbf{p}} - \dot{\hat{\mathbf{p}}}$, subtracting eqn. (11.65) from eqn. (11.66) yields the linearized position error differential equation

$$\delta\dot{\mathbf{p}} = \mathbf{F}_{pp}\delta\mathbf{p} + \mathbf{F}_{pv}\delta\mathbf{v} + \mathbf{F}_{p\rho}\boldsymbol{\rho}. \tag{11.70}$$

Eqn. (11.70) provides the first row of eqn. (11.56).

11.4.2 Attitude Error Linearization

Section 10.5.2.1 analyzed the attitude dynamics to arrive at the equation

$$\dot{\boldsymbol{\rho}} = \hat{\mathbf{R}}_b^n(\delta\boldsymbol{\omega}_{ib}^b - \delta\boldsymbol{\omega}_{in}^b). \tag{11.71}$$

The purpose of this section is to linearize eqn. (11.71) with respect to the INS error state $\delta\mathbf{x}$ and the gyro instrumentation error $\delta\boldsymbol{\omega}_{ib}^b$. Therefore, in the following derivation, second order terms in these error quantities will be dropped.

The inertial rotation rate of the geographic frame represented in the b-frame is

$$\boldsymbol{\omega}_{in}^b = \mathbf{R}_n^b\boldsymbol{\omega}_{in}^n \tag{11.72}$$

which can be manipulated as follows to derive an expression for $\delta\boldsymbol{\omega}_{in}^b$

$$\begin{aligned}\hat{\boldsymbol{\omega}}_{in}^b + \delta\boldsymbol{\omega}_{in}^b &= \hat{\mathbf{R}}_n^b(\mathbf{I}-\mathbf{P})(\hat{\boldsymbol{\omega}}_{in}^n + \delta\boldsymbol{\omega}_{in}^n)\\ \delta\boldsymbol{\omega}_{in}^b &= \hat{\mathbf{R}}_n^b(\delta\boldsymbol{\omega}_{in}^n - \mathbf{P}\hat{\boldsymbol{\omega}}_{in}^n),\end{aligned} \tag{11.73}$$

to first order. The matrix $\mathbf{P} = [\boldsymbol{\rho}\times]$ and various relationships between $\mathbf{R}_n^b$ and $\hat{\mathbf{R}}_n^b$ are defined in Section 10.5. When eqn. (11.73) is substituted into equation (11.71), the result is

$$\dot{\boldsymbol{\rho}} + \hat{\boldsymbol{\Omega}}_{in}^n\boldsymbol{\rho} = -\delta\boldsymbol{\omega}_{in}^n + \hat{\mathbf{R}}_b^n\delta\boldsymbol{\omega}_{ib}^b. \tag{11.74}$$

Section 11.6.3 discusses $\delta\boldsymbol{\omega}_{ib}^b$. The following paragraph defines $\delta\boldsymbol{\omega}_{in}^n$.

To complete the analysis, we use the facts that

$$\delta\boldsymbol{\omega}_{in}^n = \boldsymbol{\omega}_{in}^n - \hat{\boldsymbol{\omega}}_{in}^n$$

and that by Taylor's expansion

$$\boldsymbol{\omega}_{in}^n = \hat{\boldsymbol{\omega}}_{in}^n(\hat{\mathbf{p}},\hat{\mathbf{v}}) + \left[\frac{\partial\hat{\boldsymbol{\omega}}_{in}^n}{\partial\hat{\mathbf{p}}}, \frac{\partial\hat{\boldsymbol{\omega}}_{in}^n}{\partial\hat{\mathbf{v}}}\right]\left[\begin{array}{c}\mathbf{p}-\hat{\mathbf{p}}\\ \mathbf{v}-\hat{\mathbf{v}}\end{array}\right] + \ldots.$$

Combining these equations it is clear that, to first order,

$$\delta\boldsymbol{\omega}_{in}^n = \left[\frac{\partial\hat{\boldsymbol{\omega}}_{in}^n}{\partial\hat{\mathbf{p}}}, \frac{\partial\hat{\boldsymbol{\omega}}_{in}^n}{\partial\hat{\mathbf{v}}}\right]\left[\begin{array}{c}\delta\mathbf{p}\\ \delta\mathbf{v}\end{array}\right]. \tag{11.75}$$

Combining eqns. (11.74–11.75), yields

$$\dot{\boldsymbol{\rho}} = \mathbf{F}_{\rho p}\delta\mathbf{p} + \mathbf{F}_{\rho v}\delta\mathbf{v} + \mathbf{F}_{\rho\rho}\boldsymbol{\rho} + \hat{\mathbf{R}}_b^n\delta\boldsymbol{\omega}_{ib}^b \tag{11.76}$$

where

$$\mathbf{F}_{\rho p} = -\frac{\partial\hat{\boldsymbol{\omega}}_{in}^n}{\partial\hat{\mathbf{p}}} = \left[\begin{array}{ccc}\omega_{ie}\sin(\hat{\phi}) & 0 & \frac{\hat{v}_e}{(R_N+\hat{h})^2}\\ 0 & 0 & \frac{-\hat{v}_n}{(R_M+\hat{h})^2}\\ \omega_{ie}\cos(\hat{\phi}) + \frac{\hat{v}_e}{(R_N+\hat{h})\cos(\hat{\phi})^2} & 0 & \frac{-\hat{v}_e\tan(\hat{\phi})}{(R_N+\hat{h})^2}\end{array}\right]$$

$$\mathbf{F}_{\rho v} = -\frac{\partial\hat{\boldsymbol{\omega}}_{in}^n}{\partial\hat{\mathbf{v}}} = \left[\begin{array}{ccc}0 & \frac{-1}{R_N+\hat{h}} & 0\\ \frac{1}{R_M+\hat{h}} & 0 & 0\\ 0 & \frac{\tan(\hat{\phi})}{R_N+\hat{h}} & 0\end{array}\right], \text{ and}$$

$$\mathbf{F}_{\rho\rho} = -\hat{\boldsymbol{\Omega}}_{in}^n = \left[\begin{array}{ccc}0 & \omega_D & -\omega_E\\ -\omega_D & 0 & \omega_N\\ \omega_E & -\omega_N & 0\end{array}\right]$$

and $\boldsymbol{\omega}_{in}^n = [\omega_N, \omega_E, \omega_D]$ as defined in eqn. (11.55). Eqn. (11.76) provides the third row of eqn. (11.56).

11.4.3 Velocity Error Linearization

From eqn. (11.51), the mechanization equation for the Earth relative velocity vector in geographic frame is

$$\dot{\hat{\mathbf{v}}}_e^n = \hat{\mathbf{f}}^n + \hat{\mathbf{g}}^n - (\hat{\boldsymbol{\Omega}}_{en}^n + 2\hat{\boldsymbol{\Omega}}_{ie}^n)\hat{\mathbf{v}}^n$$

which is manipulated as follows to determine the velocity error dynamics:

$$\begin{aligned}
\dot{\hat{\mathbf{v}}}_e^n &= \hat{\mathbf{R}}_b^n\left(\tilde{\mathbf{f}}^b - \Delta\hat{\mathbf{f}}^b\right) + \hat{\mathbf{g}}^n - (\hat{\boldsymbol{\Omega}}_{en}^n + 2\hat{\boldsymbol{\Omega}}_{ie}^n)\hat{\mathbf{v}}^n \\
\dot{\hat{\mathbf{v}}}_e^n &= (\mathbf{I} - \mathbf{P})\,\mathbf{R}_b^n\mathbf{f}^b + \hat{\mathbf{R}}_b^n\left(\Delta\mathbf{f}^b - \Delta\hat{\mathbf{f}}^b\right) + \hat{\mathbf{g}}^n - (\hat{\boldsymbol{\Omega}}_{en}^n + 2\hat{\boldsymbol{\Omega}}_{ie}^n)\hat{\mathbf{v}}^n \\
\dot{\hat{\mathbf{v}}}_e^n &= \mathbf{f}^n - \mathbf{P}\mathbf{f}^n + \hat{\mathbf{R}}_b^n\delta\mathbf{f}^b + \mathbf{g}^n - \delta\mathbf{g}^n - (\hat{\boldsymbol{\Omega}}_{en}^n + 2\hat{\boldsymbol{\Omega}}_{ie}^n)\hat{\mathbf{v}}^n \qquad (11.77)
\end{aligned}$$

which is valid to first order. Subtracting eqn. (11.77) from eqn. (11.43), the linearized velocity error equation is

$$\delta\dot{\mathbf{v}} + \left(\hat{\boldsymbol{\Omega}}_{en}^n + 2\hat{\boldsymbol{\Omega}}_{ie}^n\right)\delta\mathbf{v} + (\delta\boldsymbol{\Omega}_{en}^n + 2\delta\boldsymbol{\Omega}_{ie}^n)\,\hat{\mathbf{v}}^n = \mathbf{P}\mathbf{f}^n - \hat{\mathbf{R}}_b^n\delta\mathbf{f}^b + \delta\mathbf{g}^n.$$

After additional algebra to compute $\left(\delta\boldsymbol{\Omega}_{eg}^g + 2\delta\boldsymbol{\Omega}_{ie}^g\right)\mathbf{v}^g$, the velocity error dynamic equation is

$$\delta\dot{\mathbf{v}} = \mathbf{F}_{vp}\delta\mathbf{p} + \mathbf{F}_{vv}\delta\mathbf{v} + \mathbf{F}_{v\rho}\boldsymbol{\rho} - \mathbf{R}_b^n\delta\mathbf{f}^b \qquad (11.78)$$

where

$$\begin{aligned}
\mathbf{F}_{vp} &= \begin{bmatrix} -2\Omega_N v_e - \frac{\rho_N v_e}{\cos^2(\phi)} & 0 & \rho_E k_D - \rho_N \rho_D \\ 2(\Omega_N v_n + \Omega_D v_d) + \frac{\rho_N v_n}{\cos(\phi)^2} & 0 & -\rho_E\rho_D - k_D\rho_N \\ -2\hat{v}_e\Omega_D & 0 & F_{63} \end{bmatrix}, \\
\mathbf{F}_{vv} &= \begin{bmatrix} k_D & 2\omega_D & -\rho_E \\ -(\omega_D + \Omega_D) & (k_D - \rho_E\tan(\phi)) & \omega_N + \Omega_N \\ 2\rho_E & -2\omega_N & 0 \end{bmatrix}, \text{ and} \\
\mathbf{F}_{v\rho} &= \begin{bmatrix} 0 & f_D & -f_E \\ -f_D & 0 & f_N \\ f_E & -f_N & 0 \end{bmatrix}.
\end{aligned}$$

Derivation of eqn. (11.78) has used the fact that $\mathbf{P}\mathbf{f}^n = -[\mathbf{f}^n\times]\,\boldsymbol{\rho}$, R_N and R_M have been replaced by R_e, and the notation of Table 11.1 has been used to simplify the presentation. Also, the hats over the computed variables have been dropped to simplified the notation.

The term $\delta\mathbf{g}^n$ in eqn. (11.77) deserves additional comment. This term, which was analyzed in Section 11.1.4, accounts for discrepancy between the actual and computed gravity vectors. The portion of this error term related to the error in altitude $\frac{\partial\mathbf{g}^n}{\partial h}$ is accounted for by the $-2\frac{g}{R_e}$ term in F_{63}. Note that this term is destabilizing since $\delta h > 0$ causes $\delta v_D < 0$, which causes the δh to grow more positive. This fact can also be verified by checking the eigenvalues of the overall INS $\mathbf{F}$ matrix in eqn. (11.56).

The fact that the matrix coefficient of $\hat{\mathbf{v}}^n$ in eqn. (11.77) is $(\hat{\mathbf{\Omega}}^n_{en} + 2\hat{\mathbf{\Omega}}^n_{ie})$ may lead to the preconception that $\mathbf{F}_{vv}$ will equal $(\hat{\mathbf{\Omega}}^n_{en} + 2\hat{\mathbf{\Omega}}^n_{ie})$, which is an antisymmetric matrix. As shown above, $\mathbf{F}_{vv} \neq (\hat{\mathbf{\Omega}}^n_{en} + 2\hat{\mathbf{\Omega}}^n_{ie})$, is due to the fact that $\hat{\mathbf{\Omega}}^n_{en}$ depends of $\hat{\mathbf{v}}^n_e$, which adds additional terms to the partial derivative.

11.5 INS Error Characteristics

The linearize dynamic model for the nine primary error states is

$$\delta\dot{\mathbf{x}}(t) = \mathbf{F}(t)\delta\mathbf{x}(t) + \mathbf{\Gamma}\mathbf{q} \tag{11.79}$$

which has the same structure as eqn. (11.56) where

$$\mathbf{F} = \tag{11.80}$$

$$\left[\begin{array}{ccc|ccc|ccc}
0 & 0 & \frac{\rho_E}{R_e} & \frac{1}{R_e} & 0 & 0 & 0 & 0 & 0 \\
\frac{-\rho_D}{\cos(\phi)} & 0 & \frac{-\rho_N}{R_e\cos(\phi)} & 0 & \frac{1}{R_e\cos(\phi)} & 0 & 0 & 0 & 0 \\
0 & 0 & 0 & 0 & 0 & -1 & 0 & 0 & 0 \\
\hline
F_{41} & 0 & F_{43} & k_D & 2\omega_D & -\rho_E & 0 & f_D & -f_E \\
F_{51} & 0 & F_{53} & F_{54} & F_{55} & F_{56} & -f_D & 0 & f_N \\
-2v_E\Omega_D & 0 & F_{63} & 2\rho_E & -2\omega_N & 0 & f_E & -f_N & 0 \\
\hline
-\Omega_D & 0 & \frac{\rho_N}{R_e} & 0 & \frac{-1}{R_e} & 0 & 0 & \omega_D & -\omega_E \\
0 & 0 & \frac{\rho_E}{R_e} & \frac{1}{R_e} & 0 & 0 & -\omega_D & 0 & \omega_N \\
F_{91} & 0 & \frac{\rho_D}{R_e} & 0 & \frac{\tan(\phi)}{R_e} & 0 & \omega_E & -\omega_N & 0
\end{array}\right],$$

$\mathbf{\Gamma} = \begin{bmatrix} \mathbf{0} & \mathbf{0} \\ -\hat{\mathbf{R}}^n_b & \mathbf{0} \\ \mathbf{0} & \hat{\mathbf{R}}^n_b \end{bmatrix}$ and $\mathbf{q} = \left[(\delta\mathbf{f}^b)^\top, (\delta\boldsymbol{\omega}^b_{ib})^\top\right]^\top$. The vector $\mathbf{q}$ is further discussed relative to eqns. (11.104–11.105).

The following two subsections consider this error model from two different perspectives. Subsection 11.5.1 considers the response of simplified single channel error models. These models are straightforward to analyze analytically which simplifies the understanding of the concepts underlying INS error propagation. Subsection 11.5.2 numerically solves eqn. (11.79) for a few specific instances to allow consideration of the full state error model.

$\Omega_N = \omega_{ie}\cos(\phi)$	$F_{41} = -2\Omega_N v_e - \frac{\rho_N v_e}{\cos^2(\phi)}$
$\Omega_D = -\omega_{ie}\sin(\phi)$	$F_{43} = \rho_E k_D - \rho_N \rho_D$
$\rho_N = \frac{v_e}{R_e}$	$F_{51} = 2(\Omega_N v_n + \Omega_D v_d) + \frac{\rho_N v_n}{\cos(\phi)^2}$
$\rho_E = \frac{-v_n}{R_e}$	$F_{53} = -\rho_E\rho_D - k_D\rho_N$
$\rho_D = \frac{-v_e\tan(\phi)}{R_e}$	$F_{54} = -(\omega_D + \Omega_D)$
$\omega_N = \Omega_N + \rho_N$	$F_{55} = k_D - \rho_E\tan(\phi)$
$\omega_E = \rho_E$	$F_{56} = \omega_N + \Omega_N$
$\omega_D = \Omega_D + \rho_D$	$F_{63} = \rho_N^2 + \rho_E^2 - 2\frac{g}{R_e}$
$k_D = \frac{v_d}{R_e}$	$F_{91} = \Omega_N + \frac{\rho_N}{\cos(\phi)^2}$

Table 11.1: Definition of notation for INS error equations.

11.5.1 Simplified Error Models

This section considers the initial condition and forced responses of the simplified error dynamics derived in Exercise 3.12 on p. 99. This simplified error analysis is useful for developing insight that is more difficult to discern from the full error analysis that is presented in Section 11.5.2.

11.5.1.1 Vertical Error Channel

From eqn. (3.104), the vertical channel error dynamics can be represented as

$$\begin{bmatrix} \dot{h} \\ \dot{v}_d \end{bmatrix} = \begin{bmatrix} 0 & -1 \\ \left(\frac{\hat{v}_n^2}{(R_e+h)^2} - \frac{2GM}{(R_e+h)^3}\right) & 0 \end{bmatrix} \begin{bmatrix} h \\ v_d \end{bmatrix}.$$

For a constant speed $\hat{v}_n$, the eigenvalues are $\pm\sqrt{\frac{2GM}{(R_e+h)^3} - \frac{\hat{v}_n^2}{(R_e+h)^2}}$. For the typical conditions where $h \ll R_e$ and $\hat{v}_n \ll (R_e + h)$, the eigenvalues are near $\pm\sqrt{\frac{2GM}{R_e^3}} = \pm\sqrt{\frac{2g}{R_e}}$. Due to the pole with positive real part, the vertical error dynamics are unstable. Some form of feedback error mechanism such as altimeter or GPS aiding is required to stabilize the vertical errors.

11.5.1.2 Lateral Error Channel

From eqn. (3.105), the lateral channel error dynamics can be represented as

$$\begin{bmatrix} \dot{\phi} \\ \dot{v} \\ \dot{\theta} \end{bmatrix} = \begin{bmatrix} 0 & \frac{1}{R_e} & 0 \\ 0 & 0 & -g \\ 0 & \frac{1}{R_e} & 0 \end{bmatrix} \begin{bmatrix} \phi \\ v \\ \theta \end{bmatrix} + \begin{bmatrix} 0 & 0 \\ 1 & 0 \\ 0 & 1 \end{bmatrix} \begin{bmatrix} \epsilon_a \\ \epsilon_g \end{bmatrix}. \tag{11.81}$$

For the discussion to follow, in this simplified model, v represents north velocity error, ϕ represents latitude error, and θ represents east tilt error.

Due to the fact that the error dynamics in this simplified model are independent of the position error ϕ, this model can be decomposed as

$$\begin{bmatrix} \dot{v} \\ \dot{\theta} \end{bmatrix} = \begin{bmatrix} 0 & -g \\ \frac{1}{R_e} & 0 \end{bmatrix} \begin{bmatrix} v \\ \theta \end{bmatrix} + \begin{bmatrix} 1 & 0 \\ 0 & 1 \end{bmatrix} \begin{bmatrix} \epsilon_a \\ \epsilon_g \end{bmatrix} \tag{11.82}$$

and

$$\dot{\phi} = \frac{1}{R_e} v. \tag{11.83}$$

In the following analysis the strategy will be to find closed-form expressions for $v(t)$ and $\theta(t)$, then to integrate $v(t)$ to determine the resulting $\phi(t)$.

For the initial condition $[v(0), \theta(0)] = [v_0, \theta_0]$, the solution to eqn. (11.82) is

$$\begin{bmatrix} v(t) \\ \theta(t) \end{bmatrix} = \begin{bmatrix} \cos(\omega_s t) & -\sqrt{gR_e}\sin(\omega_s t) \\ \frac{1}{\sqrt{gR_e}}\sin(\omega_s t) & \cos(\omega_s t) \end{bmatrix} \begin{bmatrix} v_0 \\ \theta_0 \end{bmatrix} \tag{11.84}$$
$$+ \int_0^t \begin{bmatrix} \cos\left(\omega_s(t-\tau)\right) & -\sqrt{gR_e}\sin\left(\omega_s(t-\tau)\right) \\ \frac{1}{\sqrt{gR_e}}\sin\left(\omega_s(t-\tau)\right) & \cos\left(\omega_s(t-\tau)\right) \end{bmatrix} \begin{bmatrix} \epsilon_a(\tau) \\ \epsilon_g(\tau) \end{bmatrix} d\tau$$

where ω_s represents the Schuler frequency. Eqns. (11.83) and (11.82) can now be analyzed to determine the error response under various conditions.

Initial Velocity Error. For $\epsilon_a(t) = \epsilon_g(t) = 0$ and the initial condition $[v(0), \theta(0)] = [v_0, 0]$, the solution is

$$\begin{bmatrix} v(t) \\ \theta(t) \end{bmatrix} = \begin{bmatrix} \cos(\omega_s t) \\ \frac{1}{\sqrt{gR_e}}\sin(\omega_s t) \end{bmatrix} v_0 \tag{11.85}$$

and $\phi(t) = \frac{v_0}{\sqrt{gR_e}}\sin(\omega_s t)$. Therefore, each $1.0\frac{m}{s}$ of initial lateral velocity error results in 1.3×10^{-4} rad (7×10^{-3} deg) of peak attitude error and 810 m of peak position error. The velocity and attitude errors oscillate out of phase at the Schuler frequency.

The discussion related to Figure 11.3 relates to this initial condition for the full INS error model. The position and attitude errors of this section correspond to the first and fourth subplots. The results of the analysis for the simplified model is accurate for approximately the first hour.

Initial Attitude Error. For $\epsilon_a(t) = \epsilon_g(t) = 0$ and the initial condition $[v(0), \theta(0)] = [0, \theta_0]$, the solution to eqn. (11.82) is

$$\begin{bmatrix} v(t) \\ \theta(t) \end{bmatrix} = \begin{bmatrix} -\sqrt{gR_e}\sin(\omega_s t) \\ \cos(\omega_s t) \end{bmatrix} \theta_0 \tag{11.86}$$

and $\phi(t) = \theta_0 (cos(\omega_s t) - 1)$. For the first several minutes, the position error is accurately modeled as $-\frac{1}{2}\frac{g}{R_e}\theta_0 t^2$ radians or $-\frac{g}{2}\theta_0 t^2$ meters, with velocity error $-g\theta_0 t \frac{m}{s}$. In steady-state, each milli-radian of tilt error θ results in approximately $7.9\frac{m}{s}$ of (peak) velocity error.

The discussion related to Figure 11.4 relates to this initial condition for the full INS. The position and attitude errors of this section correspond to the first and fourth subplots. The results of the analysis for the simplified model is again accurate for approximately the first hour.

Constant Accelerometer Error. For $\epsilon_g(t) = 0$, the initial condition $[v(0), \theta(0)] = [0, 0]$, and constant ϵ_a, the solution to eqn. (11.82) is

$$\begin{bmatrix} v(t) \\ \theta(t) \end{bmatrix} = \begin{bmatrix} \frac{1}{\omega_s} \sin(\omega_s t) \\ \frac{1}{g}(1 - \cos(\omega_s t)) \end{bmatrix} \epsilon_a$$

and $\phi(t) = \frac{\epsilon_a}{g}\left(1 - \cos(\omega_s t)\right)$. This shows that accelerometer bias will lead to a bias in the attitude and position errors. The initial error growth for the position and attitude errors is $\frac{\epsilon_a}{2R_e}t^2$. If the position error ϕ is expressed in meters ($x = R_e\phi$), then the initial position error grows as $\frac{\epsilon_a}{2}t^2$ which is the error growth expected for double integration of an acceleration bias; however, due to the interaction between the velocity and tilt errors, after transients decay away, each error state is oscillatory and bounded.

Constant Gyro Error. For $\epsilon_a(t) = 0$, the initial condition $[v(0), \theta(0)] = [0, 0]$, and constant ϵ_g the solution to eqn. (11.82) is

$$\begin{bmatrix} v(t) \\ \theta(t) \end{bmatrix} = \begin{bmatrix} R_e\left(\cos(\omega_s t) - 1\right) \\ \frac{1}{\omega_s}\sin(\omega_s t) \end{bmatrix} \epsilon_g \tag{11.87}$$

and $\phi(t) = \epsilon_g\left(\frac{1}{\omega_s}\sin(\omega_s t) - t\right)$. In this case, the gyro error leads to an oscillatory attitude error, but the velocity error is a biased oscillation. Due to the velocity bias, the position error grows without bound. The magnitude of the velocity error bias, $R_e\epsilon_g$, is useful for estimating the average rate of position error accumulation due to the gyro bias.

The discussion related to Figure 11.6 relates to this initial condition for the full INS. The position, north velocity, and attitude errors of this section correspond to the first, second, and fourth subplots. The results of the analysis for the simplified model is also accurate for approximately the first hour.

11.5.2 Full Error Model

This section analyzes numeric solutions of the unforced error dynamics

$$\delta\dot{\mathbf{x}}(t) = \mathbf{F}(t)\delta\mathbf{x}(t) \tag{11.88}$$

under certain special conditions described below. Since it is not practical to include and explain many different simulated scenarios, the main objective of this section is to illustrate the methodology and to illustrate how the solutions to the full system relate to the solutions of the simplified system of Section 11.5.1.

It has already been stated that the vertical error dynamics are unstable and that they must be compensate with some form of error feedback. Also, the vertical error dynamics are only weakly affected by the other states. Because the vertical error dynamics will be compensated, we assume that the vertical position and velocity errors are small and remove those states from the model. Therefore, the state vector is

$$\delta \mathbf{x} = [\delta\phi,\ \delta\lambda, \delta v_n, \delta v_e, \epsilon_N,\ \epsilon_E,\ \epsilon_D]^\top.$$

The reduced model has

$$\mathbf{F} = \left[\begin{array}{cc|cc|ccc} 0 & 0 & \frac{1}{R_e} & 0 & 0 & 0 & 0 \\ \frac{-\rho_D}{\cos(\phi)} & 0 & 0 & \frac{1}{R_e\cos(\phi)} & 0 & 0 & 0 \\ \hline F_{41} & 0 & k_D & 2\omega_D & 0 & f_D & -f_E \\ F_{51} & 0 & F_{54} & F_{55} & -f_D & 0 & f_N \\ \hline -\Omega_D & 0 & 0 & \frac{-1}{R_e} & 0 & \omega_D & -\omega_E \\ 0 & 0 & \frac{1}{R_e} & 0 & -\omega_D & 0 & \omega_N \\ F_{91} & 0 & 0 & \frac{\tan(\phi)}{R_e} & \omega_E & -\omega_N & 0 \end{array}\right]. \tag{11.89}$$

For a stationary system, the **F** matrix can be further simplified to

$$\mathbf{F} = \left[\begin{array}{cc|cc|ccc} 0 & 0 & \frac{1}{R_e} & 0 & 0 & 0 & 0 \\ 0 & 0 & 0 & \frac{1}{R_e\cos(\phi)} & 0 & 0 & 0 \\ \hline 0 & 0 & 0 & 2\Omega_D & 0 & -g & 0 \\ 0 & 0 & -2\Omega_D & 0 & g & 0 & 0 \\ \hline -\Omega_D & 0 & 0 & \frac{-1}{R_e} & 0 & \Omega_D & 0 \\ 0 & 0 & \frac{1}{R_e} & 0 & -\Omega_D & 0 & \Omega_N \\ \Omega_N & 0 & 0 & \frac{\tan(\phi)}{R_e} & 0 & -\Omega_N & 0 \end{array}\right]. \tag{11.90}$$

The stationary error dynamics are important during system initialization. It is the stationary system that will be briefly considered in this section.

Consider a stationary vehicle that is nominally level and at rest at 45 degrees north latitude. This results in eigenvalues of

$$\begin{array}{l} 0.0 \\ 0.0 \pm 7.2722 \times 10^{-5} j \\ 0.0 \pm 1.2915 \times 10^{-3} j \approx 0.0 \pm (\omega_s + \omega_e \sin(\phi))\, j \\ 0.0 \pm 1.1886 \times 10^{-3} j \approx 0.0 \pm (\omega_s - \omega_e \sin(\phi))\, j \end{array}$$

where $j = \sqrt{-1}$. The close proximity of the second two eigenvalue pairs results in solutions containing an oscillation (approximately) at the Schuler frequency

$$\omega_s = \sqrt{\frac{g}{R_e}} \tag{11.91}$$

with amplitude modulated[2] (approximately) at the Foucault frequency

$$\omega_f = \omega_e \sin(\phi). \tag{11.92}$$

The period of the Schuler oscillation is 84.4 minutes. At the latitude of the simulation, the period of the Foucault beat is 33.9 hrs.

Figure 11.3 shows the response of the error state of the stationary INS with $\mathbf{F}$ defined by eqn. (11.90) when the initial state error is $1.0\frac{m}{s}$ of north velocity error. The initial north velocity error directly causes east tilt error due to error in the calculation of the Earth relative navigation frame angular rate; east velocity error due to error in Coriolis compensation; and, latitude error by direct integration. All five of the error states that are plotted attain a steady-state oscillation that is dominated by the Foucault modulated Schuler oscillation. Due to the simulation occurring at latitude 45 degrees north, the north tilt and azimuth errors have equal magnitudes. In general, the $\mathbf{F}$ matrix (fourth column, fifth and seventh rows) of eqn. (11.90) indicates that the east velocity error will cause the azimuth error to be greater than the north tilt error by a factor of $\tan(\phi)$. The solution of the reduced error model from this same initial condition was considered in eqn. (11.85). The response for the reduced order system of eqn. (11.85) matches the full system response of Figure 11.3 well for approximately one hour, including the predicted peak errors for the north position and east tilt errors.

Figure 11.4 displays the error trajectory for the case of a one degree initial tilt about the east axis. The east tilt error directly causes error in the gravity and earth rate compensation, which results in azimuth, north tilt and north velocity errors. As predicted by eqn. (11.86), the initial rate of north position error is $-\frac{g}{2}\theta_0 t^2$ m. This estimate is valid for the first several minutes.

[2] To see why this is true, apply the sum of angles formula to $\sin\left(\omega_s t + \omega_e \sin(\phi) t\right)$.

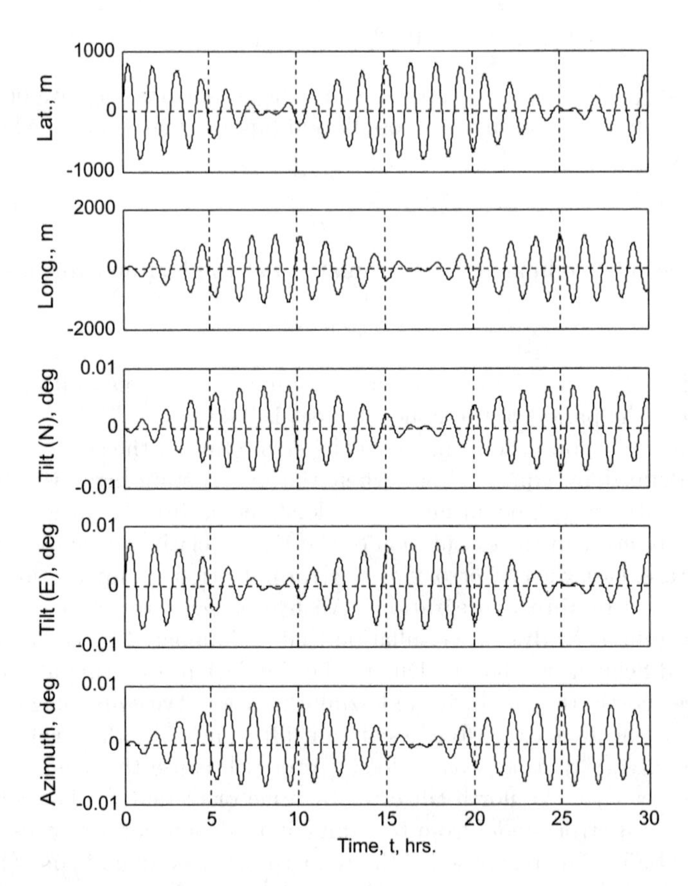

Figure 11.3: Numeric solution of eqn. (11.88) with $\mathbf{F}$ from eqn. (11.90) showing the error response for a stationary INS initialized with $1.0\frac{m}{s}$ of north velocity error.

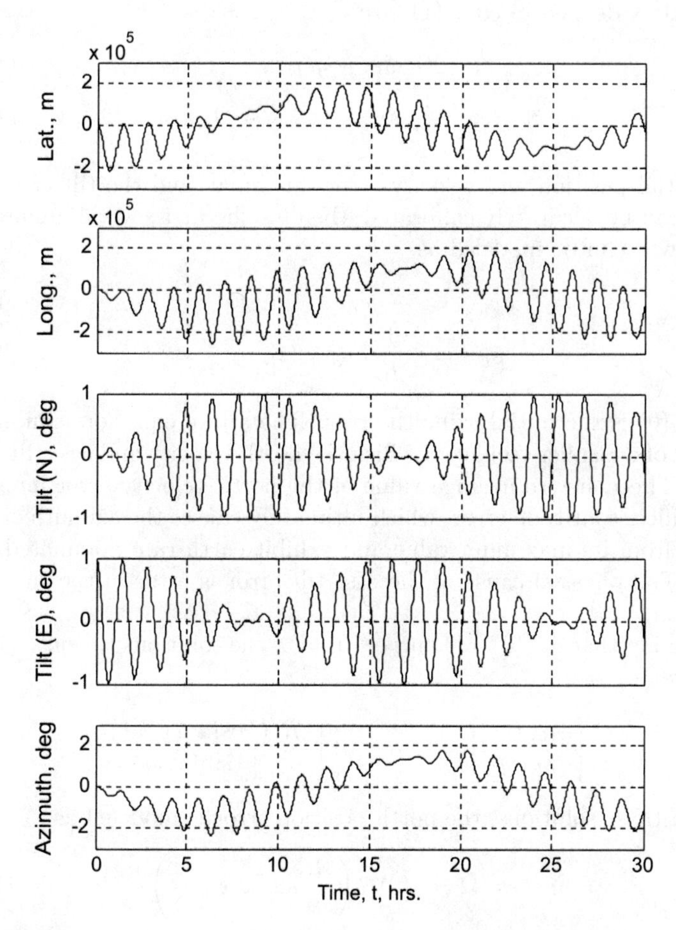

Figure 11.4: Numeric solution of eqn. (11.88) with $\mathbf{F}$ from eqn. (11.90) showing the error response for a stationary INS initialized with 1.0 degree of east tilt error.

One additional set of error conditions worth discussion is shown in Figure 11.5. This figure concerns an initial condition having 1.0 deg of azimuth error. The simplified models of Section 11.5.1 do not cover this situation. However, assuming that the system is stationary, a reduced order model can be derived by considering the third and sixth rows of the system with the **F** matrix defined in eqn. (11.90):

$$\delta \dot{v}_n = 2\Omega_D \delta v_e - g\epsilon_E \tag{11.93}$$

$$\dot{\epsilon}_E = \frac{1}{R_e}\delta v_n - \Omega_D \epsilon_N + \Omega_N \epsilon_D. \tag{11.94}$$

If the initial position and velocity errors are small and the tilt errors (ϵ_N and ϵ_E) can be accurately calibrated, then for the first several minutes the error growth can be modeled as

$$\delta \dot{v}_n = -g\epsilon_E \tag{11.95}$$

$$\dot{\epsilon}_E = \frac{1}{R_e}\delta v_n + \Omega_N \epsilon_D(0) \tag{11.96}$$

where $\epsilon_D(0)$ is the initial azimuth error. The above equation exhibits the structure of a Schuler oscillator, which the north velocity and east tilt errors exhibit. The non-zero average value of the north velocity error integrates into significant latitude error, which ultimately causes the azimuth error to decrease from its maximum value and exhibit earth rate dominated oscillations. The physical cause of the east tilt error is error in the earth rate calculation.

Assuming that $\delta v_n(0) = 0$ and $\epsilon_E(0) = 0$, the solutions to eqns. (11.95–11.96) are

$$\begin{bmatrix} \delta v_n(t) \\ \epsilon_E(t) \end{bmatrix} = \Omega_N \epsilon_D(0) \begin{bmatrix} R_e\big(\cos(\omega_s t) - 1\big) \\ \frac{1}{\omega_s}\sin(\omega_s t) \end{bmatrix}.$$

Based on these solutions, the north position error is modeled as

$$\delta n(t) = \Omega_N \epsilon_D(0) R \left(\frac{1}{\omega_s}\sin(\omega_s t) - t \right). \tag{11.97}$$

These solutions are only expected to hold for a short duration of time. Comparison with the full solution in Figure 11.5 indicates that they are valid for at least the first half of the Schuler period. Therefore, let $T = \frac{\pi}{\omega_s}$. At the end of an initialization period $t \in [0, T]$, according to the simplified model, the north position error relates to the azimuth error as

$$\delta n(T) = -\Omega_N \epsilon_D(0) R_e T.$$

Therefore, for applications where the inertial instruments are sufficiently accurate and where the system can remain stationary for approximately 40

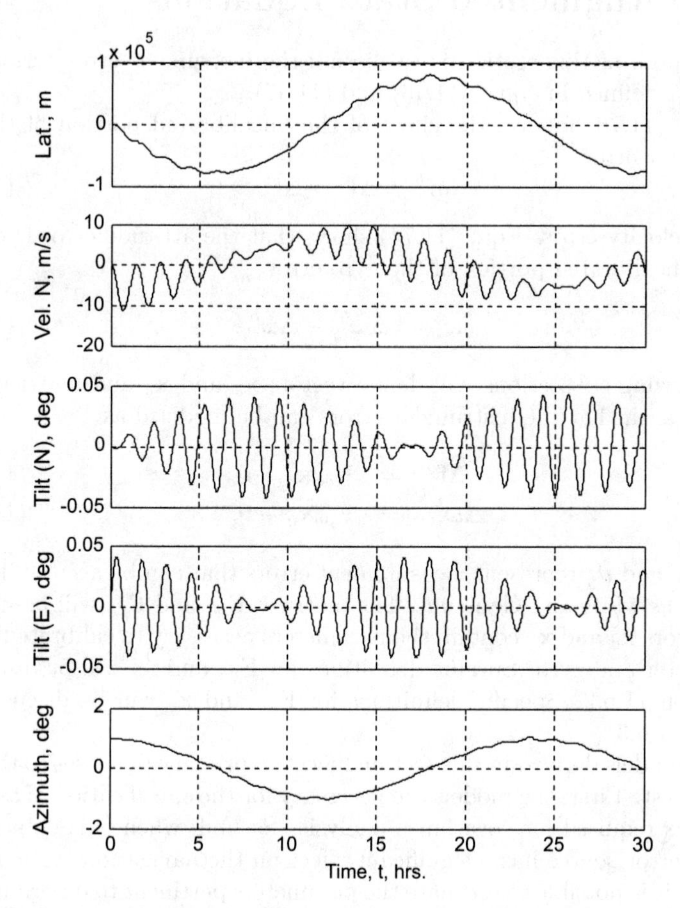

Figure 11.5: Numeric solution of eqn. (11.88) with $\mathbf{F}$ from eqn. (11.90) showing the error response for a stationary INS initialized with 1.0 degree of azimuth error.

minutes, then the initial azimuth error can be estimated as

$$\epsilon_D(0) = -\frac{1}{T}\frac{\delta n(T)}{\Omega_N R_e}.$$

11.6 Augmented State Equations

The purpose of this section is to discuss the instrument error terms $\Delta \mathbf{f}^b$ and $\Delta \boldsymbol{\omega}_{ib}^b$ defined in eqns. (11.46) and (11.47).

Eqn. (11.78) models the effect of the uncalibrated portion of the accelerometer error

$$\delta \mathbf{f}^b = \Delta \mathbf{f}^b - \Delta \hat{\mathbf{f}}^b \tag{11.98}$$

on the velocity error. Eqn. (11.74) shows that the attitude error is driven by the uncalibrated portion of the gyro error

$$\delta \boldsymbol{\omega}_{ib}^b = \Delta \boldsymbol{\omega}_{ib}^b - \Delta \hat{\boldsymbol{\omega}}_{ib}^b. \tag{11.99}$$

The following subsections will define vectors $\mathbf{x}_a$ and $\mathbf{x}_g$ and matrices $\mathbf{F}_{va}$ and $\mathbf{F}_{\rho g}$ such that the instrument errors can be modeled as

$$\Delta \mathbf{f}^b = \mathbf{F}_{va}\mathbf{x}_a + \boldsymbol{\nu}_a \tag{11.100}$$

$$\Delta \boldsymbol{\omega}_{ib}^b = \mathbf{F}_{\rho g}\mathbf{x}_g + \boldsymbol{\nu}_g \tag{11.101}$$

where $\boldsymbol{\nu}_a$ and $\boldsymbol{\nu}_g$ represent measurement errors that can be accurately represented as white and Gaussian. The matrices $\mathbf{F}_{va}$ and $\mathbf{F}_{\rho g}$ will be known. The vectors $\mathbf{x}_a$ and $\mathbf{x}_g$ contain the parameters necessary to calibrate the accelerometer and gyro. Specific definitions for $\mathbf{F}_{va}$ and $\mathbf{x}_a$ will be discussed in Section 11.6.2. Specific definitions for $\mathbf{F}_{vg}$ and $\mathbf{x}_g$ will be discussed in Section 11.6.3.

The models derived in this section have two primary purposes to the system analyst. First, the models are necessary for the specification of a 'truth model' as required for covariance analysis. Second, when it is determined that an error source has a significant effect on the navigation system error and that it is possible to estimate the parameters pertinent to describing the error source, then the parameter estimates can be augmented to the error state vector of the implemented filter. Common candidates for augmentation to the error state vector (i.e., $\mathbf{x}_a$ and $\mathbf{x}_g$) include[3]: accelerometer bias (3), accelerometer scale factor (3), accelerometer non-orthogonality (6), accelerometer nonlinearity (18), gyro bias (3), gyro scale factor (3), gyro non-orthogonality (6), gyro acceleration sensitivity (27), and calibration factors for additional sensors. The final decision as to which additional

[3]The number in parenthesis indicates the number of states that would be augmented to the nominal state vector to model the listed error. These numbers are defined in Sections 11.6.2 and 11.6.3.

error states should be included in the implemented filter involves a tradeoff between the cost of increased computation and the benefit of increased accuracy. The tradeoff results are different for each application. Quantitative decisions can be based on covariance or simulation analysis only if the 'truth model' is accurate.

Given the models of eqns. (11.100–11.101), the estimates of the instrument calibration factors are

$$\Delta\hat{\mathbf{f}}^b = \mathbf{F}_{va}\hat{\mathbf{x}}_a \quad (11.102)$$

$$\Delta\hat{\omega}^b_{ib} = \mathbf{F}_{\rho g}\hat{\mathbf{x}}_g \quad (11.103)$$

where $\hat{\mathbf{x}}_a$ and $\hat{\mathbf{x}}_g$ are initialized based on values determined off-line, but may be improved during system operation using state estimation. Based on eqns. (11.98–11.99), the instrument calibration errors are defined as

$$\delta\mathbf{f}^b = \mathbf{F}_{va}\delta\mathbf{x}_a + \boldsymbol{\nu}_a \quad (11.104)$$

$$\delta\boldsymbol{\omega}^b_{ib} = \mathbf{F}_{\rho g}\delta\mathbf{x}_g + \boldsymbol{\nu}_g \quad (11.105)$$

where $\delta\mathbf{x}_a = \mathbf{x}_a - \hat{\mathbf{x}}_a$ and $\delta\mathbf{x}_g = \mathbf{x}_g - \hat{\mathbf{x}}_g$. The models for the derivatives of $\delta\mathbf{x}_a$ and $\delta\mathbf{x}_g$ will have the structure

$$\delta\dot{\mathbf{x}}_a = \mathbf{F}_{aa}\delta\mathbf{x}_a + \boldsymbol{\omega}_a \quad (11.106)$$

$$\delta\dot{\mathbf{x}}_g = \mathbf{F}_{gg}\delta\mathbf{x}_g + \boldsymbol{\omega}_g. \quad (11.107)$$

The spectral densities for $\boldsymbol{\omega}_a$ and $\boldsymbol{\omega}_g$ are instrument specific and often specified by the manufacturer. The matrices $\mathbf{F}_{aa}$ and $\mathbf{F}_{gg}$ are also instrument specific, but are often selected such that $\hat{\mathbf{x}}_a$ and $\hat{\mathbf{x}}_g$ are either random constants or random walk plus random constants.

With the above definitions, the state augmented version of eqn. (11.56) is

$$\begin{bmatrix} \delta\dot{\mathbf{p}} \\ \delta\dot{\mathbf{v}} \\ \dot{\boldsymbol{\rho}} \\ \delta\dot{\mathbf{x}}_a \\ \delta\dot{\mathbf{x}}_g \end{bmatrix} = \begin{bmatrix} \mathbf{F}_{pp} & \mathbf{F}_{pv} & \mathbf{F}_{p\rho} & \mathbf{0} & \mathbf{0} \\ \mathbf{F}_{vp} & \mathbf{F}_{vv} & \mathbf{F}_{v\rho} & -\mathbf{R}^n_b\mathbf{F}_{va} & \mathbf{0} \\ \mathbf{F}_{\rho p} & \mathbf{F}_{\rho v} & \mathbf{F}_{\rho\rho} & \mathbf{0} & \mathbf{R}^n_b\mathbf{F}_{\rho g} \\ \mathbf{0} & \mathbf{0} & \mathbf{0} & \mathbf{F}_{aa} & \mathbf{0} \\ \mathbf{0} & \mathbf{0} & \mathbf{0} & \mathbf{0} & \mathbf{F}_{gg} \end{bmatrix} \begin{bmatrix} \delta\mathbf{p} \\ \delta\mathbf{v} \\ \boldsymbol{\rho} \\ \delta\mathbf{x}_a \\ \delta\mathbf{x}_g \end{bmatrix} + \begin{bmatrix} \mathbf{0} & \mathbf{0} & \mathbf{0} & \mathbf{0} \\ -\mathbf{R}^n_b & \mathbf{0} & \mathbf{0} & \mathbf{0} \\ \mathbf{0} & \mathbf{R}^n_b & \mathbf{0} & \mathbf{0} \\ \mathbf{0} & \mathbf{0} & \mathbf{I} & \mathbf{0} \\ \mathbf{0} & \mathbf{0} & \mathbf{0} & \mathbf{I} \end{bmatrix} \begin{bmatrix} \nu_a \\ \nu_g \\ \omega_{\mathbf{a}} \\ \omega_{\mathbf{g}} \end{bmatrix}. \quad (11.108)$$

11.6.1 Instrument Error Overview

Several forms of instrumentation error can be analyzed to determine their affect on the errors in the nine nominal error states. The major error sources to be discussed are: accelerometer bias, accelerometer scale factors, accelerometer alignment, accelerometer nonlinearity, gyro drift (bias), gyro scale factor, gyro alignment, and gyro g-sensitivity. These major sources of accelerometer and gyro error are considered in the following subsections. Enough detail is included so that the analysis can be extended to other forms of error as may be particular to specific applications.

11.6.2 Accelerometer Error Modeling

The differential equations for the actual and computed velocity (see eqns. (11.43) and (11.51)) involved the actual navigation frame specific force ($\mathbf{f}^n$) and the computed navigation frame specific force ($\hat{\mathbf{f}}^n$) respectively. The navigation frame velocity error is driven in part by the specific force error. This section considers the errors involved in measuring the acceleration frame specific force and computing its platform frame representation.

The actual platform specific force is determined from accelerometer measurements as

$$\mathbf{f}^p = \mathbf{C}_a^p \mathbf{f}^a. \tag{11.109}$$

The platform frame to accelerometer frame transformation $\mathbf{C}_p^a$ is a non-orthogonal transformation which accounts for accelerometer misalignment. This rotation has the form

$$\mathbf{C}_p^a = \mathbf{I} + \mathbf{\Delta}_p^a \tag{11.110}$$

$$\mathbf{\Delta}_p^a = \begin{bmatrix} 0 & a_{uw} & -a_{uv} \\ -a_{vw} & 0 & a_{vu} \\ a_{wv} & -a_{wu} & 0 \end{bmatrix} \tag{11.111}$$

where each element of $\mathbf{\Delta}_p^a$ is a component of a small angle. To first order, $\mathbf{C}_a^p = (\mathbf{C}_p^a)^{-1}$ can be shown by direct multiplication to yield

$$\mathbf{C}_a^p = \mathbf{I} - \mathbf{\Delta}_p^a. \tag{11.112}$$

The computed specific force equation is

$$\hat{\mathbf{f}}^p = \hat{\mathbf{C}}_a^p \tilde{\mathbf{f}}^a. \tag{11.113}$$

The accelerometer frame to platform frame transformation $\hat{\mathbf{C}}_a^p$ is often assumed to be an identity, but could also have off-diagonal elements determined through an off-line alignment process. This matrix is often determined by the manufacturer and corrected internal to the instrument. This rotation matrix has the same structure as defined for $\mathbf{C}_a^p$. Substituting the expression for $\hat{\mathbf{C}}_a^p$ into eqn. (11.113) yields

$$\hat{\mathbf{f}}^p = \left(\mathbf{I} - \hat{\mathbf{\Delta}}_p^a\right) \tilde{\mathbf{f}}^a. \tag{11.114}$$

The term $\tilde{\mathbf{f}}^a$ represents the actual accelerometer measurements accounting for all measurement errors. In a finite dimensional model, the designer can only hope to account for the major sources of measurement error. Let

$$\tilde{\mathbf{f}}^a = (\mathbf{I} - \delta\mathbf{SF}_a)(\mathbf{f}^a - \delta\mathbf{b}_a - \delta\mathbf{nl}_a - \boldsymbol{\nu}_a) \tag{11.115}$$

where $\delta\mathbf{SF}_a$ is a diagonal matrix representing uncompensated accelerometer scale factor error, $\delta\mathbf{b}_a$ represents uncompensated accelerometer bias, $\delta\mathbf{nl}_a$ represents uncompensated accelerometer nonlinearity, and $\boldsymbol{\nu}_a$ represents random measurement noise.

Combining eqns. (11.114-11.115) and linearizing the result yields the following equations for the actual navigation frame-specific force measurement and measurement error:

$$\hat{\mathbf{f}}^p = \left(\mathbf{I} - \hat{\boldsymbol{\Delta}}_p^a\right)\mathbf{f}^a - (\delta\mathbf{SF}_a)\,\mathbf{f}^a - (\delta\mathbf{b}_a + \delta\mathbf{nl}_a + \boldsymbol{\nu}_a) \tag{11.116}$$

$$\begin{aligned} \delta\mathbf{f}^p &= \mathbf{f}^p - \hat{\mathbf{f}}^p \\ &= \left(\delta\mathbf{SF}_a - \boldsymbol{\delta}_p^a\right)\mathbf{f}^p + (\delta\mathbf{b}_a + \delta\mathbf{nl}_a + \boldsymbol{\nu}_a) \end{aligned} \tag{11.117}$$

where $\boldsymbol{\delta}_p^a = \boldsymbol{\Delta}_p^a - \hat{\boldsymbol{\Delta}}_p^a$ accounts for error in the alignment process and $\mathbf{f}^a$ is approximated by $\mathbf{f}^p$ for a first order expression.

The next step is to define parameter vectors $\mathbf{x}_{b_a}$, $\mathbf{x}_{Aa}$, and $\mathbf{x}_{ka}$ such that eqn. (11.117) can be rewritten as

$$\delta\mathbf{f}^p = \frac{\partial\mathbf{f}^p}{\partial\mathbf{x}_{b_a}}\mathbf{x}_{b_a} + \frac{\partial\mathbf{f}^p}{\partial\mathbf{x}_{Aa}}\mathbf{x}_{Aa} + \frac{\partial\mathbf{f}^p}{\partial\mathbf{x}_{ka}}\mathbf{x}_{ka} + \boldsymbol{\nu}_a. \tag{11.118}$$

In this equation, $\mathbf{x}_{b_a}$ contains the states needed to model $\delta\mathbf{b}_a$, $\mathbf{x}_{Aa}$ contains the states needed to model $\delta\mathbf{SF}_a$ and $\boldsymbol{\delta}_p^a$, and $\mathbf{x}_{ka}$ contains the states needed to model $\delta\mathbf{nl}_a$. Given eqn. (11.118), it is straightforward to define the factors of eqn. (11.104) as

$$\mathbf{F}_{va} = \left[\frac{\partial\mathbf{f}^p}{\partial\mathbf{x}_{b_a}}, \frac{\partial\mathbf{f}^p}{\partial\mathbf{x}_{Aa}}, \frac{\partial\mathbf{f}^p}{\partial\mathbf{x}_{ka}}\right] \quad \text{and} \quad \delta\mathbf{x}_a = [\mathbf{x}_{b_a}^\top, \mathbf{x}_{Aa}^\top, \mathbf{x}_{ka}^\top]^\top.$$

The accelerometer error state vector $\delta\mathbf{x}_a$ discussed below has dimension thirty. In fact, the potential state order could be higher depending on the dynamic model that is appropriate for each state element (e.g., random constant, scalar Gauss-Markov). In any particular application, the entire vector $\delta\mathbf{x}_a$ may not be required; instead, the designer must determine which error terms have a significant effect on performance. Those terms must be identified either in the laboratory or during on-line operation.

Bias Error (3 States). Define the bias as $\mathbf{x}_{b_a} = \delta\mathbf{b}_a$. The effect of the bias error on the specific force error is

$$\frac{\partial\mathbf{f}^p}{\partial\mathbf{x}_{b_a}} = \mathbf{I}. \tag{11.119}$$

Scale Factor and Misalignment (9 States). Define the matrix $\delta\mathbf{A}_a = \left(\delta\mathbf{SF}_a - \delta_p^a\right)$ which has the form

$$\delta\mathbf{A}_a = \begin{bmatrix} \delta SF_u & -\delta a_{uw} & \delta a_{uv} \\ \delta a_{vw} & \delta SF_v & -\delta a_{vu} \\ -\delta a_{wv} & \delta a_{wu} & \delta SF_w \end{bmatrix}. \tag{11.120}$$

Define the auxiliary state composed of the accelerometer scale factor and misalignment matrix parameters to be

$$\mathbf{x}_{Aa} = [SF_u, SF_v, SF_w, a_{uw}, a_{uv}, a_{vw}, a_{vu}, a_{wv}, a_{wu}]. \tag{11.121}$$

The partial derivative of $\delta\mathbf{f}^p$ with respect to $\mathbf{x}_{Aa}$ is

$$\frac{\partial\mathbf{f}^p}{\partial\mathbf{x}_{Aa}} = \begin{bmatrix} f_u & 0 & 0 & -f_v & f_w & 0 & 0 & 0 & 0 \\ 0 & f_v & 0 & 0 & 0 & f_u & -f_w & 0 & 0 \\ 0 & 0 & f_w & 0 & 0 & 0 & 0 & -f_u & f_v \end{bmatrix}$$

where the elements of the platform frame specific force vector are $\mathbf{f}^p = [f_u, f_v, f_w]^\top$.

Nonlinear Effects (18 States). Finally, because $\mathbf{x}_{b_a}$ already accounts for zeroth order effects and $\mathbf{x}_{Aa}$ already accounts for first order effects, the error model accounting for nonlinear effects up to and including second order terms is

$$\delta\mathbf{nl}_a = \begin{bmatrix} k_{x1}f_u^2 + k_{x2}f_v^2 + k_{x3}f_w^2 + k_{x4}f_uf_v + k_{x5}f_vf_w + k_{x6}f_wf_x \\ k_{y1}f_u^2 + k_{y2}f_v^2 + k_{y3}f_w^2 + k_{y4}f_uf_v + k_{y5}f_vf_w + k_{y6}f_wf_x \\ k_{z1}f_u^2 + k_{z2}f_v^2 + k_{z3}f_w^2 + k_{z4}f_uf_v + k_{z5}f_vf_w + k_{z6}f_wf_x \end{bmatrix}.$$

Organizing the unknown parameters into the auxiliary state vector

$$\begin{aligned} \mathbf{x}_{ka} &= [k_{x1}, k_{x2}, k_{x3}, k_{x4}, k_{x5}, k_{x6}, k_{y1}, k_{y2}, k_{y3}, \\ &\quad k_{y4}, k_{y5}, k_{y6}, k_{z1}, k_{z2}, k_{z3}, k_{z4}, k_{z5}, k_{z6}]^\top, \end{aligned} \tag{11.122}$$

the partial derivative of $\delta\mathbf{f}^p$ with respect to $\mathbf{x}_{ka}$ is

$$\frac{\partial\mathbf{f}^p}{\partial\mathbf{x}_{ka}} = \begin{bmatrix} \mathbf{F}_{ka} & \mathbf{Z}_6 & \mathbf{Z}_6 \\ \mathbf{Z}_6 & \mathbf{F}_{ka} & \mathbf{Z}_6 \\ \mathbf{Z}_6 & \mathbf{Z}_6 & \mathbf{F}_{ka} \end{bmatrix} \tag{11.123}$$

where $\mathbf{Z}_6$ is a 1×6 vector of zeros and

$$\mathbf{F}_{ka} = [f_u^2, f_v^2, f_w^2, f_uf_v, f_vf_w, f_wf_u]. \tag{11.124}$$

11.6.3 Gyro Error Modeling

This section considers various forms of gyro error. For each form of error, this section defines a component of an augmentation state vector and a linear model relating that component to the gyro error. These gyro error state equations can be augmented to the nominal error state to model the effect of gyro error on the overall navigation solution. The derivations of this section are very closely related to those of Section 11.6.2.

The equation relating the actual platform angular rate to the angular rate measured in gyro frame is

$$\boldsymbol{\omega}_{ip}^{p} = \mathbf{C}_{g}^{p}\boldsymbol{\omega}_{ip}^{g}. \tag{11.125}$$

The platform frame to gyro frame transformation $\mathbf{C}_p^g$ is a non-orthogonal transformation which accounts for gyro misalignment. This rotation has the form

$$\mathbf{C}_{p}^{g} = \mathbf{I}+\boldsymbol{\Delta}_{p}^{g} \tag{11.126}$$

$$\boldsymbol{\Delta}_{p}^{g} = \begin{bmatrix} 0 & g_{pr} & -g_{pq} \\ -g_{qr} & 0 & g_{qp} \\ g_{rq} & -g_{rp} & 0 \end{bmatrix} \tag{11.127}$$

where each element of $\boldsymbol{\Delta}_p^g$ is a component of a small angle. To first order, $\mathbf{C}_g^p = (\mathbf{C}_p^g)^{-1}$ can be shown by direct multiplication to yield

$$\mathbf{C}_{g}^{p} = \mathbf{I} - \boldsymbol{\Delta}_{p}^{g}. \tag{11.128}$$

The computed angular rate in platform coordinates is described by

$$\hat{\boldsymbol{\omega}}_{ip}^{p} = \hat{\mathbf{C}}_{g}^{p}\tilde{\boldsymbol{\omega}}_{ip}^{g}. \tag{11.129}$$

The gyro frame to platform frame transformation $\hat{\mathbf{C}}_g^p$ is often assumed to be an identity, but could also have off-diagonal elements determined through a calibration process. This rotation matrix has the same structure as defined for $\mathbf{C}_g^p$. Substituting the expression for $\hat{\mathbf{C}}_g^p$ into eqn. (11.129) yields

$$\hat{\boldsymbol{\omega}}_{ip}^{p} = \left(\mathbf{I} - \hat{\boldsymbol{\Delta}}_{p}^{g}\right)\tilde{\boldsymbol{\omega}}_{ip}^{g}. \tag{11.130}$$

The term $\tilde{\boldsymbol{\omega}}_{ip}^g$ represents the actual gyro measurements accounting for all measurement errors. To develop a finite dimensional error model, only bias, scale factor, noise, and g-sensitive errors will be considered. Let

$$\tilde{\boldsymbol{\omega}}_{ip}^{g} = (\mathbf{I} - \delta\mathbf{SF}_{g})\left(\boldsymbol{\omega}_{ip}^{g} - \delta\mathbf{b}_{g} - \delta\mathbf{k}_{g} - \boldsymbol{\nu}_{g}\right) \tag{11.131}$$

where $\delta\mathbf{SF}_g$ is a diagonal matrix representing uncompensated gyro scale factor error, $\delta\mathbf{b}_g$ represents uncompensated gyro bias, $\delta\mathbf{k}_g$ represents uncompensated gyro g-sensitivity, and $\boldsymbol{\nu}_g$ represents random measurement noise.

Combining eqns. (11.130-11.131) and linearizing the result yields the following equations for the actual platform frame angular rate measurement and measurement error:

$$\begin{aligned}\hat{\boldsymbol{\omega}}_{ip}^{p} &= \boldsymbol{\omega}_{ip}^{p} - \hat{\Delta}_{p}^{g}\boldsymbol{\omega}_{ip}^{g} - (\delta\mathbf{SF}_{g})\,\boldsymbol{\omega}_{ip}^{g} - (\delta\mathbf{b}_{g} + \delta\mathbf{k}_{g} + \boldsymbol{\nu}_{g}) \quad (11.132)\\ \delta\boldsymbol{\omega}_{ip}^{p} &= \boldsymbol{\omega}_{ip}^{p} - \hat{\boldsymbol{\omega}}_{ip}^{p}\\ &= \left(\delta\mathbf{SF}_{g} - \boldsymbol{\delta}_{p}^{g}\right)\boldsymbol{\omega}_{ip}^{g} + (\delta\mathbf{b}_{g} + \delta\mathbf{k}_{g} + \boldsymbol{\nu}_{g}) \quad (11.133)\end{aligned}$$

where $\boldsymbol{\delta}_{p}^{g} = \Delta_{p}^{g} - \hat{\Delta}_{p}^{g}$ accounts for error in the alignment process. In the following paragraphs, to simplify the notation, we use the symbol $\mathbf{T}$ to denote the platform angular rate error vector,

$$\mathbf{T}^{p} = \delta\boldsymbol{\omega}_{ip}^{p}.$$

The next step is to define parameter vectors $\mathbf{x}_{b_g}$, $\mathbf{x}_{Ag}$, and $\mathbf{x}_{kg}$ such that eqn. (11.133) can be rewritten as

$$\mathbf{T}^{p} = \frac{\partial\mathbf{T}^{p}}{\partial\mathbf{b}_{g}}\mathbf{x}_{b_g} + \frac{\partial\mathbf{T}^{p}}{\partial\mathbf{x}_{Ag}}\mathbf{x}_{Ag} + \frac{\partial\mathbf{T}^{p}}{\partial\mathbf{x}_{kg}}\mathbf{x}_{kg} + \boldsymbol{\nu}_{a}. \quad (11.134)$$

In this equation, $\mathbf{x}_{b_g}$ contains the states needed to model $\delta\mathbf{b}_g$, $\mathbf{x}_{Ag}$ contains the states needed to model $\delta\mathbf{SF}_g$ and $\boldsymbol{\delta}_p^g$, and $\mathbf{x}_{kg}$ contains the states needed to model $\delta\mathbf{k}_g$. Given eqn. (11.134), it is straightforward to define the factors of eqn. (11.105) as

$$\mathbf{F}_{\rho g} = \left[\frac{\partial\mathbf{T}^{p}}{\partial\mathbf{b}_{g}}, \frac{\partial\mathbf{T}^{p}}{\partial\mathbf{x}_{Ag}}, \frac{\partial\mathbf{T}^{p}}{\partial\mathbf{x}_{kg}}\right] \quad \text{and} \quad \delta\mathbf{x}_{g} = [\mathbf{x}_{b_g}^{\top}, \mathbf{x}_{Ag}^{\top}, \mathbf{x}_{kg}^{\top}]^{\top}.$$

The gyro error state vector $\delta\mathbf{x}_g$ discussed below has dimension thirty-nine. The potential state order could be higher depending on the dynamic model that is appropriate for each state element (e.g., random constant, scalar Gauss-Markov). In any particular application, the entire vector $\delta\mathbf{x}_g$ may not be required; instead, the designer must determine which error terms have a significant effect on performance. Those terms must be identified either in the laboratory or during on-line operation.

Bias Error (3 States). Define the the augmentation state component to model gyro bias as $\mathbf{x}_{b_g} = \delta\mathbf{b}_g$. The partial derivative of $\mathbf{T}^p$ with respect to $\mathbf{x}_{b_g}$ is

$$\frac{\partial\mathbf{T}^{p}}{\partial\mathbf{x}_{b_g}} = \mathbf{I}. \quad (11.135)$$

Scale Factor and Misalignment Error (9 States). Define $\delta\mathbf{A}_g = (\delta\mathbf{SF}_g - \boldsymbol{\delta}_p^g)$, which has the structure

$$\delta\mathbf{A}_g = \begin{bmatrix} \delta SF_p & -\delta g_{pr} & \delta g_{pq} \\ \delta g_{qr} & \delta SF_q & -\delta g_{qp} \\ -\delta g_{rq} & \delta g_{rp} & \delta SF_r \end{bmatrix}. \tag{11.136}$$

Define the auxiliary state composed of the gyro scale factor and misalignment matrix parameters to be

$$\mathbf{x}_{Ag} = [SF_p, SF_q, SF_r, g_{pr}, g_{pq}, g_{qr}, g_{qp}, g_{rq}, g_{rp}]. \tag{11.137}$$

Then, the partial derivative of $\mathbf{T}$ with respect to $\mathbf{x}_{Ag}$ is

$$\frac{\partial \mathbf{T}^p}{\partial \mathbf{x}_{Ag}} = \mathbf{R}_b^g \begin{bmatrix} p & 0 & 0 & -q & r & 0 & 0 & 0 & 0 \\ 0 & q & 0 & 0 & 0 & p & -r & 0 & 0 \\ 0 & 0 & r & 0 & 0 & 0 & 0 & -p & q \end{bmatrix} \tag{11.138}$$

where $\boldsymbol{\omega}_{ip}^p = [p, q, r]^\top$.

Acceleration Sensitivity (27 States). Finally, the error model accounting for the gyro acceleration sensitivity (i.e., g-sensitivity) up to and including second order effects is

$$\delta\mathbf{k}_g = \begin{bmatrix} k_{p1}f_u + k_{p2}f_v + k_{p3}f_w + k_{p4}f_u^2 + k_{p5}f_v^2 \\ +k_{p6}f_w^2 + k_{p7}f_uf_v + k_{p8}f_vf_w + k_{p9}f_wf_u \\ \hline k_{q1}f_u + k_{q2}f_v + k_{q3}f_w + k_{q4}f_u^2 + k_{q5}f_v^2 \\ +k_{q6}f_w^2 + k_{q7}f_uf_v + k_{q8}f_vf_w + k_{q9}f_wf_u \\ \hline k_{p1}f_u + k_{r2}f_v + k_{r3}f_w + k_{r4}f_u^2 + k_{r5}f_v^2 \\ +k_{r6}f_w^2 + k_{r7}f_uf_v + k_{r8}f_vf_w + k_{r9}f_wf_u \end{bmatrix} \in \mathbb{R}^{3\times 1} \tag{11.139}$$

where lines have been used to clearly separate the components of the vector. Organizing the parameters into the auxiliary state $\mathbf{x}_{kg}$ yields

$$\begin{aligned} \mathbf{x}_{kg} \;=\; & [k_{p1}, k_{p2}, k_{p3}, k_{p4}, k_{p5}, k_{p6}, k_{p7}, k_{p8}, k_{p9}, \\ & k_{q1}, k_{q2}, k_{q3}, k_{q4}, k_{q5}, k_{q6}, k_{q7}, k_{q8}, k_{q9}, \\ & k_{r1}, k_{r2}, k_{r3}, k_{r4}, k_{r5}, k_{r6}, k_{r7}, k_{r8}, k_{r9}]. \end{aligned} \tag{11.140}$$

The partial derivative of $\mathbf{T}$ with respect to $\mathbf{x}_{kg}$ is

$$\frac{\partial \mathbf{T}^p}{\partial \mathbf{x}_{kg}} = \begin{bmatrix} \mathbf{F}_{kg} & \mathbf{Z}_9 & \mathbf{Z}_9 \\ \mathbf{Z}_9 & \mathbf{F}_{kg} & \mathbf{Z}_9 \\ \mathbf{Z}_9 & \mathbf{Z}_9 & \mathbf{F}_{kg} \end{bmatrix} \tag{11.141}$$

where $\mathbf{Z}_9$ is a 1×9 vector of zeros and

$$\mathbf{F}_{kg} = [f_u, f_v, f_w, f_u^2, f_v^2, f_w^2, f_uf_v, f_vf_w, f_wf_u]. \tag{11.142}$$

11.6.4 Error Characteristics

This section presents a single example of the INS forced error response.

Figure 11.6 is the response of the system described by eqn. (11.79) with all initial conditions set equal to zero (i.e., $\delta \mathbf{x} = \mathbf{0}$), and all the accelerometer and gyro error terms set equal to zero, except for the east (or pitch rate q) gyro bias having a constant value of $0.015°/hr$ (i.e., $\delta \mathbf{f}^b = \mathbf{0}$ and $\delta \boldsymbol{\omega}_{ib}^b = [0.000, 0.015, 0.000]°/hr$). For the first several minutes, this simulation corresponds to the analytic solution given by eqn. (11.87) for the reduced order system of eqn. (11.81).

Inspection of the $\mathbf{F}$ matrix in eqn. (11.90) yields the following comments on the simulation results displayed in Figure 11.6. The east gyro bias causes growth of the east tilt error. The east tilt error causes azimuth, north velocity, and north tilt error. The azimuth error develops a bias of $\frac{b_g}{\Omega_N}$ radians (0.08°) which cancels the effect of the gyro bias on the east tilt error. This azimuth bias does not propagate into the remaining states since the system is stationary and the bias is perfectly canceled in the east tilt channel. Due to the azimuth error bias canceling the gyro bias, the north velocity error is unbiased. Therefore the latitude error does not grow linearly with time for a stationary system. If the system were in motion, then the azimuth error would result in errors in the remaining states. The fact that azimuth error and east gyro bias are not separately observable is one of the limiting factors in the practical application of the gyro compassing technique. The oscillations at the earth frequency in the north velocity and azimuth error are out of phase by 180°, so that the earth frequency oscillation does not affect the east tilt dynamics.

11.7 Initialization

Inertial navigation systems integrate compensated inertial measurements to provide a position, velocity, and attitude reference trajectory. Each integration introduces one constant of integration. Therefore, initialization of a three dimensional INS requires specification of the nine initial conditions for position, velocity, and attitude plus initial conditions for the components of $\hat{\mathbf{x}}_a$ and $\hat{\mathbf{x}}_g$. This section is concerned with methods for determining these various initial conditions. Three topics are of interest:

Calibration – The process of determining various factors to calibrate the inertial instruments.

Initialization – The process of determining the INS initial position and velocity.

Alignment – The process of determining the relative orientation of the inertial system platform and the reference navigation frame axes.

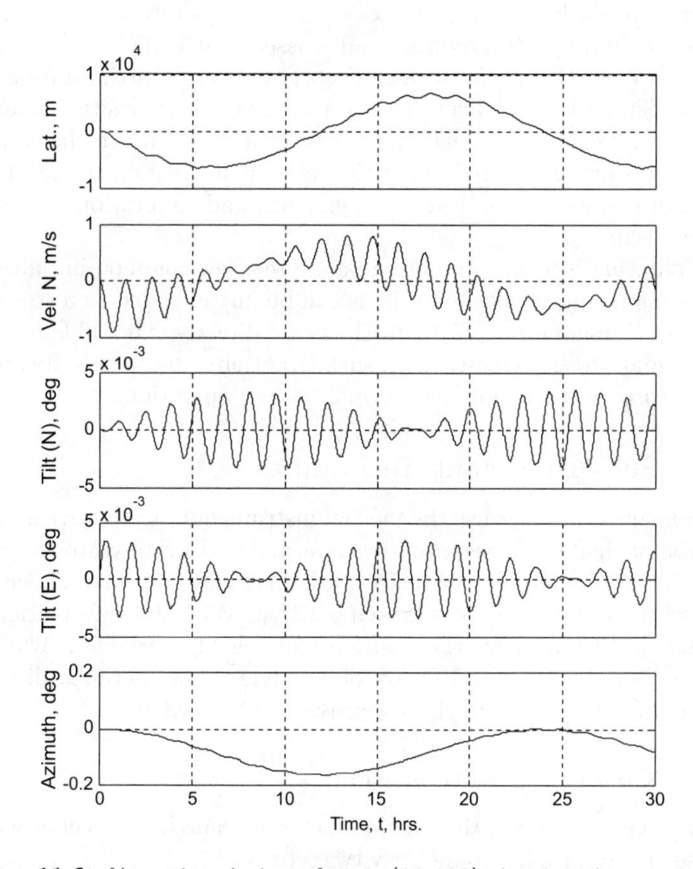

Figure 11.6: Numeric solution of eqn. (11.108) showing the error response for a stationary system. All instrument error states are zero with the exception of the east (or q) gyro bias that has a constant value of $0.015°/hr$. The vertical states are assumed to be stabilized and have been removed. The driving noise terms are zero.

Position, velocity, and attitude error estimation and correction are processes which occur at least at system start-up and may continue throughout system operation when aiding sources are available. Instrument calibration may occur off-line in a laboratory and on-line through the use of state augmentation and state estimation techniques. Since all three operations may occur simultaneously, the three topics are not completely distinct. Off-line instrument calibration techniques are discussed in [83, 126].

Inaccurate position initialization affects gravity and Earth rotation compensation. Similarly, initial velocity error integrates into position error and results in alignment error. Alignment error results in error in the transformation of platform measured quantities into the navigation frame. Therefore, accurate system initialization, alignment, and calibration are critical to system accuracy.

The following sections present and analyze a few common initialization processes. Some initialization methods can be implemented as a state estimator (e.g., Kalman filter). Such methods are directly extendible to aiding during normal system operation. In such situations, the same software can be used during initialization and normal (aided) operation.

11.7.1 Self-Alignment Techniques

Self-alignment techniques use the inertial instruments together with knowledge of the navigation frame gravity vector and earth rate vector for alignment of the inertial platform relative to the navigation frame. The system is assumed to be stationary at a known position. With this information it is straightforward to initialize the position and velocity variables. The main remaining issue is the initialization of attitude. The methods discussed below are directly related to those discussed in Section 10.3.

11.7.1.1 Coarse Self-Initialization

For a stationary system, the accelerometer measurement vector will be related to the geographic frame gravity vector by

$$\begin{bmatrix} \tilde{f}_u \\ \tilde{f}_v \\ \tilde{f}_w \end{bmatrix} = -\mathbf{R}_n^b \begin{bmatrix} 0 \\ 0 \\ g \end{bmatrix} = \begin{bmatrix} \sin(\theta) \\ -\cos(\theta)\sin(\phi) \\ -\cos(\theta)\cos(\phi) \end{bmatrix} g \tag{11.143}$$

which can be solved for θ and ϕ as follows

$$\begin{bmatrix} \hat{\phi} \\ \hat{\theta} \end{bmatrix} = \begin{bmatrix} arctan2(-\tilde{f}_v, -\tilde{f}_w) \\ arctan2(\tilde{f}_u, \sqrt{\tilde{f}_v^2 + \tilde{f}_w^2}) \end{bmatrix}. \tag{11.144}$$

The pitch and roll angles estimated by eqn. (11.144) are accurate to approximately the uncompensated accelerometer bias divided by gravity. Therefore, the level error estimation sensitivity to accelerometer bias is one milliradian per milli-g.

The alternative method described below (see [26, 27, 125]), provides an initial estimate of $\mathbf{R}_b^n$, from which roll, pitch, and yaw angles or the quaternion parameters could be determined as necessary. Let $\mathbf{u}$ and $\mathbf{v}$ be two independent vectors. Define $\mathbf{w} = \mathbf{u} \times \mathbf{v}$. Therefore, $\mathbf{w}$ is orthogonal to both $\mathbf{u}$ and $\mathbf{v}$. Assume that $\mathbf{u}$ and $\mathbf{v}$ are known in navigation frame and measured in platform frame. Then the two sets of vectors are related according to

$$[\mathbf{u}^n, \mathbf{v}^n, \mathbf{w}^n] = \mathbf{R}_b^n \left[\mathbf{u}^b, \mathbf{v}^b, \mathbf{w}^b\right] \tag{11.145}$$

$$\hat{\mathbf{R}}_b^n = [\mathbf{u}^n, \mathbf{v}^n, \mathbf{w}^n] \left[\tilde{\mathbf{u}}^b, \tilde{\mathbf{v}}^b, \tilde{\mathbf{w}}^b\right]^{-1}. \tag{11.146}$$

In particular, if $\mathbf{u}^n$ is the gravity vector and $\mathbf{v}^n$ is the earth rate vector, then

$$\hat{\mathbf{R}}_n^b = \left[\tilde{\mathbf{f}}^b, \tilde{\boldsymbol{\omega}}^b, \tilde{\mathbf{f}}^b \times \tilde{\boldsymbol{\omega}}^b\right] \begin{bmatrix} \frac{1}{g} tan(\phi) & 0 & \frac{1}{g} \\ \frac{1}{\omega_{ie}\cos(\phi)} & 0 & 0 \\ 0 & \frac{1}{\omega_{ie} g \cos(\phi)} & 0 \end{bmatrix}. \tag{11.147}$$

Non-benign environments (i.e., environments having disturbances or vibration) may result in inaccurate initialization of the attitude, which may necessitate on-line, aided error calibration. Implementation of the above approach requires high quality gyros with precision and accuracy sufficient to measure earth rate. The rotation matrix that results from the above process can be orthogonalized by the methods described in Appendix B.

When the gyros selected for a given application do not allow yaw initialization using the above methods, then alternative methods are required. Eqn. (11.146) is valid for any three independent vectors. Optical sightings are sometimes used to replace at least one of the three vectors in that equation. Otherwise, the procedure is identical. A few other common alternatives are magnetic direction sensing with local magnetic field compensation or retrieval of the last stored yaw angle for a system which is known to have been stationary between periods of operation. In either of these approaches, the accuracy will normally be low enough that on-line calibration of the azimuth error will be desirable via aiding sensors.

11.7.1.2 Fine Initialization: Physical Gyro Compassing

For a mechanized system with accurately calibrated accelerometers and gyros, alignment can be achieved by a two step process known as *gyrocompassing* [27, 36, 69]. In the first step, referred to as *leveling*, the platform is torqued to null the outputs of the north and east accelerometers. This step nominally aligns the z-axis with the local gravity vector. The resulting leveling error is directly proportional to the uncompensated accelerometer bias. Since the east component of the earth rotation rate is known to be zero, the second step of the procedure rotates the platform about the z-axis to null the output of the east gyro. The second step is referred to as *azimuth alignment*. At this point, at least theoretically, the north and down gyros could be used to estimate and correct the platform latitude as $\phi = atan2(-\Omega_D, \Omega_N)$.

11.7.1.3 Fine Initialization: Analytic Gyro-Compassing

In strap-down system applications, the platform cannot be mechanically torqued to cause alignment with the gravity and Earth rate vectors. Instead, the misalignment is estimated and used to correct the variables required to compute the direction cosine matrix.

The computed specific force vector in the body and navigation frames are related according to $\hat{\mathbf{f}}^n = \hat{\mathbf{R}}_b^n \tilde{\mathbf{f}}^b$. The corresponding error model is

$$\begin{aligned} \hat{\mathbf{f}}^n &= (\mathbf{I} - [\boldsymbol{\rho}\times])\, \mathbf{R}_b^n \left(\mathbf{f}^b + \delta\mathbf{f}^b\right) && (11.148) \\ &= \mathbf{f}^n - \mathbf{P}\mathbf{f}^n + \mathbf{R}_b^n \delta\mathbf{f}^b && (11.149) \end{aligned}$$

to first order, where $\delta\mathbf{f}^b$ represents the sum of the specific force measurement errors. For a system assumed to be stationary, the actual specific force is the sum of gravity and disturbance accelerations. In the geographic frame,

$$\mathbf{f}^n = \begin{bmatrix} 0 & 0 & -g \end{bmatrix}^\top + \mathbf{f}_d$$

where $\mathbf{f}_d$ represents the navigation frame disturbances. These disturbances can be expected to be high frequency, but the amplitude and spectral content are not accurately known and may be time varying. Disturbance motion can be decreased by mechanical isolation. The residual navigation frame specific force measurement is described by

$$\begin{aligned} \delta\mathbf{f} &= \mathbf{f}^n - \hat{\mathbf{f}}^n && (11.150) \\ &= \mathbf{P}\mathbf{f}^n - \mathbf{R}_b^n \delta\mathbf{f}^b + \mathbf{f}_d && (11.151) \\ &= \begin{bmatrix} 0 & -\epsilon_D & \epsilon_E \\ \epsilon_D & 0 & -\epsilon_N \\ -\epsilon_E & \epsilon_N & 0 \end{bmatrix} \begin{bmatrix} 0 \\ 0 \\ -g \end{bmatrix} - \mathbf{R}_b^n \delta\mathbf{f}^b + \mathbf{f}_d && (11.152) \\ &= \begin{bmatrix} 0 & -g & 0 \\ g & 0 & 0 \\ 0 & 0 & 0 \end{bmatrix} \begin{bmatrix} \epsilon_N \\ \epsilon_E \\ \epsilon_D \end{bmatrix} - \mathbf{R}_b^n \delta\mathbf{f}^b + \mathbf{f}_d. && (11.153) \end{aligned}$$

Based on eqn. (11.153), two approaches are possible for estimating the alignment error. In the first approach, the measurement

$$\delta\mathbf{f} = \mathbf{H}\delta\mathbf{x} - \mathbf{R}_b^n \boldsymbol{\nu}_a + \mathbf{f}_d$$

could be used to drive a Kalman filter with state error dynamics as previously defined, $\delta\mathbf{x} = [\delta\mathbf{p}^\top, \delta\mathbf{v}^\top, \boldsymbol{\rho}^\top, \delta\mathbf{x}_a^\top, \delta\mathbf{x}_g^\top]^\top$,

$$\mathbf{H} = \left[\begin{array}{ccccc} \mathbf{Z}_3 & \mathbf{Z}_3 & \mathbf{G} & -\mathbf{R}_b^n\mathbf{F}_{va} & \mathbf{Z}_3 \end{array}\right]$$

where $\mathbf{Z}_3$ is a three dimensional[4] zero vector, and

$$\mathbf{G} = \begin{bmatrix} 0 & -g & 0 \\ g & 0 & 0 \\ 0 & 0 & 0 \end{bmatrix}.$$

From the system description and measurement matrix, it is clear that ϵ_d is completely unobservable. In the case where the system is nearly level, the w-axis accelerometer bias is completely observable. The accuracy with which the north and east alignment errors can be estimated will depend on the accuracy of the east and north accelerometers and amount of platform motion due to disturbances. In theory, portions of the disturbance motion could be modeled through state augmentation techniques. In practice, this is difficult. Therefore, vibration and disturbances will limit the achievable accuracy. In an alternative approach, the first two components of $\delta\mathbf{f}$ could be passed through a simple low pass filter with gain $\frac{1}{g}$. The outputs are interpreted as $-\hat{\epsilon}_E$ and $\hat{\epsilon}_N$, respectively. In either case, the estimates are used to adjust $\hat{\mathbf{R}}_b^n$ (see eqn. (10.67)) until $\delta\mathbf{f}$ is zero. The accuracy of this approach will also be limited by instrumentation imperfections, vibration, and disturbances. At the conclusion of the process, the analytic version of the navigation frame is nominally in alignment with the actual navigation frame (i.e., leveling has been achieved). Again, as with the mechanized approach, leveling accuracy is directly proportional to the uncompensated accelerometer bias (1 milli-radian per 1 g bias).

Similarly, the gyro outputs can be processed to achieve azimuth alignment. Assuming that the vehicle is stationary relative to the Earth, we have that

$$\begin{aligned} \boldsymbol{\omega}_{ib} &= (\boldsymbol{\omega}_{in} + \boldsymbol{\omega}_{nb}) = (\boldsymbol{\omega}_{ie} + \boldsymbol{\omega}_d) \\ \hat{\boldsymbol{\omega}}_{ib} &= \hat{\boldsymbol{\omega}}_{in}^b = \hat{\boldsymbol{\omega}}_{ie}^b \end{aligned}$$

where the stationary assumption implies that $\boldsymbol{\omega}_{in} = \boldsymbol{\omega}_{ie}$, $\hat{\boldsymbol{\omega}}_{in} = \hat{\boldsymbol{\omega}}_{ie}$, and $\hat{\boldsymbol{\omega}}_{bn}^b = \mathbf{0}$. The symbol $\boldsymbol{\omega}_d$ is used to represent $\boldsymbol{\omega}_{nb}$ which is the disturbance angular rates of the vehicle that are in violation of the stationarity assumption. The actual and computed geographic frame rotation rates are

$$\begin{aligned} \boldsymbol{\omega}_{in}^n &= \mathbf{R}_b^n \left(\boldsymbol{\omega}_{ib}^b + \boldsymbol{\omega}_{bn}^b\right) \\ \hat{\boldsymbol{\omega}}_{in}^n &= \hat{\mathbf{R}}_b^n \hat{\boldsymbol{\omega}}_{ib}^b. \end{aligned}$$

[4] Assuming that the instrument errors are modeled as biases.

These expressions can each be expanded as follows:

$$\begin{aligned}\boldsymbol{\omega}_{in}^n &= (\mathbf{I} + [\boldsymbol{\rho}\times])\,\hat{\mathbf{R}}_b^n\left(\boldsymbol{\omega}_{ib}^b - \boldsymbol{\omega}_d^b\right)\\ &= \boldsymbol{\omega}_{ib}^n + [\boldsymbol{\rho}\times]\hat{\boldsymbol{\omega}}_{ib}^n - \boldsymbol{\omega}_d^n\end{aligned}$$

and

$$\hat{\boldsymbol{\omega}}_{in}^n = \hat{\mathbf{R}}_b^n\left(\boldsymbol{\omega}_{ib}^b + \delta\boldsymbol{\omega}_{ib}^b\right) = \boldsymbol{\omega}_{ib}^n + \hat{\mathbf{R}}_b^n\delta\boldsymbol{\omega}_{ib}^b.$$

Therefore, the residual gyro measurement in geographic frame is

$$\begin{aligned}\delta\boldsymbol{\omega} &= \boldsymbol{\omega}_{in}^n - \hat{\boldsymbol{\omega}}_{in}^n = [\boldsymbol{\rho}\times]\hat{\boldsymbol{\omega}}_{ie}^n - \hat{\mathbf{R}}_b^n\delta\boldsymbol{\omega}_{ib}^b - \boldsymbol{\omega}_d^n\\ &= \begin{bmatrix} 0 & -\epsilon_D & \epsilon_E \\ \epsilon_D & 0 & -\epsilon_N \\ -\epsilon_E & \epsilon_N & 0 \end{bmatrix}\begin{bmatrix} \Omega_N \\ 0 \\ \Omega_D \end{bmatrix} - \mathbf{R}_b^n\delta\boldsymbol{\omega}_{ib}^b - \boldsymbol{\omega}_d\\ &= \begin{bmatrix} 0 & \Omega_D & 0 \\ -\Omega_D & 0 & \Omega_N \\ 0 & -\Omega_N & 0 \end{bmatrix}\begin{bmatrix} \epsilon_N \\ \epsilon_E \\ \epsilon_D \end{bmatrix} - \mathbf{R}_b^n\delta\boldsymbol{\omega}_{ib}^b - \boldsymbol{\omega}_d. \qquad (11.154)\end{aligned}$$

Based on eqn. (11.154), two approaches are again possible. The first approach is Kalman filter based, with state error dynamics as previously defined and the measurement matrix

$$\mathbf{H} = \left[\begin{array}{ccccc} \mathbf{Z}_3 & \mathbf{Z}_3 & \mathbf{W} & \mathbf{R}_b^n & \mathbf{Z}_3 \end{array}\right] \qquad (11.155)$$

where

$$\mathbf{W} = \begin{bmatrix} 0 & \Omega_D & 0 \\ -\Omega_D & 0 & \Omega_N \\ 0 & -\Omega_N & 0 \end{bmatrix}.$$

The Kalman filter approach has the advantages that both the accelerometer based leveling and gyro based azimuth alignment can occur simultaneously, and that the approach correctly accounts for all measurement errors. In the second approach, after leveling has occurred, the second component of $\delta\boldsymbol{\omega}$ can be passed through a low pass filter to generate an estimate of ϵ_D. The utility of gyro based azimuth alignment is limited to situations where the gyros are accurate enough to measure Earth rate and the disturbances are small enough that Earth rate is discernible.

The main advantage of the low pass filter approach is its simplicity and low number of computations. Although some benefits of the Kalman filter approach have already been stated, additional benefits include:

- The initialization process uses essentially the same code (Kalman filter algorithms) as any on-line aiding processes (only the **H** and **R** matrices change). Ultimately, this can reduce the amount of code and debugging time.

- The initialization process results in the correct error covariance matrix required for on-line aiding. If the low pass filter approach is used, the designer must determine a reasonable initial value for this matrix.

11.8 Aiding Measurements

While the INS is in operation, various sensors are available to serve as aiding measurements, for example: GPS, master-slave, optical measurements, and radar. In each case, as outlined in Chapter 7, the procedure is to predict the aiding measurement using the navigation state, form the measurement residual, use the residual to estimate the navigation error state, and use the estimated error state to correct the navigation state. The following subsections present and discuss the prediction and residual equations for a few aiding scenarios.

11.8.1 Position Aiding

In position aiding methods, a measurement which is linearly related to the position is compared with the same linear combination of the INS computed positions to compute a measurement residual. The measurement residual is used to drive a Kalman filter to estimate the INS and sensor error states.

In this approach, the formulation of the measurement residual and the model relating the measurement residual to the error state appears to be straightforward. However, the designer should be very careful. Assume that the position "measurement" is the computed position estimate from a GPS receiver. There are at least two important issues to consider. First, as discussed in Section 8.5, the error in the components of the GPS computed position at a given time instant are correlated with each other. For proper Kalman filter implementation, this correlation should be known and used by the Kalman filter implementation. Unfortunately, this correlation matrix is rarely an output of the GPS receiver. Second, the GPS receiver includes various filters. The code and carrier tracking loops are filters. Also, depending on the receiver settings, the position estimate output by the receiver might itself be the state of a Kalman filter internal to the GPS receiver. Therefore, the error on the GPS estimate of the position may have significant time correlation. Without knowing the internal operation of the GPS receiver, this time correlated position error cannot be accurately modeled for INS aiding.

Often, GPS receivers will have an optional setting which causes the GPS receiver to output the position solutions pointwise. With this setting, the tracking loops still include filters, but the position measurements are computed at each time using only the GPS satellite measurements available at that time (i.e., no Kalman filtering within the GPS receiver). This setting significantly alleviates the second issue discussed in the previous paragraph. However, with this setting, there will be no position measurements available when fewer than four satellites are being tracked. The method of Section 11.8.2 allows GPS aiding of the INS even when fewer than four satellites are available.

11.8.2 GPS Pseudorange Aiding

GPS pseudorange or phase aiding of an INS is very similar to the aiding of the encoder-based dead-reckoning system discussed in Section 9.5. A major issue is the receiver clock error. Either clock error states can be augmented and estimated as discussed in Section 9.5.1 or measurements from different satellites can be differenced as discussed in Section 9.5.2. Because the majority of issues related to these approaches have already been discussed in those sections, the presentation in this section will be brief. Only the satellite differencing approach will be discussed.

11.8.2.1 Measurement Prediction Equations

Starting from the geodetic position estimate $\hat{\mathbf{p}} = [\hat{\phi}, \hat{\lambda}, \hat{h}]^\top$ from the navigation systems, this section presents the equations that would be used to predict the GPS measurements.

Using the estimated vehicle geodetic position $\hat{\mathbf{p}}$, the vehicle ECEF position $\hat{\mathbf{p}}^e$ can be computed using eqns. (2.9–2.11). The ECEF position of the i-th satellite $\hat{p}^i$ can also be computed from the ephemeris data, as discussed in Appendix C. The computed range to the i-th satellite is

$$\hat{R}(\mathbf{p}^e, \hat{\mathbf{p}}^i) = \left\|\hat{\mathbf{p}}^e - \hat{\mathbf{p}}^i\right\|_2. \tag{11.156}$$

If information is available about the common mode GPS errors, these estimates of the common-mode errors could either be added to the computed ranges or subtracted from the range measurements. The following uses the same notation as in Section 9.5.2.

Assuming that m satellites are available and that measurements from the first satellite are subtracted from all the others, the differenced pseudorange and carrier phase measurements can be predicted from the navigation state as

$$\begin{aligned}\nabla\hat{\rho}^i &= \left(\hat{R}(\hat{\mathbf{p}}^e, \hat{\mathbf{p}}^i) + \hat{\chi}^i\right) - \left(\hat{R}(\hat{\mathbf{p}}^e, \hat{\mathbf{p}}^1) + \hat{\chi}^1\right)\\ \lambda\nabla\hat{\phi}^i &= \left(\hat{R}(\hat{\mathbf{p}}^e, \hat{\mathbf{p}}^i) + \hat{\Upsilon}^i\right) - \left(\hat{R}(\hat{\mathbf{p}}^e, \hat{\mathbf{p}}^1) + \hat{\Upsilon}^1\right) + \left(\hat{N}^i - \hat{N}^1\right)\lambda\end{aligned}$$

for $i = 2, \ldots, m$. The same assumptions and comments apply as were discussed following eqns. (9.26-9.27).

11.8.2.2 Residual Measurement Equations

This section presents the GPS measurement residual equations that would be used by the Kalman filter. In this approach, the measurement residuals for the i-th satellite would be computed as

$$\begin{aligned}\delta\rho^i(t) &= \left(\tilde{\rho}^i(t) - \tilde{\rho}^1(t)\right) - \nabla\hat{\rho}^i(t) &(11.157)\\ \delta\phi^i(t) &= \left(\tilde{\phi}^i(t) - \tilde{\phi}^1(t)\right) - \nabla\hat{\phi}^i(t) &(11.158)\end{aligned}$$

which is the measured range difference between satellites i and 1 minus the predicted difference. For $i = 2, \ldots, m$, the error models for the measurement residuals are

$$\begin{aligned}\delta\rho^i(t) &= \bar{\mathbf{h}}^i_e\delta\mathbf{p}^e + \left(\eta^i - \eta^1\right) &(11.159)\\ \delta\phi^i(t) &= \bar{\mathbf{h}}^i_e\delta\mathbf{p}^e + \left(\beta^i - \beta^1\right). &(11.160)\end{aligned}$$

The quantities η^i and β^i are defined for $i = 1, \ldots, m$ in eqns. (8.83–8.91). The vector $\bar{\mathbf{h}}^i_e$ is defined as

$$\bar{\mathbf{h}}^i_e = \left(\vec{\mathbf{h}}^i_e - \vec{\mathbf{h}}^1_e\right)$$

where $\vec{\mathbf{h}}^i_e = \frac{(\hat{\mathbf{p}}^e - \hat{\mathbf{p}}^i)}{\|\hat{\mathbf{p}}^e - \hat{\mathbf{p}}^i\|_2}$ as defined in eqn. (8.12).

Currently, there is a mismatch between eqns. (11.159–11.160) which is a function of $\delta\mathbf{p}^e = [\delta x, \delta y, \delta z]^\top$ and the error state which is a function of $\delta\mathbf{p} = [\delta\phi, \delta\lambda, \delta h]^\top$. These two vectors are related according to

$$\delta\mathbf{p}^e = \mathbf{R}^e_n\mathbf{D}\delta\mathbf{p} \tag{11.161}$$

where

$$\mathbf{D} = \begin{bmatrix} (R_M + h) & 0 & 0 \\ 0 & \cos(\phi)(R_N + h) & 0 \\ 0 & 0 & -1 \end{bmatrix}.$$

Eqn. (11.161) is derived from eqns. (2.9–2.11) using the results from Exercise 2.6. Using eqn. (11.161), eqns. (11.159–11.160) become

$$\begin{aligned}\delta\rho^i(t) &= \bar{\mathbf{h}}^i_e\mathbf{R}^e_n\mathbf{D}\delta\mathbf{p} + \left(\eta^i - \eta^1\right) &(11.162)\\ \delta\phi^i(t) &= \bar{\mathbf{h}}^i_e\mathbf{R}^e_n\mathbf{D}\delta\mathbf{p} + \left(\beta^i - \beta^1\right). &(11.163)\end{aligned}$$

The residual models for use by the Kalman filter are

$$\begin{aligned}\delta\rho^i(t) &= \bar{\mathbf{H}}^i_n\delta\mathbf{x} + \left(\eta^i - \eta^1\right) &(11.164)\\ \delta\phi^i(t) &= \bar{\mathbf{H}}^i_n\delta\mathbf{x} + \left(\beta^i - \beta^1\right) &(11.165)\end{aligned}$$

where $\delta\mathbf{x} = [\delta\mathbf{p}, \delta\mathbf{v}, \boldsymbol{\rho}]^\top$ as defined following eqn. (11.56),

$$\bar{\mathbf{H}}^i_n = [\bar{\mathbf{h}}^i_e \mathbf{R}^e_n \mathbf{D}, \mathbf{0}, \mathbf{0}],$$

and $\mathbf{0}$ is the 3×1 vector.

In an alternative approach, a similarity transform could be used to transform the error state, $\mathbf{F}$ and $\boldsymbol{\Phi}$ matrices so that the error state included $\delta\mathbf{p}^e$ instead of $\delta\mathbf{p}$. Note that the north error δn and east error δe are not the horizontal error in vehicle position in geographic frame; instead, they are the error in the location of the origin of the geographic frame.

11.8.2.3 Lever Arm Compensation

Aided initialization and error estimation methods are complicated by the fact that the aiding system and INS are separated by a lever arm. The lever arm adds complication in at least two forms. First, the lever arm length and alignment will not be perfectly known due to manufacturing tolerances, different possible system configurations, lever arm flex, and vibration. Second, the measurement prediction and residual measurement models will depend on the lever arm offset vector. The first set of effects can only be accommodated by changing the mechanical structure to enhance rigidity, by augmenting calibration states, or by artificially increasing the variance of the measurement noise. Those approaches are not discussed herein. Assuming that the body frame offset vector is known, the lever arm effects should be modeled at the design state to determine their effect on accuracy and observability. Often the lever arm must be accounted for to achieve high accuracy. This section discusses the topic in general. Several specific instance are used in Chapter 12.

Consider the situation shown in Figure 11.7 where $\mathbf{p}_n$ and $\mathbf{p}_a$ denote the position vectors of the navigation system and aiding sensor, respectively. The point O denotes the navigation frame origin. The offset vector $\mathbf{r}$ between the two positions is defined as

$$\mathbf{r} = \mathbf{p}_a - \mathbf{p}_n. \tag{11.166}$$

The calibrated value is denoted $\tilde{\mathbf{r}}^b$ and the error in the calibrated value is $\delta\tilde{\mathbf{r}}^b = \mathbf{r}^b - \tilde{\mathbf{r}}^b$. We assume that both the navigation system and the aiding source are rigidly attached to a mounting structure so that $\dot{\mathbf{r}}^b \approx 0$. This assumption may not be valid if, for example, the navigation system were on a structure (e.g., a wing) that can flex relative to the structure on which an aiding source is mounted (e.g., the main body).The designer must determine the reasonableness of this assumption and the effect of the violation of this assumption on the system performance.

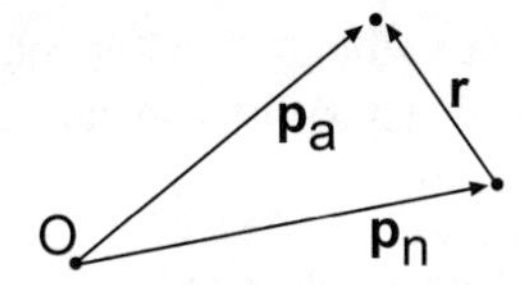

Figure 11.7: Lever Arm Compensation.

Position Aiding. Consider the hypothetical aiding sensor that provides a measurement that is linearly related to the position of the aiding sensor

$$\tilde{y}_a = \mathbf{h}\ \mathbf{p}_a^n + \nu. \tag{11.167}$$

The navigation system predicts this output by combining eqns. (11.166) and (11.167),

$$\hat{y}_a = \mathbf{h}\ \hat{\mathbf{p}}_a^n \tag{11.168}$$

$$= \mathbf{h}\left(\hat{\mathbf{p}}_n^n + \hat{\mathbf{R}}_b^n \tilde{\mathbf{r}}^b\right) \tag{11.169}$$

where $\hat{\mathbf{p}}_n^n$ and $\hat{\mathbf{R}}_b^n$ are directly computable from the state vector of the navigation system, and $\tilde{\mathbf{r}}^b$ is the specified value of the offset vector. Ideally, $\tilde{\mathbf{r}}^b$ would equal the actual value $\mathbf{r}^b$, but error $\delta\mathbf{r}^b = \mathbf{r}^b - \tilde{\mathbf{r}}^b$ between the two may exist for the reasons stated previously.

The residual aiding error, used to drive the Kalman filter, is computed as

$$\delta y_a = \tilde{y}_a - \hat{y}_a. \tag{11.170}$$

This residual error is modeled as

$$\delta y_a = \mathbf{h}\left(\mathbf{p}_n^n - \hat{\mathbf{p}}_n^n + \mathbf{R}_b^n \mathbf{r}^b - \hat{\mathbf{R}}_b^n \tilde{\mathbf{r}}^b\right) + \nu \tag{11.171}$$

$$= \mathbf{h}\left(\delta\mathbf{p}_n^n - [\hat{\mathbf{r}}^n \times]\boldsymbol{\rho}\right) + \hat{\mathbf{R}}_b^n \delta\mathbf{r}^b + \nu \tag{11.172}$$

where $\hat{\mathbf{r}}^n = \hat{\mathbf{R}}_b^n \tilde{\mathbf{r}}^b$. The final equation provides a linear relationship, suitable for Kalman filter processing, between the position measurement residual and the INS error states ($\delta\mathbf{p}_n^n$ and $\boldsymbol{\rho}$). The term containing $\delta\mathbf{r}^b$ represents the error in the measurement due to error in calibration of the lever arm $\delta\mathbf{r}^b$. This effect is time varying due to the matrix $\hat{\mathbf{R}}_b^n$ changing as the vehicle maneuvers and due to possible time variation in $\delta\mathbf{r}^b$ itself.

Velocity Aiding. Assume that an aiding sensor is available that outputs a measurement linearly related to velocity of the aiding system

$$\tilde{y}_a = \mathbf{h}\mathbf{v}_a^n + \nu. \tag{11.173}$$

By the law of Coriolis, for a vehicle rotating relative to navigation frame with angular rate $\boldsymbol{\omega}_{nb}^n$, the velocity of the navigation system and aiding source are related according to

$$\mathbf{v}_a^n = \mathbf{v}_n^n + \boldsymbol{\Omega}_{nb}^n \mathbf{r}^n. \tag{11.174}$$

The navigation system predicts the velocity aiding output according to

$$\begin{aligned} \hat{y}_a &= \mathbf{h}\hat{\mathbf{v}}_a^n & (11.175) \\ &= \mathbf{h}\left(\hat{\mathbf{v}}_n^n + \hat{\boldsymbol{\Omega}}_{nb}^n \hat{\mathbf{R}}_b^n \tilde{\mathbf{r}}^b\right) & (11.176) \end{aligned}$$

where $\hat{\mathbf{v}}_n^n$, $\hat{\boldsymbol{\Omega}}_{nb}^n$, and $\hat{\mathbf{R}}_b^n$ are directly computable from the navigation state. The residual aiding error, used to drive the Kalman filter, is computed as

$$\delta y_a = \tilde{y}_a - \hat{y}_a$$

which has the error model

$$\begin{aligned} \delta y_a &= \mathbf{h}\left(\mathbf{v}_n^n - \hat{\mathbf{v}}_n^n + \boldsymbol{\Omega}_{nb}^n \mathbf{r}^n - \hat{\boldsymbol{\Omega}}_{nb}^n \hat{\mathbf{r}}^n\right) + \nu \\ &= \mathbf{h}\left(\mathbf{v}_n^n - \hat{\mathbf{v}}_n^n + \mathbf{R}_b^n \boldsymbol{\Omega}_{nb}^b \mathbf{r}^b - \hat{\mathbf{R}}_b^n \hat{\boldsymbol{\Omega}}_{nb}^b \tilde{\mathbf{r}}^b\right) + \nu \\ &= \mathbf{h}\left(\delta\mathbf{v}_n^n + \mathbf{P}\hat{\mathbf{R}}_b^n \boldsymbol{\Omega}_{nb}^b \mathbf{r}^b + \hat{\mathbf{R}}_b^n (\delta\boldsymbol{\Omega}_{nb}^b \tilde{\mathbf{r}}^b) + \hat{\mathbf{R}}_b^n \boldsymbol{\Omega}_{nb}^b \delta\mathbf{r}^b\right) + \nu \\ &= \mathbf{h}\left(\delta\mathbf{v}_n^n - [(\boldsymbol{\Omega}_{nb}^n \mathbf{r}^n)\times]\boldsymbol{\rho} - \hat{\mathbf{R}}_b^n ([\tilde{\mathbf{r}}^b \times]\delta\boldsymbol{\omega}_{nb}^b) + \hat{\mathbf{R}}_b^n \boldsymbol{\Omega}_{nb}^b \delta\mathbf{r}^b\right) + \nu. \end{aligned}$$

The factor $\delta\boldsymbol{\omega}_{nb}^b$ could be further expanded. The final equation provides a linear relationship, suitable for Kalman filter processing, between the measurement residual and the INS error state.

Inertial Measurement Aiding. Inertial measurement matching would require another derivative which would introduce terms involving $\dot{\boldsymbol{\Omega}}_{nb}^b$. The fact that measurements of this term are not typically available limits the implementability of inertial measurement matching procedures in situations were the angular rate has significant variation.

This approach is limited to master-slave implementations. Master-slave initialization uses one accurate, previously initialized INS to calibrate a second INS. The second INS may have been recently turned on or of low enough quality that significant drift is expected to have occurred since the last calibration. For example, this scenario is applicable for the initialization of the INS state in weapon applications, where an expensive INS may not be warranted on the weapon due to the short span of time over which accurate post-launch navigation is required and due to the expectation that the INS will be destroyed with the weapon. Master-slave initialization may take a variety of forms. In one-shot alignment (or transfer alignment) the state of the master INS is used to initialize the state of the slave INS at a particular instant of time. Alternatively, the position, velocity, or inertial measurements of the master INS could be differenced with the same quantities of the slave INS. The resulting residual measurements driving a Kalman filter based error estimator [20].

The master INS specific force and angular rate vectors are processed to provide a 'correct' navigation frame version of these vectors at the slave INS location. This statement implies that the processing accounts for lever arm, flex, and non-common mode disturbance effects. This is of course not completely possible. The slave INS processes the corrected master INS vectors denoted $\mathbf{f}^n$ and $\boldsymbol{\omega}^n$ in a manner similar to that described for analytic gyro-compassing.

Inertial measurement matching can theoretically outperform analytic gyro-compassing, since the master-slave system is capable of maneuvering. Due to the maneuvers, the observability properties of the system are enhanced. However, in practice, the lever arm, flex, and disturbance effects can make inertial measurement matching techniques difficult to implement.

11.9 Observability Analysis

Consider the situation where a velocity measurement is used as an aiding signal to an INS. The full velocity measurement might be obtained from GPS, Doppler radar, a master INS, or a known velocity (e.g., a nominally stationary system).

Assuming that the platform position is accurately known during the initialization process, the error dynamics of the reduced state model incorporating lateral velocity and orientation error is

$$\begin{bmatrix} \delta\dot{V}_N \\ \delta\dot{V}_E \\ \dot{\epsilon}_N \\ \dot{\epsilon}_E \\ \dot{\epsilon}_D \end{bmatrix} = \begin{bmatrix} k_D & 2\omega_D & 0 & f_D & -f_E \\ F_{54} & F_{55} & -f_D & 0 & f_N \\ 0 & \frac{-1}{R_e} & 0 & \omega_D & -\omega_E \\ \frac{1}{R_e} & 0 & -\omega_D & 0 & \omega_N \\ 0 & \frac{\tan(\phi)}{R_e} & \omega_E & -\omega_N & 0 \end{bmatrix} \begin{bmatrix} \delta V_N \\ \delta V_E \\ \epsilon_N \\ \epsilon_E \\ \epsilon_D \end{bmatrix} + \begin{bmatrix} b_N \\ b_E \\ d_N \\ d_E \\ d_D \end{bmatrix} + \begin{bmatrix} \zeta_N \\ \zeta_E \\ \eta_N \\ \eta_E \\ \eta_D \end{bmatrix} \tag{11.177}$$

$$\mathbf{y} = \begin{bmatrix} 1 & 0 & 0 & 0 & 0 \\ 0 & 1 & 0 & 0 & 0 \end{bmatrix} \begin{bmatrix} \delta V_N \\ \delta V_E \\ \epsilon_N \\ \epsilon_E \\ \epsilon_D \end{bmatrix} + \boldsymbol{\nu} \tag{11.178}$$

where $\boldsymbol{\nu}$ accounts for measurement noise and all uncorrected lever arm or flexure effects. The instrumentation errors are represented in this model as biases.

The known position assumption is reasonable in several applications. In the master/slave calibration scenario, the offset vector between the two platforms is usually known to within a few meters (the error being due to different stored locations and structure flex). In stationary calibration applications, the calibration location may be known and programmable.

The majority of the following discussion concentrates on a nominally stationary system. The analysis is also valid for low-speed, non-accelerating vehicles. In the more general approaches, i.e., using radar, GPS, master-slave INS or other external sensors for which measurements are available while the vehicle is in motion, the approach is essentially the same. The fact that the **F** matrix is dependent on the velocity, acceleration, and rotation rate will enhance the ability to estimate the system errors (i.e., observability) relative to the performance for a stationary system that is discussed in the following.

11.9.1 Stationary, Level, Known Biases

When the velocity is near zero and the system is nearly level, the error dynamics reduce to

$$\begin{bmatrix} \delta\dot{V}_N \\ \delta\dot{V}_E \\ \dot{\epsilon}_N \\ \dot{\epsilon}_E \\ \dot{\epsilon}_D \end{bmatrix} = \begin{bmatrix} 0 & 2\omega_D & 0 & -g & 0 \\ -2\omega_D & 0 & g & 0 & 0 \\ 0 & \frac{-1}{R_e} & 0 & \omega_D & 0 \\ \frac{1}{R_e} & 0 & -\omega_D & 0 & \omega_N \\ 0 & \frac{\tan(\phi)}{R_e} & 0 & -\omega_N & 0 \end{bmatrix} \begin{bmatrix} \delta V_N \\ \delta V_E \\ \epsilon_N \\ \epsilon_E \\ \epsilon_D \end{bmatrix} + \begin{bmatrix} b_N \\ b_E \\ d_N \\ d_E \\ d_D \end{bmatrix} + \begin{bmatrix} \zeta_N \\ \zeta_E \\ \eta_N \\ \eta_E \\ \eta_D \end{bmatrix}. \tag{11.179}$$

The observability matrix is

$$\begin{bmatrix} \mathbf{H} \\ \mathbf{HF} \\ \mathbf{HF}^2 \end{bmatrix} = \begin{bmatrix} 1 & 0 & 0 & 0 & 0 \\ 0 & 1 & 0 & 0 & 0 \\ 0 & 2\omega_D & 0 & -g & 0 \\ -2\omega_D & 0 & g & 0 & 0 \\ A & 0 & -3g\omega_D & 0 & g\omega_N \\ 0 & A & 0 & -3g\omega_D & 0 \end{bmatrix} \tag{11.180}$$

where $A = (\omega_S^2 - 4\omega_D)$. This observability matrix has rank equal to 5. The five nominal states are observable from the north and east velocity measurements, denoted by $\mathbf{y}(t)$. Therefore, alignment is possible from velocity, in stationary conditions, if the IMU instrument biases are known.

11.9.2 Stationary, Level, Unknown Biases

When the instrument biases are not known, and calibration is required, then the observability analysis problem becomes more interesting. When the five instrument biases are augmented to the error state, the system dynamics are defined by

$$\begin{bmatrix} \delta\dot{\mathbf{x}} \\ \dot{\mathbf{b}} \end{bmatrix} = \begin{bmatrix} \mathbf{F} & \mathbf{I} \\ \mathbf{0} & \mathbf{0} \end{bmatrix} \begin{bmatrix} \delta\mathbf{x} \\ \mathbf{b} \end{bmatrix} + \begin{bmatrix} \zeta_x \\ \zeta_b \end{bmatrix} \tag{11.181}$$

where $\mathbf{F}$ is the dynamic matrix described for the basic five error states in eqn. (11.179). In this analysis, the augmented state is $\mathbf{b} = \mathbf{b}^n = [b_N, b_E, d_N, d_E, d_D]$ which are modeled as random constant plus random walk variables. For a strap-down system $\mathbf{b}^n$ is related to $\mathbf{b}^b$ through $\mathbf{R}_b^n$. The model above assumes that $\mathbf{R}_b^n = \mathbf{I}$, which is equivalent to assuming that the vehicle is level and north pointing. Deviations from this assumption, as long as $\mathbf{R}_b^n$ is not time varying, will change the unobservable subspace, but similar conclusions would apply. Applications in which $\mathbf{R}_b^n$ can be changed have better observability properties.

This ten state system in not completely observable, as can be shown by analysis of the observability matrix [17, 47, 74]. In fact, the observability matrix only has rank 7. Therefore, there are three directions in the ten dimensional augmented state space that are not observable. It is of interest to determine a basis for this unobservable three dimensional subspace.

Using analysis similar to that in Section 3.6.3, it is shown in [17], that a basis for the unobservable subspace $\mathbf{X}_u$ is given by

$$\begin{bmatrix} 0 \\ 0 \\ 0 \\ \frac{1}{\omega_N} \\ 0 \\ \frac{-g}{\omega_N} \\ 0 \\ \frac{-\omega_D}{\omega_N} \\ 0 \\ 1 \end{bmatrix}, \begin{bmatrix} 0 \\ 0 \\ 0 \\ 0 \\ 1 \\ 0 \\ 0 \\ 0 \\ -\omega_N \\ 0 \end{bmatrix}, \begin{bmatrix} 0 \\ 0 \\ \frac{1}{g} \\ 0 \\ \frac{\omega_D \omega_N}{g(1+\omega_N^2)} \\ 0 \\ 0 \\ 0 \\ \frac{\omega_D}{g(1+\omega_N^2)} \\ 0 \end{bmatrix}.$$

This three dimensional unobservable subspace does not correspond identically to specific states. The designer does not have the freedom to choose which states are observable. If the designer selects seven states to be estimated by fixing three variables, this does not address the fundamental issue. The seven selected variables will be estimated, but the estimates will be biased from the true values by the amount appropriate to compensate for the assumed values of the three neglected variables.

11.10 References and Further Reading

This chapter has focused on kinematic model derivations, navigation mechanization, error analysis, initialization, and INS aiding. The main references for this chapter were [27, 33, 37, 50, 73, 95, 110, 111, 116, 126, 129, 138]. The simplifying notation in Table 11.1 is based on that defined in [138]. More examples and more detailed discussion of the trajectories from the linearized inertial navigation system error dynamics is presented in [27, 138]. Gravity is discussed in [37, 67, 73, 127].

Inertial instruments have only been discussed very briefly herein. The various types of inertial instruments, their characteristics, and tradeoffs are discussed in considerable depth in, for example, [73, 83, 126]. The distinction between inertial and kinematic accelerations is discussed well in, for example, [37, 73]. Detail technical descriptions of inertial sensor modeling and design can be found in [73, 126]. The sensor error models presented herein are drawn from [27, 93, 138].

Strap-down INS specific implementation formulas are discussed in [15, 16, 115]. In particular, [115] presents numeric formulas for multi-rate implementations where the highest rate loops transform the specific force measurements to navigation frame accounting for navigation frame rotation. The lower rate loops integrate the navigation frame velocity and position differential equations. The topic of how to complete the mechanization equation computation most efficiently by performing certain subsets of the calculations in specific coordinate frames is addressed in [15, 16].

Chapter 12

LBL and Doppler Aided INS

This chapter describes a navigation system designed for a small autonomous underwater vehicle (AUV) [97, 98]. The mission for the AUV requires maneuvering underneath ships in a harbor environment. The AUV will often maneuver relative to the ship hull which may require the AUV to assume relatively large pitch and roll angles. The navigation system is responsible for maintaining accurate estimates of the AUV position, velocity, attitude, angular rates, and acceleration for use by the planning and control systems. In addition to the sensors used for navigation, the AUV is instrumented with various imaging sensors that are monitored remotely by a human operator. To enable ship hull relative maneuvering, rendezvous, and reacquisition of interesting imaged objects, the desired is to achieve sub-meter position estimation accuracy.

The navigation sensors available on the AUV included an inertial measurement unit (IMU), a Doppler velocity log (DVL), an attitude and heading sensor, a pressure sensor, and a long baseline (LBL) transceiver. The DVL measures velocity along four beam directions via acoustic Doppler measurements. The LBL measures acoustic signal round-trip travel times between a *transceiver* mounted on the AUV and four baseline *transponders* at known locations. More detailed descriptions of the characteristics of each of these sensors are presented in Section 12.2.

Given this sensor suite, two alternative navigation approaches were initially considered. One approach would use the LBL and depth measurements to aid a DVL based dead-reckoning system. However, due to the fact that the DVL requires at least three beams to have bottom-lock to resolve the velocity, this approach would fail in situations when the AUV operated with sufficiently large roll and pitch angles. In addition, the DVL update rate (7 Hz) is slow and the DVL signal is dependent on the acoustic environment; therefore, it may not always be available. The alternative approach that is presented herein uses the DVL, LBL, and depth measurements to aid an inertial solution computed from the IMU measurements. An extended Kalman filter is used to combine the available aiding measurements as they become available. The extended Kalman filter is designed to accommodate the asynchronous and delayed-state measurements that are inherent in this application due to the LBL.

12.1 Kinematics

The purpose of this section is to define the notation that will be used throughout this chapter and to derive the system kinematics.

12.1.1 Notation

The navigation variables used in this chapter are summarized in Table 12.1. As in previous chapters, superscripts will be used to identify the frame of reference in which the vectors are represented. For angular rate vectors, the notation $\boldsymbol{\omega}_{ab}^{c}$ is read as the angular rate of frame b with respect to frame a as represented in frame c. The matrix $\boldsymbol{\Omega}_{ab}^{c} = [\boldsymbol{\omega}_{ab}^{c}\times]$ represents the skew symmetric form of $\boldsymbol{\omega}_{ab}^{c}$, which is defined in Section B.15.

Symbol	Units	Description
$\mathbf{p}$	m	AUV position vector
$\mathbf{v}_e$	$\frac{m}{s}$	Earth relative velocity vector (see p. 389)
$\mathbf{a}$	$\frac{m}{s^2}$	Acceleration vector
$\mathbf{f}$	$\frac{m}{s^2}$	Specific force vector
$\mathbf{g}$	$\frac{m}{s^2}$	Gravity vector
$\boldsymbol{\omega}$	$\frac{rad}{s}$	Angular rate vector
$\boldsymbol{\theta}$	rad	Tangent to platform frame Euler angle three-tuple
$\mathbf{b}_a$	$\frac{m}{s^2}$	Accelerometer bias vector
$\mathbf{b}_g$	$\frac{rad}{s}$	Gyro bias vector
$\phi,\ \theta,\ \psi$	rad	Roll, pitch, and yaw Euler angles
$\bar{\phi}$	rad	Latitude

Table 12.1: Definition of navigation variables for Chapter 12.

12.1.2 System Kinematics

Due to the fact that the AUV will be maneuvering locally in a ship relative coordinate system defined by the LBL system, we select a fixed tangent frame implementation.

The tangent frame position $\mathbf{p}^t$ is the integral of the Earth relative velocity represented in tangent frame:

$$\dot{\mathbf{p}}^t = \mathbf{R}_p^t \mathbf{v}_e^p, \tag{12.1}$$

where the symbol $\mathbf{v}_e^p$ represents the Earth relative velocity in platform frame. By the law of Coriolis applied to $\mathbf{v}_e^p = \mathbf{R}_e^p \mathbf{v}_e^e$, the rate of change $\mathbf{v}_e^p$ is

$$\dot{\mathbf{v}}_e^p = \mathbf{R}_e^p \left(\boldsymbol{\Omega}_{pe}^e \mathbf{v}_e^e + \dot{\mathbf{v}}^e \right). \tag{12.2}$$

Using eqn. (11.32) and the fact that $\boldsymbol{\Omega}_{ep} = \boldsymbol{\Omega}_{ip} - \boldsymbol{\Omega}_{ie}$, eqn. (12.2) simplifies as follows

$$\begin{aligned} \dot{\mathbf{v}}_e^p &= \mathbf{f}^p + \mathbf{g}^p - \left(2\boldsymbol{\Omega}_{ie}^p + \boldsymbol{\Omega}_{ep}^p \right) \mathbf{v}_e^p && (12.3) \\ &= \mathbf{f}^p + \mathbf{g}^p - \left(\boldsymbol{\Omega}_{ie}^p + \boldsymbol{\Omega}_{ip}^p \right) \mathbf{v}_e^p. && (12.4) \end{aligned}$$

Combining the above equations and eqn. (2.74), the system kinematic model is

$$\begin{aligned} \dot{\mathbf{p}}^t &= \mathbf{R}_p^t \mathbf{v}_e^p && (12.5) \\ \dot{\mathbf{v}}_e^p &= \mathbf{f}^p + \mathbf{g}^p - \left(\boldsymbol{\Omega}_{ie}^p + \boldsymbol{\Omega}_{ip}^p \right) \mathbf{v}_e^p && (12.6) \\ \dot{\boldsymbol{\theta}} &= \boldsymbol{\Omega}_E^{-1} \boldsymbol{\omega}_{tp}^p && (12.7) \end{aligned}$$

where $\boldsymbol{\Omega}_E^{-1}$ is defined following eqn. (2.74). The gravity vector $\mathbf{g}^p$ is evaluated according to eqn. (11.14) at the AUV position $\mathbf{p}$. The vector $\boldsymbol{\omega}_{tp}^p$ is the angular rate of the platform frame with respect to tangent frame and satisfies $\boldsymbol{\omega}_{tp}^p = \boldsymbol{\omega}_{ip}^p - \boldsymbol{\omega}_{it}^p$, where $\boldsymbol{\omega}_{ip}^p$ is measured by the body mounted gyros and $\boldsymbol{\omega}_{it}^p = \mathbf{R}_t^p \boldsymbol{\omega}_{it}^t$ with

$$\boldsymbol{\omega}_{it}^t = \omega_{ie} [\cos(\bar{\phi}), 0, -\sin(\bar{\phi})]^\top. \tag{12.8}$$

12.2 Sensors

The following subsections discuss the characteristics, utility, and model for each onboard sensor. The sensor configuration onboard the AUV is illustrated in Figure 12.1. The IMU is at the vehicle center of gravity and defines the origin of the body frame. The DVL, LBL, and pressure sensors are offset from the body frame origin by the vectors $\mathbf{l}_D$, $\mathbf{l}_L$, and $\mathbf{l}_p$.

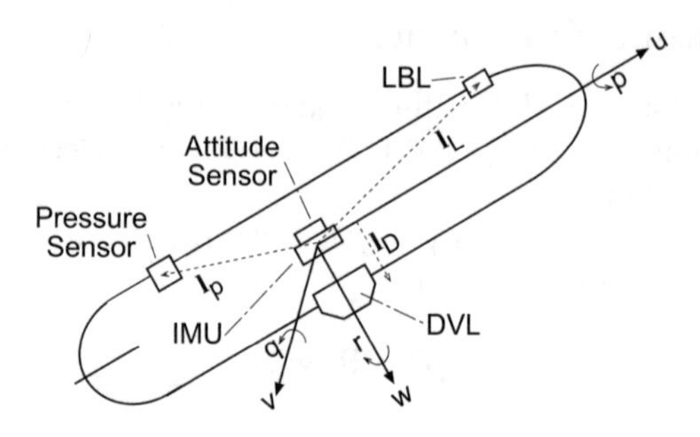

Figure 12.1: AUV sensor configuration depicting the body frame u, v, and w axes; and the sensor offsets $\mathbf{l}_D$, $\mathbf{l}_L$, $\mathbf{l}_p$.

12.2.1 Inertial Measurement Unit

The IMU provides measurements related to the acceleration and angular rates of the AUV. The IMU outputs are reliably available at a known fixed rate. The IMU outputs will be integrated through the kinematic model to provide an estimate of the state of the AUV which includes attitude, velocity, and position. Due to uncertainty in the initial conditions and imperfections in the IMU, the INS estimate of the AUV state is also imperfect. The other onboard sensor signals will be used, in a complementary filter architecture, to correct the INS state estimate.

The IMU outputs are compensated for scale factor, temperature, and non-orthogonality by the manufacturer; therefore, the gyro and accelerometer measurements are modeled as

$$\tilde{\mathbf{u}}_a = \mathbf{f}^p + \mathbf{b}_a + \boldsymbol{\eta}_a \tag{12.9}$$

$$\tilde{\mathbf{u}}_g = \boldsymbol{\omega}_{ip}^p + \mathbf{b}_g + \boldsymbol{\eta}_g \tag{12.10}$$

where $\mathbf{b}_a$ is the accelerometer bias, $\boldsymbol{\eta}_a \sim N(0, \sigma_a^2)$ represents accelerometer measurement noise, $\boldsymbol{\omega}_{ip}^p$ is the angular rate of the gyro relative to the inertial frame represented in platform frame, $\mathbf{b}_g$ represents the gyro bias, $\boldsymbol{\eta}_g \sim N(0, \sigma_g^2)$ represents gyro measurement noise, and $\mathbf{f}^p$ is the specific force vector in platform frame. As discussed in Section 11.1.2,

$$\mathbf{f}^p = \mathbf{R}_i^p \ddot{\mathbf{p}}^i - \mathbf{g}^p \tag{12.11}$$

where $\ddot{\mathbf{p}}^i$ is the acceleration of the accelerometer with respect to the inertial frame and $\mathbf{g}^p$ is the local gravity vector at the AUV location represented in platform frame, see eqn. (11.14).

The bias vectors $\mathbf{b}_a$ and $\mathbf{b}_g$ are modeled as random constants plus random walks:

$$\dot{\mathbf{b}}_a = \boldsymbol{\omega}_a \tag{12.12}$$

$$\dot{\mathbf{b}}_g = \boldsymbol{\omega}_g \tag{12.13}$$

where $\boldsymbol{\omega}_a \sim N(0, \sigma_{b_a}^2 \mathbf{I})$, $\boldsymbol{\omega}_g(0) \sim N(0, \sigma_{b_g}^2 \mathbf{I})$, and the initial bias values are distributed according to $\mathbf{b}_a(0) \sim N(0, \mathbf{P}_{b_a})$ and $\mathbf{b}_g \sim N(0, \mathbf{P}_{b_g})$. The constants σ_{b_a} and σ_{b_g} are both positive. The matrices $\mathbf{P}_{b_a}$ and $\mathbf{P}_{b_g}$ are both symmetric and positive definite.

12.2.2 Attitude and Yaw Sensor

The attitude and yaw sensor uses an inclinometer (gravity) sensor to measure attitude and a magnetometer to measure yaw. The measurements are modeled as

$$\mathbf{y}_E = \begin{bmatrix} \phi \\ \theta \end{bmatrix} + \mathbf{e}_E + \eta_E \tag{12.14}$$

$$y_\psi = \psi + e_\psi + \eta_\psi \tag{12.15}$$

where $\eta_E \sim N(0, \sigma_E^2 \mathbf{I})$ and $\eta_\psi \sim N(0, \sigma_\psi^2)$.

The symbol $\mathbf{e}_E$ represents inclinometer measurement error. The inclinometer is a specific force sensor. As such, it is incapable of distinguishing acceleration from gravity. Therefore, $\mathbf{e}_E$ is a function of the vehicle acceleration and is not stationary. During mission execution, due to the dependence of $\mathbf{e}_E$ on the acceleration, the attitude sensor is not used as an aiding sensor. The inclinometers are useful for initialization of the AUV roll and pitch angles at the start of the mission. During initialization, the AUV thrusters are off. In addition, due to the harbor environment, the currents are slowly time-varying and the waves are typically not large; therefore, the acceleration vector and $\mathbf{e}_E$ are small.

The symbol e_ψ represents non-stationary magnetometer measurement error. For example, the magnetometer will measure the vector sum of the Earth, AUV, ship, and environmental magnetic fields. Only the Earth magnetic field is the desired signal. While the AUV magnetic field can be compensated, the ship and environmental magnetic fields are not predictable and cannot be compensated. During initialization, the AUV is not near the ship. Therefore, the magnetometer is used to initialize the AUV estimate of its yaw angle. The magnetometer is also used as an aiding sensor at the beginning of the mission. When the mission planner commands the AUV to drive toward the ship, the mission planner also turns off compass aiding. This prevents the ship magnetic field from affecting navigation accuracy. The measurement model for magnetometer aiding is presented in Section 12.5.1.

12.2.3 Doppler Velocity Log

The Doppler Velocity Log (DVL) emits encoded pulses from four transducers. The instrument measures the frequency shift of the reflected pulses to determine the relative velocity between the transducer head and the reflecting surface along each beam direction.

Let $\mathbf{b}_i$ for $i = 1, \ldots, 4$ denote a unit vector in the effective direction of the i-th transducer head. These directions are known in the platform frame. The i-th Doppler measurement is

$$y_{D_i} = \left(\mathbf{v}_e^p + \boldsymbol{\omega}_{tp}^p \times \mathbf{l}_D^p\right)^\top \mathbf{b}_i^p + \eta_{D_i} \tag{12.16}$$

where the reflecting surface is assumed to be the stationary sea floor, $\eta_{D_i} \sim N(\mathbf{0}, \sigma_{D_i}^2)$, and $\mathbf{l}_D$ is the offset vector from the platform frame origin to the DVL transducer head.

The DVL will be used as an aiding sensor. The DVL is expected to provide observability for the velocity error and the accelerometer biases. A detailed discussion is presented in Section 12.7. The DVL signal is expected to be available at 7.0 Hz when the roll and pitch angles are small. When the roll and pitch angles are not small, then some DVL beams may not reflect back to the transducer.

12.2.4 Pressure Sensor

For the range of depths at which the AUV is expected to operate the Saunders-Fofonoff relationship [52] between pressure and depth is essentially linear; therefore, the pressure measurement is modeled as

$$y_p = \mathbf{s}^\top \left(\mathbf{p} + \mathbf{R}_p^t \mathbf{l}_p^p\right) + b_p + \eta_p \tag{12.17}$$

where $\mathbf{s}^\top = [0, 0, s_p]$, s_p is a known scale factor, b_p is a known bias, $\mathbf{l}_p$ is the offset from the platform frame origin to the pressure sensor, and $\eta_p \sim N(0, \sigma_p^2)$.

The pressure sensor will be used as an aiding sensor at 10 Hz rate. The pressure sensor is expected to provide observability of the depth error, the vertical velocity error, and a one dimensional subspace of the accelerometer biases. The pressure sensor aiding model is discussed in Section 12.5.3.

12.2.5 Long Baseline Transceiver

The acoustic long baseline (LBL) system precisely measures the time-of-flight of soundwaves through water. These time-of-flight measurements are used to estimate position. The principle of operation is as follows. At time t_0, the AUV, which is located at $\mathbf{p}(t_0)$, initiates the process by broadcasting an interrogation ping to a set of baseline transponders. The baseline transponders are at known locations $\mathbf{p}_i$ for $i = 1, \ldots, 4$. When the i-th baseline transponder receives the interrogation ping, it waits a fixed time interval $T_i = 250i$ milliseconds, and then emits a response ping. The AUV detects the response ping at time t_i. The location of the AUV at time t_i is denoted by $\mathbf{p}(t_i)$. The delay times T_i are unique and large enough so that the AUV can identify the responding baseline transponder by its known delay time. Figure 12.2 illustrates this operation scenario.

Given the above mode of operation, the transceiver onboard the AUV measures the total round trip transit time $y_{L_i} = t_i - t_0$. This i-th measurement is modeled as

$$y_{L_i} = \frac{1}{c(t_0)}\|\mathbf{p}_i - \mathbf{S}(t_0)\| + \frac{1}{c(t_1)}\|\mathbf{S}(t_i) - \mathbf{p}_i\| + T_i + \eta_{L_i} \tag{12.18}$$

where $c(t)$ is the speed of sound in water at time t, $\eta_{L_i} \sim N(0, \sigma_L^2)$, and the AUV transceiver location $\mathbf{S}(t)$ and vehicle location $\mathbf{p}(t)$ are related by

$$\mathbf{S}^t(t) = \mathbf{p}^t(t) + \mathbf{R}_p^t(t)\mathbf{l}_L^p.$$

The vector $\mathbf{l}_L$ denotes the offset from the platform origin to the AUV transceiver location. The speed of sound $c(t)$ is assumed to be constant over the region of interest and to be very slowly time varying. The time variation of $c(t)$ is modeled as

$$c(t) = c_0 + \delta c(t)$$

where

$$\delta\dot{c} = -\lambda_c \delta c + \omega_c \tag{12.19}$$

where c_0 and λ_c are known positive constants and $\omega_c \sim N(0, \sigma_c^2)$.

Delay is inherent in the operation of the LBL system. The measurement model of eqn. (12.18) is a function of the state at two distinct times.

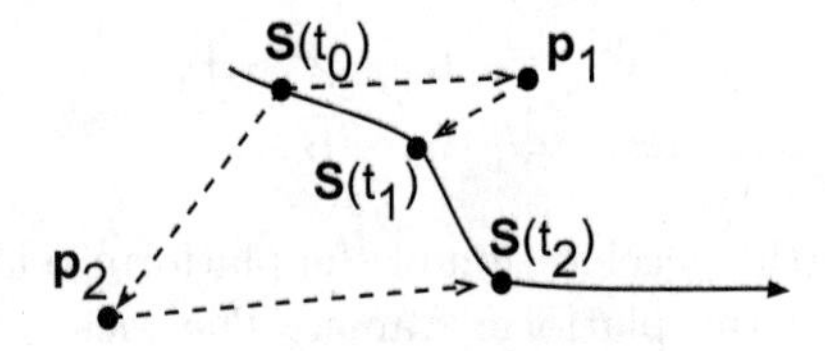

Figure 12.2: Physical setup for the long baseline transponder system. The symbol $\mathbf{S}(t_0)$ denotes the AUV transceiver position at the time at which the interrogation signal is broadcast by the AUV. The symbol $\mathbf{S}(t_i)$ denotes the AUV transceiver position at the time at which the response signal from the i-th baseline transponder is received by the AUV. The symbol $\mathbf{p}_i$ denotes the location of the i-th baseline transponder.

This delay-state estimation will requires special attention in the residual modeling.

The LBL interrogation cycle repeats at 0.5 Hz.

12.3 Mechanization and IMU Processing

The purpose of this section is to describe how the IMU outputs will be processed to provide the inertial navigation estimate of the vehicle state and how that estimate will be used to predict the outputs of the aiding sensors.

12.3.1 Mechanization Equations

Given the kinematic equations summarized in eqns. (12.5 –12.7), the inertial navigation system will propagate the state estimate through time using the equations

$$\dot{\hat{\mathbf{p}}}^t = \hat{\mathbf{R}}_p^t \hat{\mathbf{v}}_e^p \tag{12.20}$$

$$\dot{\hat{\mathbf{v}}}_e^p = \hat{\mathbf{a}}^p - \left(\hat{\mathbf{\Omega}}_{ie}^p + \hat{\mathbf{\Omega}}_{ip}^p\right) \hat{\mathbf{v}}_e^p \tag{12.21}$$

$$\dot{\hat{\boldsymbol{\theta}}} = \hat{\mathbf{\Omega}}_E^{-1} \hat{\boldsymbol{\omega}}_{tp}^p \tag{12.22}$$

where $\hat{\boldsymbol{\omega}}_{tp}^p = \hat{\boldsymbol{\omega}}_{ip}^p - \hat{\boldsymbol{\omega}}_{it}^p$, $\hat{\mathbf{a}}^p$ and $\hat{\boldsymbol{\omega}}_{ip}^p$ are defined in Section 12.3.2, and $\boldsymbol{\omega}_{it}^t = \omega_{ie}[\cos(\hat{\bar{\phi}}), 0, -\sin(\hat{\bar{\phi}})]^\top$ where $\hat{\bar{\phi}}$ represents the computed latitude. The inertial navigation system integrates at the IMU sample rate of 150 Hz.

12.3.2 IMU Processing

Given the accelerometer measurements $\tilde{\mathbf{u}}_a$ and gyro measurements $\tilde{\mathbf{u}}_g$, the platform frame acceleration and angular rate vectors are computed as

$$\hat{\mathbf{a}}^p = \tilde{\mathbf{u}}_a + \hat{\mathbf{g}}^p - \hat{\mathbf{b}}_a \tag{12.23}$$

$$\hat{\boldsymbol{\omega}}_{ip}^p = \tilde{\mathbf{u}}_g - \hat{\mathbf{b}}_g \tag{12.24}$$

where $\mathbf{a}^p = \mathbf{R}_i^p \ddot{\mathbf{p}}^i$ is the acceleration of the platform relative to the inertial frame represented in the platform frame. The local gravity vector $\hat{\mathbf{g}}^p$ is calculated at the estimated AUV location $\hat{\mathbf{p}}$. The estimated bias vectors $\hat{\mathbf{b}}_a$ and $\hat{\mathbf{b}}_g$ are computed by the Kalman filter. They are propagated through time according to the mean of eqns. (12.12–12.13):

$$\dot{\hat{\mathbf{b}}}_a = \mathbf{0} \tag{12.25}$$

$$\dot{\hat{\mathbf{b}}}_g = \mathbf{0}. \tag{12.26}$$

12.4 Error State Dynamic Model

The state of the system is

$$\mathbf{x} = \left[\mathbf{p}^t, \mathbf{v}_e^p, \boldsymbol{\theta}, \mathbf{b}_a, \mathbf{b}_g, c\right]^\top. \tag{12.27}$$

The state of the inertial navigation system is

$$\hat{\mathbf{x}} = \left[\hat{\mathbf{p}}^t, \hat{\mathbf{v}}_e^p, \hat{\boldsymbol{\theta}}, \hat{\mathbf{b}}_a, \hat{\mathbf{b}}_g, \hat{c}\right]^\top. \tag{12.28}$$

The error state vector is defined as

$$\delta\mathbf{x} = \left[\delta\mathbf{p}^\top, \delta\mathbf{v}^\top, \boldsymbol{\rho}^\top, \delta\mathbf{b}_a^\top, \delta\mathbf{b}_g^\top, \delta c\right]^\top. \tag{12.29}$$

Each delta term is defined as the true value minus the computed value (e.g., $\delta\mathbf{p} = \mathbf{p}^t - \hat{\mathbf{p}}^t$).

The attitude error is represented by the quantity $\boldsymbol{\rho}$, which is defined in Section 10.5. The vector $\boldsymbol{\rho}$ represents the tangent plane tilt error. Using the vector $\boldsymbol{\rho}$, the following useful relations between actual and computed rotation matrices can be defined:

$$\hat{\mathbf{R}}_p^t = (\mathbf{I} - [\boldsymbol{\rho}\times])\mathbf{R}_p^t \tag{12.30}$$

$$\hat{\mathbf{R}}_t^p = \mathbf{R}_t^p(\mathbf{I} + [\boldsymbol{\rho}\times]) \tag{12.31}$$

$$\mathbf{R}_t^p = \hat{\mathbf{R}}_t^p(\mathbf{I} - [\boldsymbol{\rho}\times]) \tag{12.32}$$

$$\mathbf{R}_p^t = (\mathbf{I} + [\boldsymbol{\rho}\times])\hat{\mathbf{R}}_p^t. \tag{12.33}$$

Each is accurate to first order.

12.4.1 Position Error Model

The position error dynamic model is defined by subtracting eqn. (12.20) from eqn. (12.5) and linearizing:

$$\begin{aligned}
\delta\dot{\mathbf{p}} &= \dot{\mathbf{p}} - \dot{\hat{\mathbf{p}}} \\
&= \mathbf{R}_p^t\mathbf{v}_e^p - \hat{\mathbf{R}}_p^t\hat{\mathbf{v}}_e^p \\
&= (\mathbf{I} + [\boldsymbol{\rho}\times])\,\hat{\mathbf{R}}_p^t\,(\hat{\mathbf{v}}_e^p + \delta\mathbf{v}) - \hat{\mathbf{R}}_p^t\hat{\mathbf{v}}_e^p \\
&= \hat{\mathbf{R}}_p^t\delta\mathbf{v} - [\hat{\mathbf{v}}_e^t\times]\boldsymbol{\rho}
\end{aligned} \tag{12.34}$$

where the last equality is valid to first order.

12.4.2 Velocity Error Model

The velocity error dynamic model is defined by subtracting eqn. (12.21) from eqn. (12.6) and linearizing:

$$
\begin{aligned}
\delta\dot{\mathbf{v}} &= \dot{\mathbf{v}}_e^p - \dot{\hat{\mathbf{v}}}_e^p \\
&= \mathbf{a}^p - \left(\boldsymbol{\Omega}_{ie}^p + \boldsymbol{\Omega}_{ip}^p\right)\mathbf{v}_e^p \\
&\quad - \left(\mathbf{a}^p - \mathbf{g}^p + \mathbf{b}_a + \boldsymbol{\eta}_a + \hat{\mathbf{g}}^p - \hat{\mathbf{b}}_a\right) + \left(\hat{\boldsymbol{\Omega}}_{ie}^p + \hat{\boldsymbol{\Omega}}_{ip}^p\right)\hat{\mathbf{v}}_e^p \\
&= \left(\hat{\boldsymbol{\Omega}}_{ie}^p + \hat{\boldsymbol{\Omega}}_{ip}^p\right)\hat{\mathbf{v}}_e^p - \left(\boldsymbol{\Omega}_{ie}^p + \boldsymbol{\Omega}_{ip}^p\right)\left(\hat{\mathbf{v}}_e^p + \delta\mathbf{v}\right) \\
&\quad + \hat{\mathbf{R}}_t^p\left(\mathbf{I} - [\boldsymbol{\rho}\times]\right)\mathbf{g}^t - \hat{\mathbf{R}}_t^p\hat{\mathbf{g}}^t - \delta\mathbf{b}_a - \boldsymbol{\eta}_a \\
&= \left(\hat{\boldsymbol{\Omega}}_{ip}^p - \boldsymbol{\Omega}_{ip}^p\right)\hat{\mathbf{v}}_e^p + \left(\hat{\boldsymbol{\Omega}}_{ie}^p - \boldsymbol{\Omega}_{ie}^p\right)\hat{\mathbf{v}}_e^p - \left(\boldsymbol{\Omega}_{ie}^p + \boldsymbol{\Omega}_{ip}^p\right)\delta\mathbf{v} \\
&\quad + \hat{\mathbf{R}}_t^p\left(\mathbf{g}^t - \hat{\mathbf{g}}^t\right) - \hat{\mathbf{R}}_t^p[\boldsymbol{\rho}\times]\mathbf{g}^t - \delta\mathbf{b}_a - \boldsymbol{\eta}_a \\
&= \hat{\mathbf{R}}_t^p\delta\mathbf{g}^t - \left(\boldsymbol{\Omega}_{ie}^p + \boldsymbol{\Omega}_{ip}^p\right)\delta\mathbf{v} - \hat{\mathbf{R}}_t^p[\boldsymbol{\rho}\times]\mathbf{g}^t \\
&\quad + \hat{\mathbf{R}}_t^p\hat{\boldsymbol{\Omega}}_{ie}^t\hat{\mathbf{R}}_p^t\hat{\mathbf{v}}_e^p - \hat{\mathbf{R}}_t^p\left(\mathbf{I} - [\boldsymbol{\rho}\times]\right)\left(\hat{\boldsymbol{\Omega}}_{ie}^t + \delta\boldsymbol{\Omega}_{ie}^t\right)\left(\mathbf{I} + [\boldsymbol{\rho}\times]\right)\hat{\mathbf{R}}_p^t\hat{\mathbf{v}}_e^p \\
&\quad + \left(\hat{\boldsymbol{\Omega}}_{ip}^p - \boldsymbol{\Omega}_{ip}^p\right)\hat{\mathbf{v}}_e^p - \delta\mathbf{b}_a - \boldsymbol{\eta}_a \\
&= \hat{\mathbf{R}}_t^p\left.\frac{\partial\mathbf{g}^t}{\partial\mathbf{p}}\right|_{\hat{\mathbf{p}}}\delta\mathbf{p} - \left(\hat{\boldsymbol{\Omega}}_{ie}^p + \hat{\boldsymbol{\Omega}}_{ip}^p\right)\delta\mathbf{v} + \hat{\mathbf{R}}_t^p[\mathbf{g}^t\times]\boldsymbol{\rho} \\
&\quad - \hat{\mathbf{R}}_t^p\delta\boldsymbol{\Omega}_{ie}^t\hat{\mathbf{v}}_e^t + \hat{\mathbf{R}}_t^p[\boldsymbol{\rho}\times]\hat{\boldsymbol{\Omega}}_{ie}^t\hat{\mathbf{v}}_e^t - \hat{\mathbf{R}}_t^p\hat{\boldsymbol{\Omega}}_{ie}^t[\boldsymbol{\rho}\times]\hat{\mathbf{v}}_e^t \\
&\quad - \delta\mathbf{b}_a - [\hat{\mathbf{v}}_e^p\times]\delta\mathbf{b}_g - \boldsymbol{\eta}_a - [\hat{\mathbf{v}}_e^p\times]\boldsymbol{\eta}_g \qquad (12.35)\\
&= \hat{\mathbf{R}}_t^p\left.\left(\frac{\partial\mathbf{g}^t}{\partial\mathbf{p}} + [\hat{\mathbf{v}}_e^t\times]\frac{\partial\boldsymbol{\omega}_{ie}^t}{\partial\mathbf{p}}\right)\right|_{\hat{\mathbf{p}}}\delta\mathbf{p} - \left(\hat{\boldsymbol{\Omega}}_{ie}^p + \hat{\boldsymbol{\Omega}}_{ip}^p\right)\delta\mathbf{v} \\
&\quad + \hat{\mathbf{R}}_t^p\left([\mathbf{g}^t\times] + \boldsymbol{\omega}_{ie}^t\left(\hat{\mathbf{v}}_e^t\right)^\top - \left(\boldsymbol{\omega}_{ie}^t\right)^\top\hat{\mathbf{v}}_e^t\mathbf{I}\right)\boldsymbol{\rho} \\
&\quad - \delta\mathbf{b}_a - [\hat{\mathbf{v}}_e^p\times]\delta\mathbf{b}_g - \boldsymbol{\eta}_a - [\hat{\mathbf{v}}_e^p\times]\boldsymbol{\eta}_g \qquad (12.36)
\end{aligned}
$$

where second order terms have been dropped. Eqn. (B.19) has been used to simplify the last two terms that appear in the second row of eqn. (12.35),

$$\xi = \boldsymbol{\rho}\times\left(\boldsymbol{\omega}_{ie}^t\times\hat{\mathbf{v}}_e^t\right) - \boldsymbol{\omega}_{ie}^t\times\left(\boldsymbol{\rho}\times\hat{\mathbf{v}}_e^t\right),$$

as follows:

$$
\begin{aligned}
\xi &= \left[\left(\boldsymbol{\rho}^\top\hat{\mathbf{v}}_e^t\right)\boldsymbol{\omega}_{ie}^t - \left(\boldsymbol{\rho}^\top\boldsymbol{\omega}_{ie}^t\right)\hat{\mathbf{v}}_e^t\right] - \left[\left(\left(\boldsymbol{\omega}_{ie}^t\right)^\top\hat{\mathbf{v}}_e^t\right)\boldsymbol{\rho} - \left(\left(\boldsymbol{\omega}_{ie}^t\right)^\top\boldsymbol{\rho}\right)\hat{\mathbf{v}}_e^t\right] \\
&= \left[\left(\left(\hat{\mathbf{v}}_e^t\right)^\top\boldsymbol{\rho}\right)\boldsymbol{\omega}_{ie}^t - \left(\left(\boldsymbol{\omega}_{ie}^t\right)^\top\boldsymbol{\rho}\right)\hat{\mathbf{v}}_e^t\right] \\
&\quad - \left[\left(\left(\boldsymbol{\omega}_{ie}^t\right)^\top\hat{\mathbf{v}}_e^t\right)\boldsymbol{\rho} - \left(\left(\boldsymbol{\omega}_{ie}^t\right)^\top\boldsymbol{\rho}\right)\hat{\mathbf{v}}_e^t\right] \\
&= \boldsymbol{\omega}_{ie}^t\left(\left(\hat{\mathbf{v}}_e^t\right)^\top\boldsymbol{\rho}\right) - \left(\left(\boldsymbol{\omega}_{ie}^t\right)^\top\hat{\mathbf{v}}_e^t\right)\boldsymbol{\rho}
\end{aligned}
$$

$$= \left(\boldsymbol{\omega}_{ie}^{t}\left(\hat{\mathbf{v}}_{e}^{t}\right)^{\top} - \left(\boldsymbol{\omega}_{ie}^{t}\right)^{\top}\hat{\mathbf{v}}_{e}^{t}\mathbf{I}\right)\boldsymbol{\rho}.$$

The simplified expression appears as the last two terms in the second row of eqn. (12.36).

12.4.3 Attitude Error Model

A dynamic model for the attitude error between the body and geographic frames is derived in Sections 10.5.2 and 11.4.2. This section uses an alternative derivation approach [113] to find the dynamic model for the attitude error between the platform and tangent frames.

The derivative of the actual and computed rotation matrices are

$$\dot{\mathbf{R}}_{p}^{t} = \mathbf{R}_{p}^{t}\boldsymbol{\Omega}_{tp}^{p} \quad \text{and} \quad \dot{\hat{\mathbf{R}}}_{p}^{t} = \hat{\mathbf{R}}_{p}^{t}\hat{\boldsymbol{\Omega}}_{tp}^{p}.$$

Therefore, the derivative of $\delta\mathbf{R}_{p}^{t} = \mathbf{R}_{p}^{t} - \hat{\mathbf{R}}_{p}^{t}$ is

$$\begin{aligned}
\delta\dot{\mathbf{R}}_{p}^{t} &= \dot{\mathbf{R}}_{p}^{t} - \dot{\hat{\mathbf{R}}}_{p}^{t} \\
&= \left(\mathbf{I} + [\boldsymbol{\rho}\times]\right)\hat{\mathbf{R}}_{p}^{t}\boldsymbol{\Omega}_{tp}^{p} - \hat{\mathbf{R}}_{p}^{t}\hat{\boldsymbol{\Omega}}_{tp}^{p} \\
\delta\dot{\mathbf{R}}_{p}^{t} &= \hat{\mathbf{R}}_{p}^{t}\left(\boldsymbol{\Omega}_{tp}^{p} - \hat{\boldsymbol{\Omega}}_{tp}^{p}\right) + [\boldsymbol{\rho}\times]\hat{\mathbf{R}}_{p}^{t}\boldsymbol{\Omega}_{tp}^{p}.
\end{aligned} \tag{12.37}$$

Using eqn. (12.33) the error matrix $\delta\mathbf{R}_{p}^{t} = \mathbf{R}_{p}^{t} - \hat{\mathbf{R}}_{p}^{t}$ satisfies

$$\delta\mathbf{R}_{p}^{t} = [\boldsymbol{\rho}\times]\hat{\mathbf{R}}_{p}^{t}. \tag{12.38}$$

A second equations for $\delta\dot{\mathbf{R}}_{p}^{t}$ can be found by differentiation of eqn. (12.38):

$$\delta\dot{\mathbf{R}}_{p}^{t} = [\dot{\boldsymbol{\rho}}\times]\hat{\mathbf{R}}_{p}^{t} + [\boldsymbol{\rho}\times]\hat{\mathbf{R}}_{p}^{t}\hat{\boldsymbol{\Omega}}_{tp}^{p}. \tag{12.39}$$

Combining eqns. (12.37) and (12.39) and solving for $[\dot{\boldsymbol{\rho}}\times]$, we have

$$\begin{aligned}
[\dot{\boldsymbol{\rho}}\times]\hat{\mathbf{R}}_{p}^{t} &= -[\boldsymbol{\rho}\times]\hat{\mathbf{R}}_{p}^{t}\hat{\boldsymbol{\Omega}}_{tp}^{p} + \hat{\mathbf{R}}_{p}^{t}\left(\boldsymbol{\Omega}_{tp}^{p} - \hat{\boldsymbol{\Omega}}_{tp}^{p}\right) + [\boldsymbol{\rho}\times]\hat{\mathbf{R}}_{p}^{t}\boldsymbol{\Omega}_{tp}^{p} \\
[\dot{\boldsymbol{\rho}}\times] &= -[\boldsymbol{\rho}\times]\hat{\boldsymbol{\Omega}}_{tp}^{t} + \hat{\mathbf{R}}_{p}^{t}\left(\boldsymbol{\Omega}_{tp}^{p} - \hat{\boldsymbol{\Omega}}_{tp}^{p}\right)\hat{\mathbf{R}}_{t}^{p} + [\boldsymbol{\rho}\times]\hat{\mathbf{R}}_{p}^{t}\boldsymbol{\Omega}_{tp}^{p}\hat{\mathbf{R}}_{t}^{p} \\
&= \hat{\mathbf{R}}_{p}^{t}\left(\delta\boldsymbol{\Omega}_{ip}^{p} - \delta\boldsymbol{\Omega}_{it}^{p}\right)\hat{\mathbf{R}}_{t}^{p}
\end{aligned} \tag{12.40}$$

where second order terms have been dropped and eqn. (2.23) has been used. In vector form (using eqns. (2.23) and (B.15)), the equivalent equation is

$$\dot{\boldsymbol{\rho}} = \hat{\mathbf{R}}_{p}^{t}\left(\delta\boldsymbol{\omega}_{ip}^{p} - \delta\boldsymbol{\omega}_{it}^{p}\right) \tag{12.41}$$

which is analogous to eqn. (11.71).

The gyro instrumentation error $\delta\boldsymbol{\omega}_{ip}^p$ model is

$$\delta\boldsymbol{\omega}_{ip}^p = -\delta\mathbf{b}_g - \boldsymbol{\eta}_g. \tag{12.42}$$

Eqn. (12.8) defines $\boldsymbol{\omega}_{it}^t$. An expression for $\delta\boldsymbol{\omega}_{it}^p$ can be derived as follows:

$$\begin{aligned}
\boldsymbol{\omega}_{it}^p &= \mathbf{R}_t^p \boldsymbol{\omega}_{it}^t \\
\hat{\boldsymbol{\omega}}_{it}^p + \delta\boldsymbol{\omega}_{it}^p &= \hat{\mathbf{R}}_t^p (\mathbf{I} - [\boldsymbol{\rho}\times]) \left(\hat{\boldsymbol{\omega}}_{it}^t + \delta\boldsymbol{\omega}_{it}^t\right) \\
\delta\boldsymbol{\omega}_{it}^p &= \hat{\mathbf{R}}_t^p \delta\boldsymbol{\omega}_{it}^t + \hat{\mathbf{R}}_t^p \hat{\boldsymbol{\Omega}}_{it}^t \boldsymbol{\rho} \qquad (12.43)
\end{aligned}$$

to first order. By linearization of eqn. (12.8),

$$\delta\boldsymbol{\omega}_{it}^t = -\omega_{ie} \begin{bmatrix} \sin(\bar{\phi}) \\ 0 \\ \cos(\bar{\phi}) \end{bmatrix} \frac{\partial\bar{\phi}}{\partial\mathbf{p}} \delta\mathbf{p}. \tag{12.44}$$

Combining eqns. (12.41), (12.42), and (12.43), we obtain

$$\dot{\boldsymbol{\rho}} = -\hat{\boldsymbol{\Omega}}_{it}^t \boldsymbol{\rho} - \delta\boldsymbol{\omega}_{it}^t - \hat{\mathbf{R}}_p^t \delta\mathbf{b}_g - \hat{\mathbf{R}}_p^t \boldsymbol{\eta}_g. \tag{12.45}$$

12.4.4 Calibration Parameter Error Models

The previous discussion of this chapter has introduced three instrument calibration parameters: the accelerometer bias $\mathbf{b}_a$, the gyro bias $\mathbf{b}_g$, and speed of sound in water c. Each of these is modeled as a random constant plus a random walk; therefore, the dynamic models for the error in these calibration parameters are

$$\begin{aligned}
\delta\dot{\mathbf{b}}_a &= \boldsymbol{\omega}_a \qquad (12.46) \\
\delta\dot{\mathbf{b}}_g &= \boldsymbol{\omega}_g \qquad (12.47) \\
\delta\dot{c} &= -\lambda_c \delta c + \omega_c. \qquad (12.48)
\end{aligned}$$

The characteristics of the driving noise terms $\boldsymbol{\omega}_a$ and $\boldsymbol{\omega}_g$ are defined in Section 12.2.1 while λ_c and $\boldsymbol{\omega}_c$ are discussed in Section 12.2.5.

12.4.5 Error Model Summary

Based on the derivations of the above sections, the dynamic error model is

$$\delta\dot{\mathbf{x}} = \mathbf{F}\delta\mathbf{x} + \mathbf{G}\boldsymbol{\omega} \tag{12.49}$$

where

$$\mathbf{F} = \begin{bmatrix}
\mathbf{0} & \hat{\mathbf{R}}_p^t & -[\hat{\mathbf{v}}_e^t\times] & \mathbf{0} & \mathbf{0} & 0 \\
\mathbf{F}_{vp} & -\left(\hat{\boldsymbol{\Omega}}_{ie}^p + \hat{\boldsymbol{\Omega}}_{ip}^p\right) & \mathbf{F}_{v\rho} & -\mathbf{I} & -[\hat{\mathbf{v}}_e^p\times] & 0 \\
\mathbf{F}_{\rho p} & \mathbf{0} & -\hat{\boldsymbol{\Omega}}_{it}^t & \mathbf{0} & -\hat{\mathbf{R}}_p^t & 0 \\
\mathbf{0} & \mathbf{0} & \mathbf{0} & \mathbf{0} & \mathbf{0} & 0 \\
\mathbf{0} & \mathbf{0} & \mathbf{0} & \mathbf{0} & \mathbf{0} & 0 \\
\mathbf{0} & \mathbf{0} & \mathbf{0} & \mathbf{0} & \mathbf{0} & -\lambda_c
\end{bmatrix}$$

$$\mathbf{G} = \begin{bmatrix} \mathbf{0} & \mathbf{0} & \mathbf{0} & \mathbf{0} & 0 \\ -\mathbf{I} & -[\hat{\mathbf{v}}_e^p \times] & \mathbf{0} & \mathbf{0} & 0 \\ \mathbf{0} & -\hat{\mathbf{R}}_p^t & \mathbf{0} & \mathbf{0} & 0 \\ \mathbf{0} & \mathbf{0} & \mathbf{I} & \mathbf{0} & 0 \\ \mathbf{0} & \mathbf{0} & \mathbf{0} & \mathbf{I} & 0 \\ \mathbf{0} & \mathbf{0} & \mathbf{0} & \mathbf{0} & 1 \end{bmatrix}$$

$$\boldsymbol{\omega} = \begin{bmatrix} \boldsymbol{\eta}_a & \boldsymbol{\eta}_g & \boldsymbol{\omega}_a & \boldsymbol{\omega}_g & \omega_c \end{bmatrix}^\top,$$

where

$$\mathbf{F}_{vp} = \hat{\mathbf{R}}_t^p \left(\frac{\partial \mathbf{g}^t}{\partial \mathbf{p}} + [\hat{\mathbf{v}}_e^t \times] \frac{\partial \boldsymbol{\omega}_{ie}^t}{\partial \mathbf{p}} \right) \Bigg|_{\hat{\mathbf{p}}},$$

$$\mathbf{F}_{v\rho} = \hat{\mathbf{R}}_t^p \left([\mathbf{g}^t \times] + \boldsymbol{\omega}_{ie}^t \left(\hat{\mathbf{v}}_e^t \right)^\top - \left(\boldsymbol{\omega}_{ie}^t \right)^\top \hat{\mathbf{v}}_e^t \mathbf{I} \right),$$

$$\mathbf{F}_{\rho p} = \omega_{ie} \begin{bmatrix} \sin(\bar{\phi}) \\ 0 \\ \cos(\bar{\phi}) \end{bmatrix} \frac{\partial \bar{\phi}}{\partial \mathbf{p}}.$$

The matrix $\mathbf{F}_{vp}$ is small, but the vertical term due to gravity is destabilizing. See Section 11.5.1.1. In this application, the unstable vertical error dynamics are stabilized by the depth sensor aiding. Matrix $\mathbf{F}_{\rho p}$ is derived from eqn. (12.44).

In the subsequent analysis, due to their small size, $\mathbf{F}_{\rho p}$, the last two terms in $\mathbf{F}_{v\rho}$, and $\frac{\partial \boldsymbol{\omega}_{ie}^t}{\partial \mathbf{p}}$ will be approximated as zero. It should also be noted that $\boldsymbol{\omega}_{ie} = \boldsymbol{\omega}_{it}$ because $\boldsymbol{\omega}_{te} = \mathbf{0}$.

12.5 Aiding Measurement Models

Each of the following subsections presents the model used to predict the value of an aiding measurement and derives the measurement model used in the Kalman filter error estimation approach.

12.5.1 Attitude and Yaw Prediction

The attitude sensor is only used for initializing the roll and pitch estimates:

$$\begin{bmatrix} \phi(0) \\ \theta(0) \end{bmatrix} = \mathbf{y}_E(0). \tag{12.50}$$

The yaw angle is initialized based on the output of the magnetometer as

$$\psi(0) = y_\psi(0). \tag{12.51}$$

Given $\boldsymbol{\theta}(0) = [\phi(0), \theta(0), \psi(0)]^\top$, the initial rotation matrix $\mathbf{R}_t^p$ can be computed according to the definition preceding eqn. (2.43). Let $var(\boldsymbol{\theta}(0)) = \mathbf{P}_E = diag([\sigma_E^2, \sigma_E^2, \sigma_\psi^2])$ where σ_E was defined in Section 12.2.2. Then,

$$var(\delta\rho(0)) = \mathbf{P}_\rho = \boldsymbol{\Omega}_T \mathbf{P}_E \boldsymbol{\Omega}_T^\top$$

where $\boldsymbol{\Omega}_T$ is defined in eqn. (2.80).

After initialization is complete, the magnetometer is used as an aiding signal until deactivated by the mission planner. As an aiding signal, the magnetometer output is predicted to be

$$\hat{y}_\psi = \hat{\psi}. \tag{12.52}$$

Because $\delta\boldsymbol{\theta} = \boldsymbol{\Omega}_T^{-1}\delta\boldsymbol{\rho}$ and $\delta\psi$ is the third element of $\delta\boldsymbol{\theta}$, the magnetometer error model is

$$\delta y_\psi = \mathbf{s}_\psi \boldsymbol{\Omega}_T^{-1} \delta\boldsymbol{\rho} + \eta_\psi \tag{12.53}$$

where $\mathbf{s}_\psi = [0, 0, 1]$. Therefore,

$$\mathbf{h}_\psi = \left[\begin{array}{cccccc} \mathbf{0} & \mathbf{0} & \mathbf{s}_\psi \boldsymbol{\Omega}_T^{-1} & \mathbf{0} & \mathbf{0} & 0 \end{array}\right] \tag{12.54}$$

where $\mathbf{0} = [0, 0, 0]$. The measurement variance is

$$R_\psi = \sigma_\psi^2. \tag{12.55}$$

12.5.2 Doppler Prediction

Based on the model of Section 12.2.3, the i-th Doppler measurement is predicted to be

$$\hat{y}_{D_i} = \left(\hat{\mathbf{v}}_e^p + \hat{\boldsymbol{\omega}}_{tp}^p \times \mathbf{l}_D^p\right)^\top \mathbf{b}_i^p. \tag{12.56}$$

Assuming that $\mathbf{l}_D^p$ and $\mathbf{b}_i^p$ are exactly known, the residual Doppler measurement for the i-th beam is

$$\delta y_{D_i} = (\mathbf{b}_i^p)^\top \left(\delta\mathbf{v} + [\mathbf{l}_D^p \times]\delta\mathbf{b}_g + [\mathbf{l}_D^p \times]\eta_g\right) + \eta_{D_i}. \tag{12.57}$$

Therefore,

$$\mathbf{h}_{D_i} = \left[\begin{array}{cccccc} \mathbf{0} & (\mathbf{b}_i^p)^\top & \mathbf{0} & \mathbf{0} & (\mathbf{b}_i^p)^\top [\mathbf{l}_D^p \times] & 0 \end{array}\right] \tag{12.58}$$

and the measurement variance is

$$\begin{aligned} R_{D_i} &= (\mathbf{b}_i^p)^\top [\mathbf{l}_D^p \times] \sigma_g^2 \mathbf{I} [\mathbf{l}_D^p \times]^\top (\mathbf{b}_i^p) + \sigma_{D_i}^2 \\ &= \left(\|\mathbf{l}_D\|^2 \|\mathbf{b}_i\|^2 - |\mathbf{l}_D^\top \mathbf{b}_i|^2\right) \sigma_g^2 + \sigma_{D_i}^2. \end{aligned} \tag{12.59}$$

The variance R_{D_i} is positive. The noise on the measurement δy_{D_i} is correlated to the process noise because the quantity η_g appears in both expressions. Due to the fact that $\left(\|\mathbf{l}_D\|^2 \|\mathbf{b}_i\|^2 - |\mathbf{l}_D^\top \mathbf{b}_i|^2\right) \sigma_g^2$ is significantly smaller than $\sigma_{D_i}^2$, this correlation is ignored in the Kalman filter implementation. Methods for dealing with correlation between the process and measurement noise are discussed in for example, [96].

12.5.3 Depth Prediction

Given the model of the pressure measurement from eqn. (12.17), the pressure measurement is predicted according to

$$\hat{y}_p = \mathbf{s}^\top \left(\hat{\mathbf{p}} + \hat{\mathbf{R}}_p^t \mathbf{l}_p^p \right) + b_p. \tag{12.60}$$

Therefore, the pressure measurement residual is

$$\delta y_p = \mathbf{s}^\top \delta \mathbf{p} - \mathbf{s}^\top [\mathbf{l}_p^t \times] \delta \boldsymbol{\rho} + \eta_p \tag{12.61}$$

where $\mathbf{l}_p^t = \hat{\mathbf{R}}_p^t \mathbf{l}_p^p$. This yields the measurement model

$$\mathbf{h}_p = \left[\begin{array}{cccccc} \mathbf{s}^\top & \mathbf{0} & -\mathbf{s}^\top [\mathbf{l}_p^t \times] & \mathbf{0} & \mathbf{0} & 0 \end{array} \right] \tag{12.62}$$

and the measurement variance is

$$R_p = \sigma_p^2. \tag{12.63}$$

12.5.4 LBL Prediction

Given the model from eqn. (12.18) and the measured value of t_i, the LBL measurement is predicted to be

$$\hat{y}_{L_i} = \frac{1}{\hat{c}(t_0)} \|\mathbf{p}_i - \hat{\mathbf{S}}(t_0)\| + \frac{1}{\hat{c}(t_1)} \|\hat{\mathbf{S}}(t_i) - \mathbf{p}_i\| + T_i \tag{12.64}$$

where the estimated AUV transceiver location $\hat{\mathbf{S}}(t)$ and estimated vehicle location $\hat{\mathbf{p}}(t)$ in tangent frame are related by

$$\hat{\mathbf{S}}^t(t) = \hat{\mathbf{p}}^t(t) + \hat{\mathbf{R}}_p^t(t) \mathbf{l}_L^p.$$

Let

$$d_i(t) = \frac{1}{c(t)} \|\mathbf{p}_i - \mathbf{S}(t)\| \tag{12.65}$$

$$\hat{d}_i(t) = \frac{1}{\hat{c}(t)} \|\mathbf{p}_i - \hat{\mathbf{S}}(t)\| \tag{12.66}$$

where $c(t) = c_0 + \delta c(t)$ and $\hat{c}(t) = c_0 + \delta \hat{c}(t)$. The first order Taylor's series expansion for d_i is

$$d_i(t) = \hat{d}_i(t) + \mathbf{D}_i(t) \delta \mathbf{x}(t)$$

where $\mathbf{D}_i(t) = \frac{\partial d_i}{\partial \mathbf{x}}\Big|_{\mathbf{x}=\hat{\mathbf{x}}(t)}$. With this notation, the i-th LBL residual is

$$\begin{aligned}
\delta y_{L_i} &= y_{L_i} - \hat{y}_{L_i} \\
&= \left(d_i(t_0) + d_i(t_i) + T_i + \eta_{L_i} \right) - \left(\hat{d}_i(t_0) + \hat{d}_i(t_i) + T_i \right) \\
&= \left(d_i(t_0) - \hat{d}_i(t_0) \right) + \left(d_i(t_i) - \hat{d}_i(t_i) \right) + \eta_{L_i} \\
&= \mathbf{D}_i(t_0) \delta \mathbf{x}(t_0) + \mathbf{D}_i(t_i) \delta \mathbf{x}(t_i) + \eta_{L_i} \\
&= \left(\mathbf{D}_i(t_0) \boldsymbol{\Phi}(t_0, t_i) + \mathbf{D}_i(t_i) \right) \delta \mathbf{x}(t_i) + \eta_{L_i}
\end{aligned} \tag{12.67}$$

where $\mathbf{\Phi}(t_0, t_i)$ represents the state transmission matrix from time t_i to time t_0 (i.e., backward in time). Computation of $\mathbf{\Phi}(t_i, t_0)$ is discussed in Section 12.6.3. From eqn. (12.67), the LBL output measurement matrix is

$$\mathbf{h}_{L_i} = [\mathbf{D}_i(t_0)\mathbf{\Phi}(t_0, t_i) + \mathbf{D}_i(t_i)] \tag{12.68}$$

and the measurement variance is

$$\mathbf{R}_{L_i} = \sigma_L^2. \tag{12.69}$$

The row vector $\mathbf{D}_i(t)$ is defined as

$$\mathbf{D}_i(t) = \left[\begin{array}{cccccc} \left.\frac{\partial d_i}{\partial \mathbf{p}}\right|_{\hat{\mathbf{x}}(t)} & \mathbf{0} & \left.\frac{\partial d_i}{\partial \boldsymbol{\rho}}\right|_{\hat{\mathbf{x}}(t)} & \mathbf{0} & \mathbf{0} & \left.\frac{\partial d_i}{\partial c}\right|_{\hat{\mathbf{x}}(t)} \end{array} \right]$$

where

$$\begin{aligned}
\left.\frac{\partial d_i}{\partial \mathbf{p}}\right|_{\hat{\mathbf{x}}(t)} &= -\frac{1}{\hat{c}(t)} \frac{\left(\mathbf{p}_i - \hat{\mathbf{S}}(t)\right)^\top}{\left\|\mathbf{p}_i - \hat{\mathbf{S}}(t)\right\|} \\
\left.\frac{\partial d_i}{\partial \boldsymbol{\rho}}\right|_{\hat{\mathbf{x}}(t)} &= \frac{1}{\hat{c}(t)} \frac{\left(\mathbf{p}_i - \hat{\mathbf{S}}(t)\right)^\top}{\left\|\mathbf{p}_i - \hat{\mathbf{S}}(t)\right\|} [\mathbf{l}_L^t \times] \\
\left.\frac{\partial d_i}{\partial c}\right|_{\hat{\mathbf{x}}(t)} &= -\frac{1}{\hat{c}^2(t)} \left\|\mathbf{p}_i - \hat{\mathbf{S}}(t)\right\|
\end{aligned}$$

and $\mathbf{l}_L^t = \hat{\mathbf{R}}_p^t \mathbf{l}_L^p$.

12.6 EKF Sensor Integration

The KF implementation is slightly more complicated than typical in this application due to the delayed-state issues involved in the LBL sensor.

Consider the time axis as depicted in Figure 12.3. The portion of the time axis $t \in [t_0, t_4]$ is indicated by a wide line. During this time, the LBL system has emitted an interrogation signal and is waiting for the replies from the baseline transponders. While waiting, other sensor measurements may become available. Figure 12.3 indicates availability of a magnetometer measurement at t_ψ, LBL measurements at t_i for $i = 1, \ldots, 4$, and a Doppler measurement at t_D. Figure 12.3 also indicates a pressure measurement arriving at time t_p which is outside the interrogation interval. The figure is significantly simplified because Doppler measurements for each beam arrive at 7 Hz and pressure measurements arrive at 10 Hz; therefore, because $(t_4 - t_0) > 1s$ each LBL interrogation time interval would contain at least 17

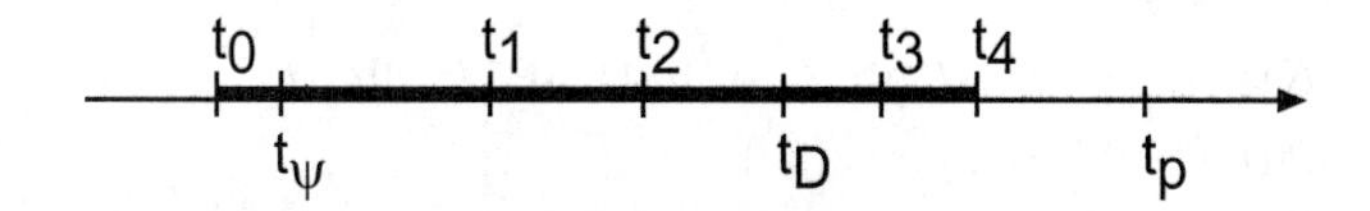

Figure 12.3: Possible sequence of measurements.

additional measurements. Nevertheless, the subsequent discussion relative to Figure 12.3 will clarify the overall measurement processing approach.

At time t_0, when the AUV emits the interrogation signal, a flag is set and the state vector $\mathbf{x}(t_0)$ is saved for prediction of the LBL measurements. While the flag is set, the navigation state is not corrected by the estimate error state vector; instead, the error state is accumulated over the interrogation interval. At time t_4 when either all four of the return signals have been received or have timed out, the flag is reset. While the flag is reset, the navigation state is corrected by the estimated error state vector as soon as it becomes available.

Section 12.6.1 will discuss measurement processing during the interrogation period (i.e., the flag is set). Section 12.6.2 will discuss measurement processing outside the interrogation period (i.e., the flag is reset). Sections 12.6.1 and 12.6.2 will not specifically discuss the time propagation of the error covariance matrix $\mathbf{P}$. Instead, those sections will assume that $\mathbf{P}$ can be computed at any time of interest. Propagation of $\mathbf{P}$ through time will be discussed in Section 12.6.3.

12.6.1 Measurement Updates for $t \in [t_0, t_4]$

For a measurement that occurs during the LBL interrogation interval $t \in [t_0, t_4]$, the estimated state error vector $\delta\hat{\mathbf{x}}$ will be accumulated. Not until time t_4 will the navigation state vector $\hat{\mathbf{x}}$ be corrected by $\delta\hat{\mathbf{x}}$. We will consider the corrections at times $t_\psi, t_1, \ldots, t_4$, and t_D to illustrate the process. During this time interval, the navigation state vector $\hat{\mathbf{x}}(t)$ is available at any time of interest because the INS is integrating eqns. (12.20–12.22) based on the IMU data.

At the beginning of the interrogation interval, it will always be the case that $\delta\hat{\mathbf{x}}^+(t_0)$. The reason that this is true will be made clear at the conclusion of Section 12.6.2.

At time t_ψ, the available information is $\hat{\mathbf{x}}(t_\psi)$, $\delta\hat{\mathbf{x}}^+(t_0) = \mathbf{0}$, and $\mathbf{P}^-(t_\psi)$. The fact that $\delta\hat{\mathbf{x}}^+(t_0) = \mathbf{0}$ implies that

$$\delta\hat{\mathbf{x}}^-(t_\psi) = \mathbf{\Phi}(t_\psi, t_0)\delta\hat{\mathbf{x}}^+(t_0) = \mathbf{0}.$$

Using $\tilde{y}_\psi(t_\psi)$ the correction valid at t_ψ is computed as

$$\begin{aligned}\mathbf{K}(t_\psi) &= \mathbf{P}^-(t_\psi)\mathbf{h}_\psi(t_\psi)\left(\mathbf{h}_\psi(t_\psi)\mathbf{P}^-(t_\psi)\mathbf{h}_\psi(t_\psi)^\top + R_\psi\right)^{-1} \\ \delta\hat{\mathbf{x}}^+(t_\psi) &= \mathbf{K}(t_\psi)\delta y_\psi(t_\psi) \\ \mathbf{P}^+(t_\psi) &= \mathbf{P}^-(t_\psi) - \mathbf{K}(t_\psi)\mathbf{h}_\psi(t_\psi)\mathbf{P}^-(t_\psi).\end{aligned} \tag{12.70}$$

Eqn. (12.70) deserves additional discussion. The standard EKF update equation for this measurement is

$$\delta\hat{\mathbf{x}}^+(t_\psi) = \delta\hat{\mathbf{x}}^-(t_\psi) + \mathbf{K}(t_\psi)\left(\delta y_\psi(t_\psi) - \mathbf{h}_\psi(t_\psi)\delta\hat{\mathbf{x}}^-(t_\psi)\right). \tag{12.71}$$

Eqn. (12.71) reduces to eqn. (12.70) because $\delta\hat{\mathbf{x}}^-(t_\psi) = \mathbf{0}$.

At time t_1, the available information is $\delta\hat{\mathbf{x}}^+(t_\psi)$, $\hat{\mathbf{x}}(t_1)$, and $\mathbf{P}^+(t_\psi)$. Eqn. (12.75) with $\tau_1 = t_\psi$ and $\tau_2 = t_1$ computes $\mathbf{P}^-(t_1)$ and supplies $\mathbf{\Phi}(t_1, t_\psi)$. The state error estimate is propagated as

$$\delta\hat{\mathbf{x}}^-(t_1) = \mathbf{\Phi}(t_1, t_\psi)\delta\hat{\mathbf{x}}^+(t_\psi).$$

Using $\tilde{y}_{L_1}$ the correction valid at t_1 is computed as

$$\begin{aligned}\mathbf{K}(t_1) &= \mathbf{P}^-(t_1)\mathbf{h}_{L_1}(t_1)\left(\mathbf{h}_{L_1}(t_1)\mathbf{P}^-(t_1)\mathbf{h}_{L_1}(t_1)^\top + R_{L_1}\right)^{-1} \\ \delta\hat{\mathbf{x}}^+(t_1) &= \delta\hat{\mathbf{x}}^-(t_1) + \mathbf{K}(t_1)\left(\delta y_{L_1}(t_1) - \mathbf{h}_{L_1}(t_1)\delta\hat{\mathbf{x}}^-(t_1)\right) \\ \mathbf{P}^+(t_1) &= \mathbf{P}^-(t_1) - \mathbf{K}(t_1)\mathbf{h}_{L_1}(t_1)\mathbf{P}^-(t_1).\end{aligned} \tag{12.72}$$

Similar processes are repeated to produce $\delta\hat{\mathbf{x}}^+(t_2)$, $\delta\hat{\mathbf{x}}^+(t_D)$, $\delta\hat{\mathbf{x}}^+(t_3)$, and $\delta\hat{\mathbf{x}}^+(t_4)$.

At time t_4, the flag is reset and the navigation state will be corrected[1]

$$\hat{\mathbf{x}}^+(t) = \hat{\mathbf{x}}^-(t) + \delta\hat{\mathbf{x}}^+(t) \tag{12.73}$$

with $t = t_4$. From the initial condition $\hat{\mathbf{x}}^+(t_4)$, eqns. (12.20–12.22) are integrated over the time interval until another measurement becomes available. For the example of Figure 12.3, the time interval is $t \in [t_4, t_p)$.

Because the navigation state vector $\hat{\mathbf{x}}^+(t_4)$ has been corrected using $\delta\hat{\mathbf{x}}^+(t_4)$, the expected value of the state error vector is now zero; therefore, $\delta\hat{\mathbf{x}}^+(t_4)$ is assigned the value zero:

$$\delta\hat{\mathbf{x}}^+(t_4) \doteq \mathbf{0}.$$

12.6.2 Measurement Updates for $t \notin [t_0, t_4]$

For a measurement that does *not* occur during the LBL interrogation interval, such as t_p in Figure 12.3, the standard EKF time update occurs. Prior

[1]This equation is applied as written for all states except those related to $\boldsymbol{\theta}$ and $\boldsymbol{\rho}$. The use of $\boldsymbol{\rho}$ to correct $\boldsymbol{\theta}$ is discussed in eqn. (10.67).

to the measurement, the available information is $\delta\hat{\mathbf{x}}^-(t_p) = \mathbf{0}$, $\hat{\mathbf{x}}^-(t_p)$, and $\mathbf{P}^-(t_p)$. When $\tilde{y}_p$ is available, the correction is computed as

$$\begin{aligned} \mathbf{K}(t_p) &= \mathbf{P}^-(t_p)\mathbf{h}_p(t_p)/\left(\mathbf{h}_p(t_p)\mathbf{P}^-(t_p)\mathbf{h}_p(t_p)^\top + R_p\right) \\ \delta\hat{\mathbf{x}}^+(t_p) &= \mathbf{K}(t_p)\delta y_p(t_p) \\ \mathbf{P}^+(t_p) &= \mathbf{P}^-(t_p) - \mathbf{K}(t_p)\mathbf{h}_p(t_p)\mathbf{P}^-(t_p) \end{aligned} \tag{12.74}$$

where eqn. (12.74) has been simplified in the same manner as discussed relative to eqn. (12.71).

Given $\delta\hat{\mathbf{x}}^+(t_p)$, the navigation state vector is corrected as described in eqn. (12.73) with $t = t_p$. From the initial condition $\hat{\mathbf{x}}^+(t_p)$, eqns. (12.20–12.22) are integrated until another measurement arrives.

Because the navigation state vector $\hat{\mathbf{x}}^+(t_p)$ has been corrected for the estimated state error $\delta\hat{\mathbf{x}}^+(t_p)$, the expected value of the state error vector is now zero; therefore, $\delta\hat{\mathbf{x}}^+(t_p)$ is assigned the value zero:

$$\delta\hat{\mathbf{x}}^+(t_p) \doteq \mathbf{0}.$$

12.6.3 Covariance Propagation

Given any two times τ_1 and τ_2 with $T = \tau_2 - \tau_1$ positive and small relative to the bandwidth of the vehicle and $\mathbf{P}(\tau_1)$ being known, the error covariance at τ_2 can be computed as

$$\mathbf{P}(\tau_2) = \mathbf{\Phi}(\tau_2, \tau_1)\mathbf{P}(\tau_1)\mathbf{\Phi}^\top(\tau_2, \tau_1) + \mathbf{Qd}(\tau_2, \tau_1). \tag{12.75}$$

In this equation, $\mathbf{\Phi}(\tau_2, \tau_1)$ and $\mathbf{Qd}(\tau_2, \tau_1)$ are computed using the method of Section 4.7.2.1. By this approach, $\mathbf{\Upsilon}$ is computed as

$$\mathbf{\Upsilon} = exp\left(\begin{bmatrix} -\mathbf{F}(t_0) & \mathbf{G}(t_0)\mathbf{Q}\mathbf{G}^\top(t_0) \\ 0 & \mathbf{F}^\top(t_0) \end{bmatrix} T\right) \tag{12.76}$$

for some $t_0 \in [\tau_1, \tau_2]$ where $\mathbf{\Phi}(\tau_2, \tau_1)$ and $\mathbf{Qd}(\tau_2, \tau_1)$ are calculated as

$$\mathbf{\Phi}(\tau_2, \tau_1) = \mathbf{\Upsilon}[(n+1:2n),(n+1):2n]^\top \tag{12.77}$$

$$\mathbf{Qd}(\tau_2, \tau_1) = \mathbf{\Phi}(t_0, t_p)\mathbf{\Upsilon}[(1:n),(n+1):2n] \tag{12.78}$$

where $\mathbf{\Upsilon}[(i:j),(k:l)]$ denotes the the sub-matrix of $\mathbf{\Upsilon}$ composed of the i through j-th rows and k through l-th columns of matrix $\mathbf{\Upsilon}$.

12.7 Observability

This section analyzes the observability of the error state under various operating conditions. The analysis makes two simplifying assumptions.

Sensors	$\mathbf{v}_e^p$ m/s	$\boldsymbol{\omega}_{ip}^p$ rad/s	$rank(\mathcal{O})$, $\mathbf{L} = 0$	$rank(\mathcal{O})$
LBL	$[0, 0, 0]$	$[0, 0, 0]$	9	11
LBL, p	$[0, 0, 0]$	$[0, 0, 0]$	12	12
LBL, p, DVL	$[0, 0, 0]$	$[0, 0, 0]$	12	13
LBL	$[1, 0, 0]$	$[0, 0, 0]$	9	12
LBL, p	$[1, 0, 0]$	$[0, 0, 0]$	12	13
LBL, p, DVL	$[1, 0, 0]$	$[0, 0, 0]$	15	15
LBL	$[1, 0, 0]$	$[0, 0, 0.1]$	9	12
LBL, p	$[1, 0, 0]$	$[0, 0, 0.1]$	12	14
LBL, p, DVL	$[1, 0, 0]$	$[0, 0, 0.1]$	15	15
LBL	$[0, 0, 0]$	$[0, 0, 0.1]$	9	11
LBL, p	$[0, 0, 0]$	$[0, 0, 0.1]$	12	12
LBL, p, DVL	$[0, 0, 0]$	$[0, 0, 0.1]$	12	13

Table 12.2: Rank of the observability matrix for various sensor combinations and operating conditions. The column marked with $\mathbf{L} = \mathbf{0}$ indicates that all sensor offset vectors have been set to zero. For full observability, the rank must be 16.

First, the stated operating conditions for each scenario are assumed to be static. Second, each set of measurements that is considered is assumed to occur simultaneously. This second assumption holds in the sense that a vector Kalman filter correction is assumed. For the model of the LBL corrections, the computation of the $\mathbf{h}_L$ matrix still accounts for different travel times and baseline transponder delay times T_i.

The structure of the $\mathbf{F}$ matrix shows that the observability may depend on the tangent plane velocity vector, the body frame velocity vector, and angular rate vector. The observability will also depend on the set of sensors that is considered.

The structure of the measurement matrix $\mathbf{h}_{L_i}$ indicates that the observability from the LBL sensor may also depend on the position of the AUV relative to the baseline transponders. In fact, given that the baseline transponders are setup on the corners of a rectangle, the placement of the AUV relative to the baseline transponders does not affect the rank of the observability matrix, although it does affect the structure of the unobservable subspace.

For various operating conditions and aiding sensor combinations, the rank of the observability matrix is summarized in Table 12.2. The first column shows which sensor combination is being used. The second and third columns show the operating condition. The fourth column shows the rank of the observability matrix $\mathcal{O}$ with all the offsets $\mathbf{l}_L$, $\mathbf{l}_p$, and $\mathbf{l}_D$ set to zero, for the stated sensor combination and operating condition. The fifth column shows the rank of the observability matrix $\mathcal{O}$ with the offsets $\mathbf{l}_L = [1.000; 0.000; -0.200]^\top$, $\mathbf{l}_p = [-0.457; 0.000; -0.070]^\top$, and $\mathbf{l}_D = [0.000; 0.000; 0.159]^\top$, for the stated sensor combination and operating condition.

From the table it is clear that the nonzero offset vectors increase the rank of the observability matrix. In fact, the situation where all the offset vectors are zero is not physically implementable, because the sensors cannot all be at the center of gravity of the vehicle. However, the structure of the unobservable subspace is easier to discuss with the offset vectors set to zero, which is the case for the discussion through the remainder of this section.

When $\mathbf{v}_e^p = [1, 0, 0]^\top$ and $\boldsymbol{\omega}_{ip}^p = \mathbf{0}$, and LBL, DVL, and pressure aiding are used, the basis vector for the unobservable space is $a\mathbf{V}_2$ where

$$\mathbf{V}_2 = \left[\begin{array}{ccc|ccc|ccc|ccc|ccc|c} 0 & 0 & 0 & 0 & 0 & 0 & \frac{1}{g} & 0 & 0 & 0 & 1 & 0 & 0 & 0 & 0 & 0 \end{array}\right].$$

and g represents the magnitude of the gravity vector. The vertical lines have been added to clearly indicate the subvectors of $\mathbf{V}_2$. This subspace indicates that for the stated sensor combination and operating condition, the system cannot discriminate between one milli-radian of *roll* error and one milli-g of lateral accelerometer bias error. This same unobservable space applies to the other two scenarios indicated in Table 12.2 to have a 15 dimensional observable subspace. Each of the scenarios indicated in Table 12.2 operates with $\psi = 0$.

When $\psi = \frac{\pi}{2}$, the tangent frame velocity changes. The rank of the observability matrix remains the same in each of the evaluated scenarios, but the unobservable subspace changes to $a\mathbf{V}_2$ where

$$\mathbf{V}_2 = \left[\begin{array}{ccc|ccc|ccc|ccc|ccc|c} 0 & 0 & 0 & 0 & 0 & 0 & 0 & \frac{1}{g} & 0 & 0 & 1 & 0 & 0 & 0 & 0 & 0 \end{array}\right].$$

This subspace indicates that for the stated sensor combination and operating condition, the system cannot discriminate between one milli-radian of *pitch* error and one milli-g of lateral accelerometer bias error. This inability to distinguish pitch error from lateral accelerometer bias appears counterintuitive, but is due to the rotation of the body frame velocity into the tangent frame to compute the position, as shown in eqn. (12.34).

Even though the lateral acceleration bias is in both subspace, all three variables (roll error, pitch error, and lateral acceleration bias) can be observed (i.e., estimated) over time intervals when the yaw angle is not constant. This statement is demonstrated in the covariance analysis contained in Figure 12.5.

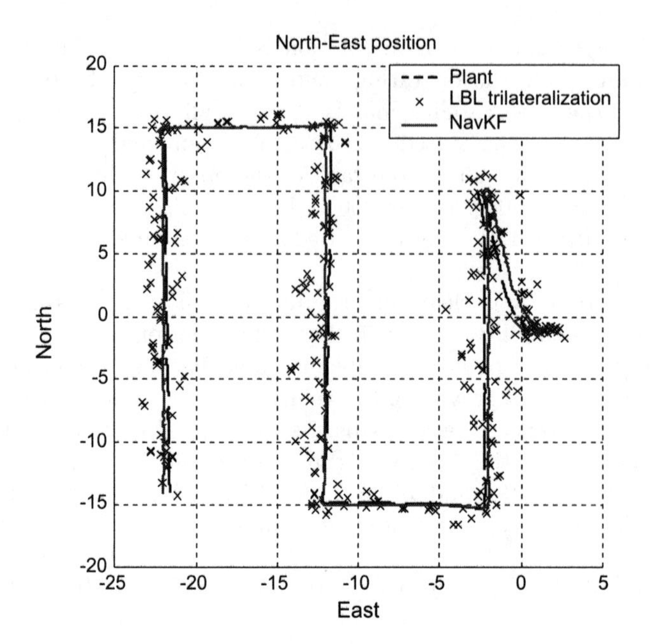

Figure 12.4: AUV trajectory (dashed), estimated trajectory (solid), and LBL-only position estimates.

12.8 Simulation Performance

This section considers the system performance in simulation. The simulation analysis is useful as both the actual and estimated states are known. Therefore, the estimation error can be directly observed.

Figure 12.4 shows the vehicle position as a function of time in the tangent plane (dashed line), the estimated vehicle position (solid line), and the LBL computed positions (x's). The sharp corners along the trajectory are achieved by commanding the forward speed u to zero at each corner while the yaw angle ψ changes by 90°. Detailed analysis of data not included herein shows the following. After approximately one minute of operation, the north and east position error standard deviations as predicted by the $\mathbf{P}$ matrix remain less that 0.15 m while the depth error standard deviation remains less that 0.01 m. The peak errors in the position errors, as computed by subtracting the estimated position from the true position, are on the order of $[0.2, 0.4, 0.01]^\top m$. Further analysis of the simulation data yields the performance indicators shown in Table 12.3.

Figure 12.5 shows the accelerometer bias $\mathbf{b}_a$ (solid line) and its estimate $\hat{\mathbf{b}}_a$ (dots) during the first 20s of the simulation. Also shown are the $\pm\sigma$, $\pm 2\sigma$, and $\pm 3\sigma$ curves, where σ represents the square root of the

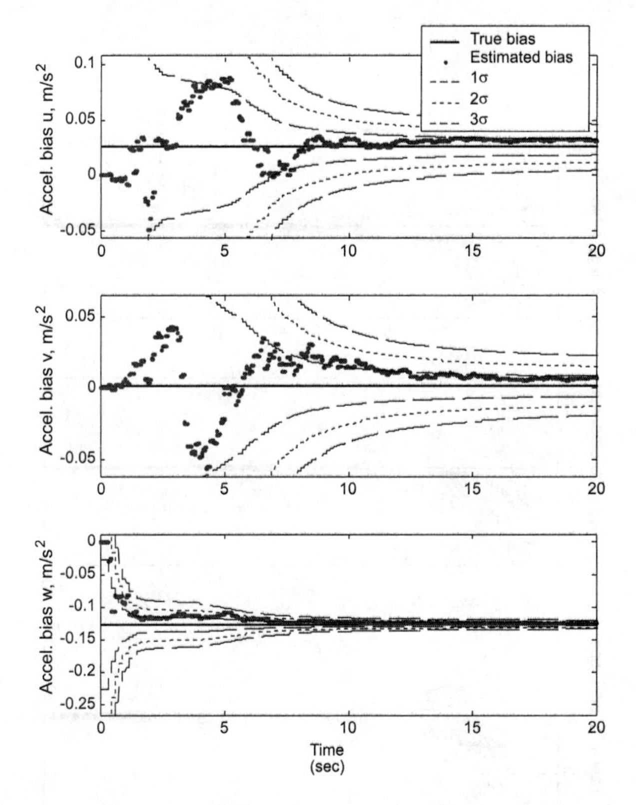

Figure 12.5: Accelerometer bias plots, b_a. Actual bias (solid); estimated bias (dotted); and $\pm\sigma$, $\pm 2\sigma$, and $\pm 3\sigma$ plots.

appropriate diagonal element of the **P** matrix. At the end of the simulation ($t = 265s$), the vector of accelerometer bias standard deviations is $[0.002, 0.002, 0.001]^\top m/s^2$. The standard deviation curves in Figure 12.5 are a covariance analysis demonstrating that the observability conclusions of Section 12.7 are valid.

Figure 12.6 shows the gyro bias $\mathbf{b}_g$ (solid line) and its estimate $\hat{\mathbf{b}}_g$ (dots) during the first $20s$ of the simulation. Also shown are the $\pm\sigma$, $\pm 2\sigma$, and $\pm 3\sigma$ curves, where σ represents the square root of the appropriate diagonal element of the **P** matrix. At the end of the simulation ($t = 265s$), the vector of gyro bias standard deviations is $[0.7, 0.7, 1.5]^\top \times 10^{-3} deg/s$.

The navigation system is operational on an AUV being used in research related to ship hull inspection. For the in-water data, the actual vehicle state is not known; therefore, the main data useful for analyzing the performance are the measurement residuals. In-water data is presented and analyzed in [97, 98].

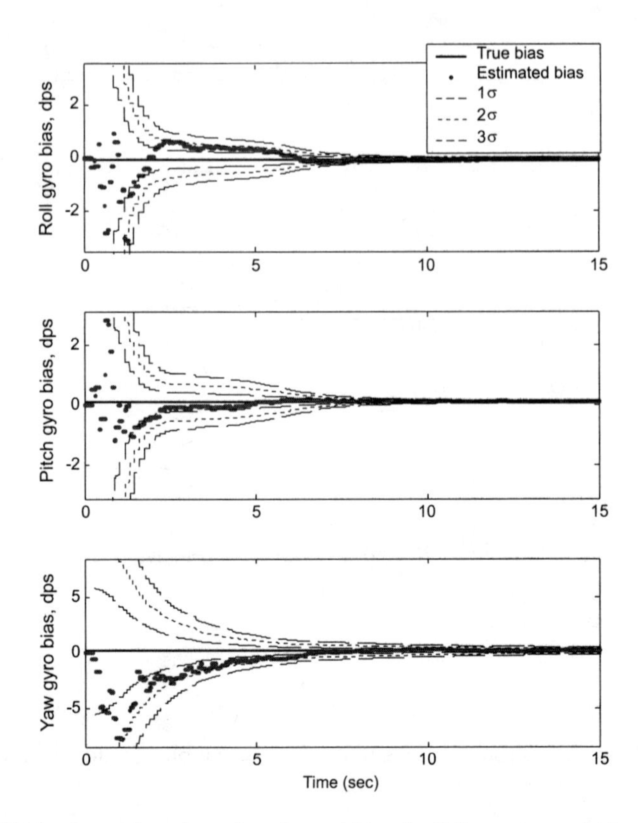

Figure 12.6: Gyro bias plots, b_g. Actual bias (solid); estimated bias (dotted); and $\pm\sigma$, $\pm 2\sigma$, and $\pm 3\sigma$ plots.

	u, $10^{-2}\frac{m}{s}$	v, $10^{-3}\frac{m}{s}$	w, $10^{-2}\frac{m}{s}$	ρ_N	ρ_E	ρ_D
$\sqrt{P}$	1.2	1.2	0.7	0.02°	0.02°	0.22°
Peak	3.0	3.0	1.0	0.01°	0.10°	0.25°

Table 12.3: Velocity and attitude metrics from the simulation of Section 12.8.

Appendix A

Notation

This appendix summarizes the notation and constants used throughout this book. It is provided for the convenience of the reader as a reference.

A.1 Notation

Table A.1 summarizes the notational conventions that are used throughout this book. A non-bold face symbol denotes a scalar quantity. A bold face symbol denotes either a vector (typically lower case) or a matrix (typically upper case). In many parts of this book, it is important to make the distinction between a *true* value, a *calculated* or *estimated* value, or a *measured* value. As shown in Table A.1, the true value has no additional mark; the calculated value has a 'hat' on it; and, the measured value has a 'tilde' above it. Further, the error between the true value and the estimated value can be defined in two different ways. In this book, the error is defined as the true value minus the estimated value. The error quantity is indicated with a δ, for example $\delta \mathbf{x} = \mathbf{x} - \hat{\mathbf{x}}$.

x	non-bold face variables denote scalars
$\mathbf{x}$	boldface denotes vector or matrix quantities
$\mathbf{x}$	denotes the true value of $\mathbf{x}$
$\hat{\mathbf{x}}$	denotes the calculated value of $\mathbf{x}$
$\tilde{\mathbf{x}}$	denotes the measured value of $\mathbf{x}$
$\delta \mathbf{x}$	denotes the error $\mathbf{x} - \hat{\mathbf{x}}$
$\mathbf{R}_a^b$	denotes the transformation matrix from reference frames a to b
$\mathbf{x}^a$	denotes vector $\mathbf{x}$ represented with respect to frame a

Table A.1: Notational conventions.

Chapter 2 introduces several different reference frames. Vector quantities are transformed between these different frames using rotational transformations which can be represented as matrices. The notation $\mathbf{R}_a^b$ denotes the transformation matrix from reference frame a to reference frame b. When a vector is being "coordinatized" in frame a, the notation $\mathbf{x}^a$ is used. All of the symbols representing the different reference frames are listed in Table A.2. As an example, using the notation defined in the Tables A.1 and A.2, the actual rotational matrix to transform a vector from body coordinates to tangent plane coordinates would be denoted $\mathbf{R}_b^t$. The calculate rotational transformation for a vector from body coordinates to tangent plane coordinates would be denoted $\hat{\mathbf{R}}_b^t$.

a	accelerometer frame (non-orthogonal)
b	body frame
e	ECEF frame
g	gyro frame (non-orthogonal) or geographic
i	inertial frame
n	navigation frame (typically geographic)
p	platform frame
t	tangent plane frame

Table A.2: Reference frame symbol definitions.

The text is concerned with estimating the state of a vehicle which evolves as a function of the continuous-time variable t. The estimation routines are implemented in discrete-time on digital computers. In addition, the estimation algorithm incorporates information from sensors that provide readings at discrete-time instants. Therefore, the estimates will evolve on the computer using a discrete-time increment T. The discrete-time increments are denoted $t_k = kT$ for positive, integer values of k. Therefore, $\mathbf{x}(t_k) = \mathbf{x}(t)|_{t=kT}$. Often, as a further shorthand, the notation $\mathbf{x}_k$ will be used instead of $\mathbf{x}(t_k)$. When the measurements do not occur with a fixed sample period T (i.e., asynchronous measurements), the symbol $\mathbf{y}_k$ simply means the k-th measurement which will have occurred at time t_k.

When considering state estimation, we are often concerned with specifying the state variables either just before or after a measurement that occurred at time t_k. We denote the value of the variable just prior to the measurement as $\mathbf{x}_k^-$ and the value of the variable just after the measurement as $\mathbf{x}_k^+$.

A.2 Useful Constants

Table A.3 presents various constants useful in navigation system design and analysis. The precision on several of the presented variables is critical to system performance. For example, the precision presented for π is that specified for the GPS system in the interface communication document [6]. Due to the radius of the satellite orbits, satellite position calculations are sensitive to this value.

Symbol	Value	Units	Description
π	3.1415926535898		pi
ω_{ie}	$7.2921151467 \times 10^{-5}$	rad/s	Earth rotation rate
μ	3.986005×10^{14}	$\frac{m^3}{s^2}$	Univ. grav. constant
c	2.99792458×10^8	$\frac{m}{s}$	Speed of light
F	$-4.442807633 \times 10^{-10}$	$\frac{s}{\sqrt{m}}$	
a	6378137.0	m	Earth semi-major axis
b	6356752.3	m	Earth semi-minor axis
f_1	1575.42	M Hz	L1 carrier frequency
f_2	1227.60	M Hz	L2 carrier frequency
λ_1	$\frac{c}{f_1} \approx 19.0$	cm	L1 wavelength
λ_2	$\frac{c}{f_2} \approx 24.4$	cm	L2 wavelength
λ_w	$\frac{c}{f_1 - f_2} \approx 86.2$	cm	wide lane wavelength
λ_n	$\frac{c}{f_1 + f_2} \approx 10.7$	cm	narrow lane wavelength

Table A.3: Important constants related to GPS and the WGS-84 world geodetic system.

A.3 Acronyms

The acronyms used in the text should be defined at their first usage which should be listed in the index.

A.4 Other Notation

The text uses two different inverse tangent functions: $arctan2(y, x)$ and $arctan(y, x) = arctan\left(\frac{y}{x}\right)$. The function $arctan(y, x)$ is a two-quadrant inverse tangent that returns the value of $\theta \in [-\frac{\pi}{2}, \frac{\pi}{2}]$ such that $\tan(\theta) = \frac{y}{x}$. The function $arctan2(y, x)$ is a four-quadrant inverse tangent that returns the value of $\theta \in (-\pi, \pi]$ such that θ is the angle of the vector $[x, y]^\top$ with respect to the positive x-axis and $\tan(\theta) = \frac{y}{x}$.

At a few locations in the book, where implementations are discussed, the symbol $\doteq$ will be used instead of $=$. Using this notation, the equation $x \doteq 3$ should be read as "x is assigned the value 3."

Appendix B

Linear Algebra Review

This appendix serves two purposes. First, it reviews basic linear algebra concepts. Second, it organizes in one place, for convenient reference, various results from linear algebra that will be used at various places in the main body of the text. Throughout this appendix, unless otherwise stated, the symbols i, j, and k are integers. The notation $i \in [1, n]$ is equivalent to $i = 1, \ldots, n$.

B.1 Definitions

A *matrix* $\mathbf{A} \in \mathbb{R}^{n \times m}$ is a tabular arrangement of real numbers with n rows and m columns. An example for $n = 2$ and $m = 3$ is

$$\mathbf{A} = \begin{bmatrix} 0.20 & 0.00 & -3.14 \\ -1.71 & 0.00 & 2.00 \end{bmatrix}.$$

Sometimes it is necessary to name the elements of a matrix individually. In this case, we state that $\mathbf{A} = [a_{i,j}]$ for $i = [1, n]$ and $j = [1, m]$. For example in the above example, $a_{2,3} = 2.00$. For matrices $\mathbf{A}$ and $\mathbf{B} \in \mathbb{R}^{n \times m}$, the notation $\mathbf{A} = \mathbf{B}$ means that $a_{i,j} = b_{i,j}$ for all $i \in [1, n]$ and $j \in [1, m]$. The *null matrix* $\mathbf{Z}_{n \times m} = \mathbf{0} \in \mathbb{R}^{n \times m}$ has $z_{i,j} = 0$ for all $i \in [1, n]$ and $j \in [1, m]$. The *identity matrix* $\mathbf{A} = \mathbf{I}_n \in \mathbb{R}^{n \times n}$ has

$$a_{i,j} = \begin{cases} 1 & \text{for } i = j \\ 0 & \text{otherwise.} \end{cases}$$

For example, $\mathbf{I}_2 = \begin{bmatrix} 1 & 0 \\ 0 & 1 \end{bmatrix}$. Normally the subscripts on $\mathbf{I}$, $\mathbf{Z}$, and $\mathbf{0}$ will be dropped, because the dimensions can be determined directly based on the context.

When a matrix has the same number of rows and columns (i.e., $n = m$), the matrix is said to be *square*. The elements $a_{i,i}$ for are referred to as the diagonal elements of $\mathbf{A}$ (i.e., $diag(\mathbf{A}) = [a_{11}, \ldots, a_{n,n}]^\top$. A square matrix $\mathbf{A}$ is called *diagonal* when $a_{ij} = 0$ for $i \neq j$. The notation $\mathbf{A} = diag([a_1, \ldots, a_n])$ will be used to state that $\mathbf{A}$ is a diagonal $n \times n$ matrix with main diagonal elements specified by the vector $[a_1, \ldots, a_n]$. The identity matrix is an example of a diagonal matrix. A square matrix $\mathbf{A}$ is *symmetric* if and only if it is square and $a_{ij} = a_{ji}$ for all $i, j \in [1, n]$. A square matrix $\mathbf{A}$ is *skew-symmetric* if and only if $a_{ij} = -a_{ji}$ for all $i, j \in [1, n]$. This implies that a skew-symmetric matrix has diagonal elements all equal to zero. Skew symmetric matrices are useful for representing the cross product as a matrix operation, see eqn. (B.14).

There are various special classes of matrices. One type of matrix is a permutation matrix. A matrix $\mathbf{P}$ is a permutation matrix when each row and column of $\mathbf{P}$ contains exactly one 1 and all other elements of $\mathbf{P}$ are 0.

The term *vector* can have both physical and algebraic meanings. For example, the line segment from point $\mathbf{P}_1$ to point $\mathbf{P}_2$ can be represented as the vector $\mathbf{d} = \mathbf{P}_2 - \mathbf{P}_1$. Note that this vector is well-defined without any discussion of reference frames. Also, the velocity vector $\mathbf{v}$ of a point has well-defined physical interpretation. The velocity vector points in the direction of travel and has magnitude equal to the speed of the point. If a frame-of-reference is specified, then either $\mathbf{d}$ or $\mathbf{v}$ can be represented algebraically as a special matrix that only has one column. When represented with respect to a specific reference frame, the i-th element of a vector $\mathbf{v}$ is the projection of $\mathbf{v}$ onto the i-th coordinate axis of the reference frame.

Alternatively, for convenience, data or variables are sometimes organized into a column matrix (or vector) without the vector having a clear physical interpretation with respect to any single frame of reference. An example of this is the vector of Euler angles and the vector of Euler angle derivatives discussed in Section 2.7.2[1].

When we want to refer to the component in the i-th row of the vector $\mathbf{v}$, we use the notation v_i instead of $v_{i,1}$. Throughout the majority of this text all vectors will be represented as column vectors. The only time that

[1]The following are formal definitions for scalars and vectors [90]:

Scalars are quantities that are invariant under coordinate transformation (e.g., mass);

Vectors are sets of quantities which if organized as $\mathbf{v} = [v_1, v_2, v_3]^\top$ transform between coordinate frames according to $\mathbf{v}^b = \mathbf{R}_a^b \mathbf{v}^a$ where $\mathbf{R}_a^b$ is a rotation matrix and the superscript on $\mathbf{v}$ indicates the frame in which it is represented.

This strict definition of vectors is required to distinguish which items arranged as $[a, b, c]^\top$ can be transformed between coordinate systems via vector transformation operations (i.e., rotations). Some quantities although convenient to organize in a vector-like notation, are not vectors in this strict sense. In this strict sense, Euler angle 3-tuple (ϕ, θ, ψ) and its derivative $\left(\frac{d\phi}{dt}, \frac{d\theta}{dt}, \frac{d\psi}{dt}\right)$ are not vectors. Note for example that each element of the Euler angle derivative 3-tuple is defined in a different reference frame.

this rule will be violated is when, in the middle of text, considerable space would be wasted by writing out the column vector.

B.2 Matrix and Vector Operations

The *transpose* operation, which is indicated by a superscript $\top$, simply exchanges the rows and columns of a matrix. Therefore, for $\mathbf{B} = [b_{i,j}] \in \mathbb{R}^{n \times m}$, $\mathbf{B}^\top = [b_{j,i}] \in \mathbb{R}^{m \times n}$. Two examples may clarify this operation. First,

$$\mathbf{A}^\top = \begin{bmatrix} 0.20 & -1.71 \\ 0.00 & 0.00 \\ -3.14 & 2.00 \end{bmatrix}$$

is a matrix in $\mathbb{R}^{3 \times 2}$. Second, for $\mathbf{v} = \begin{bmatrix} 0.00 \\ 2.00 \\ -7.14 \end{bmatrix}$,

$$\mathbf{v}^\top = \begin{bmatrix} 0.00 & 2.00 & -7.14 \end{bmatrix}$$

is a matrix in $\mathbb{R}^{1 \times 3}$ which is often referred to as a *row vector*. When a matrix is symmetric, it is straightforward to show that $\mathbf{A} = \mathbf{A}^\top$. When a matrix is skew-symmetric, $\mathbf{A}^\top = -\mathbf{A}$.

The *addition of matrices* is straightforward and is defined component-wise. When we say that $\mathbf{C} = \mathbf{A} + \mathbf{B}$ for $\mathbf{A}$, $\mathbf{B}$, $\mathbf{C} \in \mathbb{R}^{n \times m}$ we mean that $c_{i,j} = a_{i,j} + b_{i,j}$ for $i \in [1, n]$ and $j \in [1, m]$. In order to add two matrices, their dimensions must be identical. Therefore, with the matrices given previously $\mathbf{A}$ and $\mathbf{v}$ cannot be added. If we define

$$\mathbf{B} = \begin{bmatrix} 1.00 & 1.71 \\ 0.03 & 2.00 \\ 0.14 & 2.14 \end{bmatrix}$$

then $\mathbf{A}$ and $\mathbf{B}$ cannot be added or subtracted, but

$$\mathbf{A}^\top + \mathbf{B} = \begin{bmatrix} 1.20 & 0.00 \\ 0.03 & 2.00 \\ -3.00 & 4.14 \end{bmatrix}.$$

Subtraction of matrices is defined similarly.

The *symmetric* part of a square matrix $\mathbf{A}$ is defined by

$$\mathbf{A}_s = \frac{1}{2}\left(\mathbf{A} + \mathbf{A}^\top\right). \tag{B.1}$$

The *anti-symmetric* part of $\mathbf{A}$ is defined by

$$\mathbf{A}_a = \frac{1}{2}\left(\mathbf{A} - \mathbf{A}^\top\right). \tag{B.2}$$

With these definitions, any square matrix can be represented as the sum of its symmetric and anti-symmetric parts

$$\mathbf{A} = \mathbf{A}_s + \mathbf{A}_a. \tag{B.3}$$

Matrix multiplication is particularly important. When $\mathbf{A} \in \mathbb{R}^{n \times m}$ and $\mathbf{B} \in \mathbb{R}^{m \times p}$, then their product $\mathbf{C} = \mathbf{AB} \in \mathbb{R}^{n \times p}$ is defined as $c_{i,j} = \sum_{k=1}^{m} a_{i,k} b_{k,j}$. Note that the number of columns in $\mathbf{A}$ and the number of rows in $\mathbf{B}$, referred to as the inner dimension of the product, must be identical; otherwise the two matrices cannot be multiplied. Using the vectors and matrices defined above, we have the following:

$$\begin{aligned} \mathbf{AB} &= \begin{bmatrix} -0.2396 & -6.3776 \\ -1.4300 & 1.3559 \end{bmatrix} \\ \mathbf{A}^\top \mathbf{B} &= \text{not a valid operation.} \end{aligned}$$

Matrix multiplication does not commute: $\mathbf{AB} \neq \mathbf{BA}$. Matrix division is undefined. Matrix multiplication does have the following useful formulas:

$$(\mathbf{AB})\,\mathbf{C} = \mathbf{A}\,(\mathbf{BC}) \quad \text{Associative property} \tag{B.4}$$

$$(\mathbf{A} + \mathbf{B})\,\mathbf{C} = \mathbf{AC} + \mathbf{BC} \quad \text{Distributive property} \tag{B.5}$$

$$(\mathbf{ABC})^\top = \mathbf{C}^\top \mathbf{B}^\top \mathbf{A}^\top. \tag{B.6}$$

In the certain derivations it will be convenient to insert appropriately dimensioned identity matrices. The reader should be comfortable with formulas such as the following:

$$\mathbf{ABC} = \mathbf{AI}_m \mathbf{BI}_p \mathbf{C} \tag{B.7}$$

$$\mathbf{Ax} = \mathbf{AI}_m \mathbf{x}. \tag{B.8}$$

The subscript denoting the dimension will typically be dropped to simplify the notation, as it is usually clear from the context.

A square matrix $\mathbf{A}$ is an *orthogonal matrix* if and only if $\mathbf{AA}^\top = \mathbf{A}^\top \mathbf{A}$ is a diagonal matrix. A square matrix $\mathbf{A}$ is an *orthonormal matrix* if and only if $\mathbf{AA}^\top = \mathbf{A}^\top \mathbf{A} = \mathbf{I}$. Often authors use the term orthogonal to mean orthonormal. Let $\mathbf{R}_a^b$ represent the vector transformation from reference frame a to reference frame b. The inverse transformation from reference frame b to reference frame a can be shown to be $\mathbf{R}_b^a = (\mathbf{R}_a^b)^{-1} = (\mathbf{R}_a^b)^\top$. Then $(\mathbf{R}_a^b)^\top \mathbf{R}_a^b = \mathbf{R}_a^b (\mathbf{R}_a^b)^\top = \mathbf{I}$. This shows that $\mathbf{R}_a^b$ and $\mathbf{R}_b^a$ are orthonormal matrix.

The *scalar product* of two vectors $\mathbf{u},\ \mathbf{v} \in \mathbb{R}^N$ is defined to be

$$\mathbf{u} \cdot \mathbf{v} = \mathbf{u}^\top \mathbf{v} = \sum_{i=1}^{n} u_i v_i. \tag{B.9}$$

The square of the Euclidean norm of a vector is defined as the scalar product of a vector with itself: $\|\mathbf{u}\|^2 = \mathbf{u}\cdot\mathbf{u}$. The vector $\mathbf{u}$ is a *unit vector* if $\|\mathbf{u}\| = 1$. A vector has both magnitude and direction, with magnitude defined by $\|\mathbf{u}\|$ and direction defined by $\frac{\mathbf{u}}{\|\mathbf{u}\|}$. The scalar product has the following properties:

$$\mathbf{u}\cdot\mathbf{v} = \mathbf{v}\cdot\mathbf{u} = \mathbf{u}^\top\mathbf{v} = \mathbf{v}^\top\mathbf{u} \tag{B.10}$$

$$\mathbf{u}\cdot(\mathbf{v}+\mathbf{w}) = \mathbf{u}\cdot\mathbf{v} + \mathbf{u}\cdot\mathbf{w} \tag{B.11}$$

$$\mathbf{u}\cdot\mathbf{v} = \|\mathbf{u}\|\|\mathbf{v}\|\cos(\alpha) \tag{B.12}$$

where α is the angle between vectors $\mathbf{u}$ and $\mathbf{v}$. Two vectors are *orthogonal* if and only if $\mathbf{u}\cdot\mathbf{v} = 0$.

The *vector product* of two vectors $\mathbf{u}, \ \mathbf{v} \in \mathbb{R}^3$ is defined to be

$$\mathbf{u}\times\mathbf{v} = \begin{vmatrix} \mathbf{i} & \mathbf{j} & \mathbf{k} \\ u_1 & u_2 & u_3 \\ v_1 & v_2 & v_3 \end{vmatrix} \tag{B.13}$$

$$= \begin{bmatrix} u_2v_3 - u_3v_2 \\ u_3v_1 - u_1v_3 \\ u_1v_2 - u_2v_1 \end{bmatrix} = \begin{bmatrix} 0 & -u_3 & u_2 \\ u_3 & 0 & -u_1 \\ -u_2 & u_1 & 0 \end{bmatrix} \begin{bmatrix} v_1 \\ v_2 \\ v_3 \end{bmatrix}$$

where $|\cdot|$ denotes a determinant (see eqns. (B.26–B.27))and **i, j**, and **k** are unit vectors pointing along the principal axes of the reference frame. It is sometimes convenient, see for example Section 2.5.5, to express the vector product of eqn. (B.13) in the matrix form

$$\mathbf{u}\times\mathbf{v} = \mathbf{U}\mathbf{v} \tag{B.14}$$

where

$$\mathbf{U} = \begin{bmatrix} 0 & -u_3 & u_2 \\ u_3 & 0 & -u_1 \\ -u_2 & u_1 & 0 \end{bmatrix}. \tag{B.15}$$

This skew symmetric form of $\mathbf{u}$ is convenient in analysis. The equivalence expressed in eqn. (B.14) will be denoted $\mathbf{U} = [\mathbf{u}\times\ ]$. The vector product has the following properties:

$$\mathbf{u}\times\mathbf{u} = \mathbf{0} \tag{B.16}$$

$$\mathbf{u}\times\mathbf{v} = -\mathbf{v}\times\mathbf{u} \tag{B.17}$$

$$\mathbf{u}\times(\mathbf{v}\times\mathbf{w}) \neq (\mathbf{u}\times\mathbf{v})\times\mathbf{w} \tag{B.18}$$

$$\mathbf{u}\times(\mathbf{v}\times\mathbf{w}) = (\mathbf{u}\cdot\mathbf{w})\,\mathbf{v} - (\mathbf{u}\cdot\mathbf{v})\,\mathbf{w} \tag{B.19}$$

$$\mathbf{u}\cdot(\mathbf{v}\times\mathbf{w}) = \mathbf{v}\cdot(\mathbf{w}\times\mathbf{u}) \tag{B.20}$$

$$= \mathbf{w}\cdot(\mathbf{u}\times\mathbf{v}) \tag{B.21}$$

$$(\mathbf{u}\times\mathbf{v})\cdot(\mathbf{w}\times\mathbf{z}) = (\mathbf{u}\cdot\mathbf{w})\,(\mathbf{v}\cdot\mathbf{z}) - (\mathbf{v}\cdot\mathbf{w})\,(\mathbf{u}\cdot\mathbf{z}). \tag{B.22}$$

Using eqn. (B.17) and the $\mathbf{U} = [\mathbf{u}\times\ ]$ notation, we can derive the following useful formula:

$$\mathbf{u} \times \mathbf{v} = \mathbf{U}\mathbf{v} \tag{B.23}$$
$$= -\mathbf{V}\mathbf{u} \tag{B.24}$$

where $\mathbf{V} = [\mathbf{v}\times\ ]$ is the skew symmetric matrix corresponding to $\mathbf{v}$ as defined in eqn. (B.15).

The *trace* of square matrix $\mathbf{A} \in \mathbb{R}^{n\times n}$ is the sum of its diagonal elements

$$Tr(\mathbf{A}) = \sum_{i=1}^{n} a_{i,i}. \tag{B.25}$$

The trace is not defined unless its argument is square. The trace has several convenient properties:

1. $Tr\left(\mathbf{A}^\top\right) = Tr(\mathbf{A})$,
2. $Tr\left(\mathbf{A}+\mathbf{B}\right) = Tr(\mathbf{A}) + Tr(\mathbf{B})$,
3. $Tr(\mathbf{ABC}) = Tr(\mathbf{BCA}) = Tr(\mathbf{CAB})$ if all required matrix products are well defined,
4. $Tr(\mathbf{v}\mathbf{v}^\top) = Tr(\mathbf{v}^\top\mathbf{v}) = \|\mathbf{v}\|^2$,
5. $Tr(\mathbf{A}) = \sum_{i=1}^{n} \lambda_i$

where the eigenvalues λ_i are defined in Section B.6.

B.3 Independence and Determinants

Given vectors $\mathbf{u}_i \in \mathbb{R}^n$ for $i = 1, \ldots, m$, the vectors are *linearly dependent* if there exists $\alpha_i \in \mathbb{R}$, not all zero, such that $\sum_{i=1}^{m} \alpha_i \mathbf{u}_i = 0$. If the summation is *only* zero when all $\alpha_i = 0$, then the set of vectors is *linearly independent.* A set of $m > n$ vectors in $\mathbb{R}^n$ is always linearly dependent.

A set of n vectors $\mathbf{u}_i \in \mathbb{R}^n$, $i = 1, \ldots, n$ can be arranged as a square matrix $\mathbf{U} = [\mathbf{u}_1, \mathbf{u}_2, \ldots, \mathbf{u}_n] \in \mathbb{R}^{n\times n}$. Note that a matrix can always be decomposed into its component column (or row) vectors. The *rank* of a matrix, denoted $rank(\mathbf{U})$, is the number of independent column (or row) vectors in the matrix. If this set of vectors is linearly independent (i.e., $rank(\mathbf{U}) = n$), then we say that the matrix $\mathbf{U}$ is *nonsingular* (or of full rank).

A convenient tool for checking whether a square matrix is nonsingular is the determinant. The *determinant* of $\mathbf{A} \in \mathbb{R}^{n\times n}$, denoted $|\mathbf{A}|$ or $det(\mathbf{A})$, is a scalar real number that can be computed either as

$$|\mathbf{A}| = \sum_{j=1}^{n} a_{kj}c_{kj} \text{ which uses row expansion; or,} \tag{B.26}$$

$$|\mathbf{A}| = \sum_{k=1}^{n} a_{kj}c_{kj} \text{ which uses column expansion.} \tag{B.27}$$

For a square matrix $\mathbf{A}$, the *cofactor* associated with a_{kj} is

$$c_{kj} = (-1)^{k+j} M_{kj} \tag{B.28}$$

where M_{kj} is the *minor* associated with a_{kj}. M_{kj} is defined as the determinant of the matrix formed by dropping the k-th row and j-th column from $\mathbf{A}$. The determinant of $\mathbf{A} \in \mathbb{R}^{1\times 1}$ is (the scalar) $\mathbf{A}$. The determinant of $\mathbf{A} \in \mathbb{R}^{n\times n}$ is written in terms of the summation of the determinants of matrices in $\mathbb{R}^{(n-1)\times(n-1)}$. This dimension reduction process continues until it involves only scalars, for which the computation is straightforward.

If $|\mathbf{A}| \neq 0$, then $\mathbf{A}$ is nonsingular and the vectors forming the rows (and columns) of $\mathbf{A}$ are linearly independent. If $|\mathbf{A}| = 0$, then $\mathbf{A}$ is singular and the vectors forming the rows (and columns) of $\mathbf{A}$ are linearly dependent.

Determinants have the following useful properties: For $\mathbf{A}, \mathbf{B} \in \mathbb{R}^{n\times n}$

1. $|\mathbf{AB}| = |\mathbf{A}|\,|\mathbf{B}|$
2. $|\mathbf{A}| = |\mathbf{A}^\top|$
3. If any row or column of $\mathbf{A}$ is entirely zero, then $|\mathbf{A}| = 0$.
4. If any two rows (or columns) of $\mathbf{A}$ are linearly dependent, then $|\mathbf{A}| = 0$.
5. Interchanging two rows (or two columns) of $\mathbf{A}$ reverses the sign of the determinant.
6. Multiplication of a row of $\mathbf{A}$ by $\alpha \in \mathbb{R}$, yields $\alpha|\mathbf{A}|$.
7. A scaled version of one row can be added to another row without changing the determinant.

B.4 Matrix Inversion

For $\mathbf{A} \in \mathbb{R}^{n\times n}$, with $|\mathbf{A}| \neq 0$, we denote the inverse of $\mathbf{A}$ by $\mathbf{A}^{-1}$ which has the property that

$$\mathbf{A}^{-1}\mathbf{A} = \mathbf{A}\mathbf{A}^{-1} = \mathbf{I}.$$

Matrix inversion is important is both parameter and state estimation. Often problems can be manipulated into the form

$$\mathbf{y} = \mathbf{A}\boldsymbol{\theta}$$

where $\mathbf{y}$ and $\mathbf{A}$ are known, but $\boldsymbol{\theta}$ is unknown. When $\mathbf{A}$ is square and nonsingular, then its inverse exists and the unique solution is

$$\hat{\theta} = \mathbf{A}^{-1}\mathbf{y}.$$

For $\mathbf{A} \in \mathbb{R}^{m \times n}$ with $m > n$ and $rank(\mathbf{A}) = n$, then the solution

$$\hat{\theta} = \left(\mathbf{A}^\top \mathbf{A}\right)^{-1} \mathbf{A}^\top \mathbf{y}$$

minimizes the norm of the error vector $(\mathbf{y} - \mathbf{A}\hat{\boldsymbol{\theta}})$. This *least squares* solution is derived in Section 5.3.2 .

When A is a square nonsingular matrix,

$$\mathbf{A}^{-1} = \frac{\mathbf{C}^\top}{|\mathbf{A}|} \tag{B.29}$$

where $\mathbf{C}$ is the cofactor matrix for $\mathbf{A}$ and $\mathbf{C}^\top$ is called the adjoint of $\mathbf{A}$. The matrix inverse has the following properties:

1. $\left(\mathbf{A}^{-1}\right)^{-1} = \mathbf{A}$,
2. $(\mathbf{AB})^{-1} = \mathbf{B}^{-1}\mathbf{A}^{-1}$,
3. $\left|\mathbf{A}^{-1}\right| = \frac{1}{|\mathbf{A}|}$,
4. $\left(\mathbf{A}^\top\right)^{-1} = \left(\mathbf{A}^{-1}\right)^\top$ which will be denoted by $\mathbf{A}^{-\top}$, and
5. $(\mu\mathbf{A})^{-1} = \frac{1}{\mu}\mathbf{A}^{-1}$.

The inverse of an orthonormal matrix is the same as its transpose.

B.5 Matrix Inversion Lemma

Two forms of the Matrix Inversion Lemma are presented. The Lemma is useful in least squares and Kalman filter derivations. Each lemma can be proved by direct multiplication.

Lemma B.5.1 *Given four matrices $\mathbf{P}_1$, $\mathbf{P}_2, \mathbf{H}$, and $\mathbf{R}$ of compatible dimensions, if $\mathbf{P}_1$, $\mathbf{P}_2$, $\mathbf{R}$, and $\left(\mathbf{H}^\top\mathbf{P}_1\mathbf{H} + \mathbf{R}\right)$ are all invertible and*

$$\mathbf{P}_2^{-1} = \mathbf{P_1}^{-1} + \mathbf{H}\mathbf{R}^{-1}\mathbf{H}^\top, \tag{B.30}$$

then

$$\mathbf{P}_2 = \mathbf{P}_1 - \mathbf{P}_1\mathbf{H}\left(\mathbf{H}^\top\mathbf{P}_1\mathbf{H} + \mathbf{R}\right)^{-1}\mathbf{H}^\top\mathbf{P}_1. \tag{B.31}$$

Lemma B.5.2 *Given four matrices* $\mathbf{A}$, $\mathbf{B}$, $\mathbf{C}$, *and* $\mathbf{D}$ *of compatible dimensions, if* $\mathbf{A}$, $\mathbf{C}$, *and* $\mathbf{A}+\mathbf{BCD}$ *are invertible, then*

$$\left(\mathbf{A}+\mathbf{BCD}\right)^{-1} = \mathbf{A}^{-1} - \mathbf{A}^{-1}\mathbf{B}\left(\mathbf{DA}^{-1}\mathbf{B}+\mathbf{C}^{-1}\right)\mathbf{DA}^{-1}. \tag{B.32}$$

The equivalence of the two forms is shown by defining:

$$\mathbf{A} = \mathbf{P}_1^{-1}, \quad \mathbf{B} = \mathbf{H}, \quad \mathbf{C} = \mathbf{R}^{-1}, \quad \mathbf{D} = \mathbf{H}^\top,$$

and requiring $\mathbf{A}+\mathbf{BCD} = \mathbf{P}_2^{-1}$.

B.6 Eigenvalues and Eigenvectors

For $\mathbf{A} \in \mathbb{R}^{n\times n}$, the set of scalars $\lambda_i \in C$ and (nonzero) vectors $\mathbf{x}_i \in C^n$ satisfying

$$\mathbf{A}\mathbf{x}_i = \lambda_i \mathbf{x}_i \qquad \text{or} \qquad (\lambda_i \mathbf{I} - \mathbf{A})\mathbf{x}_i = \mathbf{0}_n$$

are the eigenvalues and eigenvectors of $\mathbf{A}$.

We are only interested in nontrivial solutions (i.e., solution $\mathbf{x}_i = \mathbf{0}$ is not of interest). Nontrivial solutions exist only if $(\lambda_i\mathbf{I}-\mathbf{A})$ is a singular matrix. Therefore, the eigenvalues of $\mathbf{A}$ are the values of λ such that $|\lambda\mathbf{I}-\mathbf{A}| = 0$. This yields an n-th order polynomial in λ.

If $\mathbf{A}$ is a symmetric matrix, then all of its eigenvalues and eigenvectors are real. If $\mathbf{x}_i$ and $\mathbf{x}_j$ are eigenvectors of symmetric matrix $\mathbf{A}$ and their eigenvalues are not equal (i.e., $\lambda_i \neq \lambda_j$), then the eigenvectors are orthogonal (i.e. $\mathbf{x}_i \cdot \mathbf{x}_j = 0$).

A square matrix $\mathbf{A}$ is *idempotent* if and only if $\mathbf{AA} = \mathbf{A}$. Idempotent matrices are sometimes also called *projection* matrices. Idempotent matrices have the following properties:

1. $rank(\mathbf{A}) = Tr(\mathbf{A})$;
2. the eigenvalues of $\mathbf{A}$ are all either 0 or 1;
3. the multiplicity of 1 as an eigenvalue is the $rank(\mathbf{A})$;
4. $\mathbf{A}(\mathbf{I}-\mathbf{A}) = (\mathbf{I}-\mathbf{A})\mathbf{A} = \mathbf{0}$; and,
5. $\mathbf{A}^\top$, $(\mathbf{I}-\mathbf{A})$ and $(\mathbf{I}-\mathbf{A}^\top)$ are idempotent.

Example B.1 *Select a state estimation gain vector* $\mathbf{L}$ *such that the error state model for the discrete-time system*

$$\begin{aligned}\hat{\mathbf{x}}_{k+1} &= \begin{bmatrix} 0.8187 & 0.0000 \\ 0.0906 & 1.0000 \end{bmatrix}\hat{\mathbf{x}}_k + \begin{bmatrix} 0.0906 \\ 0.0047 \end{bmatrix} f_k + \mathbf{L}(p_k - \hat{p}_k) \\ \hat{p}_k &= \begin{bmatrix} 0.0000 & 1.0000 \end{bmatrix}\hat{\mathbf{x}}_k\end{aligned}$$

has eigenvalues as $0.9 \pm 0.1j$. *The actual system is modeled as*

$$\begin{aligned} \mathbf{x}_{k+1} &= \begin{bmatrix} 0.8187 & 0.0000 \\ 0.0906 & 1.0000 \end{bmatrix} \mathbf{x}_k + \begin{bmatrix} 0.0906 \\ 0.0047 \end{bmatrix} f_k \\ p_k &= \begin{bmatrix} 0.0000 & 1.0000 \end{bmatrix} \mathbf{x}_k. \end{aligned}$$

The observability is considered in Example 3.16 and will not be repeated here.

The error state $\delta \mathbf{x}_k = \mathbf{x}_k - \hat{\mathbf{x}}_k$ *has the state space model*

$$\delta \mathbf{x}_{k+1} = \begin{bmatrix} 0.8187 & -L_1 \\ 0.0906 & (1 - L_2) \end{bmatrix} \delta \mathbf{x}_k = \mathbf{\Phi}_L \delta \mathbf{x}_k.$$

The eigenvalues of the discrete-time, error-state, transition matrix are the roots of the equation

$$\begin{aligned} 0 &= |z\mathbf{I} - \mathbf{\Phi}_L| \\ &= \begin{vmatrix} (z - 0.8187) & L1 \\ -0.0906 & (z + L_2 - 1) \end{vmatrix} \\ &= z^2 + (L_2 - 1.8187)z + (0.8187 - 0.8187L_2 + 0.0906L_1). \end{aligned}$$

To achieve the desired eigenvalues, we must have

$$\begin{aligned} -1.8 &= L_2 - 1.8187 \\ 0.82 &= 0.8187 - 0.8187L_2 + 0.0906L_1. \end{aligned}$$

The solution is $\mathbf{L} = [0.1833, 0.0187]^\top$. △

B.7 Positive Definite Matrices

For a square matrix $\mathbf{B} \in \mathbb{R}^{n \times n}$ and $\mathbf{x} \in \mathbb{R}^n$, the scalar mapping $\mathbf{x}^\top \mathbf{B} \mathbf{x}$ is called a *quadratic form.* It is straightforward using eqn. (B.3) to show that the quadratic form depends only on the symmetric portion of $\mathbf{B}$:

$$\begin{aligned} \mathbf{x}^\top \mathbf{B} \mathbf{x} &= \mathbf{x}^\top \mathbf{B}_s \mathbf{x} + \mathbf{x}^\top \mathbf{B}_a \mathbf{x} \\ &= \mathbf{x}^\top \mathbf{B}_s \mathbf{x}, \end{aligned}$$

see Exercise B.14. Therefore, it is standard practice to assume that $\mathbf{B}$ is symmetric.

When a quadratic form is defined, the designer is often interested in the sign of the scalar output, $\mathbf{x}^\top \mathbf{B} \mathbf{x}$, which depends on the properties of the symmetric matrix $\mathbf{B}$.

1. A matrix $\mathbf{B}$ is *Positive Definite* if and only if $\mathbf{x}^\top \mathbf{B}\mathbf{x} > 0$ for all non-zero $\mathbf{x}$.

2. A matrix $\mathbf{B}$ is *Positive Semi-definite* if and only if $\mathbf{x}^\top \mathbf{B}\mathbf{x} \geq 0$ for all non-zero $\mathbf{x}$.

3. A matrix $\mathbf{B}$ is *Negative Definite* if and only if $\mathbf{x}^\top \mathbf{B}\mathbf{x} < 0$ for all non-zero $\mathbf{x}$.

4. A matrix $\mathbf{B}$ is *Negative Semi-definite* if and only if $\mathbf{x}^\top \mathbf{B}\mathbf{x} \leq 0$ for all non-zero $\mathbf{x}$.

When the matrix fails to have any of the above properties, then the matrix is sign indefinite.

Properties of the eigenvalues and inverse of a positive definite matrix are discussed further in Exercise B.15. Corresponding properties apply to negative definite matrices, with appropriate changes to the statements.

B.8 Singular Value Decomposition

The singular value decomposition (SVD) of a matrix $\mathbf{A} \in \mathbb{R}^{m \times n}$ is defined as $\mathbf{A} = \mathbf{U}\mathbf{\Sigma}\mathbf{V}^\top$ where $\mathbf{U} \in \mathbb{R}^{m \times m}$ and $\mathbf{V} \in \mathbb{R}^{n \times n}$ are orthonormal matrices and $\mathbf{\Sigma} \in \mathbb{R}^{m \times n}$ is a diagonal matrix. The diagonal of $\mathbf{\Sigma}$ is defined as $(\sigma_1, \sigma_2, \ldots, \sigma_p)$ where $p = min(m, n)$. The diagonal elements of $\mathbf{\Sigma}$ are ordered so that $\sigma_1 > \sigma_2 > \ldots > \sigma_p \geq 0$. The σ_i are the singular values of the matrix $\mathbf{A}$. The rank of $\mathbf{A}$ is the smallest integer r such that $\sigma_i = 0$ for $i > r$. The first r columns of $\mathbf{U}$ form an orthonormal basis for the range space of $\mathbf{A}$ while the last $(n - r)$ columns of $\mathbf{U}$ span the null space of $\mathbf{A}^\top$. The first r columns of $\mathbf{V}$ form an orthonormal basis for the range of $\mathbf{A}^\top$, while the last $(n - r)$ columns of $\mathbf{V}$ form an orthonormal basis for the null space of $\mathbf{A}$. Additional properties of the singular value decomposition and algorithms for its computation can be found in [59].

B.9 Orthogonalization

Let the matrix $\mathbf{A}$ be orthogonal and let $\hat{\mathbf{A}}$ be a computed version of $\mathbf{A}$ that is not orthogonal due to numeric errors involved in its computation. The main topic of this section is orthogonalization of $\hat{\mathbf{A}}$. The orthogonalized version of $\hat{\mathbf{A}}$ will be denoted by $\hat{\mathbf{A}}_o$.

A traditional orthogonalization approach is calculated as

$$\hat{\mathbf{A}}_o = \hat{\mathbf{A}}\left(\hat{\mathbf{A}}^\top \hat{\mathbf{A}}\right)^{-1/2} \tag{B.33}$$

which is optimal in the sense of minimizing *trace* $\left(\left(\hat{\mathbf{A}}_o - \hat{\mathbf{A}}\right)^\top \left(\hat{\mathbf{A}}_o - \hat{\mathbf{A}}\right)\right)$ subject to the constraint that $\hat{\mathbf{A}}_o^\top \hat{\mathbf{A}}_o = \mathbf{I}$. See Exercise B.16.

Given its SVD $\hat{\mathbf{A}} = \mathbf{U}\boldsymbol{\Sigma}\mathbf{V}^\top$ with $\mathbf{U}^\top\mathbf{U} = \mathbf{I}$ and $\mathbf{V}^\top\mathbf{V} = \mathbf{I}$, the matrix $\hat{\mathbf{A}}$ can be orthogonalized as

$$\hat{\mathbf{A}}_o = \mathbf{U}\mathbf{V}^\top. \tag{B.34}$$

This SVD approach to orthogonalization is completely equivalent to the traditional orthogonalization approach of eqn. (B.33):

$$\begin{aligned}
\hat{\mathbf{A}}_o &= \hat{\mathbf{A}}\left(\hat{\mathbf{A}}^\top\hat{\mathbf{A}}\right)^{-1/2} && \text{(B.35)}\\
&= \mathbf{U}\boldsymbol{\Sigma}\mathbf{V}^\top\left(\mathbf{V}\boldsymbol{\Sigma}\mathbf{U}^\top\mathbf{U}\boldsymbol{\Sigma}\mathbf{V}^\top\right)^{-1/2}\\
&= \mathbf{U}\boldsymbol{\Sigma}\mathbf{V}^\top\Big(\left(\mathbf{V}\boldsymbol{\Sigma}\mathbf{V}^\top\right)\left(\mathbf{V}\boldsymbol{\Sigma}\mathbf{V}^\top\right)\Big)^{-1/2}\\
&= \mathbf{U}\boldsymbol{\Sigma}\mathbf{V}^\top\left(\mathbf{V}\boldsymbol{\Sigma}^{-1}\mathbf{V}^\top\right)\\
&= \mathbf{U}\mathbf{V}^\top. && \text{(B.36)}
\end{aligned}$$

A traditional approximate solution to eqn. (B.35) is

$$\hat{\mathbf{A}}_o = \hat{\mathbf{A}} + \frac{1}{2}\left(\mathbf{I} - \hat{\mathbf{A}}\hat{\mathbf{A}}^\top\right)\hat{\mathbf{A}}. \tag{B.37}$$

This expression is not obvious, but can be derived by Taylor series expansion of eqn. (B.35), see p. 39 in [113]. Because the solution is approximate, several iterations of the solution may be required to achieve convergence. It can be shown that this manipulation orthogonalizes the rows of $\hat{\mathbf{A}}_o$. The computation does not correct the asymmetric errors in $\hat{\mathbf{A}}_o$ [115].

B.10 LU Decomposition

Given a matrix $\mathbf{A} \in \mathbb{R}^{m \times m}$ the LU decomposition computes an upper triangular matrix $\mathbf{U}$ and a (unit diagonal) lower triangular matrix $\mathbf{L}$ such that $\mathbf{A} = \mathbf{L}\mathbf{U}$. In this section, we present the LU algorithm in a non-standard form to exhibit certain properties that will be required in Section 8.9.1; we will actually derive a decomposition $\mathbf{A} = \mathbf{P}\mathbf{L}\mathbf{U}$ where $\mathbf{P}$ is a permutation matrix. We assume that the matrix $\mathbf{A}$ is known to have rank equal to r.

Given that the $rank(\mathbf{A}) = r$, $\mathbf{A}$ has exactly r linearly independent rows. Select a permutation matrix $\mathbf{P}$ such that the the first r rows of the matrix $\mathbf{A}_0 = \mathbf{P}^{-1}\mathbf{A}$ are linearly independent[2]. For $i = 1, \ldots, r$ define the matrix

[2]The matrix $\mathbf{P}$ need not be known at the start of the algorithm. There are versions of the LU decomposition algorithm that compute $\mathbf{P}$ as needed during the LU decomposition.

$\mathbf{L}_i$ as

$$\mathbf{L}_i = \begin{bmatrix} 1 & \dots & 0 & \dots & 0 \\ & \ddots & & & \\ & & 1 & & \\ & & -\frac{a_{(i+1),i}}{a_{i,i}} & & \\ & & \vdots & \ddots & \\ 0 & & -\frac{a_{m,i}}{a_{i,i}} & & 1 \end{bmatrix} \tag{B.38}$$

which has a unit diagonal and nonzero elements only below the diagonal in the i-th column. The matrix

$$\mathbf{A}_i = \mathbf{L}_i \mathbf{A}_{i-1}$$

will have all zero elements below the diagonal in the first i columns. Because the last $(n-r)$ rows of $\mathbf{A}_0$ are a linear combination of the the first r rows, the last $(n-r)$ rows of $\mathbf{A}_r$ will be zero. Define

$$\mathbf{U} = \mathbf{A}_r$$

and note that $\mathbf{U}$ can be expressed as

$$\mathbf{U} = \begin{bmatrix} \mathbf{C} & \mathbf{D} \\ \mathbf{0} & \mathbf{0} \end{bmatrix} \tag{B.39}$$

where $\mathbf{C} \in \mathbb{R}^{r \times r}$ is upper diagonal and invertible, and $\mathbf{D} \in \mathbb{R}^{r \times (m-r)}$.

From the definition of each $\mathbf{L}_i$ for $i = 1, \dots, r$,

$$\begin{aligned} \mathbf{U} &= \mathbf{L}_r \mathbf{A}_{r-1} \\ &= \mathbf{L}_r \mathbf{L}_{r-1} \mathbf{A}_{r-2} \\ &\vdots \\ \mathbf{U} &= \mathbf{L}_r \mathbf{L}_{r-1} \dots \mathbf{L}_1 \mathbf{P}^{-1} \mathbf{A}; \end{aligned}$$

therefore, defining $\mathbf{L} = \mathbf{L}_1^{-1} \dots \mathbf{L}_{r-1}^{-1} \mathbf{L}_r^{-1}$ we have that

$$\mathbf{A} = \mathbf{PLU}$$

where $\mathbf{L}$ and $\mathbf{P}$ are nonsingular.

In the case where the matrix $\mathbf{A}$ has rank equal to m, then for $\mathbf{P} = \mathbf{I}$, the LU decomposition facilitates the solution for $\mathbf{x}$ to problems of the form $\mathbf{y} = \mathbf{Ax}$. The problem is decomposed into two subproblems

$$\mathbf{Lz} = \mathbf{y} \tag{B.40}$$

$$\mathbf{Ux} = \mathbf{z}. \tag{B.41}$$

Due to the lower triangular structure of $\mathbf{L}$, eqn. (B.40) is easily solved for $\mathbf{z}$, then using the upper triangular structure of $\mathbf{U}$ eqn. (B.41) is easily solved for $\mathbf{x}$.

B.11 UD Decomposition

Given a positive definite symmetric matrix $\mathbf{P}$, the UD decomposition computes an upper triangular matrix $\mathbf{U}$ and a diagonal matrix $\mathbf{D}$ such that $\mathbf{P} = \mathbf{UDU}^\top$. This section outlines the computation of the $\mathbf{UD}$ factors in the case where $\mathbf{P}$ is a 3×3 matrix. General algorithms can be found in [59]. Kalman filter implementations algorithms using the UD decomposition are presented in [60, 121].

For $\mathbf{P} \in \mathbb{R}^{3\times 3}$, the UD factorization is

$$\begin{aligned}\mathbf{P} &= \begin{bmatrix} 1 & u_{12} & u_{13} \\ 0 & 1 & u_{23} \\ 0 & 0 & 1 \end{bmatrix} \begin{bmatrix} d_{11} & 0 & 0 \\ 0 & d_{22} & 0 \\ 0 & 0 & d_{33} \end{bmatrix} \begin{bmatrix} 1 & 0 & 0 \\ u_{12} & 1 & 0 \\ u_{13} & u_{23} & 1 \end{bmatrix} \\ &= \begin{bmatrix} d_{11} + d_{22}u_{a2}^2 + d_{33}u_{13}^2 & d_{22}u_{12} + d_{33}u_{13}u_{23} & d_{33}u_{13} \\ d_{22}u_{12} + d_{33}u_{13}u_{23} & d_{22} + d_{33}u_{23}^2 & d_{33}u_{23} \\ d_{33}u_{13} & d_{33}u_{23} & d_{33} \end{bmatrix}.\end{aligned}$$

This expression allows the elements of $\mathbf{U}$ and $\mathbf{D}$ to be determined one element at a time starting at the lower right with $d_{33} = p_{33}$, working up through the third column with $u_{23} = p_{23}/d_{33}$ and $u_{13} = p_{12}/d_{33}$, then repeating a similar procedure on the next column to the left. For each column, the computations start with the diagonal element (e.g. $d_{22} = p_{22} - d_{33}u_{23}^2$) and work up through the column (e.g., $u_{12} = (p_{12} - d_{33}u_{13}u_{23})/d_{22}$).

B.12 Matrix Exponential

As should be evident from Section 3.5.4, the computation of matrix exponentials is important. Various approaches are compared in [100]. A key conclusion of that article is that there is no universally best algorithm. An algorithm that often works very well combines the Padé approximation and "scaling and squaring." That algorithm is presented well in [60]. This section presents two approaches that provide insight into the matrix exponential.

B.12.1 Power Series

The power series expansion of the scalar exponential function is

$$e^{at} = 1 + at + \frac{(at)^2}{2!} + \frac{(at)^3}{3!} + \dots$$

for $a,\ t \in \mathbb{R}$. Extension of this power series to matrix arguments serves as a definition of the matrix exponential

$$e^{\mathbf{F}t} = \mathbf{I} + \mathbf{F}t + \frac{(\mathbf{F}t)^2}{2!} + \frac{(\mathbf{F}t)^3}{3!} + \dots \tag{B.42}$$

where $t \in \mathbb{R}$, $\mathbf{F} \in \mathbb{R}^{n\times n}$, and $\mathbf{I}$ is the identity matrix in $\mathbb{R}^{n\times n}$.

Note that by the definition of the matrix exponential, it is always true that a matrix commutes with its exponential:

$$\mathbf{F}e^{\mathbf{F}t} = e^{\mathbf{F}t}\mathbf{F}.$$

Also, $e^{\mathbf{F}t}\big|_{t=0} = \mathbf{I}$.

Power series expansion is usually not the best numeric technique for the computation of matrix exponentials; however when the structure of the $\mathbf{F}$ matrix is appropriate, power series methods are one approach for determining closed form solutions for the matrix exponential of $\mathbf{F}$. An example of this is contained in Example 3.13.

B.12.2 Laplace Transform

The formula

$$e^{\mathbf{F}t} = \mathcal{L}^{-1}\left\{(s\mathbf{I}-\mathbf{F})^{-1}\right\} \tag{B.43}$$

is derived by taking the Laplace transform of both sides of eqn. (B.42),

$$\begin{aligned}
\mathcal{L}\left\{e^{\mathbf{F}t}\right\} &= \mathcal{L}\left\{\mathbf{I}+\mathbf{F}t+0.5\mathbf{F}^2t^2+\ldots\right\}\\
&= \frac{1}{s}\mathbf{I}+\frac{1}{s^2}\mathbf{F}+\frac{1}{s^3}\mathbf{F}^2+\ldots\\
&= (s\mathbf{I}-\mathbf{F})^{-1}\\
e^{\mathbf{F}t} &= \mathcal{L}^{-1}\left\{(s\mathbf{I}-\mathbf{F})^{-1}\right\}.
\end{aligned}$$

This derivation has used the fact that

$$(s\mathbf{I}-\mathbf{F})^{-1} = \left(\frac{1}{s}\mathbf{I}+\frac{1}{s^2}\mathbf{F}+\frac{1}{s^3}\mathbf{F}^2+\ldots\right)$$

which can be shown by direct multiplication of both sides of the equation by $(s\mathbf{I}-\mathbf{F})$.

Example B.2 *The purpose of this example is to clarify the application of eqn. (B.43).*

Consider computation of $e^{\mathbf{F}t}$ *for* $\mathbf{F} = \begin{bmatrix} 0 & -\omega_n \\ \omega_n & 0 \end{bmatrix}$ *where* ω_n *is a positive real number:*

$$\begin{aligned}
e^{\mathbf{F}t} &= \mathcal{L}^{-1}\left\{(s\mathbf{I}-\mathbf{F})^{-1}\right\}\\
&= \mathcal{L}^{-1}\left\{\begin{bmatrix} s & \omega_n \\ -\omega_n & s \end{bmatrix}^{-1}\right\}
\end{aligned}$$

$$
\begin{aligned}
&= \mathcal{L}^{-1}\left\{\begin{bmatrix} \frac{s}{s^2+\omega_n^2} & \frac{-\omega_n}{s^2+\omega_n^2} \\ \frac{\omega_n}{s^2+\omega_n^2} & \frac{s}{s^2+\omega_n^2} \end{bmatrix}\right\} \\
e^{\mathbf{F}t} &= \begin{bmatrix} \cos\omega_n t & -\sin\omega_n t \\ \sin\omega_n t & \cos\omega_n t \end{bmatrix}.
\end{aligned}
$$

△

B.13 Matrix Calculus

This section contains results related to the derivatives of vectors, scalar products, vector products, and matrices that will be used in the main body of the text.

B.13.1 Derivatives with Respect to Scalars

In the case where the vectors $\mathbf{u}$ and $\mathbf{v}$, the matrix $\mathbf{A}$, and the scalar μ are functions of a scalar quantity s, the derivative with respect to s has the following properties:

$$\frac{d}{ds}(\mathbf{u}+\mathbf{v}) = \frac{d\mathbf{u}}{ds} + \frac{d\mathbf{v}}{ds} \tag{B.44}$$

$$\frac{d}{ds}(\mathbf{u}\cdot\mathbf{v}) = \frac{d\mathbf{u}}{ds}\cdot\mathbf{v} + \mathbf{u}\cdot\frac{d\mathbf{v}}{ds} \tag{B.45}$$

$$\frac{d}{ds}(\mathbf{u}\times\mathbf{v}) = \frac{d\mathbf{u}}{ds}\times\mathbf{v} + \mathbf{u}\times\frac{d\mathbf{v}}{ds} \tag{B.46}$$

$$\frac{d}{ds}(\mu\mathbf{v}) = \frac{d\mu}{ds}\mathbf{v} + \mu\frac{d\mathbf{v}}{ds} \tag{B.47}$$

$$\frac{d}{ds}\mathbf{A}^{-1} = -\mathbf{A}^{-1}\left(\frac{d}{ds}\mathbf{A}\right)\mathbf{A}^{-1}. \tag{B.48}$$

B.13.2 Derivatives with Respect to Vectors

Using the convention that gradients of scalar functions are defined as row vectors,

$$\frac{d}{d\mathbf{v}}(\mathbf{u}\cdot\mathbf{v}) = \frac{d}{d\mathbf{v}}(\mathbf{v}\cdot\mathbf{u}) = \mathbf{u}^\top \tag{B.49}$$

$$\frac{d}{d\mathbf{v}}(\mathbf{A}\mathbf{v}) = \mathbf{A} \tag{B.50}$$

$$\frac{d}{d\mathbf{v}}(\mathbf{v}^\top\mathbf{A}) = \mathbf{A}^\top \tag{B.51}$$

$$\frac{d}{d\mathbf{v}}(\mathbf{v}^\top\mathbf{A}\mathbf{v}) = \mathbf{v}^\top(\mathbf{A}+\mathbf{A}^\top) \tag{B.52}$$

$$= 2\mathbf{v}^\top\mathbf{A},\ \text{if } \mathbf{A} \text{ is symmetric.} \tag{B.53}$$

B.13.3 Derivatives with Respect to Matrices

The derivative of the scalar μ with respect to the matrix $\mathbf{A}$ is defined by

$$\frac{d\mu}{d\mathbf{A}} = \begin{bmatrix} \frac{d\mu}{da_{11}} & \frac{d\mu}{da_{12}} & \cdots & \frac{d\mu}{da_{1n}} \\ \frac{d\mu}{da_{21}} & \frac{d\mu}{da_{22}} & \cdots & \frac{d\mu}{da_{2n}} \\ \vdots & \vdots & \ddots & \vdots \\ \frac{d\mu}{da_{m1}} & \frac{d\mu}{da_{m2}} & \cdots & \frac{d\mu}{da_{mn}} \end{bmatrix}. \tag{B.54}$$

For the scalar operation defined by the trace, the following differentiation formulas are useful:

$$\frac{d(trace(\mathbf{AB}))}{d\mathbf{A}} = \mathbf{B}^\top \quad (\mathbf{AB} \text{ must be square}) \tag{B.55}$$

$$\frac{d(trace(\mathbf{ABA}^\top))}{d\mathbf{A}} = 2\mathbf{AB}^\top \quad (\mathbf{B} \text{ must be symmetric}). \tag{B.56}$$

B.14 Numeric Zero Finding & Optimization

This section reviews the topics on numeric zero finding and numeric optimization. Section B.14.1 uses linear algebra to derive iterative algorithms to approximate the value of $\mathbf{x}$ such that $\mathbf{y} = \mathbf{g}(\mathbf{x}) = \mathbf{0}$ when $\mathbf{x}, \mathbf{y} \in \mathbb{R}^n$. This is referred to as *zero finding.* Section B.14.2 uses linear algebra to extend the results of Section B.14.1 to derive iterative algorithms to approximate the value of $\mathbf{x}$ that minimizes a cost function $J(\mathbf{x})$. This is referred to as *numeric optimization.*

The result derived in Section B.14.2 is directly related to solving the GPS measurement equations and will be used in Section 8.2.2 and Example 8.1.

B.14.1 Numerical Zero Finding

Newton's method is a procedure for finding a value of $\mathbf{x}$ such that $\mathbf{g}(\mathbf{x}) = \mathbf{0}$. This value of $\mathbf{x}$ is called a zero of $\mathbf{g}$. The function $\mathbf{g}$ is assumed to be at least differentiable.

Let $\mathbf{y} = \mathbf{g}(\mathbf{x})$ for $\mathbf{x}, \mathbf{y} \in \mathbb{R}^n$. The Taylor series linear approximation to $\mathbf{g}(\mathbf{x})$ at the point $\mathbf{x}_k$ is

$$\mathbf{g}(\mathbf{x}) = \mathbf{g}(\mathbf{x}_k) + \mathbf{G}_k (\mathbf{x} - \mathbf{x}_k) \tag{B.57}$$

where $\mathbf{G}_k = \left.\frac{\partial \mathbf{g}}{\partial \mathbf{x}}\right|_{\mathbf{x}=\mathbf{x}_k}$. Newton's method solves eqn. (B.57) for the value of $\mathbf{x}$ that causes the linearized function to be zero. This solution is an

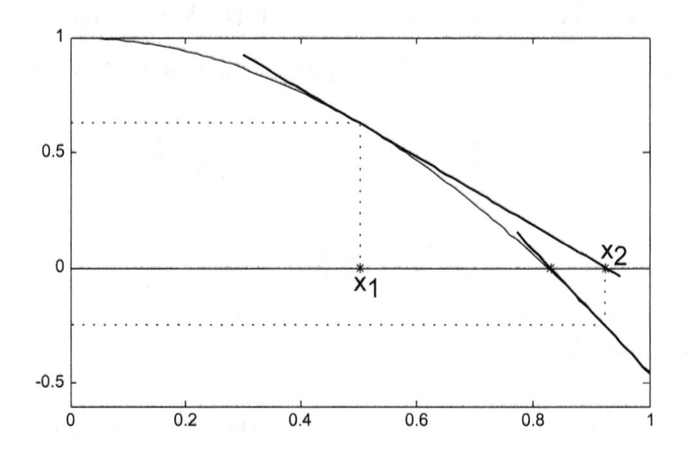

Figure B.1: The function $g(x)$ and the linear approximations to $g(x)$ at x_1 and x_2 as determined by Newton's method in Example B.3.

improved estimate of the zero of $\mathbf{g}$. Due to the nonzero higher order terms (h.o.t.'s), the result may not immediately yield a zero of $\mathbf{g}$, but iteration of

$$\mathbf{x}_{k+1} = \mathbf{x}_k - \mathbf{G}_k^{-1}\mathbf{g}(\mathbf{x}_k) \tag{B.58}$$

results in quadratic convergence of $\mathbf{x}_k$ toward a (simple) zero of $\mathbf{g}$.

Example B.3 *Figure B.1 shows the first three iterations of eqn. (B.58) for*

$$g(x) = \cos(x) - x^2$$

with a starting condition of $x = 0.5$. The figure shows the function $g(x)$, the linear approximation at $x = x_1$, and the linear approximation at $x = x_2$. The sequence of estimates of the zero of $g(x)$ is as follows:

k	x_k	y_k
1	0.50000000000000	0.62758256189037
2	0.92420692729320	-0.25169067707687
3	0.82910575599742	-0.01188097378152
4	0.82414613172820	-0.00003292121442
5	0.82413231240991	-0.00000000025583
6	0.82413231230252	0.00000000000000.

The convergence is faster near the final result due to the accuracy of the Taylor approximation of eqn. (B.57) improving as x approaches the location of the zero of the g.

△

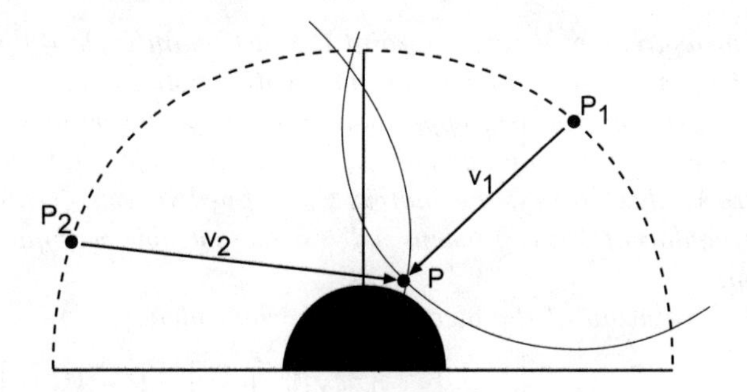

Figure B.2: Illustration of the scenario for Example B.4. The position of two satellites $\mathbf{P}_1$ and $\mathbf{P}_2$ and measurements of the ranges $\|\mathbf{v}_1\|$ and $\|\mathbf{v}_2\|$ are available.

B.14.2 Numeric Optimization

If we have a situation where $\mathbf{y} = \mathbf{f}(\mathbf{x})$ for $\mathbf{x} \in \mathbb{R}^n$ and $\mathbf{y} \in \mathbb{R}^m$ where $m \geq n$, then in general no exact solution will exist. As in Newton's method we can use iteration to find the solution that minimizes $\|\mathbf{y} - \mathbf{f}(\mathbf{x})\|$. Starting at an estimate $\mathbf{x}_k$ the Taylor's approximation to $\mathbf{f}(x)$ near $\mathbf{x}_k$ is

$$\mathbf{f}(\mathbf{x}) = \mathbf{f}(\mathbf{x}_k) + \mathbf{F}_k\left(\mathbf{x} - \mathbf{x}_k\right). \tag{B.59}$$

where $\mathbf{F}_k = \frac{\partial \mathbf{f}}{\partial \mathbf{x}}\big|_{\mathbf{x}=\mathbf{x}_k}$ is the *Jacobian* matrix for $\mathbf{f}(\mathbf{x})$ evaluated at $\mathbf{x}_k$. The approximate cost function is then

$$J(\mathbf{x}) = \left\|\mathbf{y} - \left(\mathbf{f}(\mathbf{x}_k) + \mathbf{F}_k\left(\mathbf{x} - \mathbf{x}_k\right)\right)\right\|^2.$$

Taking the partial derivative of J with respect to $\mathbf{x}$ and setting it equal to zero yields

$$\frac{\partial J}{\partial \mathbf{x}} = \mathbf{F}_k^\top\Big(\mathbf{y} - \left(\mathbf{f}(\mathbf{x}_k) + \mathbf{F}_k\left(\mathbf{x} - \mathbf{x}_k\right)\right)\Big) = \mathbf{0}.$$

Using the solution of this equation as the next estimate of the minimum $\mathbf{x}_{k+1}$ yields the algorithm

$$\mathbf{x}_{k+1} = \mathbf{x}_k + \left(\mathbf{F}_k^\top \mathbf{F}_k\right)^{-1} \mathbf{F}_k^\top\left(\mathbf{y} - \mathbf{f}(\mathbf{x}_k)\right) \tag{B.60}$$

assuming that the inverse matrix exists.

Example B.4 *Consider the scenario depicted in Figure B.2. The goal is to find the location of the point* $\mathbf{P}$ *when the ranges* $\|\mathbf{v}_1\|$ *and* $\|\mathbf{v}_2\|$ *between the point* $\mathbf{P}$ *and the two satellites at locations* $\mathbf{P}_1$ *and* $\mathbf{P}_2$*, respectively, are known.*

For each satellite, the figure includes a narrow solid line representing the set of points that are at the measured range from the satellite. In the absence of measurement error, the point $\mathbf{P}$ *must lie on each of these circles; therefore, there are two possible solutions. Notice that as the vectors* $\mathbf{v}_1$ *and* $\mathbf{v}_2$ *become nearly the linearly dependent, that is as the point* $\mathbf{P}$ *moves toward the line connecting the two points, the two possible solutions approach each other. As this occurs, the matrix* $\mathbf{F}_k$ *defined in eqn. (B.61) will become nonsingular and the problem will not be solvable without additional information.*

Using the notation of the algorithm described above,

$$\mathbf{y} = \boldsymbol{\rho} + \boldsymbol{\nu}, \quad and \quad \mathbf{f}(\mathbf{x}) = \begin{bmatrix} \|\mathbf{v}_1\| \\ \|\mathbf{v}_2\| \end{bmatrix} = \begin{bmatrix} \|\mathbf{P} - \mathbf{P}_1\| \\ \|\mathbf{P} - \mathbf{P}_2\| \end{bmatrix}$$

where $\mathbf{v}_1 = \mathbf{P} - \mathbf{P}_1$ *and* $\mathbf{v}_2 = \mathbf{P} - \mathbf{P}_2$ *and* $\boldsymbol{\nu}$ *represents measurement noise. If we let* $\hat{\mathbf{P}}_k$ *denote the current estimate of the minimizing value of* $\mathbf{P}$*, then*

$$\mathbf{F}_k = \left.\frac{\partial \mathbf{f}}{\partial \mathbf{P}}\right|_{\mathbf{P}=\hat{\mathbf{P}}_k} = \begin{bmatrix} \frac{\mathbf{v}_1^\top}{\|\mathbf{v}_1^\top\|} \\ \frac{\mathbf{v}_2^\top}{\|\mathbf{v}_2^\top\|} \end{bmatrix} \tag{B.61}$$

where $\mathbf{v}_1$ *and* $\mathbf{v}_2$ *are represented as column vectors. All terms in eqn. (B.60) are now defined and the equation can be iterated until convergence.*

Let $\bar{\mathbf{P}}$ *denote the final result of the iteration of eqn. (B.60). As shown in eqn. (5.26), when* $\boldsymbol{\nu}$ *is zero mean Gaussian noise with covariance* $E\langle \boldsymbol{\nu}\boldsymbol{\nu}^\top \rangle = \mathbf{R} = \sigma^2 \mathbf{I}$*, then the error* $\delta\mathbf{P} = \mathbf{P} - \bar{\mathbf{P}}$ *is a zero mean Gaussian random variable with covariance* $\mathbf{Q} = E\langle (\delta\mathbf{P})(\delta\mathbf{P})^\top \rangle = \sigma^2 \left(\mathbf{F}^\top\mathbf{F}\right)^{-1}$ *where* $\mathbf{F}$ *denotes the final value of* $\mathbf{F}_k$ *at the conclusion of the iteration. Figure B.3 shows theoretical and simulation results for two instances of the scenario of Figure B.2 that are each further described below. In both cases,* $\sigma = 3m$*, the radius of the satellite orbit was* $R = 20 \times 10^6 m$*,* $\mathbf{P} = [0, 6]^\top \times 10^6 m$*, and the iteration was terminated when the estimated position had converged to 0.001m. Each figure shows both a scatter plot of the position estimation error for 400 realizations repetitions of the experiment and the ellipse that is defined in Section 4.9.1.2 that should contain 50% of the samples.*

For the simulation results shown in the plot on the left, the satellite positions were

$$\mathbf{P}_1 = \begin{bmatrix} -19,156,525.704 \\ 5,746,957.711 \end{bmatrix} m \quad and \quad \mathbf{P}_2 = \begin{bmatrix} 0.000 \\ 20,000,000.000 \end{bmatrix} m.$$

This situation, with one satellite at a high elevation and another satellite at a significantly lower elevation results in the matrix $\mathbf{F}^\top\mathbf{F}$ *being well-conditioned. The iteration converges rapidly to millimeter accuracy. The position estimation error covariance matrix is*

$$\mathbf{Q} = \begin{bmatrix} 9.00 & -0.12 \\ -0.12 & 9.00 \end{bmatrix} m^2$$

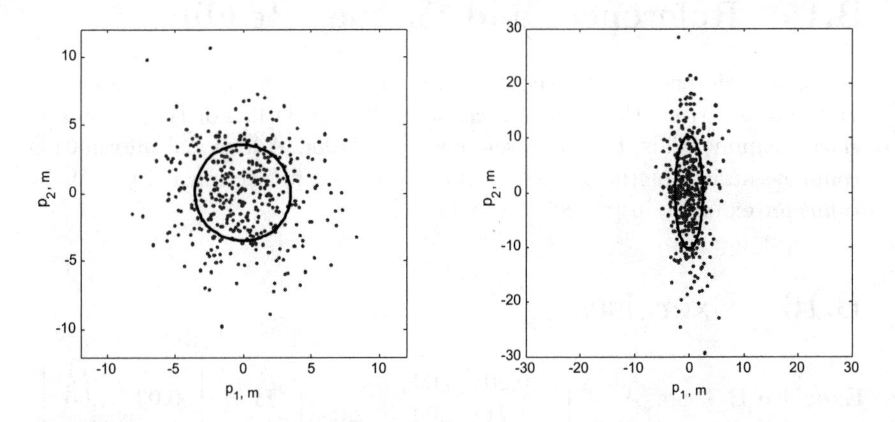

Figure B.3: Estimation error plots for Example B.4. The dots are 400 realizations of the position estimation error from simulation. The curve is the ellipse expected to contain 50% of the position estimation errors.

which shows that the errors in the components of $\mathbf{P}$ *are expected to be almost the same in magnitude and only weakly correlated.*

For the simulation results shown in the plot on the right, the satellite positions were

$$\mathbf{P}_1 = \begin{bmatrix} -17,149,858.514 \\ 10,289,915.108 \end{bmatrix} m \quad and \quad \mathbf{P}_2 = \begin{bmatrix} 17,149,858.514 \\ 10,289,915.108 \end{bmatrix} m.$$

This situation, with both satellites at very low elevation, results in the matrix $\mathbf{F}^\top\mathbf{F}$ *being poorly-conditioned; therefore, the iteration may converge slowly. In this case, the position estimation error covariance matrix is*

$$\mathbf{Q} = \begin{bmatrix} 4.78 & 0.00 \\ 0.00 & 76.42 \end{bmatrix} m^2.$$

which shows significant magnification of the measurement noise in the estimation of the second component of the position vector. This is due to some portion of the state, in this case the second component of the position, only being weakly observable. Note that the components of the position along the vectors $\mathbf{v}_1$ *and* $\mathbf{v}_2$ *are accurately estimated while those orthogonal to both* $\mathbf{v}_1$ *and* $\mathbf{v}_2$ *are not.*

The fact that the accuracy of the position estimate can be reliably expressed in terms of the Jacobian matrix $\mathbf{F}$ *is used in GPS applications as is discussed in Section 8.5.* △

B.15 References and Further Reading

The main reference for this chapter was [59]. On the topic of orthogonalization of matrices, the main sources were [27, 49, 113]. For the Section on matrix exponentials, the main reference was [100]. Additional information about numerical methods such as that described in Section B.14.1 can be found for example in [86, 88].

B.16 Exercises

Exercise B.1 For $\mathbf{A} = \begin{bmatrix} 0.20 & 0.00 & -3.14 \\ -1.71 & 0.00 & 2.00 \end{bmatrix}$, $\mathbf{B} = \begin{bmatrix} 1.00 & 1.71 \\ 0.03 & 2.00 \\ 0.14 & 2.14 \end{bmatrix}$, and $\mathbf{v} = \begin{bmatrix} 0.00 \\ 2.00 \\ -7.14 \end{bmatrix}$, find the following products or state why the product is not valid: $\mathbf{Av}$; $\mathbf{B}^\top\mathbf{v}$; $\mathbf{Bv}$.

Exercise B.2 Define matrices $\mathbf{A} \in \mathbb{R}^{1\times n}$ and $\mathbf{x} \in \mathbb{R}^{n\times 1}$ so that the summation $\sum_{i=1}^{n} a_i x_i$ can be written as a matrix product $\mathbf{Ax}$.

Exercise B.3 A set of linear equations can be written in matrix format. For example, define matrices $\mathbf{Y} \in \mathbb{R}^{4\times 1}$, $\mathbf{A} \in \mathbb{R}^{4\times 4}$ and $\mathbf{x} \in \mathbb{R}^{4\times 1}$ so that

$$\begin{aligned} 4 &= 23x - 4y - 8s \\ 8 &= 42s + 15t + 16x \\ 15 &= 23s - 23t \\ 16 &= 42y - 8s \end{aligned}$$

is equivalent to $\mathbf{Y} = \mathbf{Ax}$. Equations of this form are easily solvable using software packages such as MATLAB.

Exercise B.4 A set of first order ODE's can be conveniently written in matrix form. For example, define a state vector $\mathbf{x}$ and a matrix $\mathbf{A}$ such that the set of three scalar ordinary differential equations

$$\begin{aligned} \dot{v} &= 23v - 4y - 8z \\ \dot{y} &= 42z + 15y + 16v \\ \dot{z} &= 42y - 8z \end{aligned}$$

can be written in the state space form $\dot{\mathbf{x}} = \mathbf{Ax}$ with order $n = 3$.

Exercise B.5 For $\mathbf{A} = \begin{bmatrix} 1 & 5 & 4 \\ 0 & 7 & -1 \\ 3 & 2 & 7 \end{bmatrix}$, the minor $M_{1,1}$ is

$$M_{11} = \begin{vmatrix} 7 & -1 \\ 2 & 7 \end{vmatrix} = (7)(7) - (2)(-1) = 51.$$

1. Find the other eight minors.
2. Use the row expansion formula to compute the determinant of **A**.
3. Use the column expansion formula to compute the determinant of **A**.

Exercise B.6 Use the determinant to decide whether the matrices **A** and **B** are nonsingular.

$$\mathbf{A} = \begin{bmatrix} 1 & 0 & 0 \\ 1 & 1 & 0 \\ 1 & 1 & 1 \end{bmatrix} \qquad \mathbf{B} = \begin{bmatrix} 1 & 0 & 1 \\ 1 & 1 & 0 \\ 1 & 1 & 0 \end{bmatrix}.$$

Exercise B.7 Show that the determinant of a permutation matrix is 1.

Exercise B.8 For $\mathbf{A} = \begin{bmatrix} 1 & 5 & 4 \\ 0 & 7 & -1 \\ 3 & 2 & 7 \end{bmatrix}$, using the cofactor formula in eqn. (B.28):

$$\begin{aligned} c_{11} &= (-1)^{1+1} 51 = 51. \\ c_{21} &= (-1)^{2+1} 27 = -27. \end{aligned}$$

Compute the remaining seven elements of the cofactor matrix.

Exercise B.9 Let $\mathbf{A} = \begin{bmatrix} 1 & 0 & 6 \\ 0 & 4 & -3 \\ 3 & 5 & -1 \end{bmatrix}$.

1. Compute the cofactor matrix.
2. Compute the inverse matrix.

Exercise B.10 Use the properties of determinants to show that

$$|\mathbf{A}^{-1}| = \frac{1}{|\mathbf{A}|}.$$

Exercise B.11 Let $(s\mathbf{I} - \mathbf{A}) = \begin{bmatrix} s & -1 & 0 \\ a & s+b & -c \\ -d & 0 & s+f \end{bmatrix}$. Find $(s\mathbf{I} - \mathbf{A})^{-1}$.

Exercise B.12 Verify Lemma B.5.1 by direct multiplication of the right-hand sides of eqns. (B.30) and (B.31).

Exercise B.13

1. Let λ_i and $\mathbf{v}_i$ for $i \in [1, n]$ denote the eigenvalues and eigenvectors of $\mathbf{A} \in \mathbb{R}^{n \times n}$. Determine the eigenvalues and eigenvectors of $\mathbf{A}^2$.

2. Determine the eigenvalues and eigenvectors of $\mathbf{A}^m$ for any positive integer m.

3. Use the above property of eigenvalues to show that the only possible eigenvalues of an idempotent matrix are 0 and 1.

Exercise B.14 For the special case of $\mathbf{x} \in \mathbb{R}^n$, either use direct multiplication or the properties of scalar and vector products to show that $\mathbf{x}^\top \mathbf{B}_a \mathbf{x} = 0$, where $\mathbf{B}_a$ is an antisymmetric matrix.

Exercise B.15 Prove the following:

1. If $\mathbf{B}$ is a positive definite matrix, then all the eigenvalues of $\mathbf{B}$ are strictly positive real numbers.

2. If $\mathbf{B}$ is a positive semi-definite matrix, then all the eigenvalues of $\mathbf{B}$ are non-negative real numbers.

3. If $\mathbf{B}$ is a positive definite matrix, then $\mathbf{B}^{-1}$ exists and is also positive definite.

Exercise B.16 Prove eqn. (B.33) by following the outline below. Note that the hats on the variables have been dropped to simplify the notation.

1. Consider the cost function

$$J(\mathbf{A}_o, \mathbf{\Lambda}) = tr\left((\mathbf{A}_o - \mathbf{A})^\top (\mathbf{A}_o - \mathbf{A}) + \mathbf{\Lambda}\left(\mathbf{A}_o^\top \mathbf{A}_o - \mathbf{I}\right)\right)$$

where the Lagrange multiplier $\mathbf{\Lambda}$ is a symmetric matrix. Take the derivative of J with respect to $\mathbf{A}_o$ and show that the matrix $\mathbf{A}_o$ that minimizes this expression is

$$\mathbf{A}_o = \mathbf{A}\left(\mathbf{I} + \mathbf{\Lambda}\right)^{-1}. \tag{B.62}$$

2. Substitute the result of eqn. (B.62) into the constraint

$$\mathbf{A}_o^\top \mathbf{A}_o = \mathbf{I}$$

to show that

$$(\mathbf{I} + \mathbf{\Lambda}) = \pm\left(\mathbf{A}^\top \mathbf{A}\right)^{\frac{1}{2}}. \tag{B.63}$$

3. Complete the proof by combining eqns. (B.62) and (B.63) to obtain

$$\hat{\mathbf{A}}_o = \hat{\mathbf{A}} \left(\hat{\mathbf{A}}^\top \hat{\mathbf{A}} \right)^{-1/2}.$$

Exercise B.17 Rederive the algorithm of Section B.14.2 for the case where

$$J(\mathbf{x}) = \left(\mathbf{y} - \left(\mathbf{f}(\mathbf{x}_k) + \mathbf{F}_k\left(\mathbf{x} - \mathbf{x}_k\right)\right)\right)^\top \mathbf{R}^{-1} \left(\mathbf{y} - \left(\mathbf{f}(\mathbf{x}_k) + \mathbf{F}_k\left(\mathbf{x} - \mathbf{x}_k\right)\right)\right).$$

Appendix C

Calculation of GPS Satellite Position & Velocity

The interface between the GPS *space* and *user* segments consists of a two frequency (L1 = 1575.42 MHz and L2 = 1227.60 MHz) radio link. Utilizing these links, the satellites that comprise the *space* segment provide continuous Earth coverage, transmitting signals which provide to the *user* segment the ranging codes and system data needed to accomplish the GPS navigation mission. The transmitted satellite parameter set is computed and controlled by the *control* segment and uploaded to the satellite for broadcast to the user. The satellite signals are received and decoded by the user GPS receiver. The decoded data is available from the receiver usually via a serial link.

The data decoded from the satellite signal include satellite clock calibration parameters, satellite position calculation parameters, atmospheric correction parameters, reference time, and almanac and health information for all space segment satellites. The satellite message parameter set is displayed in Table C.1. This book assumes that the reader will download the satellite parameter set into a local data structure. This appendix describes how the time variables and ephemeris data set are processed to determine clock corrections, satellite positions, and atmospheric corrections.

To clearly present the main ideas, it is important to distinguish between the following time variables:

t'_r – the pseudo-time at which the receiver takes a measurement,

t_{sv} – the pseudo-time at which the satellite broadcast the signal that is measured at the receiver,

t'_p – the propagation pseudo-time of the signal from the satellite antenna to the receiver antenna.

Δt_{sv} – the satellite clock bias,

t – the GPS time at which the satellite broadcast the signal that is measured at the receiver,

b – the receiver clock bias,

t_r – the GPS time at which the receiver takes a measurement,

t_p – the propagation time of the signal from the satellite antenna to the receiver antenna.

The variables t and t_{sv} are related by $t_{sv} = t + \Delta t_{sv}$. The variable t_{sv} is called the *channel time* because the receiver maintains a distinct value for each channel. The variables t_r and t'_r are related by $t'_r = t_r + b$. The variable t'_r is the *user* or *receiver* time and is common for all channels. The propagation time is $t_p = t_r - t$. For a receiver on the surface of the Earth, typical values for t_p are 60 to 90 ms.

The above variables are listed in the order in which they could actually be computed by the receiver. An overview of the process is as follows, assuming that b is initially unknown:

1. At time t'_r, the receiver samples all channels that are tracking a signal.

2. For each channel, based on various channel variables (z-count, bit count, number of C/A repetitions, number of C/A code chips, and code phase)[1] the receiver determines t_{sv}.

3. For each channel, the receiver computes the propagation pseudo-time and pseudorange

$$t'_p = t'_r - t_{sv} \tag{C.1}$$

$$\rho = ct'_p. \tag{C.2}$$

4. Based on eqn. (C.4) in Section C.1, the receiver computes Δt_{sv}.

5. Knowledge of Δt_{sv} allows computation of t by eqn. (C.6) in Section C.1. Knowledge of t allows calculation of the satellite positions as discussed in Sections C.2 and C.3. With the satellite positions and the pseudorange measurements from each channel the receiver can compute its position and the clock bias b.

6. At this point, the receive could compute t_r and t_p.

[1]See p. 215 in [76].

The details are described in the following two subsections.

Within the receiver the above process works well, but for users working with reported observables outside the receiver, the channel time variables are usually not reported. When the receiver output stream includes a single time stamp t'_r for all pseudorange observables, then the channel time is recovered as

$$t_{sv} = t'_r - \frac{\rho}{c} \tag{C.3}$$

where ρ is computed as in eqn. (C.2). Since t_{sv} is the first variable needed by the user, receivers frequently output the pseudorange in the form of t'_p instead of ρ, which avoids a unit transformation on the receiver that would just be undone by the user.

C.1 Satellite Clock Corrections

Assume that a set of simultaneous pseudoranges are available. For each channel, determine the channel time t_{sv}.

The correction Δt_{sv} is calculated by the polynomial correction[2]

$$\Delta t_{sv} = a_{f0} + a_{f1}(t_{sv} - t_{oc}) + a_{f2}(t_{sv} - t_{oc})^2 + \Delta t_r \tag{C.4}$$

where t_{oc} and the polynomial coefficients are a portion of the navigation message. The term

$$\Delta t_r = FeA^{1/2}\sin(E_k) \tag{C.5}$$

is a relativistic correction, F is a constant (see Appendix A), e and A are defined in Table C.1, and E_k will be defined in Section C.2. Then the actual GPS system time at which the signal was sent is

$$t = t_{sv} - \Delta t_{sv}. \tag{C.6}$$

The ground segment uses eqns. (C.6) and (C.4) as written to determine the polynomial based estimate for the clock correction; however, this results in a coupled set of equations for the user. Sensitivity analysis has shown [6] that it is sufficiently accurate to approximate t by t_{sv} in eqn. (C.4).

The parameter t_{oc} is the reference time for the applicability of the clock correction data. The difference $(t_{sv} - t_{oc})$ is not expected to be large. In fact, the difference is assumed to lie in the interval $[-302400, 302400]$ seconds, whose length corresponds to the number of seconds in a GPS week. However, since t_{oc} is expressed in seconds of a given GPS week, it may happen near the beginning or end of a week that t_{sv} is referenced to one week while t_{oc} is referenced to a different week. In such cases $|t_{sv} - t_{oc}|$

[2]In [6] the following equation is written in a slightly different format that implies an iterative implementation is required if Δt_{sv} were large. Because Δt_{sv} is normally small, the iteration is not required and the following version is typically sufficient.

would be greater than 302400 s. When this occurs the user is responsible for reflecting $(t_{sv} - t_{oc})$ back into the proper range by adding or subtracting 604800 s.

The parameter IODC in Table C.1 is an integer which allows the user to detect when new clock model parameters are available. By monitoring IODC of the incoming satellite signal relative to the IODC of the clock parameters currently in use, the user can determine when to update the clock model parameters.

C.2 Satellite Position Calculations

The ephemeris parameters describe the orbit during a given interval of time (at least one hour). The ephemeris parameters are the parameters for an extension of the orbital model predicted by Kepler. The extension is necessary to account for non-uniformities in the Earth gravitational field. The ephemeris parameters are determined by the control segment as a curve fit to the measured satellite orbit. The parameter t_{oe} is the reference time of applicability (local origin) of the ephemeris parameters.

The ephemeris parameters are defined in Table C.1 [6]. These parameters are broadcast by the satellites and available after decoding on-board the receiver. The units of the broadcast parameters are also defined in the table.

The ECEF coordinates for the phase center of the satellite antenna can be calculated using a variation of the equations shown in Table C.2 [131]. The main inputs to these equations are the variable t_{sv} and the satellite clock correction and ephemeris parameters defined in Table C.1.

The satellite antenna phase center position is very sensitive to small perturbations in most ephemeris parameters. The sensitivity of position to the parameters $\sqrt{A}$, C_{rc}, and C_{rs} is about one meter/meter. The sensitivity to angular parameters is on the order of 10^8 meters/semi-circle, and to the angular rate parameters is on the order of 10^{12} meters/semi-circle/second. Because of this extreme sensitivity to angular perturbations, the values of all constants should exactly match those stated in Appendix A, from [6].

The parameter IODE is an integer which allows the user to detect when a new set of ephemeris parameters is available. By monitoring the IODE value for the incoming satellite signal relative to the IODE of the ephemeris data currently in use, the user can determine when to update the ephemeris model parameters.

A few comments are appropriate concerning Table C.2. The variable t_k is the actual total time difference between the time t and the ephemeris reference time t_{oe}. The calculation must account for beginning or end of week crossovers. That is, if t_k is greater than 302,400 seconds, subtract 604,800 seconds from t_k. If t_k is less than -302,400 seconds, add 604,800

Parameter	Description
T_{GD}	Group Delay, sec
IODC	Issue of Data, Clock
t_{oc}	Clock data reference time, sec
a_{f2}	Second order correction to satellite clock, $\frac{sec}{sec^2}$
a_{f1}	First order correction to satellite clock, $\frac{sec}{sec}$
a_{f0}	Constant correction to satellite clock, sec
M_0	Mean anomaly at reference time, semi-circles
Δn	Mean motion difference from computed value, semi-circles/sec
e	Eccentricity, dimensionless
$A^{1/2}$	Square root of the semi-major axis, $\sqrt{m}$
Ω_0	Longitude of ascension node at reference time, semi-circle
i_0	Inclination angle at reference time, semi-circle
ω	Argument of perigee, semi-circle
$\dot{\Omega}$	Rate of right ascension, semi-circle/sec
$IDOT$	Rate of inclination angle, semi-circle/sec
C_{uc}	Amplitude of the cosine harmonic correction to the argument of latitude, rad
C_{us}	Amplitude of the sine harmonic correction to the argument of latitude, rad
C_{rc}	Amplitude of the cosine harmonic correction to the orbit radius, m
C_{rs}	Amplitude of the sine harmonic correction to the orbit radius, m
C_{ic}	Amplitude of the cosine harmonic correction to the angle of inclination, rad
C_{is}	Amplitude of the sine harmonic correction to the angle of inclination, rad
t_{oe}	Ephemeris reference time, sec
$IODE$	Issue of Data, Ephemeris

Table C.1: Clock and ephemeris parameter definitions.

seconds to t_k. The equation for E_k must be solved iteratively. Various iterative solution techniques are considered in [131].

Table C.3 [44] contains a complete example set of ephemeris parameters. By using these ephemeris parameters in the equations of Table C.2 the space vehicle (satellite) time offset and ECEF satellite antenna coordinates can be determined. The results of these calculations are shown in Table C.4.

Equation	Units	Description
$\Delta t_r = Fe\sqrt{A}\sin(E_k)$	s	Rel. corr. term
$\Delta t_{sv} = a_{f0} + a_{f1}(t_{sv} - t_{oc}) + a_{f2}(t_{sv} - t_{oc})^2 + \Delta t_r$	s	Corr. to SV clock
$t = t_{sv} - \Delta t_{sv}$	s	Corr. mess. trans. time
$A = (\sqrt{A})^2$	m	Orbit semi-major axis
$n_0 = \sqrt{\frac{\mu}{A^3}}$	rps	Comp'd mean motion
$t_k = t - t_{oe}$	s	Time from ref. epoch
$n = n_o + \Delta n$	rps	Corr. mean motion
$M_k = M_0 + t_k n$	rad	Mean anomaly
$E_k = M_k + e\sin(E_k)$	rad	Kepler's ecc. anom. eqn.
$v_k = \text{atan}\left(\frac{\sqrt{1-e^2}\sin(E_k)}{1-e\cos(E_k)}, \frac{\cos(E_k)-e}{1-e\cos(E_k)}\right)$	rad	True anomaly
$\phi_k = v_k + \omega$	rad	Argument of lat.
$\delta u_k = C_{us}\sin(2\phi_k) + C_{uc}\cos(2\phi_k)$	rad	Arg. of lat. corr.
$\delta r_k = C_{rs}\sin(2\phi_k) + C_{rc}\cos(2\phi_k)$	m	Radius corr.
$\delta i_k = C_{is}\sin(2\phi_k) + C_{ic}\cos(2\phi_k)$	rad	Inclination corr.
$u_k = \phi_k + \delta u_k$	rad	Corr. arg. of lat.
$r_k = A(1 - e\cos(E_k)) + \delta r_k$	m	Corr. radius
$i_k = i_0 + \delta i_k + (IDOT)t_k$	rad	Corr. inclination
$X_k = r_k\cos(u_k)$	m	Orb. plane x pos.
$Y_k = r_k\sin(u_k)$	m	Orb. plane y pos.
$\Omega_k = \Omega_0 + (\dot{\Omega} - \dot{\Omega}_e)t_k - \dot{\Omega}_e t_{oe}$	rad	Corr. long. of asc. node
$x_k = X_k\cos(\Omega_k) - Y_k\cos(i_k)\sin(\Omega_k)$	m	SV ECEF x coord.
$y_k = X_k\sin(\Omega_k) + Y_k\cos(i_k)\cos(\Omega_k)$	m	SV ECEF y coord.
$z_k = Y_k\sin(i_k)$	m	SV ECEF z coord.

Table C.2: Equations for calculating satellite ECEF position based on navigation message ephemeris parameters and time. The abbreviation rps stands for radians per second.

Parameter	Value	Units
t_{sv}	4.03272930×10^5	sec.
w_n	910	GPS week number
t_{ow}	403230	Seconds of GPS week
t_{gd}	2.3283×10^{-9}	sec. (Group delay)
$aodc$	409	(Clock data issue)
t_{oc}	410400	sec
a_{f2}	0.00	sec/sec^2
a_{f1}	1.819×10^{-12}	sec/sec
a_{f0}	$3.29776667 \times 10^{-5}$	sec
$AODE$	153	(Orbit data issue)
Δn	4.3123×10^{-9}	rad/sec
M_0	2.24295542	rad
e	$4.27323824 \times 10^{-3}$	–
$A^{1/2}$	5.15353571×10^3	$\sqrt{\text{m}}$
t_{oe}	410400	sec
C_{ic}	9.8720193×10^{-8}	rad
C_{rc}	282.28125	meters
C_{is}	$-3.9115548 \times 10^{-8}$	rad
C_{rs}	-132.71875	meters
C_{uc}	$-6.60121440 \times 10^{-6}$	rad
C_{us}	$5.31412661 \times 10^{-6}$	rad
Ω_0	2.29116688	rad
ω	-0.88396725	rad
i_0	0.97477102	rad
$\dot{\Omega}$	-8.025691×10^{-9}	rad/sec
$IDOT$	-4.23946×10^{-10}	rad/sec

Table C.3: An example of received ephemeris information for a sample satellite.

Variable	Value	Units
A	2.6559×10^7	m
n_0	1.4587×10^{-4}	rad/s
Δt_{sv}	3.2965×10^{-5}	sec
t	4.0327×10^5	sec
t_k	-7.1271×10^3	sec
n	1.4587×10^{-4}	rad/s
M_k	1.2033	rad
E_k	1.2073	rad
v_k	1.2113	rad
ϕ_k	3.2735×10^{-1}	rad
δu_k	-2.0003×10^{-6}	rad
δr_k	1.4310×10^2	m
δi_k	5.4489×10^{-8}	rad
u_k	3.2735×10^{-1}	rad
r_k	2.6519×10^7	m
i_k	9.7477×10^{-1}	rad
X_k	2.5111×10^7	m
Y_k	8.5267×10^6	m
Ω_k	-2.7116×10^1	rad
x_k	-5.67841101×10^6	m
y_k	-2.49239629×10^7	m
z_k	7.05651887×10^6	m

Table C.4: Calculated Satellite Position in ECEF Coordinates

C.3 Reference Frame Consistency

Section C.2 presented the standard equations use to compute the position $\mathbf{p}^i(t^i)$ of the i-th satellite vehicle at the time of transmission t^i as represented in the ECEF frame at time t^i.

A subtlety in the equations presented in Table C.2 is the fact that the computed position for each satellite is in a distinct frame-of-reference and are therefore not compatible. The ECEF is not an inertial frame. Due to the rotation of the frame, the ECEF frame a time t_1 is distinct from the ECEF frame at time t_2. Let $[\mathbf{P}_i]^{e(t)}$ denote the coordinates of the point $\mathbf{P}_i$ with respect to the ECEF frame at time t. Even in the case that $\mathbf{P}_1$ is fixed in inertial space, in general $[\mathbf{P}_1]^{e(t_1)} \neq [\mathbf{P}_1]^{e(t_2)}$ for $t_1 \neq t_2$.

Use of time-of-transit to measure range is premised on the fact that the speed-of-light through a vacuum is constant relative to inertial reference frames. Therefore, if the coordinates of two points were known in an inertial frame $[\mathbf{P}_1]^i$ and $[\mathbf{P}_2]^i$ the time for light to transit between the points through a vacuum would be

$$\tau = \frac{\left\| [\mathbf{P}_1]^i - [\mathbf{P}_2]^i \right\|}{c}.$$

When the coordinates of the two points are known in distinct reference frames a and b, $[\mathbf{P}_1]^a$ and $[\mathbf{P}_2]^b$, neither of which is an inertial frame, then they could be transformed to an inertial frame:

$$\tau = \frac{\left\| \mathbf{R}_a^i [\mathbf{P}_1]^a - \mathbf{R}_a^i \mathbf{R}_b^a [\mathbf{P}_2]^b \right\|}{c}.$$

Because the norm of a vector is invariant under rotational transformation, this is equivalent to

$$\tau = \frac{\left\| [\mathbf{P}_1]^a - \mathbf{R}_b^a [\mathbf{P}_2]^b \right\|}{c}.$$

There are two important points. First, to calculate the distance between these two points $\mathbf{P}_1$ and $\mathbf{P}_2$, they must first be represented in the same reference frame. Second, there is not a unique best frame. The following discussion will use the ECEF frame at an arbitrary time denoted as t_a. Often the selected time is t_r' the time at which the receiver measured the pseudoranges.

Given $[\mathbf{P}_1]^{e(t_1)}$ and $[\mathbf{P}_2]^{e(t_2)}$ the coordinates of $\mathbf{P}_1$ and $\mathbf{P}_2$ with respect to the ECEF frame at time t_a are

$$\begin{aligned} [\mathbf{P}_1]^{e(t_a)} &= \mathbf{R}_{e(t_1)}^{e(t_a)} [\mathbf{P}_1]^{e(t_1)} \\ [\mathbf{P}_2]^{e(t_a)} &= \mathbf{R}_{e(t_2)}^{e(t_a)} [\mathbf{P}_2]^{e(t_2)}. \end{aligned}$$

The coordinate transformation matrix

$$\mathbf{R}_{e(t_i)}^{e(t_a)} = [\omega_{ie}^e(t_a - t_i)]_3 = \begin{bmatrix} \cos(\omega_{ie}^e(t_a - t_i)) & \sin(\omega_{ie}^e(t_a - t_i)) & 0 \\ -\sin(\omega_{ie}^e(t_a - t_i)) & \cos(\omega_{ie}^e(t_a - t_i)) & 0 \\ 0 & 0 & 1 \end{bmatrix}$$

is a plane rotation that accounts for the the rotation of the Earth over the time span $(t_a - t_i)$ at angular rate ω_{ie} around the ECEF z-axis.

In the case where $t_a = t_r'$, then for the i-th satellite, its position is rotated by the angle $(\omega_{ie}^e t_p^i) = \omega_{ie}^e(t_r' - t_{sv}^i)$ where t_p^i and t_{sv}^i are the quantities defined in eqn. (C.1) referred to the i-th satellite. Neglecting to transform all satellite positions to the same frame-of-reference can result in range errors of up to $\pm 40m$.

C.4 Satellite Velocity

Use of the GPS Doppler measurements requires knowledge of the velocity of the satellite. That issue is addressed in this section.

C.4.1 Equations from Ephemeris

Equations for the satellite position as a function of time are shown in Table C.2. The satellite velocity can be computed as the derivative of the satellite position as summarized in Table C.5.

$$\begin{aligned}
\dot{E}_k &= \frac{n_0 + \Delta n}{1 - e\cos(E_k)} \\
\dot{\phi}_k &= \frac{\sqrt{1-e^2}}{1 - e\cos(E_k)} \dot{E}_k \\
\dot{u}_k &= (1 + 2C_{us}\cos(2\phi_k) - 2C_{uc}\sin(2\phi_k))\,\dot{\phi}_k \\
\dot{r}_k &= 2\,(C_{rs}\cos(2\phi_k) - C_{rc}\sin(2\phi_k))\,\dot{\phi}_k + Ae\sin(E_k)\dot{E}_k \\
\dot{X}_k &= \dot{r}_k\cos(u_k) - r_k\sin(u_k)\dot{u}_k \\
\dot{Y}_k &= \dot{r}_k\sin(u_k) + r_k\cos(u_k)\dot{u}_k \\
\frac{di_k}{dt} &= 2\,(C_{is}\cos(2\phi_k) - C_{ic}\sin(2\phi_k))\,\dot{\phi}_k + IDOT \\
\dot{\Omega}_k &= \dot{\Omega} - \dot{\Omega}_e \\
\dot{x}_k &= \dot{X}_k\cos(\Omega_k) - \dot{Y}_k\cos(i_k)\sin(\Omega_k) + Y_k\sin(i_k)\sin(\Omega_k)\frac{di_k}{dt} - y_k\dot{\Omega}_k \\
\dot{y}_k &= \dot{X}_k\sin(\Omega_k) + \dot{Y}_k\cos(i_k)\cos(\Omega_k) - Y_k\sin(i_k)\cos(\Omega_k)\frac{di_k}{dt} + x_k\dot{\Omega}_k \\
\dot{z}_k &= \dot{Y}_k\sin(i_k) + Y_k\cos(i_k)\frac{di_k}{dt}
\end{aligned}$$

Table C.5: Equations for calculating satellite ECEF velocity based on navigation message ephemeris parameters and time.

The equations of Table C.5 are derived by starting with the last three rows of Table C.2 and working up through that table computing the necessary derivatives to complete the computation. The only step requiring clarification is the derivation of $\dot{\phi}_k$.

From Table C.2 it is clear that $\dot{\phi}_k = \dot{\upsilon}_k$ and

$$\upsilon_k = \operatorname{atan}\left(\frac{\sqrt{1-e^2}\sin(E_k)}{1-e\cos(E_k)}, \frac{\cos(E_k)-e}{1-e\cos(E_k)}\right).$$

By the chain and quotient rule for differentiation it is straightforward to derive

$$\dot{\upsilon}_k = \frac{\sqrt{1-e^2}\cos(E_k)\left(\cos(E_k)-e\right)+\sqrt{1-e^2}\sin^2(E_k)}{\left(\cos(E_k)-e\right)^2+\left(\sqrt{1-e^2}\sin(E_k)\right)^2}\dot{E}_k$$

which simplifies as follows:

$$\begin{aligned}
\dot{\upsilon}_k &= \frac{\sqrt{1-e^2}\left(1-e\cos(E_k)\right)}{e^2-2e\cos(E_k)+\cos^2(E_k)+\sin^2(E_k)-e^2\sin^2(E_k)}\dot{E}_k \\
&= \frac{\sqrt{1-e^2}\left(1-e\cos(E_k)\right)}{\left(1-e\cos(E_k)\right)^2}\dot{E}_k \\
&= \left(\frac{\sqrt{1-e^2}}{1-e\cos(E_k)}\right)\dot{E}_k.
\end{aligned}$$

C.4.2 Practical Issues

The GPS Doppler measurement is constructed as the change in the carrier phase angle over a stated time interval. Therefore the GPS Doppler measurement is the change in the distance between the receiver and satellite antennae (plus clock drift) over the stated period of time. Dividing this pseudorange change by the length of the interval yields a measurement of the average pseudorange rate of change over the time interval.

To remove the effects of satellite motion from the Doppler measurement accurately, the satellite velocity should be compatible with the average pseudorange rate. This requires the average satellite velocity over the same time interval. One approach is to compute the satellite velocity at the center of the interval. Another approach is to compute the satellite velocity at the beginning and end of the interval and to average their values. A third approach is to difference the satellite positions at the beginning and end of the time interval. This third approach eliminates the need to compute the equations in Table C.5.

C.5 Ionospheric Model

This section presents the Klobuchar ionospheric correction model. The variable definitions and units are summarized in Table C.6. This model should only be used by users with single frequency receivers that are not operating in differential mode. Dual frequency users should refer to Section 8.6. Differential operation requires consideration of the DGPS protocol and base-to-user distance. If the RTCM-104 standard is being used, then the corrections should include an atmospheric error correction.

The Klobuchar model was defined under the constraints of using only eight coefficients and approximately one daily model update, to provide a worldwide correction for approximately 50% of the ionospheric delay at mid-latitudes. The selected form of the model was a bias plus half-cosine:

$$t_{ion} = F\left(b + AMP\cos\left(\frac{2\pi(t-\zeta)}{PER}\right)\right). \tag{C.7}$$

The F in this equation is a slant correction factor and is distinct from the constant used in eqn. (C.5). Due to the eight coefficient constraint, a study was performed to determine how to best allocate the eight GPS message coefficients to the four parameters of eqn. (C.7). The model was found to be most sensitive to the amplitude AMP and period PER terms. Therefore, b and ζ are represented by constant terms in the model. Third order polynomial expansions are used for AMP and PER.

The inputs to the model are the user geodetic latitude ϕ and longitude λ, the GPS time t_{gps}, and the user relative azimuth A and elevation E of the satellite. The GPS message supplies four α and four β parameters. The algorithm proceeds as follows:

1. Based on the satellite elevation, compute the Earth central angle between the user position and the Earth projection of the ionospheric intersection point:

$$\psi = \frac{0.0137}{E+0.11} - 0.022. \tag{C.8}$$

2. Based on the user location and the satellite azimuth, calculate the geodetic latitude and longitude of the Earth projection of the ionospheric intersection point:

$$\phi_i = \begin{cases} \phi_u + \psi\cos(A) & \text{if } |\phi_i| \le 0.416 \\ 0.416 & \text{if } \phi_i > 0.416 \\ -0.416 & \text{if } \phi_i < -0.416 \end{cases} \tag{C.9}$$

$$\lambda_i = \lambda_u + \frac{\psi\sin(A)}{\cos(\phi_i)}. \tag{C.10}$$

3. Calculate the local time at the Earth projection of the ionospheric intersection point:

$$t = 4.32 \times 10^4 \lambda_i + t_{gps}. \tag{C.11}$$

The variable t in the model is assumed to lie in the stated range. If the computation results in an out-of-range value, the user is responsible for reflecting it back in to the specified range by adding or subtracting 86400.

4. Calculate the geomagnetic latitude of the Earth projection of the ionospheric intersection point:

$$\phi_m = \phi_i + 0.064 \cos(\lambda_i - 1.617). \tag{C.12}$$

5. Used the β terms from the GPS message to calculate the period:

$$PER = \begin{cases} \sum_{n=0}^{3} \beta_n (\phi_m)^n & \text{if } PER \geq 72000 \\ 72000 & \text{if } PER < 72000. \end{cases} \tag{C.13}$$

6. Calculate the argument of the cosine term:

$$x = \frac{2\pi(t - 50400)}{PER}. \tag{C.14}$$

7. Calculate the slant factor:

$$F = 1.0 + 16.0(0.53 - E)^3. \tag{C.15}$$

8. Use the α terms from the GPS message to calculate the amplitude:

$$AMP = \begin{cases} \sum_{n=0}^{3} \alpha_n (\phi_m)^n & \text{if } AMP \geq 0 \\ 0 & \text{if } AMP < 0. \end{cases} \tag{C.16}$$

9. Calculate the L1 ionospheric correction:

$$t1_{ion} = \begin{cases} F\left(5 \times 10^{-9} + AMP(1 - \frac{x^2}{2} + \frac{x^4}{24})\right) & \text{if } |x| < 1.57 \\ 5F \times 10^{-9} & \text{if } |x| \geq 1.57. \end{cases}$$

A three term (fourth order) expansion of the cosine has been used. The condition $|x| \geq 1.57$ indicates the time t is in the night.

10. If needed, calculate the L2 ionospheric correction:

$$t2_{ion} = \frac{77^2}{60^2} t1_{ion}. \tag{C.17}$$

Symbol	Units	Description
α_n		Coefficients of cubic fit to the amplitude of vertical delay
β_n		Coefficients of cubic fit to the period of the model
E	semi-circles	User to SV elevation angle
A	semi-circles	User to SV azimuth (clockwise from true north)
ϕ	semi-circles	User geodetic WGS-84 latitude
λ	semi-circles	User geodetic WGS-84 longitude
t_{gps}	sec	GPS system time
ψ	semi-circles	Earth central angle between the user position and the Earth projection of the ionospheric intersection point
ϕ_i	semi-circles	Geodetic latitude of the Earth projection of the ionospheric intersection point
λ_i	semi-circles	Geodetic longitude of the Earth projection of the ionospheric intersection point
t	sec	Local time, range = $[0, 86400)$
ϕ_m	semi-circles	Geomagnetic latitude of the Earth projection of the ionospheric intersection point
PER	sec	Period
x	radians	Phase
F		Slant factor
AMP	sec	Amplitude of cos term
$t1_{ion}$	sec	L1 ionospheric correction
$t2_{ion}$	sec	L2 ionospheric correction

Table C.6: Ionospheric correction variable definitions.

C.6 References and Further Reading

Readers interested in the details of the satellite signal format should see [6, 131]. The main source for Sections C.1–C.2 was [6]. The main reference for Section C.3 was [13]. At the time of publication, the author is not aware of a published source for the information in Section C.4; although the information is well known. The main sources for Section C.5 were [6, 79, 80].

Appendix D

Quaternions

While the orientation between any two frames may result from a sequence of rotations, a theorem due to Euler states that the transformation between any two frames can be represented as a single rotation about a single fixed vector [61]. In the following, we will consider frames a and b. Frame b results from rotation of frame a by the angle ζ about the unit vector $\mathbf{E} \in \mathbb{R}^3$.

For a vehicle experiencing arbitrary angular rotations, the effective axis $\mathbf{E}$ and rotation angle ζ that represent the rotational transformation between the body and navigation frames will evolve over time. This section introduces the quaternion method for parameterizing the effective axis and rotation angle. This section reviews the properties of quaternions [50, 61, 103, 118, 126, 129]. It presents the methods to compute the direction cosine matrix R_b^n and the roll ϕ, pitch θ, and yaw ψ angles. Finally, it presents algorithms for computing quaternions based on the body frame angular rate vector. Quaternion parameterizations may appear less intuitively appealing, but the quaternion approach is often the preferred implementation approach due to the linearity of the quaternion differential equations, the lack of singularities, the lack of trigonometric functions in the integration routine (in comparison to Euler angle integration), and the small number of parameters (relative to direction cosine integration).

D.1 Quaternions Basics

A complex number has two parameters $(a_1, a_2) \in \mathbb{R}^2$ and can be expressed as

$$\mathbf{z} = a_1 + \mathbf{i}a_2$$

where $\mathbf{i}$ satisfies $\mathbf{i}^2 = -1$. The complex number z is the real linear combination of the basis 1 and $\mathbf{i}$. Complex numbers are a convenient means for expressing rotations of vectors in a two-dimensional space.

The quaternion has four parameters $\mathbf{b} = (b_1, b_2, b_3, b_4) \in \mathbb{R}^4$ and can be represented by a generalized (four component) complex number:

$$\mathbf{b} = b_1 + b_2\mathbf{i} + b_3\mathbf{j} + b_4\mathbf{k} \tag{D.1}$$

where 1, $\mathbf{i}, \mathbf{j}, \mathbf{k}$ are the quaternion basis. The symbol $\circ$ will be used to denote the quaternion product. The product of two quaternions yields a third quaternion. The quaternion product has the following properties:

$$\begin{array}{lll} \mathbf{i} \circ \mathbf{i} = -1, & \mathbf{i} \circ \mathbf{j} = \mathbf{k}, & \mathbf{i} \circ \mathbf{k} = -\mathbf{j}, \\ \mathbf{j} \circ \mathbf{j} = -1, & \mathbf{j} \circ \mathbf{k} = \mathbf{i}, & \mathbf{j} \circ \mathbf{i} = -\mathbf{k}, \\ \mathbf{k} \circ \mathbf{k} = -1, & \mathbf{k} \circ \mathbf{i} = \mathbf{j}, & \mathbf{k} \circ \mathbf{j} = -\mathbf{i}. \end{array}$$

The conjugate or adjoint of $\mathbf{b}$ is

$$\bar{\mathbf{b}} = b_1 - b_2\mathbf{i} - b_3\mathbf{j} - b_4\mathbf{k}. \tag{D.2}$$

Addition or subtraction of quaternions is defined as the addition and subtraction of the corresponding components of the quaternions.

By the distributive properties of multiplication and the above properties, the product of quaternions $\mathbf{b}$ and $\mathbf{c}$ is

$$\begin{aligned} \mathbf{b} \circ \mathbf{c} &= (b_1c_1 - b_2c_2 - b_3c_3 - b_4c_4) + (b_1c_2 + b_2c_1 + b_3c_4 - b_4c_3)\,\mathbf{i} \\ &\quad + (b_1c_3 - b_2c_4 + b_4c_2 + b_3c_1)\,\mathbf{j} + (b_1c_4 + b_2c_3 - b_3c_2 + b_4c_1)\,\mathbf{k} \\ &= \begin{bmatrix} b_1 & -b_2 & -b_3 & -b_4 \\ b_2 & b_1 & -b_4 & b_3 \\ b_3 & b_4 & b_1 & -b_2 \\ b_4 & -b_3 & b_2 & b_1 \end{bmatrix} \begin{bmatrix} c_1 \\ c_2 \\ c_3 \\ c_4 \end{bmatrix} \qquad \text{(D.3)} \\ &= \begin{bmatrix} c_1 & -c_2 & -c_3 & -c_4 \\ c_2 & c_1 & c_4 & -c_3 \\ c_3 & -c_4 & c_1 & c_2 \\ c_4 & c_3 & -c_2 & c_1 \end{bmatrix} \begin{bmatrix} b_1 \\ b_2 \\ b_3 \\ b_4 \end{bmatrix}. \qquad \text{(D.4)} \end{aligned}$$

It is important to note that quaternion multiplication is not commutative: $\mathbf{b} \circ \mathbf{c} \neq \mathbf{c} \circ \mathbf{b}$; but is associative: $\mathbf{a} \circ (\mathbf{b} \circ \mathbf{c}) = (\mathbf{a} \circ \mathbf{b}) \circ \mathbf{c}$. The norm of a quaternion is

$$\|\mathbf{b}\| = \mathbf{b} \circ \bar{\mathbf{b}} = b_1^2 + b_2^2 + b_3^2 + b_4^2. \tag{D.5}$$

This inverse of quaternion $\mathbf{b}$ is

$$\mathbf{b}^{-1} = \frac{\bar{\mathbf{b}}}{\|\mathbf{b}\|}.$$

The quaternion $\mathbf{b}$ can also be expressed in the *vector form*

$$\mathbf{b} = b_1 + \vec{\mathbf{b}} \tag{D.6}$$

where $\vec{\mathbf{b}} = [b_2, b_3, b_4]^\top$. The vector form allows compact representation of quaternion operations. For example, the quaternion conjugate is

$$\bar{\mathbf{b}} = b_1 - \vec{\mathbf{b}}.$$

The quaternion product can be written as

$$\mathbf{b} \circ \mathbf{c} = b_1 c_1 - \vec{\mathbf{b}} \cdot \vec{\mathbf{c}} + b_1 \vec{\mathbf{c}} + c_1 \vec{\mathbf{b}} + \vec{\mathbf{b}} \times \vec{\mathbf{c}}. \tag{D.7}$$

Based on the quaternion $\mathbf{b}$ we can form the matrices

$$\mathbf{Q}_b = \begin{bmatrix} b_1 & -\vec{\mathbf{b}}^\top \\ \vec{\mathbf{b}} & (b_1 \mathbf{I} + [\vec{\mathbf{b}}\times]) \end{bmatrix} \text{ and } \bar{\mathbf{Q}}_b = \begin{bmatrix} b_1 & -\vec{\mathbf{b}}^\top \\ \vec{\mathbf{b}} & (b_1 \mathbf{I} - [\vec{\mathbf{b}}\times]) \end{bmatrix}. \tag{D.8}$$

From which it is clear that $\mathbf{Q}_{\bar{\mathbf{b}}} = \mathbf{Q}_b^\top$ and $\bar{\mathbf{Q}}_{\bar{\mathbf{b}}} = \bar{\mathbf{Q}}_b^\top$. Using the matrices $\mathbf{Q}_b$ and $\bar{\mathbf{Q}}_b$, the quaternion product can be expressed as

$$\mathbf{b} \circ \mathbf{c} = \mathbf{Q}_b \mathbf{c} \tag{D.9}$$

$$= \bar{\mathbf{Q}}_c \mathbf{b} \tag{D.10}$$

which are the same matrices written in component form in eqns. (D.3–D.4). It can also be shown, by direct multiplication, that $\mathbf{Q}_b$ and $\bar{\mathbf{Q}}_c$ commute, $\mathbf{Q}_b \bar{\mathbf{Q}}_c = \bar{\mathbf{Q}}_c \mathbf{Q}_b$.

D.2 Rotations

Let frame a be aligned with frame b by rotating frame a by ζ radians about unit vector $\mathbf{E}$. The quaternion $\mathbf{b}$ that represents the rotational transformation from frame a to frame b is

$$\mathbf{b} = \begin{bmatrix} \cos(\zeta/2) \\ \mathbf{E}\sin(\zeta/2) \end{bmatrix}.$$

Note that $\mathbf{b}$ has the normality property that $\|\mathbf{b}\| = 1$. Therefore, for the representation of rotational transformations, the quaternion $\mathbf{b}$ has only three degrees of freedom.

Let $\mathbf{z} = \mathbf{R}_a^b \mathbf{v}$ where $\mathbf{v}$ is coordinatized in frame a and $\mathbf{z}$ is the representation of $\mathbf{v}$ when coordinatized in frame b. Each vector can be expressed as the quaternion form as

$$\mathbf{q}_v = \begin{bmatrix} 0 \\ \mathbf{v} \end{bmatrix} \text{ and } \mathbf{q}_z = \begin{bmatrix} 0 \\ \mathbf{z} \end{bmatrix}.$$

Using quaternions, the transformation of the vector quantity $\mathbf{v}$ from frame a to frame b is

$$\mathbf{q}_z = \mathbf{b} \circ \mathbf{q}_v \circ \mathbf{b}^{-1} = \mathbf{b} \circ \mathbf{q}_v \circ \bar{\mathbf{b}} \tag{D.11}$$

which can be written using eqn. (D.10) as

$$\begin{aligned}\mathbf{q}_z &= \mathbf{Q}_b\bar{\mathbf{Q}}_{\bar{b}}\mathbf{q}_v \\ &= \begin{bmatrix} b_1 & -b_2 & -b_3 & -b_4 \\ b_2 & b_1 & -b_4 & b_3 \\ b_3 & b_4 & b_1 & -b_2 \\ b_4 & -b_3 & b_2 & b_1 \end{bmatrix} \begin{bmatrix} b_1 & b_2 & b_3 & b_4 \\ -b_2 & b_1 & -b_4 & b_3 \\ -b_3 & b_4 & b_1 & -b_2 \\ -b_4 & -b_3 & b_2 & b_1 \end{bmatrix} \begin{bmatrix} 0 \\ \mathbf{v} \end{bmatrix}.\end{aligned} \tag{D.12}$$

The product matrix $\mathbf{Q}_b\bar{\mathbf{Q}}_{\bar{b}}$ is

$$\begin{bmatrix} 1 & 0 & 0 & 0 \\ 0 & b_1^2+b_2^2-b_3^2-b_4^2 & 2(b_2b_3-b_1b_4) & 2(b_1b_3+b_2b_4) \\ 0 & 2(b_2b_3+b_1b_4) & b_1^2-b_2^2+b_3^2-b_4^2 & 2(-b_1b_2+b_3b_4) \\ 0 & 2(-b_1b_3+b_2b_4) & 2(b_1b_2+b_3b_4) & b_1^2-b_2^2-b_3^2+b_4^2 \end{bmatrix},$$

with the desired rotation matrix being the lower right 3×3 matrix. Based on the above analysis, the rotation matrix for transforming vectors from a frame to b frame can be computed from the quaternion $\mathbf{b}$ using the expression

$$\mathbf{R}(\mathbf{b}) = \begin{bmatrix} b_1^2+b_2^2-b_3^2-b_4^2 & 2(b_2b_3-b_1b_4) & 2(b_1b_3+b_2b_4) \\ 2(b_2b_3+b_1b_4) & b_1^2-b_2^2+b_3^2-b_4^2 & 2(b_3b_4-b_1b_2) \\ 2(b_2b_4-b_1b_3) & 2(b_1b_2+b_3b_4) & b_1^2-b_2^2-b_3^2+b_4^2 \end{bmatrix} \tag{D.13}$$

where $\mathbf{R}(\mathbf{b})$ is used as a shorthand for $\mathbf{R}_a^b(\mathbf{b})$ to simplify the notation below. Using the vector form of the quaternion, Eqn. (D.13) can be compactly expressed as

$$\mathbf{R}(\mathbf{b}) = \left(b_1^2 - \vec{\mathbf{b}} \cdot \vec{\mathbf{b}}\right)\mathbf{I} + 2\vec{\mathbf{b}}\vec{\mathbf{b}}^\top + 2b_1[\vec{\mathbf{b}}\times] \tag{D.14}$$

where $\mathbf{I}$ is the identity matrix for $\mathbb{R}^3$ and the final simplification uses the fact that

$$[\vec{\mathbf{b}}\times][\vec{\mathbf{b}}\times] = \vec{\mathbf{b}}\vec{\mathbf{b}}^\top - \vec{\mathbf{b}}^\top\vec{\mathbf{b}}\mathbf{I}.$$

Typical applications integrate the quaternion at a high rate and only compute $\mathbf{R}_a^b$ at a lower rate as needed for other computations.

D.2.1 Direction Cosine to Quaternion

If the matrix $\mathbf{R}_a^b$ is known, then the quaternion can be computed from eqn. (D.13) as

$$\mathbf{b} = \begin{bmatrix} \frac{1}{2}\sqrt{1+\mathbf{R}_a^b[1,1]+\mathbf{R}_a^b[2,2]+\mathbf{R}_a^b[3,3]} \\ \frac{\mathbf{R}_a^b[3,2]-\mathbf{R}_a^b[2,3]}{4b_1} \\ \frac{\mathbf{R}_a^b[1,3]-\mathbf{R}_a^b[3,1]}{4b_1} \\ \frac{\mathbf{R}_a^b[2,1]-\mathbf{R}_a^b[1,2]}{4b_1} \end{bmatrix}. \tag{D.15}$$

Eqn. (D.15) can be useful to determine initial conditions for the quaternion differential equation. When the quantity under the square root is near zero, the numeric properties of eqn. (D.15) can be improved, see eqns. (166-168) in [120].

D.2.2 Quaternions to Euler Angles

In the particular case when the quaternion $\mathbf{b}$ represents the rotation from tangent to body frame, the Euler angles can be calculated when required (e.g., for control) from the components of the direction cosine matrix as, by comparison of eqns. (2.43) and (D.13),

$$\sin(\theta) = -2(b_2 b_4 + b_1 b_3) \tag{D.16}$$
$$\phi = atan2\left(2(b_3 b_4 - b_1 b_2), 1 - 2(b_2^2 + b_3^2)\right) \tag{D.17}$$
$$\psi = atan2\left(2(b_2 b_3 - b_1 b_4), 1 - 2(b_3^2 + b_4^2)\right). \tag{D.18}$$

D.3 Quaternion Derivative

The purpose of this section is to derive an expression relating the derivative of the quaternion to the angular rate vector $\boldsymbol{\omega}$. The derivation will be derived between generic a and c frames. Then, the result is specialized to the body to tangent plane application.

D.3.1 General Derivation

Let $\mathbf{r}$ be an arbitrary constant position vector in frame a and let c be another frame. The origins of frames a and c are coincident and the angular velocity of frame a with respect to frame c coordinatized in frame c is $\boldsymbol{\omega}_{ca}^{c}$. Let $\mathbf{z} = \mathbf{R}_a^c \mathbf{r}$ denote then representation of the vector with respect to frame c. The rate of change of the vector $\mathbf{z}$ in frame c is

$$\begin{aligned} \dot{\mathbf{z}} &= \mathbf{R}_a^c \boldsymbol{\Omega}_{ca}^a \mathbf{r} \\ &= \boldsymbol{\Omega}_{ca}^c \mathbf{R}_a^c \mathbf{r} \\ &= \boldsymbol{\omega}_{ca}^c \times \mathbf{z}. \end{aligned} \tag{D.19}$$

Letting $\mathbf{b}$ denote the unit quaternion that transforms vectors represented in frame a to representations in frame c, by eqn. (D.11)

$$\mathbf{q}_z = \mathbf{b} \circ \mathbf{q}_r \circ \bar{\mathbf{b}}$$

the derivative of the quaternion $\mathbf{q}_z$ is

$$\dot{\mathbf{q}}_z = \dot{\mathbf{b}} \circ \mathbf{q}_r \circ \bar{\mathbf{b}} + \mathbf{b} \circ \mathbf{q}_r \circ \dot{\bar{\mathbf{b}}}.$$

Using the facts that $\mathbf{q}_r = \bar{\mathbf{b}} \circ \mathbf{q}_z \circ \mathbf{b}$ and that $\mathbf{b} \circ \bar{\mathbf{b}}$ is the identity quaternions, we have that

$$\begin{aligned} \dot{\mathbf{q}}_z &= \dot{\mathbf{b}} \circ \bar{\mathbf{b}} \circ \mathbf{q}_z \circ \mathbf{b} \circ \bar{\mathbf{b}} + \mathbf{b} \circ \bar{\mathbf{b}} \circ \mathbf{q}_z \circ \mathbf{b} \circ \dot{\bar{\mathbf{b}}} && \text{(D.20)} \\ &= \dot{\mathbf{b}} \circ \bar{\mathbf{b}} \circ \mathbf{q}_z + \mathbf{q}_z \circ \mathbf{b} \circ \dot{\bar{\mathbf{b}}}. && \text{(D.21)} \end{aligned}$$

Using the fact that $\mathbf{b}$ is a unit vector, it can be shown that the scalar portion of both $\dot{\mathbf{b}} \circ \bar{\mathbf{b}}$ and $\mathbf{b} \circ \dot{\bar{\mathbf{b}}}$ is zero. Therefore, as expected, the scalar portion of eqn. (D.21) is zero. It can also be shown that the vector portions of these quaternions are equal in magnitude and opposite in sign. Therefore, if we let $\dot{\mathbf{b}} \circ \bar{\mathbf{b}} = 0 + \vec{\mathbf{w}}$, then $\mathbf{b} \circ \dot{\bar{\mathbf{b}}} = 0 - \vec{\mathbf{w}}$. The vector portion of eqn. (D.21) is

$$\dot{\mathbf{z}} = 2\vec{\mathbf{w}} \times \mathbf{z}. \tag{D.22}$$

Comparing eqns. (D.19) and (D.22) we conclude that $\boldsymbol{\omega}_{ca}^{c} = 2\vec{\mathbf{w}}$. Writing this in quaternion form yields

$$\mathbf{q}_{\omega_{ca}^{c}} = 2\dot{\mathbf{b}} \circ \bar{\mathbf{b}} \tag{D.23}$$

where $\mathbf{q}_{\omega_{ca}^{c}} = 0 + \boldsymbol{\omega}_{ca}^{c}$. Solving eqn. (D.23) for $\dot{\mathbf{b}}$, by multiplying on the right by $\mathbf{b}$, yields

$$\begin{aligned} \dot{\mathbf{b}} &= \frac{1}{2}\mathbf{q}_{\omega_{ca}^{c}} \circ \mathbf{b} && \text{(D.24)} \\ &= \frac{1}{2}\mathbf{Q}_{\omega_{ca}^{c}}\mathbf{b} && \text{(D.25)} \\ &= \frac{1}{2}\bar{\mathbf{Q}}_b \mathbf{q}_{\omega_{ca}^{c}}. && \text{(D.26)} \end{aligned}$$

In particular, because the scalar part of $\mathbf{q}_{\omega_{ca}^{c}}$ is zero, based on the definition of $\bar{\mathbf{Q}}_b$ in eqn. (D.8) it is possible to reduce eqn. (D.26) to

$$\dot{\mathbf{b}} = \frac{1}{2}\begin{bmatrix} -\vec{\mathbf{b}}^{\top} \\ (b_1\mathbf{I} - [\vec{\mathbf{b}}\times]) \end{bmatrix} \boldsymbol{\omega}_{ca}^{c}. \tag{D.27}$$

D.3.2 Body to Navigation Frame Result

The special case where $\mathbf{b}$ represents the rotation from navigation to body frame is of interest. In this case,

$$\begin{aligned} \boldsymbol{\omega}_{bn}^{b} &= \boldsymbol{\omega}_{bi}^{b} + \boldsymbol{\omega}_{in}^{b} \\ &= \boldsymbol{\omega}_{in}^{b} - \boldsymbol{\omega}_{ib}^{b} \end{aligned}$$

where $\boldsymbol{\omega}_{ib}^{b} = [p, q, r]^{\top}$ are the measured body rates. For the construction of the matrices to follow, we will use the following notation for the components of $\boldsymbol{\omega}_{bn}^{b} = [\omega_1, \omega_2, \omega_3]^{\top}$.

Eqn. (D.27) specializes to

$$\dot{\mathbf{b}} = \frac{1}{2}\boldsymbol{\Omega}_b \begin{bmatrix} \omega_1 \\ \omega_2 \\ \omega_3 \end{bmatrix} \tag{D.28}$$

where

$$\boldsymbol{\Omega}_b = \begin{bmatrix} -b_2 & -b_3 & -b_4 \\ \hline b_1 & b_4 & -b_3 \\ -b_4 & b_1 & b_2 \\ b_3 & -b_2 & b_1 \end{bmatrix} \tag{D.29}$$

and the horizontal line is inserted in the matrix to facilitate comparison with eqn. (D.27). Similarly, eqn. (D.25) specializes to

$$\dot{\mathbf{b}} = \frac{1}{2}\mathbf{Q}_{\omega^b_{bn}}\mathbf{b}, \text{ where} \tag{D.30}$$

$$\mathbf{Q}_{\omega^b_{bn}} = \left[\begin{array}{c|ccc} 0 & -\omega_1 & -\omega_2 & -\omega_3 \\ \hline \omega_1 & 0 & -\omega_3 & \omega_2 \\ \omega_2 & \omega_3 & 0 & -\omega_1 \\ \omega_3 & -\omega_2 & \omega_1 & 0 \end{array}\right] \tag{D.31}$$

where the horizontal and vertical lines are included to facilitate comparison with the definition of $\mathbf{Q}_{\omega^b_{bn}}$ in eqn. (D.8).

D.4 Summary

Given the results derived above, the quaternion approach can be defined by the following sequence of actions.

1. At $t = 0$, the initial attitude is estimated and $\mathbf{R}^b_n(0)$ is computed according to eqns. (2.43).

2. The initial value of the quaternion $\mathbf{b}(0)$ is computed according to eqn. (D.15).

3. Given $\boldsymbol{\omega}^b_{bn}(t)$, the quaternion $\mathbf{b}(t)$ for $t \geq 0$ is computed using either eqn. (D.28) or (D.30).

4. The rotation matrix $\mathbf{R}^b_n(t)$ is computed from $\mathbf{b}(t)$ using eqn. (D.13) when required. Alternatively, vectors can be transformed between the frames-of-reference directly by use of eqn. (D.11).

5. The attitude is computed according to eqns. (D.16-D.18) at any time that it is needed.

D.5 Quaternion Integration

The quaternion approach using a four parameter vector with a unity magnitude constraint allows three degrees of freedom. Due to the unity magnitude constraint, quaternions are elements of a unit sphere in four space. This unit sphere is not a Euclidean vector space in which the usual definitions of vector addition and scaling apply; therefore, care must be taken in the integration of the quaternion differential equation. Since the equations are linear, if the sampling rate is high enough that it is accurate to consider the rotation rate as constant over the sample interval T, then it is possible to find and implement a closed form solution to eqn. (D.30). Such a solution is presented below.

To simplify notation, let $\mathbf{Q}(t) = \frac{1}{2}\mathbf{Q}_{\omega^b_{bn}}(t)$ so that eqn. (D.30) can be written as

$$\dot{\mathbf{b}}(t) = \mathbf{Q}(t)\mathbf{b}(t). \tag{D.32}$$

Our objective is to find $\mathbf{b}(t_2)$ when $\mathbf{b}(t_1)$ is known and $\mathbf{Q}(t)$ is constant for $t \in (t_1, t_2]$ where $T = t_2 - t_1$.

Given these assumptions, for $t \in (t_1, t_2]$ we have that

$$\mathbf{Q}(t) = \mathbf{Q} = \frac{1}{2}\left[\begin{array}{c|ccc} 0 & -\omega_1 & -\omega_2 & -\omega_3 \\ \hline \omega_1 & 0 & -\omega_3 & \omega_2 \\ \omega_2 & \omega_3 & 0 & -\omega_1 \\ \omega_3 & -\omega_2 & \omega_1 & 0 \end{array}\right].$$

and we define the integrating factor

$$\Xi(t) = e^{-\int_{t_1}^{t} \mathbf{Q}(\tau)d\tau}. \tag{D.33}$$

Multiplying the integrating factor into eqn. (D.32) from the left and simplifying proceeds as follows:

$$\begin{aligned} e^{-\int_{t_1}^{t} \mathbf{Q}d\tau}\dot{\mathbf{b}} - e^{-\int_{t_1}^{t} \mathbf{Q}d\tau}\mathbf{Q}\mathbf{b} &= 0 \\ \frac{d}{dt}\left(e^{-\int_{t_1}^{t} \mathbf{Q}d\tau}\mathbf{b}(t)\right) &= 0. \end{aligned}$$

Integrating both sides over the interval yields

$$\begin{aligned} e^{-\int_{t_1}^{t_2} \mathbf{Q}d\tau}\mathbf{b}(t_2) &= e^{-\int_{t_1}^{t_1} \mathbf{Q}d\tau}\mathbf{b}(t_1) \\ \mathbf{b}(t_2) &= e^{\int_{t_1}^{t_2} \mathbf{Q}d\tau}\mathbf{b}(t_1) \end{aligned} \tag{D.34}$$

where it has been recognized that $\left(e^{-\int_{t_1}^{t} \mathbf{Q}d\tau}\right)^{-1} = e^{\int_{t_1}^{t} \mathbf{Q}d\tau}$.

To write the solution of eqn. (D.34) in a more convenient form, we define $\mathbf{w} = [W_1, W_2, W_3]^\top$ and

$$\mathbf{W} = \begin{bmatrix} 0 & -\mathbf{w}^\top \\ \mathbf{w} & ([\mathbf{w}\times]) \end{bmatrix} = \begin{bmatrix} 0 & -W_1 & -W_2 & -W_3 \\ W_1 & 0 & -W_3 & W_2 \\ W_2 & W_3 & 0 & -W_1 \\ W_3 & -W_2 & W_1 & 0 \end{bmatrix},$$

where

$$W_1 = \tfrac{1}{2}\int_{t_1}^{t} \omega_1(\tau)d\tau, \quad W_2 = \tfrac{1}{2}\int_{t_1}^{t} \omega_2(\tau)d\tau, \quad W_3 = \tfrac{1}{2}\int_{t_1}^{t} \omega_3(\tau)d\tau.$$

With these definitions, eqn. (D.34) is equivalent to

$$\mathbf{b}(t_2) = e^{\mathbf{W}}\mathbf{b}(t_1). \tag{D.35}$$

Fortunately, the state transition matrix $e^{\mathbf{W}}$ allows further simplification.

First, it is easily verified by direct multiplication that

$$\mathbf{W}^2 = -\|\mathbf{w}\|^2\mathbf{I}.$$

Expanding $e^{\mathbf{W}}$ using a power series

$$\begin{aligned} e^{\mathbf{W}} &= \mathbf{I} + \mathbf{W} + \frac{\mathbf{W}^2}{2!} + \frac{\mathbf{W}^3}{3!} + \dots \\ &= \left(\mathbf{I} + \frac{\mathbf{W}^2}{2!} + \frac{(\mathbf{W}^2)^2}{4!} + \dots\right) + \mathbf{W}\left(\mathbf{I} + \frac{\mathbf{W}^2}{3!} + \frac{(\mathbf{W}^2)^2}{5!} + \dots\right) \\ &= \cos(\|\mathbf{w}\|)\,\mathbf{I} + \frac{\sin(\|\mathbf{w}\|)}{\|\mathbf{w}\|}\mathbf{W}. \end{aligned}$$

Therefore, because the above derivation can be repeated over any interval of duration T for which the assumption of a constant angular rate is valid, we have the general equation

$$\mathbf{b}(t_k) = \left(\cos(\|\mathbf{w}\|)\,\mathbf{I} + \frac{\sin(\|\mathbf{w}\|)}{\|\mathbf{w}\|}\mathbf{W}\right)\mathbf{b}(t_{k-1}) \tag{D.36}$$

where $\mathbf{w}$, $\|\mathbf{w}\|$ and $\mathbf{W}$ involve the integral of the angular rates over the k-th sampling interval $t \in (t_{k-1}, t_k]$.

Note that no approximations were made in this derivation, so the solution is in closed form given the assumption that $\mathbf{Q}(t)$ is constant over the interval of length T. The solution is easily verified to be norm preserving by showing that the norm of the right side of eqn. (D.36) is equal to $\|\mathbf{b}(t_{k-1})\|$. The algorithm is not an exact solution to the differential equation over any interval where the angular rates are not constant. In [73] the algorithm of eqn. (D.36) is shown to be second order and a third order algorithm is also presented. The algorithm being second order means the algorithm error contains terms proportional to T^3 and higher powers. Because T is small, higher order algorithms have smaller errors; however, the tradeoff is that they require additional computation.

D.6 Attitude Representation Comparison

The following example clarifies the singularity that occurs with the Euler angle representation of attitude as $\theta \to \frac{\pi}{2}$.

Example D.1 *Consider the tangent to body direction cosine matrix corresponding to $\theta = \frac{\pi}{2}$:*

$$\mathbf{R}_t^b = \begin{bmatrix} 0 & 0 & -1 \\ -s\psi c\phi + c\psi s\phi & c\psi c\phi + s\psi s\phi & 0 \\ s\psi s\phi + c\psi c\phi & -c\psi s\phi + s\psi c\phi & 0 \end{bmatrix}$$

which is well defined for any ψ and ϕ. However, there are multiple values of $\psi, \phi \in (-\pi, \pi]$ that yield the same orientation.

Consider two sequences of rotations. For each, assume that the tangent and body frames are initially aligned. The first rotation sequence is defined by a rotation of $\frac{\pi}{2}$ rads. about the navigation frame y-axis to give $(\phi, \theta, \psi) = (0, \frac{\pi}{2}, 0)$. The second rotation sequence is defined by a rotation about the navigation frame z-axis by $\frac{\pi}{2}$ rads., rotation about the body frame y-axis by $\frac{\pi}{2}$ rads., and then rotation about the vehicle x-axis by $\frac{\pi}{2}$ rads to yield $(\phi, \theta, \psi) = (\frac{\pi}{2}, \frac{\pi}{2}, \frac{\pi}{2})$. Both of these rotation sequences result in the same vehicle orientation.

For each of these rotation sequences the angular rate vectors in body frame are well defined. In the first case, e.g., $(p, q, r) = (0, 90, 0) deg/s$ for 1.0s. In the second case, e.g.:

$$\begin{aligned} (p, q, r) &= \left(0,\ 0, \frac{\pi}{2}\right) rad/s \text{ for } t \in [0, 1)s \\ (p, q, r) &= \left(0, \frac{\pi}{2},\ 0\right) rad/s \text{ for } t \in [1, 2)s \\ (p, q, r) &= \left(\frac{\pi}{2},\ 0,\ 0\right) rad/s \text{ for } t \in [2, 3)s. \end{aligned}$$

In both cases, the matrix $\mathbf{\Omega}_E$ involved in the computation of the Euler angle derivatives of eqn. (2.74) on page 57 becomes singular. For $\theta \approx \frac{\pi}{2}$ the matrix $\mathbf{\Omega}_E^{-1}$ is large (potentially infinite) which can greatly magnify sensor errors in the computation of the Euler angle derivatives. △

The following example considers the motion similar to that of the previous example in the sense that the vehicle is maneuvering near $\theta = \frac{\pi}{2} rad$, but using a quaternion implementation.

Example D.2 *For the initial condition $(\phi, \theta, \psi) = (0, 0, 0)$, the initial quaternion is $\mathbf{b}(0) = [1, 0, 0, 0]$.*

Figure D.1 shows the body frame angular rates $[p, q, r]$. The quaternion that results from the integration of either eqn. (D.28) or (D.30) is plotted

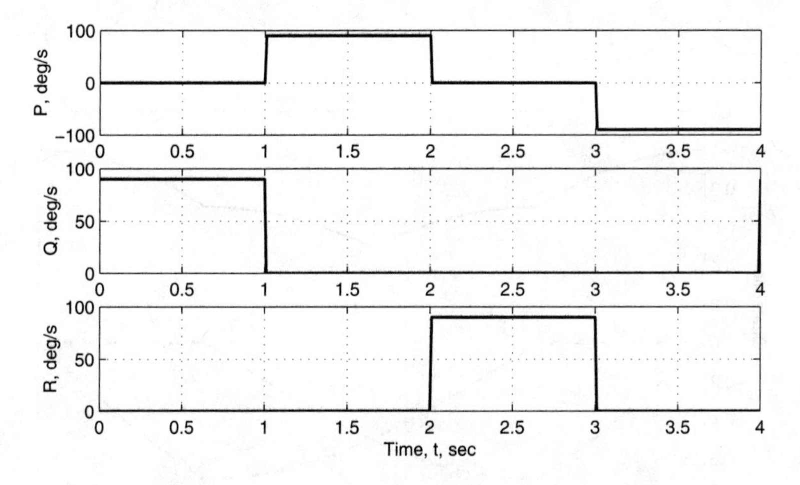

Figure D.1: Angular rates for Example D.2.

in Figure D.2. The Euler angles computed from the quaternion are plotted in Figure D.3. Throughout the simulation, the gain defined by the matrix $\mathbf{\Omega}_b$ *in eqn. (D.29) is small. For example, due to* $\mathbf{b}$ *being a unit quaternion each row has magnitude less than 1.0.*

△

For an additional example of the use of quaternions, especially for attitudes with θ near 90 degrees, see the discussion related to Figure 10.5.

D.7 References and Further Reading

The main sources for the material in this appendix were [23, 45, 50, 103, 115, 118, 137]. Attitude representations are compared in [120].

D.8 Exercises

Exercise D.1 Show that eqn. (D.7) yields the same result as eqn. (D.3).

Exercise D.2 Use the definitions of eqns. (D.6) and (D.7) with the definition of quaternion products in eqn. (D.7) to verify that eqn. (D.5).

Exercise D.3 Verify that eqn. (D.14) yields the same result as eqn. (D.13).

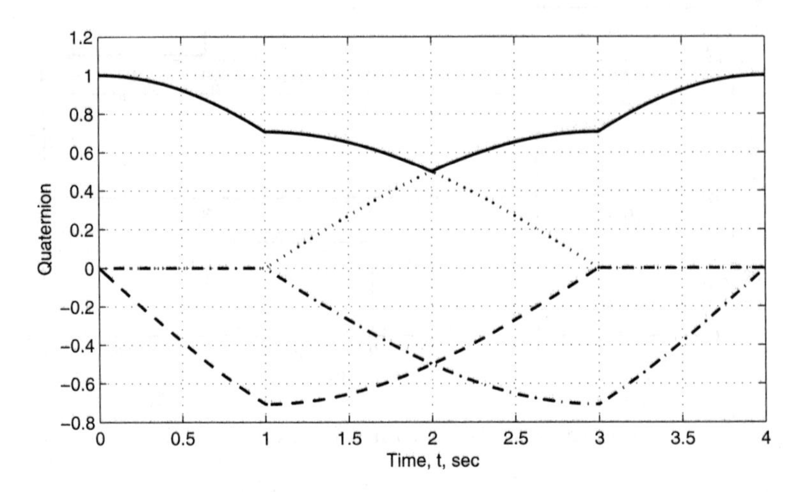

Figure D.2: Quaternion $\mathbf{b}(t)$ for Example D.2: b_1 is solid, b_2 is dash-dotted, b_3 is dashed, and b_4 is dotted.

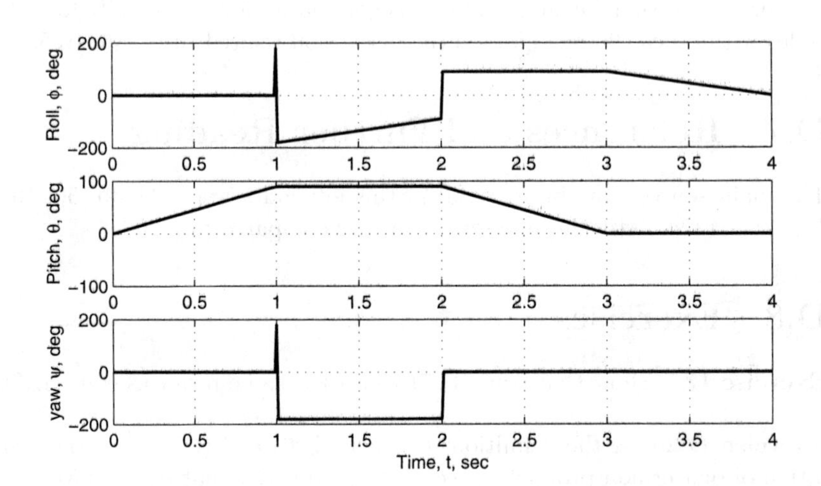

Figure D.3: Computed attitude by eqns. (D.16-D.18) for Example D.2.

Exercise D.4 Let $\mathbf{b}$ and $\hat{\mathbf{b}}$ represent a quaternion and its computed value where $\dot{\mathbf{b}} = \frac{1}{2}\mathbf{q}_\omega \circ \mathbf{b}$ and $\dot{\hat{\mathbf{b}}} = \frac{1}{2}\tilde{\mathbf{q}}_\omega \circ \hat{\mathbf{b}}$ with $\mathbf{q}_\omega = [0, \boldsymbol{\omega}^\top]^\top$ being the quaternion representation of the angular rate vector $\boldsymbol{\omega}$ and $\tilde{\mathbf{q}}_\omega = [0, \tilde{\boldsymbol{\omega}}^\top]^\top$ being the quaternion representation of the measured angular rate vector $\tilde{\boldsymbol{\omega}} = \boldsymbol{\omega} + \mathbf{x}_g + \boldsymbol{\nu}_g$.

The quaternion error can be define as $\delta\mathbf{b} = \mathbf{b} \circ \hat{\mathbf{b}}^{-1}$ where $\delta\mathbf{b} \approx [1, \delta\boldsymbol{\xi}^\top]^\top$.

1. Use eqn. (D.14) to show that the small angle rotation corresponding to $\delta\mathbf{b}$ is

$$\mathbf{R}(\delta\mathbf{b}) = (\mathbf{I} + 2[\delta\boldsymbol{\xi}\times]) \tag{D.37}$$

   to first order.

2. Given the product $\delta\mathbf{b} = \mathbf{b} \circ \hat{\mathbf{b}}^{-1}$ and eqn. (D.37), show that

$$\mathbf{R}(\mathbf{b}) = (\mathbf{I} + 2[\delta\boldsymbol{\xi}\times])\,\mathbf{R}(\hat{\mathbf{b}}).$$

3. Starting from the equation

$$\delta\dot{\mathbf{b}} = \dot{\mathbf{b}} \circ \hat{\mathbf{b}}^{-1} + \mathbf{b} \circ \dot{\hat{\mathbf{b}}}^{-1}$$

   use the fact that $\dot{\hat{\mathbf{b}}}^{-1} = -\hat{\mathbf{b}}^{-1} \circ \dot{\hat{\mathbf{b}}} \circ \hat{\mathbf{b}}^{-1}$ to show that

$$\delta\dot{\mathbf{b}} = \frac{1}{2}\left(\mathbf{q}_\omega \circ \delta\mathbf{b} - \delta\mathbf{b} \circ \tilde{\mathbf{q}}_\omega\right). \tag{D.38}$$

4. Show that the vector portion of eqn. (D.38) reduces to

$$\delta\dot{\boldsymbol{\xi}} = [\boldsymbol{\omega}\times]\delta\boldsymbol{\xi} - \frac{1}{2}\mathbf{x}_g - \frac{1}{2}\boldsymbol{\nu}_g \tag{D.39}$$

   after all higher order terms have been dropped.

Exercise D.4 Let [illegible] quaternion and [illegible] [illegible]

[illegible]

[illegible] to show that [illegible] small angle rotation [illegible]

[illegible]

[illegible]

[illegible] (D.38)

[illegible] (D.39)

after all [illegible] terms have been dropped.

Bibliography

[1] N. Ackroyd and R. Lorimer. *Global navigation: a GPS user's guide.* Lloyd's of London Press, London; NewYork, 1990.

[2] D. W. Allan. Statistics of atomic frequency standards. *Proceedings of the IEEE*, 54(2):221–230, 1966.

[3] W. G. Anderson and E. H. Fritze. Instrument approach system steering computer. *Proceedings of Institute of Radio Engineers*, 41(2):219–228, February 1953.

[4] Anonymous. Phase I NAVSTAR/GPS major field test objective report thermostatic correction. Technical report, Navstar/GPS Joint Program Office, Space & Missile Systems Organization, Los Angeles Air Force Station, Los Angeles, California, May 4 1979.

[5] Anonymous. Recommended practice for atmospheric and space flight vehicle coordinate systems. Technical Report AIAA/ANSI R-004-1992, American Institute of Aeronautics and Astronautics, Reston, VA, 1992.

[6] Anonymous. NAVSTAR GPS space segment/navigation user interfaces. Technical Report ICD-GPS-200, ARINC Research Corporation, April 1993.

[7] Anonymous. RTCM recommended standards for differential navstar GPS service. Technical Report 104, RTCM Special Committee, January 1994.

[8] Anonymous. The global positioning system: A shared national asset. Technical report, National Research Council, Washington, D.C., 1995.

[9] Anonymous. IEEE standard specification format guide and test procedure for single-axis laser gyros. Technical Report IEEE Std. 647-1995, IEEE, New York, NY, 1995.

[10] Anonymous. IEEE standard specification format guide and test procedure for single-axis interferometric fiber optic gyros. Technical Report IEEE Std. 952-1997, IEEE, New York, NY, 1997.

[11] Anonymous. IEEE standard specification format guide and test procedure for single-axis, nongyroscopic accelerometers. Technical Report IEEE Std. 1293-1998, IEEE, New York, NY, 1998.

[12] Anonymous. World Geodetic System 1984 (WGS-84)–Its definition and relationships with local geoderic systems. Technical Report NIMA TR 8350.2, Defense Mapping Agency, Fairfax, VA, 2000.

[13] N. Ashby and M. Weiss. Global position system receivers and relativity. Technical Report NIST TN 1385, National Institute of Standards and Technology, Boulder, CO, March 1999.

[14] I. Y. Bar-Itzack, P. Y. Montgomery, and J. C. Garrick. Algorithms for attitiude determination using GPS. *Journal of Guidance, Control, and Dynamics*, 21(6):846–852, 1998.

[15] I. Y. Bar-Itzhack. Navigation computation in terrestrial strapdown inertial navigation systems. *IEEE Transactions on Aerospace and Electronic Systems*, 13(6):679–689, 1977.

[16] I. Y. Bar-Itzhack. Corrections to 'navigation computation in terrestrial strapdown inertial navigation systems'. *IEEE Transactions on Aerospace and Electronic Systems*, 14(3):542–544, 1978.

[17] I. Y. Bar-Itzhack and N. Bergman. Control theoretic approach to inertial navigation systems. *Journal Guidance*, 11(3):237–245, 1988.

[18] T. Barnes. Selective availability via Levinson predictor. In *Proceedings of the Institute of Navigation (ION) GPS–95*, 1995.

[19] J. S. Bay. *Fundamentals of Linear State Space Systems.* WCB McGraw-Hill, Boston, MA, 1999.

[20] J. Baziw and C. T. Leondes. In-flight alignment and calibration of inertial measurement units–part 1: General formulation. *IEEE Transactions on Aerospace and Electronic Systems*, 8(4):439–449, 1972.

[21] H. Blomenhofer, G. Hein, E. Blomenhofer, and W. Werner. Development of a real-time DGPS system in the centimeter range. In *Proc. of the IEEE 1994 Position, Location, and Navigation Symposium*, pages 532–539, 1994.

[22] Y. Bock and N. Leppard. *Global Positioning System: An Overview.* International Association of Geodesy Symposia, No. 102, Springer-Verlag, New York, 1989.

[23] J. E. Bortz. A new mathematical formulation for strapdown inertial navigation. *IEEE Transactions on Aerospace and Electronic Systems*, 7(1):61–66, 1971.

[24] B. Bowring. Transformation from spatial to geographical coordinates. *Survey Review*, XXXIII:323–327, 1976.

[25] M. S. Braasch, A. M. Fink, and K. Duffus. Improved modeling of GPS selective availability. In *Proceedings of the Institute of Navigation National Technical Meeting*, pages 121–130, 1993.

[26] K. R. Britting. Self alignment techniques for strapdown inertial navigation systems with aircraft applications. *Journal of Aircraft*, 7(4):302–307, 1970.

[27] K. R. Britting. *Inertial Navigation Systems Analysis*. Wiley-Interscience, New York, 1971.

[28] W. L. Brogan. *Modern Control Theory*. Prentice Hall, Englewood Cliffs, NJ, 3rd edition, 1991.

[29] R. G. Brown. Kalman filter modeling. In *Proceedings of the 16th Annual Precise Time and Time Interval (PTTI) Applications and Planning Meeting*, pages 261–272, 1984.

[30] R. G. Brown. Integrated navigation systems and Kalman filtering: A perspective. *Navigation: Journal of the Institute of Navigation*, 19(4):355–362, Winter 1972-73.

[31] R. G. Brown and Y. C. Hwang. *Introduction to random signals and applied Kalman filtering*. J. Wiley, New York, 2nd edition, 1992.

[32] R. G. Brown and D. J. Winger. Error analysis of an integrated inertial/Doppler-satellite system with continuous and multiple satellite coverage. Technical report, Engineering Research Institute, Iowa State University, January 1971.

[33] C. Broxmeyer. *Inertial navigation systems*. McGraw-Hill electronic sciences. McGraw-Hill, New York, 1964.

[34] R. S. Bucy and P. D. Joseph. *Filtering for Stochastic Processes with Applications to Guidance*. Wiley, New York, 1968.

[35] E. Cannon. High-accuracy GPS semikinematic positioning: Modeling and results. *Navigation: Journal of the Institute of Navigation*, 37(1):53–64, Summer 1990.

[36] R. H. Cannon. Alignment of inertial guidance systems by gyrocompassing — linear theory. *Journal of Aerospace Science*, 28(11):885–895, 912, 1961.

[37] A. B. Chatfield. *Fundamentals of High Accuracy Inertial Navigation*, volume 171 of *Progress in Astronautics and Aeronautics.* AIAA, Reston Virginia, 1997.

[38] C.-T. Chen. *Linear Systems Theory and Design.* Oxford University Press, New York, 3rd edition, 1999.

[39] D. R. Childs, D. M. Coffey, and S. P. Travis. Error statistics for normal random variables. Technical Report AD-A011 430, Naval Underwater Systems Center, May 1975.

[40] C. E. Cohen. *Attitude Determination Using GPS.* Ph.D. dissertation, Stanford Univ., Stanford, CA, Depart. of Aeronautics and Astronautics, December 1992.

[41] J. L. Crassidis and F. L. Markley. New algorithm for attitude determination using global positioning system signals. *Journal of Guidance, Control, and Dynamics*, 20(5):891–896, 1997.

[42] J. L. Crassidis and F. L. Markley. Predictive filtering for attitude estimation without rate sensors. *Journal of Guidance, Control, and Dynamics*, 20(3):522–527, 1997.

[43] J. J. DeAzzo and C. H. Houpis. *Linear Control System Analysis and Design: Conventional and Modern.* McGraw-Hill, New York, 2nd edition, 1981.

[44] M. Djodat. Comparison of various differential global positioning systems. Master's thesis, California State University, Fullerton, July 1996.

[45] W. L. Elbert. Estimating the Euler attitudes. Technical Report JHU/APL-T-G-1329, John Hopkins Applied Physics lab, November 1981.

[46] H. J. Euler and C. H. Hill. Attitude determination: Exploring all information for optimal ambiguity resolution. In *Proc. of the ION GPS-95*, pages 1751–1757, 1995.

[47] J. C. Fang and D. J. Wan. A fast initial alignment method for strapdown inertial navigation system on stationary base. *IEEE Transactions on Aerospace and Electronic Systems*, 32(4):1501–1505, 1996.

[48] J. A. Farrell and M. Barth. *The Global Positioning and Inertial Navigation.* McGraw-Hill, New York, 1999.

[49] J. L. Farrell. Performance of strapdown inertial attitude reference systems. *Journal of Spacecraft and Rockets*, 3(9):1340–1347, 1966.

[50] J. L. Farrell. *Integrated Aircraft Navigation.* Academic Press, New York, 1976.

[51] K. Feigl, R. King, and T. Herring. A scheme for reducing the effect of selective availability on precise geodetic measurements from the global positioning system. *Geophysical Research Letters*, 18(7):1289–1292, 1991.

[52] N. P. Fofonoff and R. C. Millard Jr. Algorithms for computation of fundamental properties of seawater. Technical Report 44, Unesco, Paris, France, 1983.

[53] G. F. Franklin, J. D. Powell, and A. Emami-Naeini. *Feedback Control of Dynamic Systems.* Addison-Wesley, Reading MA, 3rd edition, 1994.

[54] G. T. French. *Understanding the GPS.* GeoResearch, 1997.

[55] S. Frodge, S. Deloach, B. Remondi, D. Lapucha, and R. Barker. Real-time on-the-fly kinematic GPS system results. *Navigation: Journal of the Institute of Navigation*, 41(2):175–186, 1994.

[56] G. J. Geier. Corrections to GPSPFP memo 03-76. Technical Report GPSPFP Memo 03-76, Intermetrics Inc., Cambridge, MA, January 1976.

[57] G. J. Geier. Delayed state Kalman filter equations for delta range measurement processing. Technical Report GPSPFP Memo 03-76, Intermetrics Inc., Cambridge, MA, 1976.

[58] A. Gelb. *Applied Optimal Estimation.* MIT Press, Cambridge, MA, 1974.

[59] G. H. Golub and C. F. Van Loan. *Matrix Computations.* The Johns Hopkins University Press, Baltimore, MD, 2nd edition, 1989.

[60] M. S. Grewal and A. P. Andrews. *Kalman Filtering: Theory and Practice using Matlab.* John Wiley, New York, 2nd edition, 2001.

[61] C. Grubin. Derivation of the quaternion scheme via the Euler axis and angle. *J. Spacecraft*, 7(10):1261–1263, 1970.

[62] S. Han and C. Rizos. Integrated method for instantaneous ambiguity resolution using new generation GPS receivers. In *IEEE PLANS*, pages 254–261, 1996.

[63] R. Hatch. The synergism of GPS code and carrier measurements. In *International Geodetic Symposium on Satellite Doppler Positioning, 3rd*, volume 2, pages 1213–1231, 1983.

[64] R. Hatch. Instantaneous ambiguity resolution. In *Symposium No. 107, Kinematic Systems in Geodesy, Surveying and Remote Sensing*, pages 299–308. Springer Verlag, 1990.

[65] R. Hatch, J. Jung, P. Enge, and B. Pervan. Civilian GPS: The benefits of three frequencies. *GPS Solutions*, 3(4):1–9, 2000.

[66] M. Heikkinen. Geschlossene formeln zur berechnung räumlicher geodäticher koordinaten aus rechtwinkligen koordinaten. *Zeitschrift für Vermessungswesen*, 5:207–211, 1982.

[67] W. A. Heiskanen and H. Moritz. *Physical Geodesy.* W. H. Freeman, San Francisco, 1967.

[68] B. Hofmann-Wellenhof, H. Lichtenegger, and J. Collins. *Global Positioning System: Theory and Practice.* Springer-Verlag, New York, 3rd edition, 1994.

[69] J. C. Hung and H. V. White. Self-alignment techniques for inertial measurement units. *IEEE Trans. on Aerospace and Electronic Systems*, 11(6):1232–1247, 1975.

[70] P. Hwang and R. Brown. GPS navigation: Combining pseudorange with continuous carrier phase using a Kalman filter. *Navigation: Journal of the Institute of Navigation*, 37(2):181–196, 1990.

[71] K. Itô. *Lectures on Stochastic Processes.* TATA Institute of Fundamental Research, Bombay, India, 1961.

[72] A. H. Jazwinski. *Stochastic Processes and Filtering Theory*, volume 4 of *Mathematics in Science and Engineering Series.* Academic Press, San Diego, 1970.

[73] C. Jekeli. *Inertial Navigation Systems with Geodetic Applications.* Walter de Gruyter, Berlin, 2001.

[74] Y. F. Jiang and Y. P. Lin. Error estimation of INS ground alignment through observability analysis. *IEEE Transactions on Aerospace and Electronic Systems*, 28(1):92–96, 1992.

[75] P. S. Jorgensen. Navstar/Global Positioning System 18-satellite constellation. In *Global Positioning System, Papers published in Navigation*, volume II, pages 1–12. The Institute of Navigation, 1984.

[76] E. D. Kaplan, editor. *Understanding GPS: Principles and Applications.* Artech House, Boston, MA, 2nd edition, 2006.

[77] S. Kay. *Intuitive Probability and Random Processes using MATLAB.* Springer, 2005.

[78] M. Kayton and W. R. Fried. *Avionics Navigation Systems.* John Wiley & Sons, 2nd edition, 1997.

[79] J. A. Klobuchar. Ionospheric time-delay algorithm for single frequency GPS users. *IEEE Transactions on Aerospace and Electronic Systems*, 23(3):325–331, 1987.

[80] J. A. Klobuchar. Ionospheric effects on GPS. In B. W. Parkinson and J. J. Spilker, Jr., editors, *The Global Positioning System: Theory and applications*, chapter 12, pages 485–516. American Institute of Aeronautics and Astronautics, Washington, DC, 1996.

[81] G. Lachapelle, M. Cannon, and G. Lu. High-precision GPS navigation with emphasis on carrier-phase ambiguity resolution. *Marine Geodesy*, 15:253–269, 1992.

[82] G. Lachapelle, M. E. Cannon, G. Lu, and B. Loncarevic. Shipborne GPS attitude determination during MMST-93. *IEEE Journal of Oceanic Engineering*, 21(1):100–105, 1996.

[83] A. Lawrence. *Modern Inertial Technology: Navigation, Guidance and Control.* Springer-Verlag, New York, 1993.

[84] A. Leon-Garcia. *Probability and Random Processes for Electrical Engineering.* Prentice Hall, 1993.

[85] G. Lu, M. E. Cannon, and G. Lachapelle. Attitude determination using dedicated and nondedicated multiantenna GPS sensors. *IEEE Transactions on Aerospace and Electronic Systems*, 30(4):1053–1058, 1994.

[86] D. G. Luenberger. *Optimization by Vector Space Methods.* Wiley Interscience, New York, NY, 1969.

[87] D. G. Luenberger. *Introduction to Dynamic Systems: Theory, Models, and Applications.* John Wiley & Sons, New York, NY, 1979.

[88] D. G. Luenberger. *Linear and Nonlinear Programming.* Addison-Wesley, Menlo Park, CA, 2nd edition, 1984.

[89] G. Mader. Dynamic positioning using GPS carrier phase measurements. *Manuscripta Geodetica*, 11:272–277, 1986.

[90] J. B. Marion. *Classical Dynamics of Particles and Systems.* Academic Press, Orlando, 2nd edition, 1970.

[91] F. L. Markley. Attitude determination using vector observations and the singular value decomposition. *Journal of the Astronautical Sciences*, 36(3):245–258, 1988.

[92] C. W. Marquis. Integration of differential GPS and inertial navigation using a complementary Kalman filter. Master's thesis, Naval Postgraduate School, September 1993.

[93] P. S. Maybeck. *Stochastic Models, Estimation, and Control.* Academic Press, New York, 1979.

[94] R. J. Mayhan. *Discrete-time and Continuous-time Linear Systems.* Addison-Wesley, Reading, MA, 1984.

[95] C. L. McClure. *Theory of Inertial Guidance.* Prentice-Hall, Englewood Cliffs, New Jersey, 1960.

[96] J. M. Mendel. *Lessons in Estimation Theory for Signal Processing, Communications, and Control.* Prentice Hall, Englewood Cliffs, NJ, 1995.

[97] P. A. Miller, J. A. Farrell, Y. Zhao, and V. Djapic. Autonomous underwater vehicle navigation. Technical Report 1968, SSC San Diego, February 2008.

[98] P. A. Miller, Y. Zhao, V. Djapic, and J. A. Farrell. Autonomous underwater vehicle navigation. In *Proc. of the 15th Int'l. Symp. on Unmanned Untethered Submersibles Technologies*, Durham, NH, August 2007.

[99] P. Misra and P. Enge. *Global Positioning System: Signals, Measurements, and Performance.* Ganga-Jamuna Press, Lincoln, MA, 2001.

[100] C. Moler and C. Van Loan. Nineteen dubious ways to compute the exponential of a matrix. *SIAM Rev.*, 20:801–836, 1978.

[101] B. Moore. Principal component analysis in linear systems: Controllability, observability, and model reduction. *IEEE Transactions on Automatic Control*, 26(1):17–32, 1981.

[102] H. Moritz. Bull. Geod. *Geodetic Reference System 1980*, 58(3):388–398, 1984.

[103] R. E. Mortenson. Strapdown guidance error analysis. *IEEE Transactions on Aerospace and Electronic Systems*, 10(4):451–457, 1994.

[104] A. E. Niell. Global mapping functions for the atmosphere delay at radio wavelengths. *Journal of Geophysical Research*, 101(B2):3227–3246, February 1996.

[105] N. S. Nise. *Control Systems Engineering*. John Wiley & Sons, Hoboken, NJ, 4th edition, 2004.

[106] B. K. Oksendal. *Stochastic Differential Equations: An Introduction with Applications*. Springer, Berlin, 2003.

[107] A. Papoulis. *Probability, Random Variables, and Stochastic Processes*. McGraw-Hill, New York, 2nd edition, 1984.

[108] B. W. Parkinson and J. J. Spilker, Jr., editors. *The Global Positioning System: Theory and applications*. American Institute of Aeronautics and Astronautics, Washington, DC, 1996.

[109] D. Pietraszewski, J. Spalding, C. Viehweg, and L. Luft. U.S. Coast Guard differential GPS navigation field test findings. *Navigation: Journal of the Institute of Navigation*, 35(1):55–72, 1988.

[110] J. C. Pinson. Inertial guidance for cruise vehicles. In C. T. Leondes, editor, *Guidance and Control of Aerospace Vehicles*, chapter 4. McGraw-Hill, 1963.

[111] G. R. Pitman. *Inertial guidance*. Wiley, New York, 1962.

[112] M. Pratt, B. Burke, and P. Misra. Single-epoch integer ambiguity resolution with GPS L1-L2 carrier phase measurements. In *Proceedings ION GPS-97*, pages 1737–1746, 1997.

[113] R. M. Rogers. *Applied Mathematics in Integrated Navigation Systems*. AIAA Educational Series. American Institute of Aeronautics and Astronautics, Inc, Reston, VA, 2nd edition, 2003.

[114] C. E. Rohrs, J. L. Melsa, and D. G. Schultz. *Linear Control Systems*. McGraw-Hill, New York, 1993.

[115] P. Savage. *Strapdown System Algorithms*. Advances in Strapdown Inertial Systems. AGARD Lecture Series 133, 1984.

[116] C. J. Savant. *Principles of inertial navigation*. McGraw-Hill, New York, 1961.

[117] R. Scherrer. *The WM GPS Primer*. Wild Heerbrugg Ltd, Heerbrugg, Switzerland, 1985.

[118] A. L. Schwab. Quaternions, finite rotation, and Euler parameters. May 2002.

[119] J. Shockley. Consideration of troposheric model corrections for differential GPS. Technical report, SRI International, February 1984.

[120] M. D. Shuster. Survey of attitude representations. *Journal of the Astronautical Sciences*, 41(4):439–517, 1993.

[121] D. Simon. *Optimal State Estimation: Kalman, H_∞, and Nonlinear Approaches.* John Wiley & Sons, Hoboken, NJ, 2006.

[122] J. J. Spilker. Tropospheric effects on GPS. In B. Parkinson, J. Spilker, P. Axelrad, and P. Enge, editors, *Global Positioning System: Theory and Applications*, volume 1, pages 517–546. AIAA, 1996.

[123] P. Teunissen. A new method for fast carrier phase ambiguity estimation. In *Proceedings of IEEE PLANS*, pages 562–573, 1994.

[124] P. J. G. Teunissen. GPS carrier phase ambiguity fixing concepts. In A Kleusberg and P. Teunissen, editors, *GPS for Geodesy*, Lecture Notes for Earth Sciences, pages 263–335. Springer, 1996.

[125] E. H. Thompson, J. L. Farrell, and J. W. Knight. Alignment methods for strapdown inertial systems. *AIAA JSR*, 3(9):1432–1434, 1966.

[126] D. H. Titterton and J. L. Weston. *Strapdown inertial navigation technology.* Peter Peregrinis Ltd. on behalf of the Institution of Electrical Engineers, London, UK, 1997.

[127] C. Tsuboi. *Gravity.* George Allen & Unwin, London, Boston, Sydney, 1983.

[128] J. B.-Y. Tsui. *Fundamentals of Global Positioning System Receivers: A Software Radio Approach.* Wiley-Interscience, 2004.

[129] A. Van Bronkhorst. *Strapdown System Algorithms*, volume 95 of *AGARD Lecture Series.* 1978.

[130] A. Van Dierendonck, J. McGraw, and R. Brown. Relationship between Allan variances and Kalman filter parameters. In *Proceedings of the 16th Annual Precise Time and Time Interval (PTTI) Applications and Planning Meeting*, pages 273–293, 1984.

[131] A. Van Dierendonck, S. Russell, E. Kopitzke, and M. Birnbaum. The GPS navigation message. In *Global Positioning System: Papers published in Navigation, Vol. 1*, pages 55–73. Institute of Navigation, 1980.

[132] F. Van Graas and M. Braasch. GPS interferometric attitude and heading determination: Initial flight test results. *Navigation: Journal of the Institute of Navigation*, 38(4):297–316, 1991-1992.

[133] C. Van Loan. Computing integrals involving the matrix exponential. *IEEE Transactions on Automatic Control*, 23:395–404, 1978.

[134] R. D. J. Van Nee. GPS multipath and satellite interference. In *Proceedings of the Forty-eigth Annual Meeting of the Institute of Navigation*, pages 167–178, 1992.

[135] G. Wahba. A least squares estimate of spacecraft attitude. *SIAM Review*, 7(3):409, 1965.

[136] J. R. Wertz, editor. *Spacecraft Attitude Determination and Control.* Kluwer Academic Publishers, Boston, 1978.

[137] S. A. Whitmore, M. J. Fife, and L. A. Brashear. Development of a closed-loop strapdown attitude system for the high altitude aerodynamic performance experiment. In *Proc. of the 35th AIAA Aerospace Sciences Conference*, 1997.

[138] W. S. Widnall and P. A. Grundy. Inertial navigations system error models. Technical Report TR-03-73, Intermetrics Inc., May 1973.

[139] N. Wiener. *Extrapolation, Interpolation, and Smoothing of Stationary Time Series.* Wiley, New York, 1949.

[140] J. Williams. *From Sails to Satellites: The Origin and Development of Navigation Science.* Oxford Press, Oxford, 1992.

[141] W. H. Wirkler. Aircraft course stabilization means. U.S. Patent 2,548,278, April 10, 1951.

[142] S. Wu, W. Bertiger, and J. Wu. Minimizing selective availability error on satellite and ground global positioning system measurements. *Journal Guidance*, 15(5):1306–1309, 1992.

[143] Y. Yang. Personnal Communication, 2007.

[144] Y. Yang. Method and apparatus for adaptive filter based attitude updating. U.S. Patent 2005/0240347 A1, October 27, 2005.

[145] Y. Yang and J. A. Farrell. Two antennas GPS-aided INS for attitude determination. *IEEE Transactions on Control Systems Technology*, 11(6):905–918, 2003.

Index